Nelson Yuan-sheng Kiang

rcvd 6/83

SYSTEMS PHYSIOLOGY

WILEY–INTERSCIENCE SERIES ON BIOMEDICAL ENGINEERING

Advisory Editor: **JOHN H. MILSUM, University of British Columbia**

L. A. Geddes and L. E. Baker

Principles of Applied Biomedical Instrumentation

George H. Myers and Victor Parsonnet

Engineering in the Heart and Blood Vessels

Robert Rosen

Dynamical System Theory in Biology

Stanley Middleman

Transport Phenomena in the Cardiovascular System

Samuel A. Talbot and Urs Gessner

Systems Physiology

SYSTEMS PHYSIOLOGY

SAMUEL ARMSTRONG TALBOT (1903–67)
Associate Professor of Medicine (Biomedical Engineering)
The Johns Hopkins University
Baltimore, Maryland

URS GESSNER
Interdisciplinary Research Center
Department of Health of St. Gallen
St. Gallen, Switzerland

A WILEY INTERSCIENCE PUBLICATION

JOHN WILEY & SONS, New York · London · Sydney · Toronto

Library of Congress Cataloging in Publication Data:

Talbot, Samuel Armstrong, 1903–67.
Systems physiology.

(Wiley–Interscience series in biomedical engineering)
"A Wiley–Interscience publication."
Includes bibliographies.
1. Physiology. 2. Biomedical engineering.
I. Gessner, Urs, joint author. II. Title.

[DNLM: 1. Biomedical engineering. 2. Physiology.
QT 34 T142s 1973]
QP33.5.T34 1973 612 72–10536
ISBN 0–471–84415–2

Printed in the United States of America

10 9 8 7 6 5 4 3 2 1

Series Preface

"To love,
is not to gaze steadfast
at one another:
it is to look together
in the same direction."

St. Exupery, *Terre des Hommes*

Biomedical engineering is but one of many new interdisciplinary endeavors having great importance for us. These endeavors are important because their study and exploitation provide almost our only hope for finding acceptable compromise solutions to increasingly complex sociotechnological problems. Of course the different disciplines of science have always cross-fertilized one another to some extent, but until very recently science's explosive growth has tended necessarily to fragment knowledge into somewhat isolated compartments. However, this specialization has been so successful in generating new knowledge and technological capacity that now one of our desperate needs is for synthesizing this knowledge in socially useful groupings, for example, groupings to tackle the problems of political systems, pollution of resources, ecological systems, urban and transportation systems, and health systems.

The synthesizing process seldom continues for long without raising some practical difficulties. Each discipline has inevitably acquired its own language and a set of basic assumptions and constraints and, of course, some mystique. Furthermore, there may be considerable resistance against the retrenchment of individual disciplines necessary for new ones to be synthesized from some of their parts. Indeed sound arguments may be advanced also to show that experts from several disciplines must work in concert in complex problems such as those already mentioned, thus achieving a crude form of "intelligence amplification." It would seem that both methods will be found necessary;

that is, interdisciplinary collaboration will fruitfully coexist with new synthetic disciplines.

A basic necessary attribute of either method is expressed in the words of St. Exupery quoted above—" to look together in the same direction," even if from different points of view. The present series in Biomedical Engineering is addressed to this need.

The Biomedical Engineering series aims both to educate persons from one discipline in what they need to know of other disciplines and to catalyze a synthesis in the core subject generally known as biomedical engineering. Thus one set of volumes will introduce the classically trained biologist and medical scientist to the quantitatively based analytical theories and techniques of the engineer and physical scientist. Instrumentation, mathematical modeling, signal and system analysis, communication and control theory, and computer and simulation techniques will be covered.

A second set is intended for engineers and physical scientists and will cover basic material on biological and medical systems with as quantitative and compact a presentation as is possible. The omissions and simplifications necessitated by this approach should be justified by the increased ease of transferring the information and, more subtly, by the increased pressure this can bring to bear upon the search for unifying quantitative principles.

Some subjects of biomedical engineering interest will be treated in texts of considerable depth, while fast-changing subjects will be treated in monographs. Typical areas include biomaterials, prosthetic systems, ecological modeling, systems identification, and computer-aided diagnosis.

Medicine is being forced increasingly into the public spotlight, in direct relation to its increasing power to affect the quality and quantity of human life. Biomedical engineering is contributing significantly to these developments and must be responsibly concerned with the inevitable social questions of what, to whom, when, and where. This series therefore will also focus on the social correlates of biomedical engineering, for only in this way, I believe, can the theoretical benefits of technological advance be assured of transformation into real, practical benefits for humans in their complex, imperfect societies.

JOHN H. MILSUM

Préville, Quebec,
November 1971

Preface

This book presents a modest beginning in systems physiology—a synthesis of the many aspects and principles important in the complex organization of processes and organ systems. Our approach emphasizes quantitative, functional, that is, performance-oriented analyses of systems, most of them at supracellular levels. However, not all of the ideas, models, and concepts that are advanced can be regarded as firmly established, although they have proven to be useful, if not indispensable, for the early developments of systems physiology. Moreover, we would like to analyze many more organ systems and their interactions, but are hindered by lack of data as well as by insufficient understanding in numerous areas within physiology.

The task of synthesizing according to the rules and tools of systems theory is all important but enormous. Dr. S. A. Talbot, the senior author, had the wide scientific experience and understanding of many disciplines so essential for such a synthesis. His death in 1967 was a tragic loss; not only had he completed far too few chapters before his death, but I found myself without a guide. Although the road had been well defined by much cooperative work and discussion, the many steps leading to the final manuscript were arduous. A level that would do full justice to the profound insights that Dr. Talbot had developed during his distinguished career could hardly be maintained. His main concern was to create order; I can only hope that my own contribution helps in preparing the first steps toward that goal.

We also believe that many pathological states of higher systems should be analyzed by the methods presented in this book. An entire section on external processing of biological information, including a chapter on symptom-disease logic, might have been added. The state of knowledge and order in this area seems not yet ripe for textbook presentation, however.

It is a pleasure to acknowledge our indebtedness to the many faculty colleagues, students, and administrators and to other biophysical scientists whose ideas and assistance were used in developing this book. Any clarity of exposition is the result of graduate seminar courses given over seven years at The Johns Hopkins University and the University of Pennsylvania Engineering Schools. Dr. David A. Robinson is responsible for a number of the more unusual treatments given here, some of them unpublished. We are beholden to Dr. Herman P. Schwan for many discussions of bioelectricity and teaching in this area. Dr. W. H. Huggins helped greatly in interpreting the relations between physiology and both systems analysis and signal theory.

Reaching further back, the years of research with Dr. W. H. Marshall afforded Dr. Talbot first insights into image processing in the visual system. Above all, Dr. Talbot wanted to express appreciation of the twenty-year support and encouragement of Professor A. McGehee Harvey toward the development, on a scientific basis, of Biomedical Engineering at the Johns Hopkins' Department of Medicine.

I am greatly indebted to Professor Richard J. Johns for sustained interest and help during the years after the death of Dr. Talbot. Grants from the Commonwealth Fund were important for the early work on this book, earning the thanks of both authors. The cooperation given to me by the firm of F. Hoffmann–LaRoche & Co. during the last phases of manuscript preparation is also deeply appreciated.

I also express my gratitude to Belle Otto Talbot (Mrs. S. A. Talbot), whose help has been invaluable. She not only kept an eye on our discussions of chemical systems, but also corrected my writings, teaching me the fine points of English grammar and trying to eliminate all traces of Swiss-German from the text. This must have been a trying job. She thus is, effectively, a coauthor.

Most of the material was developed with the secretarial help of Miss Mary Franke. The manuscript was typed by Mrs. Vivian Cassel, who also prepared the drawings, thereby earning our profound gratitude. Without her painstaking work, this book could never have been finished. We also thank Mr. F. Schweizer for his assistance with the illustrations.

URS GESSNER

Basel, Switzerland
October 1972

Contents

Part III Sensory–Motor Systems

Abbreviations

ACTH	adrenocorticotropic hormone
ATP	adenosin-5′-triphosphate
A-V	atrial-ventricular
BVP	Bonhoeffer-Van der Pol model
CE	contractile element (of muscle)
CFF	critical flicker fusion frequency
CNS	central nervous system
cps	cycles per second (in illustrations; equals Hz)
CRF	corticotropin releasing factor
CSF	cerebrospinal fluid
db	decibel (in illustrations; equals dB)
D.C.	direct current
deg	degree (of angle)
ECG	electrocardiogram
EPSP	excitatory postsynaptic potential
g*	gram force (equals 981 dynes)
HH	Hodgkin–Huxley model
HR	Hodgkin–Rushton model
IC	initial condition
Im	imaginary part of . . .
IPSP	inhibitory postsynaptic potential
jnd	just noticeable difference
LCR	L = inductance, R = resistance, C = capacitance
MAF	minimum audible field
PSP	postsynaptic potential
RC	R = resistance, C = capacitance
Re	real part of . . .
RLC	see *LRC*
SA	sinoatrial (node)
sc	scala (e.g., scala tympani)
SE	series-elastic element (of muscle)
URS	unidirectional rate sensitivity

Mathematical symbols are defined in the text.

SYSTEMS PHYSIOLOGY

Introduction

This book is a text in the general field of mathematical physiology. It evolved from a course of study on biological processes and systems for graduate engineers.* These students, who were competent at the graduate level in applied mathematics, could focus on the use of systems analysis in the biological situation without stopping to develop the mathematical or physical apparatus and concepts. For these students, however, a detailed review of the biological mechanisms was required. This review, carefully selected and guided, was ancillary to the course and is not included here.

Taking for granted an ability to follow the mathematics needed, it became possible to examine critically the applicability and adequacy of the models, and to ask whether we were doing problems in curve fitting or finding "intrinsic" (i.e., mechanistically or functionally particular) models. This issue we discuss in Chapter 1. The concept and need of modeling are, of course, already understood by such students.

However, several M.D.'s with basic engineering training also found this course useful. At least one medical student in twenty now has such a background, and one need hardly dilate on the value of this factor in medical research. This trend toward engineering is enhanced by the many biomedical researchers who are now studying mathematics, physics, and basic engineering theory, essential also in the use of computers as an aid to research. So we have reason to believe that there is an extensive body of scientists who can use a book that focuses on mathematical physiology rather than on development of the requisite biomathematics.

Several sources for materials serve as background to mathematical physiology. For those who seek such an introduction, perhaps the first

* Both authors taught this material to graduate biomedical engineers at The Johns Hopkins University and collaborated for many years in teaching and research. The Introduction and Chapter 1 are Dr. Talbot's, the rest of the text is a joint effort.

problem is to translate the biological situation into a model that can be represented mathematically. This is taught admirably by Riggs [0.1]. He emphasizes the simple but important rate models of physiology and biochemistry. Mathematically, these take the form of linear first-order differential equations and their representation and solution. In the vegetative functions (e.g., respiration) the equations for transport, reaction, conversion, diffusion, and distribution of materials, hormonal levels, and so on are so numerous and coupled that matrix methods are needed. For these methods Searle [0.2] gives a good introduction as well as a foundation for the present book.

The restricted utility of compartmental analysis and of linear first-order equations in physiology is brought out by Yamamoto [0.3]. He cites more recent references and surveys the mathematics required by more complex and realistic physiological models of coupled rate processes.

Other varieties of mathematical modeling per se are surveyed in readable fashion by Stibitz [0.4]. This work suggests that the biologist may wish to ally himself with an applied mathematician who knows the limitations and pitfalls of the more advanced methods so often required in physiology, and which we illustrate. Stibitz introduces computer methods by several examples of numerical (difference) analysis alongside the usual treatment by continuous functions. His emphasis is still on biological problems describable by simple physical models, rather than complex sequential coupled systems. Books on biophysics and systems engineering for medical students (Randall [0.5], Ackerman [0.6], and Blesser [0.7]) tend also to keep to the same introductory level of complexity.

Systems physiology transcends control systems physiology; about these distinctions more is said in Chapter 1. Yet it is true that much of modern mathematical physiology illustrates how intermediate as well as advanced *control theory* illuminates not only homeostasis but also the dynamics of performing circulatory and sensory-motor tasks. Excellent introductions to this have been published, building on Riggs' "primer." Among these is Milhorn's *Application of Control Theory to Physiological Systems* [0.8], which emphasizes the mathematical and engineering basics. The author discusses several dynamic systems, notably the respiratory carbon dioxide control, in a treatment that for many years was the most comprehensive one available. He also discusses static physiological equilibria of several back-coupled systems (which lack discrete sensing or comparator mechanisms) and outlines others ripe for research from this point of view.

Neither Milhorn nor Riggs, however, discuss analytically the steady states of flow systems—irreversible thermodynamics being the analytical foundation of such treatment—nor the dynamical approach to such states (Chapter 2).

More advanced and rigorous mathematically is Milsum's *Biological Control Systems Analysis* [0.9], written for the graduate engineer as well as for the mathematically sophisticated physiologist. This work introduces models requiring nonlinear and stochastic methods. Like Milhorn, Milsum deals with control aspects and transient behavior, a small fraction of the subject matter of our text. Unlike Milhorn, Milsum's main competence is in what might be called "exterior ballistics": systems whose dynamics can be studied in terms of nonpenetrating input and output measurements. Schwan's book [0.10] includes, besides rigorous discussions of nerve excitation, tissue properties, and hemodynamics, an excellent treatment of biocontrol.

The control systems viewpoint also has several assumptions which are severely limiting, as Yamamoto, in *Physiological Controls and Regulations* [0.3], points out. Many factors make straightforward application of technologic control theory more complex than present texts would suggest. These factors include the inadequacy of linear theory, the basic conflict between a reaction or materials-flow graph and a signal-flow graph, the biological rarity of simple summing at junctions, the presence of large transport delays, and a limited nonlinear memory. The problem becomes even more interesting when changes in control behavior can be used to characterize disease or to quantify therapy. To illustrate some of these cases is one of our aims.

Although control is highly important for stability and for excellent performance of tasks at every level of bioorganization, it is but a part of our study of biosystems and processes. For example, *communication* is only partially linked with control. In addition to its many control functions, the central nervous system also has more subtle and important functions at the human level in affording communication with the wellsprings of human interaction, exploration, and invention. Therefore the science of biocommunications transcends that of biocontrol, especially if one goes beyond the engineering definition of communication, as Weaver has done [0.11]. This also is illustrated.

Inseparable from study of biosystems, and distinct from control and communication, is the study of how we get information about such systems. The informational content of the electrocardiogram exemplifies this area. Of similar challenge and biomedical importance are phono- and ballistocardiograms from the viewpoint of maximal information and diagnostic decision making. A book frankly oriented to biomedical aspects might generalize into the new field of formal diagnostic theory and logical processes. We stop considerably short of covering this, however.

Instead of developing a text on mathematical procedures, we attempt to survey and especially to interrelate the hierarchic relations of biosystems, using a mathematical approach. We emphasize that *concepts* (biological, physicochemical, and engineering) change according to level of organization, whereas analytic *forms* remain more or less similar. This brings out the intrinsic changeover in biosystems and processes, from dealing with matter (flows, concentrations, reactions, etc.) and energetics at the simpler physiological levels, to a major emphasis on transformation and production of information (negentropy) and organization per se at the higher levels of bioorganization. This represents important conceptual shifts too, from the older ideas of chemistry and physics to those of systems engineering, and from viewing the organism as vegetative and self-contained to viewing it as not only adaptive but also creative, purposive, and interactive at conscious human and social levels. Here communication and control as such become minor elements, means to an end, rather than the focus of interest and the key to systems design.

In biophysical sciences, there are two main trends. *Basic biophysics* emphasizes the study of isolated structures and functions: ribosome, membrane, muscle fiber, blood vessel, cochlea, eye. *Bioengineering* emphasizes how these are coupled into systems—the problems of measuring, characterizing, and quantifying the coupling and systems behavior—to accomplish some purpose of the organism. Beyond these, *biomedical engineering* and medical science (especially from the professional point of view) are further interested in how such systems fail. In this book are many references to problems of dysfunction, and observations from these provide both biomedical engineers and medical scientists with much of their knowledge.

However, this is hardly a textbook of biomedical engineering in general. As a rule, practicing biomedical engineers spend much of their time on problems of physical methods and measurements for the living system, especially the human, in various stages of health and activity. Another large portion of time is devoted to the major area of recording, reducing, analyzing, displaying, and interpreting multivariate data. Neither of these areas is included here; however, the book by Clynes and Milsum [0.12] presents an excellent cross section of the many facets involved.

The first subject, bioengineering methods, measurements, and instrumentation, constitutes an extensive and rapidly growing field, applying to biomedical problems all the recognized divisions of engineering science. It is covered by several books [0.13] and journals [0.14]. The specialized biomedical computer hardware, application, techniques, and programming are likewise too extensive [0.15] and too much subject to change for inclusion in a textbook of lasting worth.

Many biomedical measurements are made indirectly through intervening complexity and make sense only in terms of mechanistic models (e.g., ECG, PCG, BCG*). Understanding the mechanism should help distinguish artifact

* Electrocardiography, phonocardiography, ballistocardiography.

from information in instrumental records, as well as leading to clearer thinking about which measurements are necessary and significant or most relevant to the particular problem and which are not. (The era is passing of measuring everything technically measurable that might have bearing; one must weight for significance to avoid taking excessive data.)

For design and applications of equipment, experience shows that the biomedical engineer needs a quantitative understanding of the biological phenomena and mechanisms he is measuring, assisting, or controlling. The quantitative study of biosystems offered by this book would seem to be basic to the designer of biomedical experiments as well, regardless of the type of engineering involved. It also underlies the mathematical models implied in the use of computers.

Biosystems are so interacting that we can hardly confine our knowledge to hemodynamics (for instance) in disregard of neural, hormonal, or muscular factors in the results. Nor can the bioengineer or physiologist, studying the nervous system or its performance, disregard the physiologic and psychophysical content or *meaning* of the information being processed. It is in this spirit that the subject matter of this book has been presented to bioengineering graduate students, as basic to both instrumentation and computer applications, which change so rapidly from year to year.

The *area* this book covers reflects the growth of research on biosystems in the last 10 to 15 years, primarily at the macrolevel. We have selected topics that illustrate our major themes:

1. That systems principles and analysis occur at every level of physiology, psychophysiology, and pathophysiology, whereas at each level the analytical elements change as each system becomes a subsystem of that above.
2. That as we approach higher systems, we deal increasingly with organization and information, their modification and production.

Formulating and analyzing this organization moves toward engineering sciences, and away from physics and chemistry. We wish to emphasize the corresponding change in our conceptual framework of such systems.

The *selection* and distribution of topics is revealed in the table of contents. The field is growing so rapidly that the treatment of many topics will be outdated by the time of publication, as indeed proved true during the development of the seminar course. Both instructor and students revised and updated it annually. However, most of the original treatments here selected may qualify as classic fundamental references, worthy of exposition in a textbook. It has been the practice of the authors to provide students with copies of the originals of these, on which to base written reports. Indeed, an offset-printed compendium of such original papers would be valuable. Should the reader wish to refresh his mathematics, the textbook by Wylie [0.16] is recommended.

Furthermore, when this work was started in 1962, much of the present source material had not been published, and systems physiology in this broad sense was a slim subject. Recent research has opened so many new areas of mathematical physiology from the systems viewpoint that a book twice this size could now be written. In selecting topics from this richer domain, it has seemed wise to adhere to the original motivation. That is to maintain an overview of the entire field in order to preserve a sense of relationships and so to help the student locate his area of interest. The specialist will note lacunae, especially in stochastic and adaptive systems, still regarded by the authors as in the competitive modeling stage.

The future may bring specialized texts on mathematical systems analysis of many kinds: in the exterosensory and communicative systems, in respiratory and circulatory systems dynamics, in organismic control by hormonal mechanisms, in biosystems design for complex muscular performance, and in biological controls, internal and extrapersonal. As research progresses these will go far beyond the individual sections of this book. But we would still beware the process of reduction and fractionation—"knowing more and more about less and less"—by which one loses understanding of how subsystems couple to contribute to task-oriented systems, weaving all their functions for human purposes.

The biomedical engineer or the medical scientist can hardly afford to lose sight of the human as an abstracting, purposive, emotional, responsive, and original being. This consideration enters into all realistic scientific research with humans. The thrust of this book is therefore integrative rather than analytical; we hope to interest and attract students into this area of integrated human performance in the real world. One can and should study subsystems primarily to understand their *contribution* to the whole, as well as in their own right.

It is hoped that from this work the student will gain some critical feeling for theories and models as devices for bringing conceptual order. This is less to render biology more tractable for calculation (for which curve fitting is adequate), than to bring the performance of biological tasks at every level into the general framework of scientific understanding. Models therefore must be criticized not only for sufficiency or adequacy for the data at hand, but also for fidelity to nature, completeness, and relation to larger models of which they are a part.

Models are indeed aids to thinking about a particular biosystem or function. But more than this, they help to build a framework or fabric of science whereby general principles are discerned as operating throughout nature—not merely in a mathematical formalism, but as organizing nexi. Some principles may find their main importance in *inanimate* nature and structures, others are mainly exemplified in *living* nature and performance. We will remark on these. The quest for principles, which interests the practitioner as well as the scientist, is served by comparisons at all levels of organization and among a variety of special systems.

This book is written, therefore, from the generalist's viewpoint. It is more than a survey or collection of special topics; rather, it stresses their assembly into a pattern, both for wider understanding and for disclosing principles that underlie the performance of biosystems as a whole. By study of this purposiveness of ever-broadening biosystems, we may be able to extrapolate to the purpose of a human society.

A textbook may represent the content of a course of systematic study or provide a framework about which to build a course. This book is intended to be the latter: a guide to the self-motivated student who prefers to read the original work with full mathematical detail, but who desires a guide for organization, critique, comparison, and perspective.

REFERENCES

0.1. Riggs, D. S. *The Mathematical Approach to Physiological Problems.* Williams and Wilkins Co., Baltimore, 1963.

0.2. Searle, S. R. *Matrix Algebra for the Biological Sciences.* John Wiley and Sons, New York, 1966.

0.3. Yamamoto, W. S. and J. R. Brobeck (Eds.). *Physiological Controls and Regulations.* W. B. Saunders Co., Philadelphia, 1965.

0.4. Stibitz, G. R. *Mathematics in Medicine and the Life Sciences.* Yearbook Medical Publishers, Chicago, 1966.

0.5. Randall, J. E. *Elements of Biophysics,* 2nd ed. Yearbook Medical Publishers, Chicago, 1962.

0.6. Ackerman, E. *Biophysical Science.* Prentice-Hall, Englewood Cliffs, N.J., 1962.

0.7. Blesser, W. B. *A Systems Approach to Biomedicine.* McGraw-Hill Book Co., New York, 1969.

0.8. Milhorn, H. T., Jr. *The Application of Control Theory to Physiological Systems.* W. B. Saunders Co., Philadelphia, 1966.

0.9. Milsum, J. H. *Biological Control Systems Analysis.* McGraw-Hill Book Co., New York, 1966.

0.10. Schwan, H. P. (Ed.). *Biological Engineering.* McGraw-Hill Book Co., New York, 1969.

0.11. Shannon, C. E. and W. Weaver. *The Mathematical Theory of Communication.* University of Illinois Press, Urbana, 1949.

0.12. Clynes, M. and J. H. Milsum. *Biomedical Engineering Systems.* McGraw-Hill Book Co., New York, 1970.

0.13. (a) Donaldson, P. E. K. *Electronic Apparatus for Biological Research.* Academic Press, London, 1958. (*b*) Lion, K. S. *Instrumentation in Scientific Research—Electrical Input Transducers.* McGraw-Hill Book Co., New York, 1959. (c) Stacy, R. W. *Biological and Medical Electronics.* McGraw-Hill Book Co., New York, 1960. (d) Potter, V. R. (Ed.).

Methods in Medical Research, Vols. 1–12. Yearbook Medical Publishers, Chicago, 1948–1970. (e) Hill, D. W. *Principles of Electronics in Medical Research.* Butterworth and Co., London, 1965. (f) Geddes, L. A. and L. E. Baker. *Principles of Applied Biomedical Instrumentation.* John Wiley and Sons, New York, 1968.

0.14. Journals on Instrumentation
(a) *IEEE Transactions on Bio-Medical Engineering.* Institute of Electrical and Electronics Engineers, New York. (b) *Medical and Biological Engineering.* Pergamon Press, New York. (c) *Medical Research Engineering.* Medical Research Technology, Little Falls, N.J. (d) *Bio-Medical Engineering.* United Trade Press, Ltd., London. (e) *Medical Equipment.* Milton Publishing Co., London. (f) *Medical Instrumentation* (Journal of the Association for the Advancement of Medical Instrumentation, JAAMI). Williams and Wilkins Co., Baltimore. (g) *Medical Progress through Technology.* Springer-Verlag, New York.

0.15. Computer Methods: Books (a selection)
(a) Enslein, K. (Ed.). *Data Acquisition and Processing in Biology and Medicine*, Vol. 3. Pergamon Press, London, 1964. (b) Enslein, K. and J. F. Kinslow (Eds.). *Data Acquisition and Processing in Biology and Medicine*, Vol. 4. Pergamon Press, New York, 1966. (c) Stacy, R. W. and B. D. Waxman (Eds.). *Computers in Biomedical Research*, Vol. 2. Academic Press, New York, 1965. (d) Ledley, R. S. *Use of Computers in Biology and Medicine.* McGraw-Hill Book Co., New York, 1965. (e) Taylor, T. R. *The Principles of Medical Computing.* Blackwell Scientific Publishers, Oxford, 1967.

Computer Methods: Journals
(a) *Methods of Information in Medicine.* Schattauer Verlag, Stuttgart. (b) *Computers and Biomedical Research.* Academic Press, New York. (c) *Computer Progress in Biomedicine.* North Holland Publishing Co., Amsterdam. (d) *Computers in Biology and Medicine.* Pergamon Press, New York. (e) *International Journal of Bio-Medical Computing.* Elsevier Publishing Co., Amsterdam.

0.16. Wylie, C. R., Jr. *Advanced Engineering Mathematics*, 3rd ed. McGraw-Hill Book Co., New York, 1966.

CHAPTER ONE

The Concept of Systems Physiology

A. MEANING AND SCOPE OF SYSTEMS PHYSIOLOGY

Systems physiology is physiology from the viewpoint of systems science. *Systems science* is a recent branch which attempts to abstract and generalize the *systems aspect* from functioning systems of all kinds. It aims to formulate principles in "organized complexity," ranging from simple organisms to the human tasks of organizing and communicating.

Of course, the quantitative study of physiological systems (as seen at the medical level) has been proceeding for many decades; numerous textbooks describe their operation, both normal and abnormal. As methods of measurement have improved, these descriptions have become increasingly precise. Modeling is now generally accepted as a necessary framework for analysis. The ranges of normal and abnormal variations have been studied for many important parameters and measures of function, for example, temperature, metabolic rate, contractile force, and arousal. In all this, the major emphasis and intent has been traditionally on the particular *mechanism*: just how it works, and under what conditions and limitations.

1. The Purpose of a System

Systems are studied with quite different viewpoints and approaches. Like a traffic bridge, a *static* system is studied in terms of its purposes: Where should the bridge be located in reference to regional traffic flow? What is the minimum design for the maximum stresses it will bear—cars per days, storms, tides, erosion of piers? Will its appearance be pleasing? Which design will satisfy all requirements at least cost? From these considerations of the task result the details of how the bridge is designed, built, and worked; its stress distribution, struts, cables, bents, and structural materials.

A *dynamic* time-varying system seeks to fulfill its task optimally, according to certain criteria, by adaptation. For example, a publishing house providing societal and cultural satisfactions must minimize costs. The successful publisher perceives and molds these many subtle changing satisfactions and cost factors, and flexibly designs his organization to deal with them. Examples of systems study are endless, ranging in complexity from nonlinear bottle openers and systems of personal transportation to governments.

Physiologically, we may mention human systems designed for: (1) internal *maintenance* of gravitational equilibrium, temperature, blood pressure, glucose level, red cells, blood volume, and osmotic pressure, as well as presently less objective parameters such as alertness or equanimity; (2) *protection* against infection, foreign proteins, and noxious experiences; (3) normal and extraordinary external *activity*, both physical and mental; (4) *communication* by hearing, speech, reading, and writing, by music, by mutual recognition of ideas; (5) *storage*, retrieval, and novel recombination of experiences and ideas; and (6) creative and representational *expression*. Obviously, systems physiology as such has barely begun, and is still preoccupied with animal and existential levels.

The first unifying viewpoint of systems science then is the concept of the *task* and of its satisfactory performance: a value judgment. Most of its analytical framework to date has developed in relation to engineering systems (chemical, electrical, management); however, identical optimizing task-oriented concepts have been springing up in the study of societal, economic, government, military, and hospital operations. In general, systems physiology seeks to understand the individual tasks and problems of the human. Are they biochemical, physical, psychosensory, emotional, mental, social? Do the methods of their solution resemble, analytically, task solving in other systems sciences?

More specifically, systems physiology is the study of particular physiological systems in terms of the design of and interaction between their subsystems for optimal task performance. Thus study of the performance of the human temperature regulator depends on a sufficiently comprehensive analysis of range and distribution of load (heat and cold), adaptive reset, diurnal reset, speed of response, mixed commands, and emergency demands (Chapter 19). These task considerations must be expressed in details of subsystem structure and function and they must be consistent with other tasks of the circulatory system.

The muscles with their controls comprise another task-oriented system. In particular, brain and hand, in their linked design and development, optimize concurrently several different tasks of great sophistication, including delicate manipulation of fine objects in complex sequences of motion, grasping in least time, and positioning in six dimensions with compensation for load dynamics. In terms like these, not only can we describe and model quantitatively the design of observed neuromuscular structures and measured performances, but the design also becomes meaningful—we learn "why?", as well as "how?" and "how well?" (Chapter 11). The visuoperceptive system is one that is even more fully studied. Its immediate tasks are discrimination, comprehension, and interpretation of what is looked upon. This broad view of physiology is hardly novel.

The philosophically inclined reader will at once set up the equation: "systems physiology = teleological physiology"; for task or purpose brings in the "final cause" of Aristotle, rather than the antecedent causation of determinism. (*Purpose* here is not intent, but design objective deduced from performance.) A reading of this text also shows that the viewpoint is holistic or compositionist [1.1] rather than reductionist. This teleology is what distinguishes synthetic from analytic chemistry, engineering from physics. Indeed, for a physicist final cause is meaningless. His systems have no task and so cannot be judged on optimum performance of it. Systems science asks, in addition: Under what circumstances does the system do its best? The element of optimal value judgment is present.

From a horde of examples, take the eye: What is the best it can be made to do for the human user with respect to light sensitivity (the two-photon problem)? Next, what combinations of physical, chemical, physiological, psychophysical, and cognitive mechanisms and environmental circumstances are needed by the human to so optimize this task? Until the question was framed this way, the complexity of the exquisite performance of the visual system was not disclosed (Chapter 5). By consideration of overall performance, we assign subtasks to submechanisms. Many physiologists have long thought this way, but some "basic" scientists, working with "bricks and mortar," have insisted that the study of final cause meant circular reasoning, having no explanatory value. The systems physiologist then, like the engineer, is distinguished by the questions he asks.

Our position is the heuristic one. If the teleological approach yields more penetrating explanations, and the compositionist view uncovers observations otherwise missed, let them be used. The reader may make his own choice.

Systems physiology starts with the system as a whole, in relation to its *design* objectives. Observing the tendency of the human to reorganize his environment, to interpret human relations, to explore and create, to become involved and to enjoy, we may wonder if there is not something fundamentally inadequate in a viewpoint restricted to internal mechanisms per se.

As long as "physiology" is confined to the study of lower animals, mainly in an inert condition, how can we appreciate the intricate system that gives rise not only to the extraordinary range of dynamic performance of the animal or human in full, complex, purposeful activity, but also to its motivational parameters and flexibility in adjusting that performance to the wide variety of tasks it conceives? Physiology can raise its sights, given the conceptual and analytical tools becoming available from systems science for dealing with such problems. Just as the classical chemistry of closed systems is inadequate to describe intracellular phenomena, so the classical viewpoint of internal systems excludes major variables in human physiology.

The history of phylogeny has been encephalization, producing eventually a form of life that revolves not around bare survival, as in Darwin's concept, but around production of a world that is rich emotionally, intellectually, and in social justice. Why are these factors "physiology"? Do certain innate biological processes bring about these effects? The encephalization of man, the function of the forebrain, and man's mechanisms for sensitivity, sympathy, and empathy appear indeed to be products of lawful biological evolution, transcending and unrelated to the areas that have preoccupied mammalian physiology, namely, the demands and apparatus of vegetative operation or survival. Medical physiology, especially, must analyze the function and dysfunction of control systems which themselves change with the task; it must also measure the role of motivation in the effectiveness of physiological performance. The closed and mindless mechanism is not enough in the cooperative struggle for health.

2. The "Operator" in Systems Science

Systems science has a second defining emphasis besides that of fixation on objectives: "a system for . . . ." It is the interpretation of subsystems as an assemblage of *operators*. A system (seen at a given level) is conceived as comprised of a network of such elementary operations performed by its subsystems.

Many of these operations can be stated mathematically as (1) summating—addition, subtraction, comparison, integrations over time or space; (2) multiplying by a coefficient or a variable, scaling, sampling, modulation, powering; division (inverse multiplication); (3) exponentiating—exp, log; (4) delaying, weighting with delay (convoluting); (5) projecting—rotating, transforming, or remapping (e.g., time to complex frequency domain); (6) differentiating—taking slope, rate or gradient, and/or curvature, spatially or temporally. Control subsystems (of types recognized in engineering) may be described as sequences or *networks* of such mathematical operations, coupled appropriately.

These operations are applied at every level of organization, not to the physical substrate on which a mechanism works, but to the physical *measures* that are made and interpreted technically as "signals." Examples of signals are: concentrations, temperatures, diffusional flow rates; nerve frequencies, spatial excitation patterns, carotid pressures; angular velocity of eye or head, position of hand; electrocardiographic lead voltage, phonocardiographic spectrum. We readily note the distinction between these objective signals and "information." The latter in its biological meaning is relative to sender and receiver and, like "organization," its measure depends on the dimensions recognized. (In its engineering meaning, as in a message, this is merely bit content or a measure of its probability, e.g., "$P \ln P^{-1}$.")

In this operational framework, control networks, for example, are studied in a very general way, as applicable to a variety of specific systems (inanimate, animate, social, industrial, etc.), justifying the name "systems science" apart from subject matter. Today physiological processes are being reformulated in terms of such typical engineering subsystems, many of them included here. This mode of representation or modeling can also be used erroneously, as we shall see.

In addition, many operations are poorly amenable to disciplined mathematical statement because of essential nonlinearity, nonstationarity, or plain nonrandom ignorance. Such behavior among the subsystems at a given level of organization may still be generalizable among biosystems, even if rare—or not yet observed—in inanimate systems. In this general class are threshold effects, discontinuous derivatives, destruction, and synthesis. A schematized representation calls for special new operators to keep such "happenings" within systems science. We present many examples of this.

What is important is not the possibility of stating mathematically the relations and contributions of subsystems, but rather the generalization that similar combinations of operations are recognizable at every level of organization and field of application, so that principles may be sought. Systems physiology tries to state the processes and corresponding operations whereby the task is best accomplished at a particular level of organization, be it mitochondrial or relating to the control of nutrition at the organ level or a skilled psychomotor performance (e.g., playing basketball).

3. Modeling in Biological Systems

The systems engineering concepts just outlined must still be applied to biosystems with reservations. It is one thing to design a physical or organizational system to fulfill certain optimal performance criteria, using in this design subsystem components of known "operational" characteristics. The biological problem asks something further: Given a system already designed for excellent performance of certain tasks, how can we best *model* its existing design at this level? We do not request just any model that will perform similarly; this property is necessary but not sufficient for models acceptable biologically. A certain homology or correspondence is needed for credibility.

In the elucidation of biological design, this factor is so important, and so often overlooked by theorists in engineering systems, that we must examine it further.

The model in the first place largely expresses the viewpoint, interests, and concepts of its maker. If mathematical, it may be analytical, as in differential equations of synaptic summation (Chapter 3); geometrical, as in the vector space of signal theory; stochastic, as in modeling temporal neural threshold behavior (Chapter 4); or statistical, as in Bayesian diagnostic theory. Those who think better in physical models may prefer the electrical (e.g., the *RLC* models of circulation discussed in Chapter 16), thermodynamic (potential, flux), thermal, or hydrodynamic models. Bimolecular or autocatalytic models may clarify simpler subsystems. All these are proposals for orderly thinking about a complex problem, process, or task performance in a way that depends partly on the structure of the data and partly on personal knowledge or preference.

Second, such models are not unique because they are usually constructed to accommodate only the data the author elects to include. Adding new data, dimensions, or constraints may force drastic changes in the model. If our aim is deeper biological understanding as well as mathematical representation for practical purposes, we will wish to transcend description or taxonomy by models of scope wider than their data. Implied in this book is the hope that, by studying critically a wide range of such abstractions, one may achieve a flexibility and feeling for alternatives which will lead to more powerful and general models. This brings up the need to consider intrinsic models, which is discussed below.

4. The Hierarchic Approach to Biosystems

By what procedure can we implement such a study of "systems physiology"? The answer would seem to evolve from the perspective strongly advocated for years by the great physiologist, R. W. Gerard: to keep in mind at all times the successive *levels of organization* which are inextricably interrelated. To this, systems physiology would add task orientation at each level. The nerve membrane is depolarized "in order to" carry messages. The dendrites, soma, and nuclei are there to transform and process these messages; the primary sensory levels have the task of translating the messages into analytic and synthetic patterns on which consciousness operates; the inner nuclei color and associate these patterns with sympathetic system activity; the higher networks associate the enriched patterns with learned experience and meaning and with external decision and action, as well as with imagination and attitudes affecting these decisions.

At each level we find a set of subsystem concepts suited to that level, quite meaningless at levels below and only confusing to the study of levels above (a thesis attributable to P. Bridgman). Thus in studying the performance of a locomotive or a watch, we hardly need to think of the metallurgy of the bearings; in studying the mobilization of increased muscular coordination in exercise, we may forget the role of molecular orbitals in phosphate-bond energy. At each higher level, new and extraneous considerations enter and the subsystems must be reformulated. Thus the viscosity of blood as a tissue is not fixed by a determinate intrinsic structure, but by the extrinsic relation of that structure to capillary walls, to impressed velocity, to rouleaux formation determined by hormone activation, and to other external factors. Viscosity cannot be stated in purely molecular terms, determined from below, or even by boundary conditions in a system that is open, like the living one. Hence it must be redefined *operationally* for each level of observation (Chapters 13 and 14).

As we climb the hierarchy in "open systems," the extraneous nongenetic factors that enter become more numerous and complex, as well as more informational and less physicochemical. Finally the *external* interactions of the system itself, both outward and mindward, become the governing considerations. We reach the level where control systems modulated by human motivation and intention become the focus of study, modeling, and formulation. This finding has long been recognized in the physiology of fatigue, and it enters strongly in pathophysiology.

What is gained by use of such task language as "in order to do," rather than "which does," in the foregoing? The latter is merely descriptive, but the former raises the question "how well?" For example, nerve depolarization carries messages very well because it is nondecrementing, is locally maintained, has a high safety factor, and is noninterfering in bundles (little external field, electric or magnetic). Thus task language directs the scientist's attention to departures from an ideal or goal and so forces him to define the goal for that function. This proves to be salutary. By defining a goal, we examine the properties of that goal to see if they are also present in the biological system.

5. Plan of this Book

The table of contents indicates that this book follows analytically the design and operation of a number of physiological systems from their lower subsystems to the level of interaction with their tasks.

Many of these are presented as classic examples of fortunate mathematical modeling of an "intrinsic" type, as well as of major mathematical apparatus that will be important more than once to the student of advanced physiological systems. In all cases adequate mathematical references are given in the source papers, which it is taken for granted the student will wish to procure. However, it will be clear to any student of physiology that this is a thin sampling indeed of physiological systems. Each year sees a few more contributions at the level of mathematical power we have tried to represent in this book, and which we believe to be required by the complexity of medical physiology. The capable teacher can tie in such additions with the pattern of systems physiology.

Obviously systems physiology in its broadest sense is a program in its early stages, especially if the goal is to handle it analytically rather than descriptively. We are still working largely at the level of the simpler mechanisms. However, enough material has developed (mainly since 1960) to make a beginning in systems physiology. The first analytic models are rapidly being improved; several used here are being superseded by models more powerful and accurate, which are "intrinsic" rather than "fitted" (see below). New areas of physiologic, psychophysiologic, and pathophysiologic function are fast being formulated quantitatively, as the conceptual grasp of physiologists becomes more sophisticated and as there become available new applied mathematical tools better adapted to physiology. Indeed, this book may be "dated" before it is issued. In this sense it may be proffered as an "introduction to topics in systems physiology" in the broader context of that field.

Clearly also, developments at the higher levels of physiologic function are still nascent, in that the quantitative mechanisms of biocontrol systems and especially environmental interaction are only now being explored. Large gaps remain in our knowledge of control of arousal, tonicity, and interaction inside the central nervous system. Even in the hormonal control systems, sensitive methods of quantitation are scarce and slow. Indeed, for this very reason, the dynamical analysis of such systems, and their interaction with the nervous system, is perhaps the major current barrier to developing systems physiology.

B. CONTROL IN BIOSYSTEMS

One important aspect of modern systems physiology is the study of control. Classically, this has mainly meant homeostasis (maintenance of *internal* levels) in its manifold forms from intracellular to organismic. Studies of external behavior (motor, speech, emotion) have been mainly in stimulus–response context, and recently in the *en route* dynamics (time course) of control. The sensory system is now being analyzed in terms of data processing, although not yet dynamically, and also in relation to dynamical control of motor action. This research area (Chapters 5 and 6) also is still rather internally oriented.

However, modern study of psychophysiology in primates and in man has opened new fields of *external* control functions. These comprise not only the psychophysics of threshold and suprathreshold discriminations, but also the dynamics of man's behavior and perceptions as links in a complex external task or process, and of man as a performer, problem solver, inventor, and value judge. The detailed physiology of this broad area is new and little developed analytically; its organized exposition is not quite due. We give some indication of such analytic methods in Chapters 4, 10, and 12.

Broadly speaking, control has *pattern* aspects which are temporal and spatial as well as functional and structural. Most current engineering control theory deals with temporal control: the study of transient errors in approaching a goal. Spatial and temporal control are trivially related in that it is possible to simultaneously time-control several coordinates (e.g., $x\ y\ z$, $\theta\ \phi\ \psi$, the position and attitude of a spacecraft). But without reference to any time factor (dynamics) whatsoever, we can also speak of minimizing or controlling the departures of several features of a geometric or functional pattern from a norm. Spatial rather than time gradients and time integrals are involved, as in visual contrast, relationship, or color. An example of a functional "pattern" would be the set of steady-state operation points of several flows or concentrations in the respiratory system.

Quality control, as another example of pattern control, may be applied to many kinds of human performance, including sports, and has its systems physiology aspects. This book develops topics in pattern control in the study of visuosensory performance and in spatiotemporal control through the visuomuscular system (Chapters 5 and 11, respectively).

Furthermore, we will find it important in systems physiology to distinguish clearly between *control* mechanisms and *stability* mechanisms. In the homeostatic category both are used, and they are superficially similar. A damped pendulum or the relaxed human arm may approach its rest position with the same oscillatory path whereby a pulsed speed controller or the blood sugar level after glucose injection approaches its steady state. In each case, restoring force varies with displacement, but the first two have no specialized error-sensing mechanism or control substance, and the others do. Conversely, for the "control mechanism," it would be silly to invent a "potential" as in the molecule, whose gradient expresses the restoring force toward a stable position. Indeed, when a nonlinear feedback system is used, especially if positive, it may restore toward no "stable" configuration at all (Chapter 3) but rather toward a stable cyclic path. (Such rhythms are in fact necessary to biological homeostasis, but analytically separate from it.)

An even closer distinction must be made between stability and control in open systems. The distinction between static and dynamic (steady-state) equilibria is detailed in Chapter 2.

1. Equilibrium

Static equilibrium, which deals with potential (work) minima, is familiar from physics and physical chemistry. It is the core of most physical and chemical reasoning, and it is implied in Newton's and D'Alembert's laws of mechanics and classical hydro- and thermodynamics. Classical engineering rests on it, as do biophysics and physical chemistry at the molecular level. But above this level, examples of static equilibrium are hard to find in biology, or in *in situ* biochemistry for that matter. Thus the law of thermal death (entropy increase), which holds for closed equilibrium systems, has been widely cited as a biological principle. In living systems that never approach static equilibrium, however, such thermo"dynamics" is operative but does not dominate. Physicists and chemists in the older tradition find it difficult to accept this limited applicability in biosystems of much classical theory, although nonequilibrium thermodynamics was first developed by physical chemists. We develop this topic more thoroughly in Chapter 2.

2. Steady State

Dynamic (flow) equilibrium deals with minimum entropy production (power or effort). It is becoming familiar from open-system (irreversible) thermodynamics, which is encountered especially in the more complex processes of biochemistry and in the production of organization (negentropy) at every level. Dynamic equilibrium describes the steady-state "operating points" of open, coupled systems: chains, circuits, or cross-coupled networks of concentration and flow, of materials, of reactions and, indeed, of information. A recent brief discussion by Engelberg [1.2] gives, as examples of dynamic equilibrium, the steady-state operating concentrations of cortisol and of thyroxin. Although these subsystems may indeed form part of a controller system, they are really not themselves controlled in the engineering sense. Guyton extends this graphical analysis of dynamic equilibrium to the circulation (Chapter 16), again part of a controlled system. (The "load diagram" of a motor or a vacuum tube circuit is an electrical engineering example of such dynamic "operating points" in a flow situation. Similarly, there are many simple hydraulic analogs in which a control system is not involved in steady state.)

Such dynamical flow systems, with distributed sources and sinks, usually form closed circuits and exhibit many intercouplings. These may be backward, forward, negative, or positive in effect. Although perhaps an obvious example of a "feedback system," such a functional network is by no means as obviously a control system. No particular quantity or value is being either maintained or made to move with minimum error toward a given objective. Indeed, all the variables of such a system are being controlled or stabilized by the cross-couplings; but with many reentries it is hard to say, from the linkages alone, which variables are controlled for what specific purpose.

This situation applies also to information flow networks of nerve centers in the brain stem and cortex, where thousands of facilitating and inhibiting activities are associated with the simplest sensory–motor control. It is a most unlikely situation for applying engineering control theory, if one starts from a structure–function viewpoint. Too many systems and subsystems are being flexibly controlled at once, for too many simultaneous purposes, to allow successful association of all observed feedback junctions with a specific control purpose. Even a well-designed engineering controller of a single variable may have half a dozen control loops: forward, backward, positive, and negative. One can only start with task performance at a given level and elucidate the layers of subsystems—a major task for "systems physiology, control systems division." Add to this the striking feature of the many alternative backup substitution networks and load-adaptive switching of control networks, and we confront a whole new field of physiology just to specify the normal and pathological dynamics.

Other criteria help us distinguish clearly between feedback and control. Feedback may be used to tighten up a sloppy performance with no particular control system involved. Likewise, quite independently of control, it may be used to extend a dynamic range, to generate or alter the operating relation (characteristic) between variables by introducing or removing nonlinearities, or to set operating limits for one subsystem in sequential or parallel relation to another. (Illustrations are to be found in the visual system and in many other places.) So, although control engineering theory deals competently with many problems, mainly dynamic, it does not explain all biological feedback, nor cross-couplings in general. These problems must be studied in the context of the individual physiological tasks at all levels of organization, and they must be modeled in their own terms. This is systems physiology because, when reduced to flow graphs (see below), these models have many analytically identical components or operators, which not only simplify the programming of a working model for verification, but also give a basis for generalizations.

3. Nonlinear Regulation

In addition, what the engineer calls good regulation and the physiologist homeostasis may result simply from the quite *curvilinear* functional relations which abound in physiological systems, although they in no way involve identifiable specialized feedback, sensing, reference, or comparator elements. Moreover, neither storage (capacitative) elements nor rate government is necessary, although generally each is present. The dynamic equilibrium of such nonlinear coupled systems may show high stability against external disturbance, especially when one component operates on the production or destruction rate of another—a coupling exceedingly common in physiology. However, the locus of this stability is sensitive to many elements in the system. There is no reference value or comparator involved, other than the location of the characteristic in its coordinates. Milhorn [0.8] develops some quantitative aspects of this kind of nonlinear regulation for several systems.

Much like true control systems, such dynamic equilibria exhibit oscillatory *transient* response to a disturbance (Chapter 2). But these are still systems without the standard apparatus of control theory: input summing junctions, error amplifiers, or feedback transfer functions. Although mathematical formalisms and physical models (analogs) using dynamic analysis of standard control systems may be invented to accurately mimic ("predict") their behavior, such models really in no way "explain," correspond to, or help understand the open-system mechanisms in question. Engineers have much to learn from physiology about varieties of system stability. One purpose of this book is to illustrate and clarify these distinctions and the validity of related models.

4. Stable Oscillation

Still another type of "dynamic equilibrium," again standing apart from control systems theory, is the maintenance of *steady* oscillation or rhythmicity. Rhythmicity may be weak (easily disturbed) or highly stable. Physiology exhibits several of these besides the cardiac and respiratory examples we discuss. In linear closed systems (to which the beginning student is usually restricted), a fixed oscillation is the unstable case between a dying and a runaway transient. But steady physical oscillators like clocks or electronic relaxation oscillators, and biological oscillators like the nerve cell membrane, are all open (energy-fed) nonlinear, nonequilibrium systems. If disturbed, their oscillatory frequency and amplitude return monotonically to stable values which, in general, are rather insensitive thermally and so not dependent on chemical rates. The analysis of such stable rhythms is most comprehensible by a graphical (topologic) formalism called the phase plane, which we develop in Chapter 3 in relation to the axon. Higgins [1.3] has worked out the necessary and sufficient conditions for steady oscillations in complex coupled chemical systems (a special case).

Finally, having separated several physiological homeostatic systems that do *not* fit the analysis of feedback control engineering, we can examine more closely those which *do* include the canonical elements for error sensing, control function, effector unit, and so on. For example, temperature control (Chapter 19), carotid pressure control (Chapter 16), and perhaps certain pursuit mechanisms (Chapters 11 and 12) are not only modeled or fitted by, but also correspond mechanistically to, standard engineering control systems. However, the latter must be quite sophisticated to suffice, that is, rate sensitive, adaptive, and sampling systems must be present. Since the recent texts of Milsum [0.9] and Milhorn [0.8] develop adequately the mathematical apparatus of several such systems, we need not present detailed coverage here.

Other nonstandard feedback control mechanisms, indeed, use the output to modulate the input, but the mechanisms are strange to engineering. The comparator may be at the output, and the feedback operating not on input magnitude, but on an input velocity (rate) (glucose, Chapter 20), or on gain (muscle, Chapter 10), or on an output velocity (flux) element.

In modern physiology, we may consider mental as well as anatomically identifiable control systems; for even at the cognitive level, information (signal) processing systems may be tested experimentally (Chapter 4). In human relations (sociology), quantitative feedback control is well recognized; it is a general phenomenon of higher levels of organization of every sort.

Frequently then the transfer function (ratio of output to input) does not obey the elementary control theory, described in most physiology and biophysics texts, which depends on subtractive control and so nicely fits the ordinary analog computer. We are warned thereby not to blithely assume that all feedback is a function of the output alone and is fed into an input summing junction. Rather we must look carefully into the physiology to see not only what is the phenomenological law of control [1.4], but also what kind of intercoupling is actually present. It may not be additive at all, but multiplicative (gain, fractionation), variable impedance (conductance, permeability), or a variable coefficient of a derivative, and of novel insertion in the system. Without such caution, we may generate a plausible mathematical or physical model that will fit and so "predict" the data if we adjust the parameters. Such a model would have no scientific significance, however, and would in fact mislead us about what to measure and study next.

Terminology in the area of biological control systems is still evolving. Thus "activating feed-forward" is a stabilizing factor [1.3]. The necessary phase inversion of control processes may depend on the entry: inhibition may be "activated." In biosystems the generators (energy inputs) are often distributed, so that the customary gain (lumped-amplifier) elements are unusual, disguised, or missing. Gain also may be obtained from chemical processing of a substance that is, by itself, irrelevant, although it determines a signal or a property of the system. Matter–energy conversion is ubiquitous.

As regards coupling, feedback has been called a "point of view" (Milsum [0.9]), and truly so if all stabilization is attributed to it.

In the interest of encouraging careful distinctions, although we cannot change developing usage, we can point out, in examples exhibited hereafter, the class of control involved: (1) "*passive*," which depends on *equilibrium* of forces and concentrations, as well as on *steady states* of flow systems under their respective laws; (2) "*active*," wherein identifiable control mechanisms mix in information from upstream or downstream; and (3) "*regulatory*," wherein a system element is designed to be locally autolimiting (or, by contrast, autocatalytic).

This moves us into the broader area of biomathematics.

C. THE ROLE OF BIOMATHEMATICS

In what sense can there be a biomathematics? Is not mathematics unified, but with different applications?

Long usage and developments over the last century have established a *physical mathematics*, with several compendia (e.g., Courant-Hilbert, *Methods of Mathematical Physics*). This branch, which was originated and developed by physicists for physics, was subsequently polished by pure and applied mathematicians. Every branch of physics is represented. Physicists from Helmholtz and Maxwell to Einstein, Dirac, and Feynman have developed new formalisms.

The mathematics of physics as such has paid little attention to stationarily coupled systems of high complexity. On the one hand, modern (quantum) physics has been interested in collisions and the transient interactions thereof. On the other, the interaction of particles orbiting in each other's quantized fields (as in molecules, atoms, and nuclei) deals with statistical and resonance patterns of equilibria, rather than with the relatively fixed characteristic configurations seen in dynamic mechanisms at the macro level, or the steady-state couplings of open-flow systems. So in physical mathematics we find neither the conceptual models nor the mathematical forms needed to cope with the highly complex, but also highly specific, organization that is encountered in biosystems and used for processing both matter and information. In addition, the "boundary value problems" of physical dynamics (mechano, electro, and thermo) become meaningless when the boundaries themselves are not fixed but dynamic and externally modulated, and when the coefficients are variable. In general, therefore, physicists have left coupled systems to the engineers, as mere special cases of the more "fundamental study of elements" rather than of linkage relations. In fact, the prestige and authority of physics and biophysics (which to a large extent shares this view) have perhaps influenced some biologists to prefer these simpler "fundamental" problems.

Similarly, there is a typical body of *chemical mathematics* used in theoretical and physical chemistry, developed by such men as Gibbs, Kirkwood, Pauling, Debye, and Prigogine. It covers the micro and macro levels of gas and solution properties, but its power extends little beyond the macromolecule. Typically, it relies on thermostatics and dynamics, on quantum mathematics of bond structures, on modifications of Boltzmann's statistics, and on systems of linear differential equations rather simply coupled. The advent of open-system (irreversible) thermodynamics and chemical (process) engineering is rapidly extending this body.

From physics and chemistry, new mathematical applications have developed in many fields of engineering, so that it is possible to speak meaningfully of *engineering mathematics.* Algebraically, the application of vector, tensor, and matrix methods is greatly expanded in engineering mechanics and electricity. Multidimensional geometric and topological representations also appear, especially in nonlinear differential equations. Coupled systems become so complex that reduction by algebra is too cumbersome, and shortcut graphical methods are used to transform the system into simpler equivalent graphs (signal-flow, Chapter 10) before solution. Here the logic of algebra is replaced by a network topology which is easier to follow, calculate, and verify. Such mathematics, now constrained to correspond with the actual (physical) system, unifies many alternative formalisms. Many systems of differential equations can be reduced to algebra by the Laplace transform and then solved by such graphical reduction without loss of accuracy. We will discuss and use this technique.

Since engineering problems approach biologic ones in complexity, engineering mathematics is the best preparation for quantitative analysis and formulation of the design of biosystems. Engineering mathematics deals extensively with many kinds of stability, and types and degrees of instability, along with quantitation; it also considers yields, adaptive changes of parameters, wide dynamic range, and nonstationary operators (changing system structure). It is concerned with analyzing systems in dysfunction, with reserve systems, and with factors of safety. All these areas are common in biology and medicine, as soon as one addresses biosystems at work.

Biological mathematics is developing slowly, and means many things to many people, depending on the individual's approach. It is the next step after "quantitative biology": the use of objective scaled measurements of biostructures and processes, expressed in tabular or graphical form. Four aspects merit consideration, namely, curve fitting, concept fitting, intrinsic biomathematics, and the signal flow diagram technique.

1. Curve Fitting

To the bioscientist and computer man, biomathematics may mean simply curve fitting: finding the simplest or otherwise optimal mathematical expression to describe an observed relation between variables. We might call this mathematical taxonomy. Such reduction to symbolic shorthand makes possible complex calculations and helps to ascertain the relative importance of certain terms in order to simplify a calculation. As an example, we might mention the optimal "basis functions" for the electrocardiographic (ECG) wave form [1.5].

Curve fitting is also used with stochastic data to find the distribution function that most simply describes the observations without a mechanistic basis for this distribution, as in nerve impulse physiology. Indeed, till the 1950s, biomathematics was predominantly statistics and statistical decisions. This book may help demonstrate that this situation has been reversed now.

Characteristic of the curve-fitting approach is the adjustment of parameters to fit the data. In one sense this permits "prediction." The results of an experiment can be "mathematically predicted" from previous experiments of the *same* kind. Although this may astound and impress the biologist who is weak mathematically or the engineer who is weak biologically, it contributes little to understanding of mechanism (i.e., scientific insight). The mathematical model and its parameters may not be related (homologous) to the biological system involved, so that the result of a *new* kind of experiment cannot be predicted; it has no generalizing value. Many different mathematical models may fit equally well, the more complicated ones generally affording scope for other variables as they are introduced, and so corresponding better with nature. Thus, "Occam's razor" (William of Occam's principle that a theory should be parsimonious) is a poor guide; the objects instead are to understand the living mechanism and to incorporate in the mathematics as many properties of its components as one needs. This is one reason for collaboration of biomedical and bioengineering scientists. Nevertheless, putting an observed relationship into mathematical form need not (as Kelvin suggested) help us to understand it, but perhaps only to manipulate it. We grant that this alone is a worthy end, as some mathematical transformations may indeed suggest or reveal a mechanism.

At any one level of biological organization, it is desirable to relate to submechanisms by finding models that exhibit homology as well as fit. But it is obviously impractical, intractable, and confusing to express cardiovascular control in subcellular terms or cognitive processes in terms of membrane depolarization. Just because metastable structures occur at the macromolecular level is no indication that these are implicated in the learning mechanism. Such structures occur at every level of organization. Nor should the science of human function be delayed while subcellular possibilities are explored. Thus mathematical models, first descriptive and then homologous, are needed at every level of organization to deal with the complexity that occurs at that level. The equations formulating the chemistry of the mitochondrion are no less complex or more accessible than the logical relations between cognitive processes and communicative behavior; we need to know both.

Curve fitting, then, is an initial procedure toward a homologous systems model at a given level of organization. It is also a source of examples of interest to biologists in learning mathematical procedures such as the Laplace transform. It serves as shorthand for calculations. But to seek an explicit analytic expression to fit a curve, if it does only that, is hardly needed for calculation, since both digital and analog computers deal quite readily with the curves themselves (or their estimates, if stochastic).

2. Concept Fitting

Another approach to biological mathematics is concept fitting. An engineer, observing a homeostatic or regulated variable, may jump to the conclusion that this involves a "feedback mechanism." And sometimes he will be right, especially if all such stability is defined as feedback (Milsum [0.9]). This term, introduced in the 1940s by control engineers for a specific error-minimizing mechanism, is now applied to many other modes of stabilization or limiting. Our point here is that "feedback" is used in connection with so many mechanisms in biology with which a control engineer is unfamiliar, that forcing all stabilization into the concept of a negative feedback control, without regard to the actual mechanism used, may be more misleading than enlightening.

An equally important concept recently developed in the engineering mathematics of systems is that of "memory" in the form of the weighting function of filters. This irreversible dying or dissipation of information and of the form and span of operator action gives mathematical format to important biological phenomena. Thus there are advantages as well as dangers in applying engineering concepts biomedically. These conceptual advances from the science of systems are perhaps the most important modern development of biological mathematics, although new bioapplications of statistical mathematics (nonparametric) are advancing rapidly too.

Overstress on the important concept of "linear" models (often sufficient for engineering design) may also be misleading, especially now that nonlinear analysis is available topologically, analytically, and by computer (Chapters 3, 4, 7, 15, 16, and 20). An important analytical and conceptual tool is the piecewise linear and lumped model, which has special power in branched systems.

A final example of concept fitting is the quest for electrical analogs of biologic systems. A linear analysis of differential equations of temporal behavior (through the Laplace transform) at once suggests an electrical model; these models indeed have great power for visualizing or even solving a reduced equivalent system (e.g., in the circulation). However, many features of living hydromechanical systems such as plasticity or variable capacity (versus capacitance) seem to lack an electrical equivalent. In biology, functions and derivatives discontinuous at zero are common. Nevertheless, on the whole both the linear mathematical formalism and electrophysical models are valuable enough to become familiar to biologists. We use them freely in this book, pointing out their weaknesses as well as strengths.

3. Intrinsic Biomathematics

The term intrinsic biomathematics denotes formulations or formalisms that describe in some singular sense the inner mechanisms of a biosystem, its processes, or behavior. An intrinsic representation or mathematical model of the control of blood sugar, on the hyperglycemic side, is the set of six highly coupled equations whose terms identify with the component reactions and hormones in the liver–pancreas–blood–muscle system. Less intrinsic formalisms are also possible [20.2]. We point out intrinsic modeling as it occurs.

Another kind of intrinsic mathematics deals with procedures for finding observables or descriptors that are informationally independent with respect to a given scientific question. Thus one peculiar property of the biological system represented by the normal ECG is its set of four intrinsic (eigen) components, an orthogonal frame or four-space basis in terms of which all "leads" can be described [1.5].

Stochastic (mathematical) forms may or may not be "intrinsic" rather than "curve fits." The variability of nerve-firing thresholds is a borderline example. Stochastic formulations here may be intrinsic if attributable to intrinsically random mechanisms, such as Brownian molecular collisions, or if they are due to a multiplicity of decoupled factors, as we find in agricultural or disease situations. But superficial variability can so often be sharply reduced by identifying systematic factors ("intrinsic axes" in probability space) that to some degree "probabilistic behavior" may be an artifact of the reference frame. This is true of nerve firing, where synapses are highly organized. In biomathematics our task is to find such axes and factors. We discuss this more fully in threshold theory (Chapter 4).

Historically, the works of Helmholtz on eye and ear are monumental early contributions to biomathematics. Lotka [1.6] and Volterra [1.7] treated ecological relations and their stability by systems of linear differential equations. In more highly variate situations, the relevant mathematics was early developed by Thurstone and R. A. Fisher [1.8]. Rashevsky and his colleagues pioneered mathematical modeling at a variety of organizational levels, with rather idealized assumptions [1.9]. C. E. Shannon [0.10] perceived the possibilities of mathematically expressing organization per se (which increases *pari-passu* with levels of life)—a proposal investigated at the simpler levels by Yockey [1.10].

In the 1950s came a revolution in engineering mathematics, brought about by the discovery and rationalization of multidimensional control systems, the advent of computers to handle their calculation, the use of tracers to delineate subsystems, and an outburst of dynamic models of biosystems, described here. A general theory of systems, independent of mechanisms and first suggested by Bertalanffy [1.11], became recognized as equally applicable to chemical, electrical, biological, industrial, or social systems. One of its very powerful formalisms, the signal-flow diagram, was developed [1.12] by electrical engineers, and in certain applications it might be called the mathematics of systems.

4. The Signal-Flow Diagram

In the signal-flow approach, matter and energetics (the cornerstones of chemistry and physics) become irrelevant or subsidiary to information (negentropy) changes and transforms. We focus on the quantitative aspects of the changing organization of informational structure, on behavioral mechanisms, and on the performance of tasks.

Such mathematics is intrinsically suited to the nature of living systems, which constantly and spontaneously (Chapter 2, Appendix) increase organization, simultaneously at many levels, internally and externally, environmentally, socially, and culturally. Because this kind of mathematics is expressly developed for systems designed to serve a changing and adapting purpose (goal-seeking inventive and exploring operations), it is well suited to describe and handle the behavior of complex biosystems. The signal-flow diagrams and the associated formalism are presented in Chapter 10. We may expect their much wider use in biology as they become more generally known and used by bioengineers. They have an important directional property next to be discussed.

A set of first-order equations may represent, for instance,

1. A system of chemical reactions involved in a biological relationship (e.g., sugar, insulin, fats in the blood, muscle, liver).
2. A set of relations between the observables of a coupled system (e.g., in the circulatory system—pressure, flow or velocity, resistance, capacity, elastance or capacitance, concentration or quantity).

But this is not enough. In a bio (open) system there is a defined *direction* of reaction, flow, or throughput, and in most systems this cannot be reversed. (Inanimate examples are worm gear trains and amplifiers.) Thus stimulus and response are not interchangeable—a feature missing in the mathematical statement $y = ax$ or $x = y/a$. We can bear this directionality in mind, but the purely analytic equation alone is inadequate as a biological model.

In the signal-flow diagram, or flow graph, input and output are specified, both in detail and for the system. The flow graph resembles a "dual" of the physical anatomy (Fig. 1.1) in that it deals with local processes P which

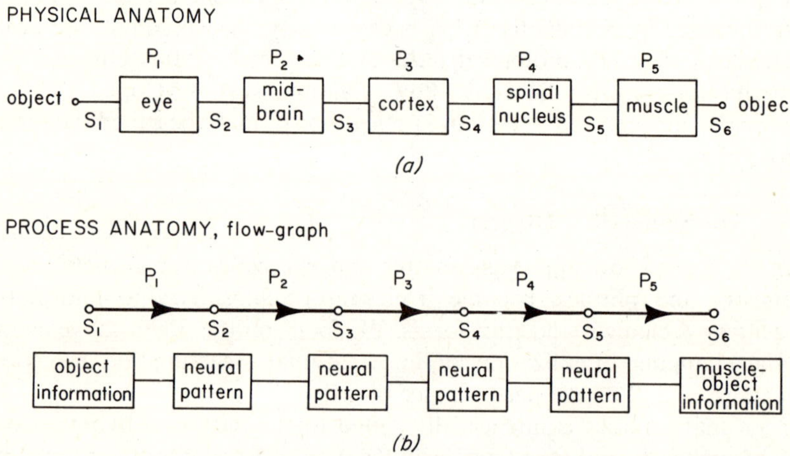

Fig. 1.1. Two different graphical representations of a system, illustrated with the example of muscle performance (output) under visual control (input). (*a*) Physical anatomy: the subsystems stand out, their characteristics defined by the operators P_1 to P_5. The signals processed by the subsystems are called S_1 to S_6. (*b*) Process anatomy: in this signal-flow diagram (or flow graph), emphasis is placed on the results (S's) of the processes. It may be considered the "dual" of the block diagram in the upper panel. Here, the functional relations between the signals S are conspicuous. Block diagrams adhere more closely to the physical anatomy. Both block diagrams and flow graphs are used in this book. Signal-flow diagram techniques are discussed in detail in Chapter 10.

transform one signal (information) pattern S into the next. Each connection P is considered to be an *operator* in both time and extent, expressible graphically or mathematically. The emphasis is on the result of the process rather than the mechanism for it. In this way systems analysis focuses on the *task*, overall and at each step. It is equally applicable to neural information flow, transport chains (pressure–velocity, flux of electrons, material, or concentration), or in combinations of these, as happens in respiration and circulation. The mathematics is carried "piggyback" as it were, in the transfer functions or operations P. These in turn (ideally) summarize the *effect* of mechanisms as seen at that level of function.

The flow graph also is apprehended at a glance topographically as a pattern, rather than logically as a set of algebraic relations. To the visually minded this is another advantage. When flow graphs are used, however, the operations and solutions of linear algebraic relations can be performed more simply graphically than by analytic algebra.

In addition, the flow graph (with due precaution) can handle nonlinear and nonstationary problems. The latter implies that the operators themselves may change, as in adaptive systems, which biosystems are *par excellence*. In particular, nonlinear systems may be noncommutative. A rectifier preceding a multiplier (both encountered biologically) gives a result differing from that of a rectifier after. Algebra postulates that $AB = BA$ in solving problems. If this is untrue, as it is in biosystems, one should use a formalism like the flow graph, which preserves the functional relations but also in a simple way preserves the *order* of operations. Although this restriction excludes some of the reductive tricks that can be used in graphical solution, it preserves the real mathematical properties of the biosystem.

In the flow graph, which is equivalent to a system of algebraic or differential equations, transforming calculus to algebra by the Laplace transform may of course be used. This is particularly true where, as in viscoelastic tissues, the time derivative of displacement (or in mass-loaded systems, of velocity) must be expressed. Such derivative terms also occur in much neural processing. Here again, the algebraic formalism and its flow graph equivalent become the simplest formulation of input–output relations. In this case linearity, commutativity, and stationarity are required.

Finally, it is possible to draw a particular flow graph that confines its attention to the particular observables (e.g., stimulus–response) being used experimentally; it expresses, illustrates, and solves the system in terms of these, explicitly. Other observables are left implicit. A flow graph also makes possible more simple and stable *models* of systems in analog computers, by use of the time-inversion transform. This does away with nonconverging solutions. In fact, the flow graph is essentially a wiring diagram for such a computer (Chapter 10). All these properties of the flow graph have been simply described and explained in recent references [1.13].

Simple systems formulable by one or two algebraic equations may not be advantageously represented in the form of a flow graph. The algebraic form is of course more familiar to most people who have not had to deal with complex systems. But if we are to cope with and analyze complex systems, as engineers and biomedical scientists must, it is well to study the mathematical forms best suited. Flow graph methods are only now being developed for systems whose couplings are non-additive (i.e., by way of parametric modulations, the commonest biologic case).

However, control and signal theory, which together gave rise to the flow graph, are not yet fully suitable for biology. Both the mathematics of information theory and its biological applications have been hampered by the narrow definitions used in communications or message engineering. Applied more broadly to the dynamics of spontaneous increase of organization, to the problems of meaning and of exploratory behavior, in a "thermodynamics" primarily of negentropic change (i.e., in systems in which the flux of energy becomes negligible or irrelevant), a quantitative formalism of information forces and flows appears to be possible (Chapter 2). Another weakness of control and signal theory lies in the restriction to time-series or temporal variables. In the biological system a great deal of emphasis must be laid on spatial variables and on the transformation and interaction between the spatial and the temporal.

Engineers have been impressed by the fact that in the sophisticated design and cognate mathematics of living systems (bionics) there occur mathematical operations that are still uncommon in engineering science. Some of these are cumbersome mathematically but easily modeled physically, as shown in nonlinear controls. Of special interest is the weakness in biological applications of Aristotelian logic: the binary decision or excluded middle. So the term "biomathematics" might be applied to those operations and mathematical systems-formulations which mainly occur in biosystems—just as was true for physical or chemical mathematics. The chapters of this book collect and critically discuss many examples of such biomathematics.

In summary, biomathematics goes beyond curve fitting and concept fitting to mathematical approaches intrinsically suited to specific biological situations and often uncommon elsewhere. It selects freely from the tools of linear and nonlinear analysis, systems theory, the formalisms of open-system (negentropic) thermodynamics, and it mixes topological with algebraic representation. The spatial domain, usually multidimensional, is as important as the time or frequency domain. These factors are illustrated by the text, not as a course in biomathematics, but as a map and critique of current models from the broadest area possible in an introductory book at this level.

D. CONCLUSION

Systems science has a somewhat different appearance to every scientist, in terms of his thought categories. Systems physiology, in the authors' view, sees the human in health and disease as a whole person, complete with motives, emotions, reflection, and physical and mental enjoyment—as an individual who functions at changing tasks or goals in a complex social framework. In this context, his functioning is to be studied and described by scientific methods. But we assume to exist and seek to formulate principles by which his performance (as a whole and in detail) can be made intelligible.

First of these principles is that he may be viewed in terms of levels and kinds of organization that are more or less statable but are not predetermined, because they are constantly changing and in fact synthesizing.* The lower or internal levels are somewhat more fixed in anatomical (structural) aspects, though with much functional variation between individuals and within the individual according to the tasks cognate to that level. Much of this lability resides in the recombinations possible in the nervous system, both in learning new optimizing patterns and in compensating for deficiencies (referred to various tasks). The higher levels, involving for example interpersonal psychophysiology, are relatively unexplored and are outside this book's scope. But the principle of continuing change, specifically this continual increase of organization, is of the essence. Much of a person's effort may be spent in reorganizing for new tasks or problems, especially in ageing, dysfunction, and misfortune.

Second, this reorganization occurs in a direction that serves to minimize effort: entropy production rate (Prigogine's principle). This is accomplished both by finding functional patterns of least waste power and by optimizing information flow by way of prediction, feed-forward, and preprogramming—that is, by minimizing useless information flow and processing.

It is particularly this developing *organization* or reorganization-for-task which occupies the systems scientist. How is coordination developed between organs, or between mind and body? In what does it consist, specifically? Can its processes be analyzed in an orderly way and are they composed of elements that, when expressed quantitatively, turn up quite generally in all kinds of systems? These are questions for systems physiology. Recognizing that organization is task directed and that optimal success with each task is only approximate, the approach to success must be studied and measured.

The dynamical aspects of this optimization constitute biocontrol theory. But a controller is not a performer, rather it is a submechanism toward performance; and dynamics is not all-important. No amount of error correction will make a maestro. The system as a whole must be organized for high performance, and this total organization (and reorganization) we would study, rationalize and, where convenient and possible for generalizing purposes, mathematize. For the aim of all science is generalization—the quest for principles. (Be it said at once that all principles are *not* statable in

* See Rothstein's proof ([2.7] and in Chapter 2, p. 60). Also, this continual increase in organization has additional dimensions noted by Teilhard de Chardin [1.15].

mathematical form; indeed, many are qualitative, and mathematics is often limited and inflexible for this purpose.)

Clearly we start where we are, with familiar physiological systems and subsystems, and try to formulate their organization framework, called a model. (A more thorough discussion of the theory of physiological models is given by Harmon and Lewis [1.14].) One model will not suffice for all possible tasks of the system, for that would require a model as complex as the system itself. But by studying models and their deficiencies, we force ourselves to focus accurately and quantitatively on detailed performance and its control. The classical physiologist studies each system for its own sake. The systems physiologist, by study of diverse engineering systems, has come to recognize the analytic commonality of models, and so primarily seeks the commonality or generalization of diverse physiological (and psycho- and pathophysiological) systems. A reasonable way to uncover this is to compare large numbers of models of diverse physiological systems—which explains this enterprise, although it be limited, tentative, and in its beginning stages.

The biosystems scientist has another problem. His models are not only inescapably hierarchic, but also may need to be set up independently for the many levels of organization. Unlike the models of the physical systems scientist, each of the models of the investigator of biosystems need *not* be inherently expressible in terms of models at lower levels. This is because (1) in open systems new extrinsic factors enter as complexity grows, so that the model must be reformulated in new variables, and (2) the model suffices only for a limited task or goal for which it is optimal; that is, it is goal directed. In addition, the systems physiologist, the biomedical engineer, and the physician view this challenge of modeling in different ways—the last two being also under constraints of time and penalty for failure. In view of this hierarchic diversity a systems physiologist will likely restrict his research to certain levels of organization, especially if we confine ourselves to human systems. How then is the student to learn of these many opportunities for study, since all knowledge cannot be his field? Again we propose that in this book, as in medical science, the examination of a large number of representative models opens the best route to an appreciation of the systems view of physiology.

E. REFERENCES*

1.1. Simpson, G. G. Status of the study of organisms. *Amer Sci* **50**: 36–45, 1962.

1.2. Engelberg, J. Physiological regulation: the steady state. *Physiologist* **9**: 69–88, 1966.

1.3. Higgins, J. Chapter 1 in *Control of Energy Metabolism*. B. Chance, R. Estabrook, and J. Williamson (Eds.). Academic Press, New York, 1965.

1.4. Trimmer, J. D. *Response of Physical Systems.* John Wiley and Sons, New York, 1950.

1.5. Young, T. Y. and W. H. Huggins. On the representation of electrocardiograms. *IEEE Trans Bio-Med Electron* **BME-10**: 86–95, 1963.

1.6. Lotka, A. J. *Elements of Physical Biology.* Williams and Wilkins Co., Baltimore, 1925. Reprinted as *Elements of Mathematical Biology.* Dover Publications, New York, 1965.

1.7. Volterra, V. *Leçons sur la Théorie Mathématique de la Lutte pour la Vie.* Villars, Paris, 1931.

1.8. Fisher, R. A. *The Design of Experiments*, 8th ed. Hafner Publishing Co., Darien, Conn., 1966.

1.9. Rashevsky, N. *Mathematical Biophysics*, 3rd ed. Dover Publications, New York, (1938) 1960.

1.10. Yockey, H. P. (Ed.). *Symposium on Information Theory in Biology.* Pergamon Press, New York, 1958.

1.11. v. Bertalanffy, L. *General System Theory.* G. Braziller, New York, 1968.

1.12. Mason, S. J. and H. J. Zimmerman. *Electronic Circuits, Signals and Systems.* John Wiley and Sons, New York, 1960. Chapter 4.

1.13. Huggins, W. H. and D. R. Entwisle. *Introductory Systems and Design.* Blaisdell Publishing Co., Waltham, Mass., 1968.

1.14. Harmon, L. D. and E. R. Lewis. Neural modeling. Physiol Rev **46**: 513–591, 1966.

1.15. Teilhard de Chardin, P. *The Phenomenon of Man.* Harper and Brothers, New York, 1959.

* Journal references follow the style of *Index Medicus.*

Part I

NEUROSENSORY SYSTEMS

CHAPTER TWO

Nerve Membrane: Statics and Steady-flow Processes

This chapter deals with the origin of nerve membrane potentials, an example of a complex physiological system at a very low (subcellular) level of biological organization. Using chemical thermodynamics, we calculate the total diffusion potential resulting from three ions (Na^+, K^+, and Cl^-). The standard derivation implies a model of components electrically in parallel [2.1–2.3]. Finkelstein and Mauro [2.4] show that this is incompatible with a homogeneous (versus mosaic) membrane and thus also incompatible with the dynamic (series) models used in older nerve impedance studies.

After sorting out these difficulties, we grapple with the large discrepancy between this simple passive electrochemical model and the actual living membrane, part of an open thermodynamic system. The difference involves active transport and irreversible thermodynamics. For the latter we outline the conceptual foundations and analytic methods, with examples. These enable us to illustrate how active chemical processes couple quantitatively with the diffusion potentials to account for the total potential observed [2.5].

A. ORIGIN OF POTENTIALS

Qualitatively, a potential exists across the nerve membrane because it is impermeable to large anions of protein P^- (mainly creatine) located inside, while it is permeable to water and to cations K^+ and Na^+, as well as to small anions Cl^-. Such a structure approximates a cationic exchange membrane. To calculate a potential we must turn to work theory (thermodynamics) and assume a model. The simplest model begins with a membrane permeable to the single cation K^+ drawn to a higher concentration inside by

P^-. Approximate electroneutrality inside results. The concentration or density difference of K^+ across the membrane gives rise to differential Brownian or entropic forces, but this mechanical force is balanced by the electric force resulting from an opposing double layer of ions. That is, a cation C^+ is pushed out of the axoplasm mechanically by the cation density and pulled in to the axoplasm electrically by the slight internal excess of anions (P^- and Cl^-) whereby the membrane becomes an electric doublet, inside negative. At equilibrium, in a *closed system*, we would find that any ionized salt CA is partitioned according to the Gibbs-Donnan ratio, enabling us to calculate a potential determined primarily by P^-. This would approximate the very low potential of a metabolically inactive nerve fiber.

Quantitatively, an incremental concentration ∂n of particles of species i, inside, increases the Gibbs potential G, and so the chemical potential, μ_i, defined as

$$\mu_i = \frac{\partial G}{\partial n_i}\bigg|_{p,\,T}$$

The increase of μ can be calculated by considering the solute a perfect gas, compressed by the small volume dV. This is so because experimentally in dilute solutions $pV/T = nR$, where R is the gas constant. At constant p, T the work done goes into increased density or "chemical" potential:

$$d\mu_i = p\,dV = RT\,\frac{dp}{p} \qquad \text{using} \quad pV = RT = \text{a constant}$$

In a solution, the partial pressure of species i is

$$p_i = \frac{n_i}{V}\,RT = c_i RT$$

where $\qquad c_i =$ molar concentration

Hence

$$d\mu_i\big|_{p,\,T} = RT\,\frac{dc_i}{c_i} \qquad \text{so that} \quad \mu_i = \mu_{i_0} + RT\,\ln c_i \qquad (2.1)$$

that is, "chemical potential" increases logarithmically with concentration. If several species are present, $G\big|_{p,T} = \sum \mu_i\,dn_i$. In a concentration gradient independent of charge, force per unit of material (mole) due to density alone is proportional to the spatial gradient:

$$f = -\frac{d\mu}{dx} = -RT\,\frac{1}{c}\frac{dc}{dx}$$

which varies directly with dc/dx and inversely with c; that is, force per mole increases in dilute solutions (curiously).

If eq. 2.1 is integrated from a point inside (′) to a point outside (″) the membrane, the "chemical potential" difference for uncharged particles of species i, regardless of membrane thickness, is

$$\Delta\mu = \mu' - \mu'' = RT \ln \frac{c'}{c''} \tag{2.2}$$

in calories (cal) of work or joules per mole (J/mole).

Since the solute is ionic, an electrical potential Ψ is present also. The sum of μ and Ψ is the "electrochemical potential"

$$\bar{\mu} = \mu + ZF\Psi \tag{2.3}$$

where Z = ionic charge number
F = Faraday (9.65×10^4 coulomb/mole)

The net electrochemical *force* (J/cm) on a mole is the (negative) spatial gradient of this:

$$\frac{d\bar{\mu}}{dx} = -RT\frac{1}{c}\frac{dc}{dx} - ZF\frac{d\Psi}{dx} \tag{2.4a}$$

Particles in this force field tend to migrate with the concentration gradient, but only until the associated electrical force cancels the diffusional one; then $\Delta\bar{\mu} = 0$. Integrating eq. 2.4a across the membrane, the potential difference (EMF) is given by Nernst's equation, assuming *equilibrium*:

$$E = \Psi'' - \Psi' = \frac{RT}{ZF}\ln\frac{c'}{c''} \tag{2.4b}$$

where $$\frac{RT}{F} = 0.026\ V \text{ at } 25°\text{C}$$

for each component of each solution, regardless of thickness or gradient in membrane. To determine polarity for cations ($Z > 0$) consider that, as for K^+ due to excess P^-, $c' > c''$, then $\Psi'' > \Psi'$. The inside becomes negative, as would be expected.

If the ion is one of a salt pair, for example, K^+ and Cl^-, in equal numbers (an additional constraint), then the impermeable anion P^- also determines the *magnitude* of the potential. In this case, besides the electrochemical equilibrium, *chemical* equilibrium alone requires for a uniunivalent electrolyte CA that $\Delta G = 0$ so that $\mu' = \mu''$. From eq. 2.1.

$$RT \ln [\text{CA}]' = RT \ln [\text{CA}]'' \tag{2.5a}$$

or

$$\frac{[\mathrm{C}]'}{[\mathrm{C}]''} = \frac{[\mathrm{A}]''}{[\mathrm{A}]'} \equiv r \tag{2.5b}$$

where $[\mathrm{X}] =$ concentration of species X (moles/liter)

This is the Gibbs-Donnan ratio. But if there is electroneutrality as well, then inside

$$[\mathrm{C}]' = [\mathrm{A}]' + Z_p[\mathrm{P}] \tag{2.6a}$$

where $[\mathrm{P}] =$ protein anion concentration

$Z_p =$ charge on protein ion (negative)

$$[\mathrm{C}]'' = [\mathrm{A}]'' \text{ outside} \tag{2.6b}$$

Introducing r from eq. 2.5b and eliminating $[\mathrm{C}]'$ and $[\mathrm{C}]''$, we have

$$r = \frac{1}{r} + Z_p \frac{[\mathrm{P}]}{[\mathrm{A}]''} \qquad \text{a quadratic in } r \tag{2.7a}$$

But if $[\mathrm{P}] \ll 1$

$$r \simeq 1 + \frac{Z_p[\mathrm{P}]}{2[\mathrm{A}]''} = 1 - \frac{[\mathrm{A}]' - [\mathrm{C}]'}{2[\mathrm{A}]''} \tag{2.7b}$$

Applying this to the squid nerve which we are analyzing (see table of concentrations, p. 42), we find

$$r = 1 - \frac{356}{1080} \simeq \frac{2}{3}$$

Consequently

$$E = \frac{RT}{F} \ln r \quad \text{from eqs. 2.4b and 2.5b}$$

$= 0.058 \log_{10} \frac{2}{3} = -11$ mV, the squid nerve potential without oxygen versus -62 mV measured with oxygen.

This chemical equilibrium model, then, has very little relation to the nerve membrane. All we gain from it is the Nernst expression for equilibrium voltage due to an ion-concentration difference. The Donnan ratio does not determine this difference in live nerve, because now the outside is *open* and has unlimited numbers of $Na^+ + Cl^-$, and the system is not in equilibrium. Instead, a metabolizing nerve actively pumps Na^+ out and K^+ in, making the $[\mathrm{K}]'/[\mathrm{K}]''$ ratio about 30, so that $E_{K^+} = 90$ mV. The pump is strong enough to reverse the Na^+ ratio while the Cl^- ratio falls out of electroneutrality. A good quantitative description of the pump's action and possible mechanism is given in Ruch-Patton [2.2].

Pump-out of Na^+ proceeding as quickly as Na^+ leaks in, and pump-in of K^+ against its leak, give rise to ion *currents* in the membrane, and we must consider the potentials and the concentration profiles of these currents. For any cation, the current density through the membrane is

$$I_C = \bar{C}Fv \tag{2.8}$$

where $\bar{C}$ = average concentration of cation within membrane
v = velocity of cation = uf (mobility × force)

Hence

$$I_C = \bar{C}Fu\left[\frac{RT}{F}\frac{1}{C}\frac{dC}{dx} + \frac{dE}{dx}\right]$$

where the square bracket is the force (eq. 2.4a) and $dE = Zd\Psi$.

If, as Goldman proposed, the field profile within the membrane is assumed to be *uniform*, then $dE/dx = E/a$ (a = membrane thickness), and the equation becomes

$$\frac{dC}{dx} + \frac{FE}{RTa}C = \frac{I_C}{RTu}$$

This integrates from inside ″ to outside ′, giving

$$\ln\frac{[C]'' - I_Ca/uFE}{[C]' - I_Ca/uFE} = \frac{-EF}{RT}$$

Solving for the cation current

$$I_C = P_C\frac{EF^2}{RT}\frac{[C]'' - [C]'\exp(-EF/RT)}{1 - \exp(-EF/RT)}$$

where permeability $P_C \equiv uRT/aF$.

Similarly, for the anion current in the opposite direction we have

$$I_A = P_A\frac{EF^2}{RT}\frac{[A]' - [A]''\exp(-EF/RT)}{1 - \exp(-EF/RT)} \quad \text{reversing limits}$$

where $P_A \equiv vRT/aF$

The sum of these two in the *steady state* is

$$0 = \frac{F^2E}{RT}\frac{(P_C[C]'' + P_A[A]') - (P_C[C]' + P_A[A]'')\exp(-EF/RT)}{1 - \exp(-EF/RT)}$$

Solving for E by taking the logarithm results in

$$E = \frac{RT}{F}\ln\frac{P_C[C]' + P_A[A]''}{P_C[C]'' + P_A[A]'} \tag{2.9}$$

This is Goldman's linear gradient equation for the combined potential of two ions. Extending this to the three ions K^+, Na^+, and Cl^- (Goldman-Hodgkin-Katz, see [2.2]) with no net current

$$E = \frac{RT}{F} \ln \frac{P_K[K]' + P_{Na}[Na]' + P_{Cl}[Cl]''}{P_K[K]'' + P_{Na}[Na]'' + P_{Cl}[Cl]'} \tag{2.10}$$

Experimentally, this physical model with linear membrane gradient gives good calculated values for static E, using values for permeability and the concentrations measured from squid axon [2.3]:

conc. inside

$[K]' = 345$ $[Na]' = 72$ $[Cl]' = 61$ (mmole/liter)

conc. outside

$[K]'' = 10$ $[Na]'' = 455$ $[Cl]'' = 540$ (mmole/liter)

permeability

$P_K = 1.8 \times 10^{-6}$ $P_{Na} = 0.07 \times 10^{-6}$ $P_{Cl} = 0.8 \times 10^{-6}$ (cm/sec)

Nernst potentials

$E_K = 89$ mV $E_{Na} = -46$ mV $E_{Cl} = 55$ mV

Setting these in eq. 2.10, the calculated membrane potential is 60 mV versus 62 mV experimentally. Similar accuracy holds for frog sartorius muscle [2.3]. We note that K^+ dominates the calculated resting potential

$$E_K \bigg|_{20^\circ} = 0.058 \log_{10} \frac{345}{10} = 89 \text{ mV}$$

but is strongly modified by the reverse potential of Na^+ caused by the metabolic pump. We now have a satisfactory model for calculating the combined potential produced by the three permeating ions' concentration ratios, but we still have no model for what determines those concentrations. Although no components of the pump have been identified, its qualitative and quantitative properties have been well described [2.2, 2.3]. Its coupling with diffusion transport and potentials is discussed in Section 2.E.

B. NERVE MEMBRANE IMPEDANCE

To understand the depolarization process of nerve and its propagation (the impulse), it is necessary to study and model its impedance and to show how greatly this *changes* with excitation. For 20 years research was focused on the total impedance (i.e., the relation of impressed potential to total current). Only in the 1950s did it become clear that the membrane exhibits static and dynamic impedance changes that differ drastically for each ion, especially

Na^+ and K^+, both naturally and under control by drugs. In explanation, it appears that the membrane is spatially inhomogeneous, having different patches of high permeability for each ion; it is not even necessary that each species act in unison. This phenomenon also affects threshold behavior for both driven and spontaneous firing. In addition, it involves temporal and spatial patterning of patches whose randomness is only partial.

The Goldman model does well with the *membrane potential* of several ions. How does it handle the problem of *ion currents and impedance*? The following critique of Finkelstein and Mauro [2.4] focuses on how the parallel ionic model considered in Section 2.A agrees with external impedance measurements.

A quantitative expression for impedance follows from the Nernst force eq. 2.4. The flux density ϕ of the jth cation is

$$\phi_j = -u_j C_j \left[\frac{RT}{C_j} \frac{dC_j}{dx} + F \frac{d\Psi}{dx} \right] \tag{2.11}$$

where u = mobility
Ψ = membrane potential

Integrating over the thickness of the membrane (outside to inside) gives

$$\phi_j \int_o^i \frac{dx}{F u_j C_j} = -\frac{RT}{F} \int_{C''}^{C'} d \ln C_j - \int_0^{-\Psi} d\Psi = -\frac{RT}{F} \ln \frac{C'}{C''} + \Psi$$

or, since

$$I_j = F\phi_j$$

$$IR = -E + \Psi$$

as shown in Fig. 2.1, with

$$R = \int_o^i \frac{dx}{F^2 u_j C_j}$$

$$E = \frac{RT}{F} \ln \frac{C'}{C''}$$

Evidently the resistance R varies with total membrane potential Ψ_T because of C and u profiles in the membrane, but the diffusion battery E does not depend on Ψ_T. Moreover, with zero external current, there may be strong *circulating* currents of anions and cations; this hypothesis is compatible with a mosaic structure corresponding to this physical model, but not with a homogeneous membrane.

The total membrane potential Ψ also depends in a simple way on the *mobilities* u and v of cations C and anions A inside the membrane. If the currents are equal

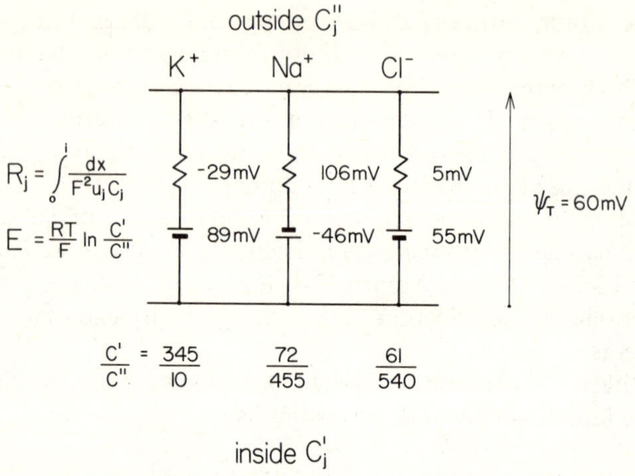

Fig. 2.1. Mixed ionic resistance model of squid axon membrane, where C_j' and C_j'' are the concentrations of the jth ion on the inside and the outside, respectively, and u_j is the mobility of the jth ion. The voltages are given as calculated across the resistances and for the "Nernst batteries." After Finkelstein and Mauro [2.4].

$$FCu\left[-\frac{RT}{F}\frac{1}{C}\frac{dC}{dx} - \frac{dE}{dx}\right] = FAv\left[-\frac{RT}{F}\frac{1}{A}\frac{dA}{dx} + \frac{dE}{dx}\right]$$

Solving for dE and integrating over the thickness a, assuming [C] = [A] within the membrane (but not $dE/dx = E/a$), the total membrane potential depends on u and v. In terms of the ionic concentrations of the two solutions

$$\Psi_T = \int_o^i dE = \int_{C''}^{C'} -\frac{u-v}{u+v}\frac{RT}{F}\frac{1}{C}\,dC = \frac{u-v}{u+v}\frac{RT}{F}\ln\frac{C'}{C''} \tag{2.12}$$

Thus if the ion pairs have nearly equal mobility ($u \simeq v$) (as in a K^+Cl^- membrane), indeed the total membrane potential $\Psi_T = 0$ experimentally.* The membrane here is a pure resistance. However, in the model (Fig. 2.1) the batteries E still work; that is, there must be *circulating* currents instead of *canceling* currents. So the usual parallel circuit potential model fails for a homogeneous membrane. Moreover, the resistances (conductances) of this model are independent of impressed current, by the following argument (at least in this single-salt case).

Defining I^* as current/mobility in the membrane from eq. 2.11, the net value is

* This contrasts with Goldman's eq. (2.9): $E = RT/F \ln (uC' + vA'')/(uC'' + vA')$ where a potential results in a linear-gradient model, even if $u = v$.

$$I^*(x) = \frac{I^-}{v} - \frac{I^+}{u} = \left[-RT\frac{dC}{dx} - FC\frac{d\Psi}{dx}\right] - \left[RT\frac{dC}{dx} - FC\frac{d\Psi}{dx}\right] = -2RT\frac{dC}{dx} \quad (2.13)$$

Integrating over thickness a yields

$$I^*a = 2RT(C' - C'')$$

At a point x in the membrane, then,

$$\int_0^x I^*\,dx \equiv \int_0^x \frac{2RT}{a}(C' - C'')\,dx = \int_0^C 2RT\,dC$$

giving the variation of C in the membrane with position

$$C = \frac{C' - C''}{a}x + C' \qquad \text{a linear change of concentration}$$

Hence the resistance in the cation branch can now be integrated:

$$R_C = \int_o^i \frac{dx}{F^2uC} = \int_o^i \frac{dC}{F^2u}\frac{a}{C' - C''} = \frac{a\ln(C'/C'')}{F^2u(C' - C'')} \qquad \text{similarly for } R_A$$

Since this depends only on external concentrations, rather than on *currents* in the membrane, a depolarizing current should not change these resistances (or conductances); they depend on membrane potential only. This proves important in the dynamics of membranes (see Chapter 3).

Pure Electrical Model. A quite different equivalent circuit of the nerve membrane may also be derived. If an external test current is impressed to determine impedance, all ion conductances act in parallel, rather than individually. This experimental method has long been used in relating nerve membrane reactance to its general impedance properties as a function of frequency (Cole [3.6]) and in studying passive propagation. It is interesting to compare the properties of components of a *series* model that fits such experiments with those of the parallel ionic networks we have been describing.

Analytically we can derive the steady-state properties of the membrane components modeled as a series network by writing eq. 2.13 as a sum rather than a difference. The currents of *all* cations and *all* anions are added:

$$\phi = -\sum_j\left[RTu_j\frac{dC_j}{dx} - Fu_jC_j\frac{d\Psi}{dx}\right] + \sum_k\left[RTv_k\frac{dA_k}{dx} - Fv_kA_k\frac{d\Psi}{dx}\right]$$

Rearranging, the total impressed flux is

$$\phi = RT\left(\frac{dV}{dx} - \frac{dU}{dx}\right) - F\frac{d\Psi}{dx}(U + V)$$

where

$$U(x) = \sum u_j C_j \quad \text{and} \quad dU = \sum u_j dC_j$$
$$V(x) = \sum v_k A_k \quad \text{and} \quad dV = \sum v_k dA_k$$

Rearranging again and substituting $I = F\phi$ gives

$$-d\Psi = I \frac{dx}{F^2(U + V)} + RT \frac{d(U - V)}{F(U + V)}$$

Integrating over x, the potential difference from outside to inside is

$$\Psi_T = I \int_o^i \frac{dx}{F^2(U + V)} + \frac{RT}{F} \int_o^i \frac{d(U - V)}{U + V} = I\bar{R} + \Psi_D \quad \text{(see Fig. 2.2)} \quad (2.14)$$

Note that the diffusion EMF Ψ_D now depends on the terminal voltage Ψ_T through its effects on the concentration gradients dC_j/dx *within* the membrane itself, instead of only on C' and C'' external to the membrane as in Nernst's equation.

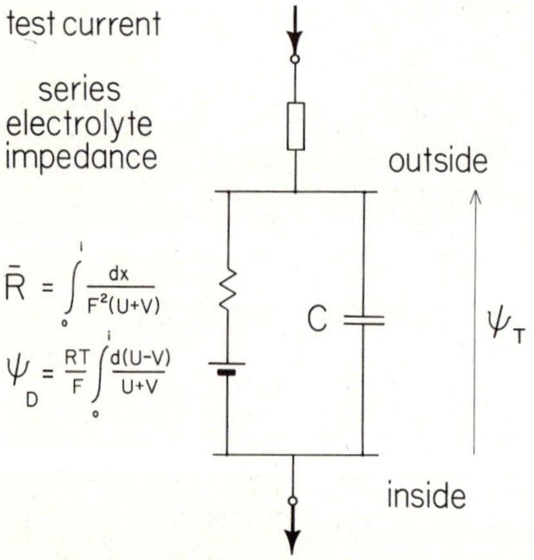

Fig. 2.2. "Pure electrical" model of nerve membrane. Both the apparent resistance $\bar{R}$ (as obtained experimentally by imposing test current) and the diffusion potential Ψ_D depend on mobilities in this model. For the applicability of the model, see text. After Finkelstein and Mauro [2.4].

This is a more realistic model because impedance measurements do show Ψ_D to have a reactive component (Fig. 3.1); that is a frequency dependence on a time-varying intramembrane profile of $C_j(x)$, even with zero net current. Because of the large membrane capacitance in parallel with these generator–resistance assemblies, however, it is difficult to design an electrical experiment that distinguishes between a mosaic and a homogeneous model [2.4].

In summary, we presently have both series and parallel models for the nerve membrane. The former type gives no information about ionic components, whose *differential* behavior must be known if we are to account for depolarization. The latter, unless it represents a mosaic of specific ion patches, fails to depict the intramembrane behavior, either for alternating current injected or in the simple case of a potassium chloride membrane interface. It appears that the choice of models depends on small versus large signal behavior; for large signals (depolarization) only a nonlinear, parallel model will suffice.

C. THE PROBLEM OF INTERACTION BETWEEN DIFFUSION AND CHEMICAL PROCESSES

Since most of the measured membrane potential arises from the large excess K^+ inside, produced by the "active" metabolic pump (chemical reactions) rather than from equilibrium diffusion processes, we must look for the connection or coupling between such *chemical* reactions and *diffusional* transport. A naïve theory would assume them to be linear and independent. But to investigate this question quantitatively requires a formalism for precise symbolic thinking about these quite different kinds of flux processes.

Classical (equilibrium) thermodynamics suffices to describe the free-energy aspects of these "active" chemical processes [2.6]. Thus we may picture a system (Fig. 2.3) in which the first change of Gibbs free energy $\Delta\mu' = \Delta G|_{p,T}$, that is, where the sodium ion Na^+ at low concentration $[Na]_i$ just inside the boundary is combined by chemical reaction with carrier C by an enzyme residing at interface I–II, using ATP bonding energy $\tilde{P}$

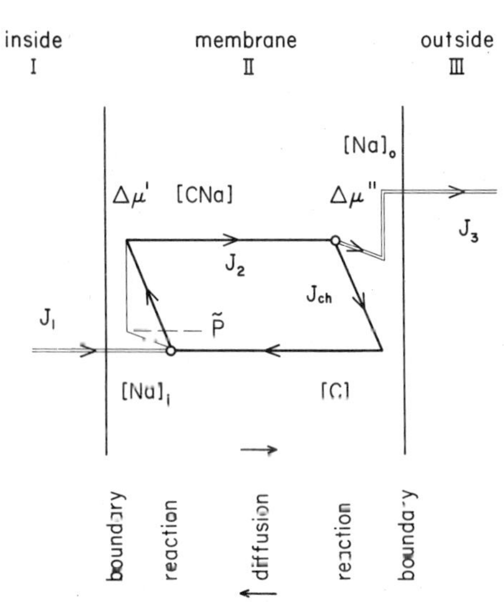

Fig. 2.3. Coupling of chemical and diffusion processes. The free energy change $\Delta\mu'$ is due to the reaction of carrier C with Na^+ at interface I–II. The compound [CNa] is driven toward the outside by its large diffusion gradient. At interface II–III, a second enzyme frees energy $\Delta\mu''$, which forces Na^+ out of the membrane. This frees carrier C to diffuse back toward the inside of the cell. Energy gained from the chemical reaction is represented by $\tilde{P}$. J_{ch} is given by the rate of the reaction within the membrane.

produced in solution I. The compound CNa, now having a high diffusion potential referred to the other side, moves outward. At interface II–III it is then dissociated by a second enzyme freeing energy $\Delta\mu''$, which is thereupon utilized to force out Na^+ at its higher concentration $[Na]_O$. Carrier C then diffuses inward cyclically, carrying K^+ by a similar process up *its* concentration gradient. This whole outward process can occur at a rate proportional to $[Na]_i^3$ [2.2]—a rate that is much higher than the slow inward *diffusion* of Na^+ (explained as a low sodium permeability P_{Na} when the membrane potential is high). A steady state of flow is finally reached when the amount of Na^+ pumped outward equals the amount of Na^+ diffusing inward. Although flow proceeds, the system is regarded as "near equilibrium" energetically.

The Na^+ pumping rate of the inactive nerve [3 pM/cm^2 sec] would need to increase in a nerve firing at 100 impulses/sec by 100-fold to maintain polarization [2.2, p. 60]. The control mechanism whereby the pump holds the $[Na]_i$ at its leakage value is not known. But since such membranes can fire continuously at a high rate for the lifetime of the animal, we know that the pump rate is variable.

However, classical equilibrium thermodynamics gives no way to formulate several important aspects of this active transport process, which occurs indeed in many other living membranes. These are:

1. It occurs at a rapid but steady rate, involving flow conductances (resistances) whose values are significant and must be expressed explicitly.
2. At steady states, the rate of the *chemical reaction* in II is limited by the rate of *diffusion* from I.
3. Within the anisotropic (structured) membrane II there may be entrainment or drag phenomena, whereby the *degree of advancement* of the chemical reaction (not even defined in equilibrium theory) depends on the counterflows of K^+, Na^+, C, and CNa (in this example).

Recent theory of nonequilibrium thermodynamics deals mainly with the frictional interaction or interference factors that accompany the transport of materials, electrons, entropy, or information. The reader is referred to Katchalsky [2.5] for a readable and detailed review of the thermodynamics involved in this "irreversible" formalism and its applications to passive and active transport, as far as these have been carried.

Moreover, the "carrier" molecule just described is still hypothetical and has not been identified yet in nerve or in the many other biological membranes where active transport occurs. Mechanisms of diffusional coupling and their reactions are unknown. Evidence for them is circumstantial, but strong. In view of this situation, only generalized models have been worked out, and the formalism developed for four possible cases. These are:

1. The chemical (diffusion) potential drop $\Delta\mu_i$ across a membrane depends on the rates (J_{ch}) of chemical *reactions* within membrane II (Fig. 2.3), because substances diffusing from regions I and III participate actively in this reaction.

2. The chemical potential drop across the membrane involves one component that is *coupled* diffusively (entrained) with another component participating in the reaction.

3. The passive rate of transport (e.g., Na^+ outward) is increased because it is combined chemically with a carrier within the membrane. Each carrier molecule bears several of the molecules being extruded—a "facilitation" theory. The only example proved so far is an artificial membrane containing hemoglobin Hb, which (as HbO_2) rapidly transports O_2 (although not *up* an O_2 gradient, as required by nerve membrane theory).

4. If the membrane boundaries are complementary ion exchangers (semipermeable diffusers for anions and cations respectively), then the decomposition (a chemical reaction) of a salt within it will produce a potential across it.

It seems evident that we still do not have a satisfactory model describing not only the results (as do the Nernst and Goldman models), but also the mechanisms of producing the observed (mainly active) potential of the nerve membrane. It is also clear that to deal with this situation we need a *formalism* (physicochemical and mathematical) that is capable of expressing quantitatively the *nonequilibrium* situation, ubiquitous in both normal and abnormal physiology.

Hence the next section presents a short introduction to this formalism, essential even for reading current literature on membrane potentials and transport. We must also describe membrane *dynamics*, which depends on the nonlinear resistance of the membrane: its resistance to combined Na^+ and K^+ flow changing transiently from positive to negative during the nerve impulse. The formalism is also needed for nerve models involving such elements. Finally, we study nonequilibrium thermodynamics because it treats quantitatively the rate of increase of both entropy and negentropy, using the extremum principle of Prigogine (see [2.5]) and the synthesis principle of Rothstein [2.7], which have wide application and great importance in systems physiology at every level.

D. INTRODUCTION TO NONEQUILIBRIUM, OPEN-SYSTEM THERMODYNAMICS: ENTRAINMENT

Thermodynamics differs from energetics (as used in mechanics or electricity) in its emphasis on the entropy term S—the dissipation or randomization accompanying natural processes of potential loss. In thermodynamics, however, S is restricted to *thermal* randomization. Randomizations of pattern, of system couplings, or of kinds of organization in a complex system of message information or of meantent* transmission (Colby [2.8]) have not yet been formalized. However, for simple physicochemical models pertinent at low levels of biological organization, thermal entropy is important and adequate. Like mechanical work $p\ dV$, heat ($dQ = T\ dS$) has an intensive factor $T \sim p$ and an extensive factor (entropy) $S \sim V$. Classical equilibrium thermodynamics (thermostatics) exhibits this thermal dissipation dS as a composite resulting from many kinds of work. Thus a small increment in the total energy of a system can contain a number of generalized "work" terms. The total internal energy may be written, holding the intensive factors constant, as follows:

$$dU = T\ dS - p\ dV + \sigma\ dA + f\ dl + \Psi\ de + \sum_i \mu_i\ dn_i \text{ (diffusive or "chemical")} + \sum_i A_i\ d\xi_i \text{ (chemical reaction)} + \text{gravitational and centrifugal diffusion equilibria} + \text{thermoelectric, Joulean, Fourier heats} + \text{etc.} \tag{2.15}$$

where dS contains the dissipative ("incurred waste") contributions from all these.

Many examples using a variety of nonchemical variables are shown by Tribus [2.9]. For membrane phenomena (diffusion, reactions) at constant temperature and pressure, we need only a few of the terms of eq. 2.15:

$$-T\ dS = -dU - p\ dV + \Psi\ de + \sum_i \mu_i\ dn_i + \sum_i A_i\ d\xi_i \tag{2.16a}$$

where (as in the general case) each "work" term has an intensive (potential or force) factor and an extensive (displacement) factor, their product being work or energy dimensionally.

If the system is open, so that (*a*) flowing through it are matter, energy, and entropy, while (*b*) locally within it (at sources and sinks) both heat and entropy are produced or consumed and matter is chemically converted, then we must deal instead with the flow or *rate* of work: force flux (dimensionally, power). Taking the time derivative of eq. 2.16a, the "dissipation," as defined by Rayleigh, is

$$\Phi = T\frac{dS}{dt} = p\dot{V} + \Psi i + \sum_i \mu_i \frac{dn_i}{dt} + \sum_i A_i \frac{d\xi_i}{dt} \tag{2.16b}$$

$$= \sum_i X_i J_i \qquad \text{in general}$$

where X_i are forces, J_i are fluxes (of $\dot{S}$, entropy), $\dot{V}$ is volume flow, i is current, n is mass flow, ξ is relative speed of reaction.

* This is the intentional component of the meaning of a sentence, or of an utterance [2.8].

A simple hydraulic example (Franck [2.10]) using only the first term serves to illustrate the stability (minimization) principle that is applicable to this flow regime. Instead of the familiar minimization of various *potentials* as G in thermostatics, h ("pressure head") in hydromechanics, or Ψ in electricity with maximization of entropy S, Prigogine [2.5] showed that in open systems at steady state there is a minimization of *waste power* ($\dot{S}$), that is, the rate of generating random (unorganized) activity, or conversely, a maximum rate of generating organization (negentropy). That this generation of organization is a natural self-acting process was proved for *cyclic* open systems by Rothstein [2.7] (see Section 2.F). Most biosystems are cyclic, having evolved under solar drive on a rotating planet.

Consider the system of tanks in Fig. 2.4. This has two dissipative elements $k = 1/R$ and a capacitance C defined by

$$I_1 - I_2 = C\dot{h} = a\rho\dot{h} \tag{2.17}$$

$$C\dot{h} = [k_1 h_1 + k_2 h_2] - (k_1 + k_2)h \qquad \text{hydraulically}$$

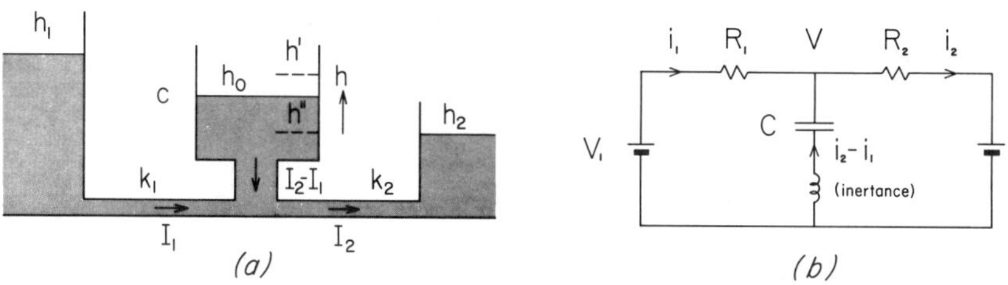

Fig. 2.4. Examples illustrating the minimization of dS/dt in open systems: (a) A hydraulic system in which the levels h_1 and h_2 are held constant; the inertial forces are neglected (see eq. 2.17). Inspection of the function $h(t)$ in the $(h, \dot{h})$ phase plane shows that the system is stable; dS/dt is minimized, see Fig. 2.5. The level h_0 is reached in steady state; h' and h'' represent the steady-state levels as would be found in two intermediate reservoirs (see text). (b) The electrical analog of (a). The inductance demonstrates how fluid inertia in a hydraulic system would have to be taken into account. After Franck [2.10].

or analogously

$$C\frac{dV}{dt} = [g_1 V_1 + g_2 V_2] - (g_1 + g_2)V \qquad \text{electrically}$$

where

h = fluid level in tank
a = area
ρ = density
$k = g$ = conductance

The dissipation function Φ just defined thermodynamically in this simple case is the frictional power dissipated in R_1 and R_2.

$$\Phi = \phi_1 + \phi_2 = I_1(h_1 - h) + I_2(h - h_2) = k_1(h_1 - h)^2 + k_2(h - h_2)^2 \quad (2.18)$$

Expanding and rearranging gives

$$\Phi = (k_1 + k_2)(h - h_0)^2 + \Phi_0 \quad (2.19)$$

where $h_0 = (k_1 h_1 + k_2 h_2)/(k_1 + k_2)$.

At the steady state, eq. 2.19 becomes

$$\Phi_0 = \frac{k_1 k_2}{k_1 + k_2}(h_1 - h_2)^2 = \frac{(\text{driving force})^2}{\text{Total } R}$$

This (eq. 2.19) is a right parabola with vertex at (h_0, Φ_0) (Fig. 2.5*a*). The system is clearly stable, in the sense that fluid (disturbance) added to h would seek the level h_0 by a damped oscillation (with fluid inertia), the restoring force depending on $(h - h_0)$ and the damping on R or K. This is indicated by the arrows in the phase plane (Fig. 2.5*a*). It is, however, an open flow system, nonequilibrium; its input and output driving forces are constrained, and its internal forces h are a function of the coefficients k. Although it approaches a "set point" h_0 determined by h_1, it is not a control system by any ordinary definition.

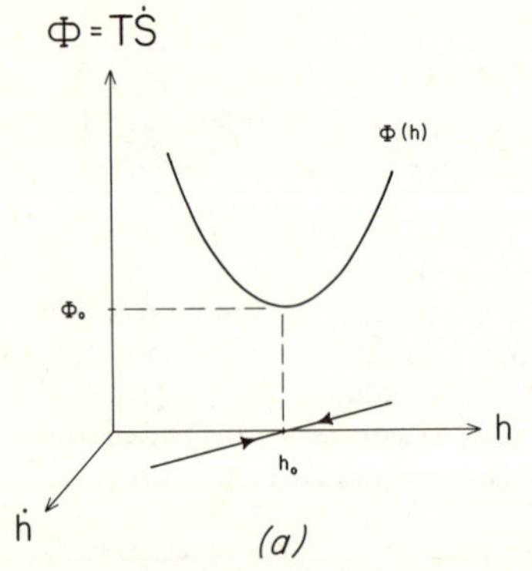

Fig. 2.5*a*. Minimization of rate of entropy production in a stable open system. The dissipation function $\Phi = T\dot{S}$ is plotted versus h, the level in the intermediate reservoir of Fig. 2.4. The trajectories in the (h, $\dot{h}$) phase plane indicate stability if R's (or k's) are positive.

Most important, at the point of stability for this particular dimension X ($= h$), the rate Φ of entropy production (waste effort) is minimal. That is, an open system, regardless of how many and varied, unconstrained forces X are acting, adjusts them automatically in such a way that $\dot{S} = \min$. For instance, if there were two intermediate tanks showing potentials h' and h'',

Fig. 2.5*a* would show a paraboloid in (h', h''), having a two-dimensional minimum. Thus each unconstrained system force contributes an extremum to Φ space. (Some of these may be locally unstable, showing negative curvature, provided the system is stable as a whole.) Note that by substituting concentrations for the forces h and letting flows be proportional to these, we obtain equally well a model of diffusion ($XJ = \mu\ dn/dt$), of reaction kinetics (where $XJ = A\ d\xi/dt$), or of first-order chemical reactions.

The next conceptual contribution of nonequilibrium thermodynamics is that of the "phenomenological coefficients" L or system conductances which are *combinations* of the k's. In particular, it is postulated that *each* flux J depends linearly on *all* the forces X acting on the system (e.g., P, T, μ, A). Physically, this means that all flows are assumed to be coupled or to entrain each other. Mathematically this is stated as follows:

$$J_j = \sum_i L_{ij} X_i = L_{1j} X_1 + L_{2j} X_2 + \dots \tag{2.20}$$

Thus if there are *two* forces, X_1 and X_2, Φ in eq. 2.16b becomes quadratic in the X's:

$$\Phi = \sum_i X_i J_i - \sum_i \sum_j L_{ij} X_i X_j = L_{11} X_1^2 + (L_{12} + L_{21}) X_1 X_2 + L_{22} X_2^2 \tag{2.21}$$

Applying this to our hydraulic example, it would be reasonable to choose, as *system forces*, the potentials referred to h_2:

$$X_1 = h_1 - h_2 \qquad X_2 = h - h_2 \tag{2.22a}$$

The corresponding system fluxes (from Fig. 2.4) are:

$$J_1 = I_1 = k_1(h_1 - h) \qquad \text{flowing out of } h_1$$
$$J_2 = I_2 - I_1 = (k_1 + k_2)h - (k_1 h_1 + k_2 h_2) \qquad \text{flowing out of } h \tag{2.22b}$$

The conductances, or phenomenological coefficients L are defined as partial derivatives [linear terms of Taylor series expansion of general $J_i(X_i)$ relations]:

$$L_{ij} = \frac{\partial J_i}{\partial X_j} \tag{2.23}$$

In our hydraulic example, then, given eq. 2.22b, we have

$$L_{11} = k_1 \qquad L_{12} = -k_1$$
$$L_{21} = -k_1 \qquad L_{22} = k_1 + k_2$$

Thus the L's (system conductances) must be calculated from the J's and X's. Note that, agreeing with Onsager's rule: $L_{21} = L_{12}$ [2.5]; the cross-conductance L_{12} here is $-k_1$, the *negative* of the physical conductance. This results merely from our choice of X_1 and X_2—an important matter, as Katchalsky shows.

The dissipation function is unchanged in this phenomenological form:

$$\Phi = T\dot{S} = \sum_i X_i J_i = L_{11}X_1^2 + (L_{12} + L_{21})\, X_1 X_2 + L_{22} X_2^2$$

$$\Phi = k_1(h_1 - h_2)^2 - 2k_1\,(h_1 - h_2)\,(h - h_2) + (k_1 + k_2)\,(h - h_2)^2$$

which rearranges to

$$\Phi = (k_1 + k_2)\,(h - h_0)^2 + \Phi_0$$

again the parabola of eq. 2.19.

In Fig. 2.5*a* the plot in the phase plane $(h, \dot{h})$ gives further information on the *stability* of this system, which we later apply to the unstable nerve membrane's dynamics. We have

$$\dot{h} = \frac{k_1 + k_2}{C}\left[\frac{k_1 h_1 + k_2 h_2}{k_1 + k_2} - h\right] = \frac{1}{\tau}\,(h_0 - h) \tag{2.24}$$

where time constant $\tau = C/(k_1 + k_2)$.

We note a negative slope for $h > h_0$; the velocity $\dot{h}$ is negative, a property of stability. But if $k_1 < 0$ (negative conductance) and also $<k_2$, the $\dot{h}$ versus h slope would be reversed, and the system would depart (unstably) from X_0 for a small displacement, instead of approaching it (Fig. 2.5*b*).

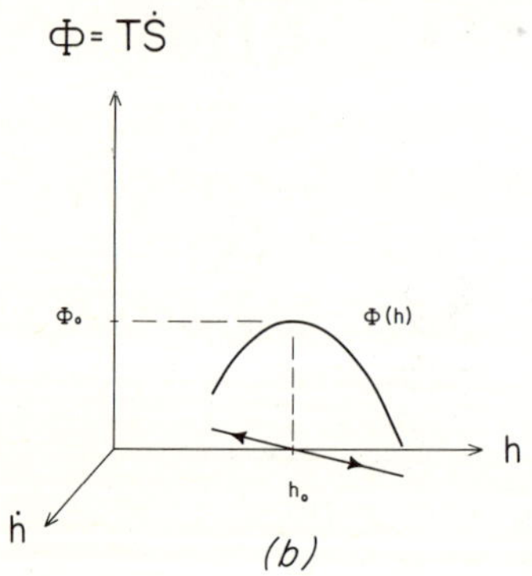

Fig. 2.5*b*. Maximum dissipation in the steady state. If *R*'s are negative, Φ_0 is maximum and the slope of the diverging trajectories in the phase plane is opposite the one in Fig. 2.5*a*. The system is unstable.

Similarly, the parabola would be inverted (Fig. 2.5*b*) with a *maximum* $\Phi = T\dot{S}$ at the vertex, as the result of a negative conductance or resistance. In a closed system such elements would be sites of continuously decreasing entropy, which is impossible. But in open systems, negative resistors are easily designed by engineers and are common in physics and biology (anabolism). In our physical model (Fig. 2.4) we need only back-couple h and k_1 positively by a float and valve, to create a negative resistance (R_1). Such a conductance dependent on a potential (as found in nerve) is not only sufficient for "excitability," but also, as we later show, can give trains of pulsatile oscillations stable in form, amplitude, and frequency.

Probably the most important feature of nonequilibrium thermodynamics is that the phenomenological equations with coefficients L_{ij} provide a formalism for quantitative analysis of flow phenomena that are known experimentally or postulated to be interdependent. This is in contrast to most classical physical and engineering analysis in which independence is postulated. The interdependence shown in eq. 2.20 is inescapable in biological analysis, although many examples are well known in inanimate systems, too.

The most familiar example in physics is the cross-coupling between electrical and heat flow in thermocouples, in the form of the Thompson, Peltier, and Seebeck coefficients (all expressible as L_{ij} [2.5]). Electrons obviously carry thermal motion also. In the physical chemistry of complex diffusion, the important process of cross-coupling or entrainment of the flows of solutes with each other and with the solvent is readily formulated as L_{ij} [2.5], and it accounts also for several cross-osmotic phenomena. Coupling of the flow of charged or polarizable fluids with the walls of a tube or a porous plug produces several familiar electrohydraulic effects. The corresponding ζ-potential coefficients may also be expressed explicitly in terms of L_{ij} [2.5]. Finally, we again mention the coupling in the nerve membrane between ionic flows and metabolic reactions, our main reason for including this subject.

Quite generally, while linear sums of independent terms may yield tractable mathematics, interdependence is so common in nature that the existence of a generalized formalism should be better known to theorists working with nonchemical phenomena. The formalism clearly may be applied to any phenomenon, process, or reorganization wherein a rate of change may be related to a force, and when waste effort, confusion, or disorganization accompanies the process. Thermal effects or mechanical motion are unnecessary—any phenomenon or pattern changing with time is sufficient. (See the elegant application of irreversible thermodynamics to a description of the tumor cell membrane by Hempling [2.11].) Further thinking along these lines appears in Section 2.F, which deals with the generalization of thermodynamics to nonthermal problems.

E. APPLICATION TO THE COUPLING BETWEEN DIFFUSION AND CHEMICAL REACTIONS IN MEMBRANES

Although the particular energetics of coupling Na^+ into the carrier molecule of a nerve membrane is not yet understood, Katchalsky describes succinctly a model by Snell which points to specific measurements needed to quantitate this coupling.

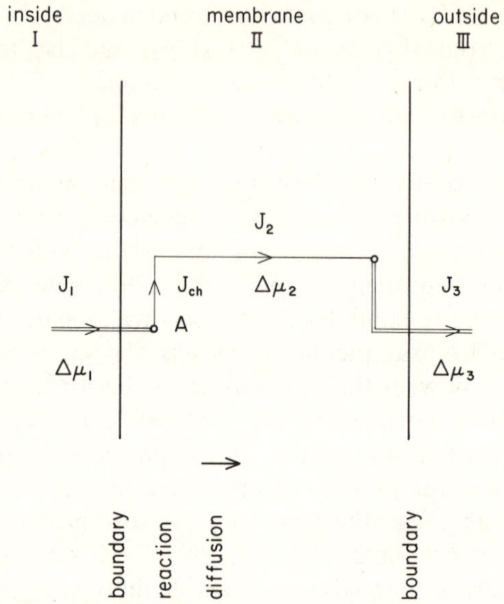

Fig 2.6. Linkage between diffusion and chemical reaction. The potentials $\Delta\mu_1$, A (driving flux J_{ch} of chemical reaction), $\Delta\mu_2$, J_1, and J_2 appear only in the first part of a carrier mechanism in a membrane.

If, for instance, near interface I–II (Fig. 2.6) Na^+ diffuses into the membrane at rate J_1, is converted chemically in a complex 2 (Na^+ + carrier C) at rate J_{ch}, and then diffuses down its gradient $\Delta\mu_2$ across the membrane at rate J_2, the flows will be (phenomenologically)

$$\begin{aligned} J_1 &= L_{11}\Delta\mu_1 + L_{12}\Delta\mu_2 \\ J_2 &= L_{21}\Delta\mu_1 - L_{22}\Delta\mu_2 \\ J_{ch} &= L_{ch}A \end{aligned} \tag{2.25}$$

where $\Delta\mu_1$ = driving density-potential difference for J_1
$\Delta\mu_2$ = same for J_2
A = affinity of the chemical reaction
$J_{ch} = \nu\, d\xi/dt$ (stoichiometric coefficient × rate of advancement)

In eq. 2.25 the flows J_1 and J_2 are assumed to be affected by both the "chemical" (density) potentials $\Delta\mu$ but not by the chemical reaction potential A. Note that the potential A (first defined in irreversible thermodynamics) deals with the progress of chemical reactions. This is unlike the "chemical potential " μ, which signifies only concentration, and so indirectly reaction probability, in an *in vitro* situation. Rewriting eq. 2.25 explicitly in the potentials (inverse phenomenological equations), we use "phenomenological" or *system resistances* rather than conductances:

$$\begin{aligned} \Delta\mu_1 &= R_{11}J_1 + R_{12}J_2 \\ \Delta\mu_2 &= R_{21}J_1 + R_{22}J_2 \\ A &= R_{ch}J_{ch} \end{aligned} \tag{2.26}$$

This is the "Ohm's law" form, where the R's may be derived from the L matrix of eq. 2.25 in the usual way, and $R_{ch} = 1/L_{ch}$ (the R's here are preferred, as determinable experimentally). But if this system is at steady state, the materials for and from the internal *reaction* J_{ch} must be supplied by ion flows J_1 and J_2, so that J_{ch} should be governed by J_1 and J_2, even if not entrained to them, in the following quantitative way. Total increase of material of species i, in the membrane at steady state, is at the rate

$$\frac{dn_i}{dt} = \int_A J_i\, dA + \int_V \nu_i J_{ch}\, dV = 0$$

where A = surface area
V = volume

Integrating over unit volume inside and unit area of membrane, we have

$$-J_i + \nu_i J_{ch} = 0 \qquad \text{so that} \quad J_i = \nu_i J_{ch}$$

Then $J_1 = J_{ch}$ and $J_2 = -J_{ch}$ assuming $\nu_1 = 1$, $\nu_2 = -1$

Thus from eq. 2.26 we see that the membrane potentials will depend on the metabolic reactions as follows:

$$\begin{aligned} \Delta\mu_1 &= (R_{11} - R_{12})\, J_{ch} \\ \Delta\mu_2 &= (R_{21} - R_{22})\, J_{ch} \end{aligned} \tag{2.27}$$

where $J_{ch} \equiv A/R_{ch}$

Clearly, the enzymes controlling metabolic reaction *rates* J_{ch} determine the nerve *potential*, regardless of whether diffusion entrainment R_{12} also occurs.

This model describes only the first half of the transport reaction at the I–II boundary. At the II–III boundary the disassociation J''_{ch} and outward diffusion J_3 of Na^+ at high concentration would have a similar description. Since the backward permeability for Na^+ is very low in the fully polarized membrane, the conductivities L_1 and L_3 would be those of rectifiers—

nonlinear and asymmetric. This is known experimentally. Without more specific information about the carrier complex, this is about as far as thermodynamic analysis will take us at the present time.

The active transport mechanism itself is somewhat complex. Its genetic, hormonal, protein-synthetic, metabolic, and other facets have been reviewed by Tosteson [2.12]. For Na^+, kidney studies indicate that at least three enzymatic mechanisms are involved, suggesting that more than one Na^+ carrier exists in different parts of the kidney tubule. We are dealing here with rather inaccessible intramolecular mechanisms, not yet modeled. Artificial bimolecular layers of complex proteolipids seem to be a promising approach. But the dynamic and stochastic aspects of active membrane permeability are still quite unexplored, although some theoretical beginnings have been made (Katchalsky et al. [2.13]).

From the systems viewpoint, accounting for the value of nerve potential as presented here is quite incomplete. The rest potential clearly depends mainly on passive Na^+ conductivity. What determines this? Why does the pump reach a steady state at the exact level at which Na^+ leaks without a pump? Does this imply a comparator? Is the upper bound of pumping limited by ATP supply? How can an eviscerated squid axon maintain a normal resting potential? How does rest potential relate to specific membrane functions exhibited locally: synaptic response, train generation, propagation, integration, storage? These are all systems questions at the subcellular level.

F. APPENDIX: TAXODYNAMICS

Taxodynamics: The (Thermo)dynamics of Living Negentropic Systems

Classical or equilibrium thermodynamics ("thermostatics") generalizes the conservation of mechanical energy to include other forms of energy (e.g., thermal, concentration, chemical, electric, surface) by introducing a special term for the unavoidably dissipated energy converted to random motion ($T\ dS$, $S =$ entropy) in a *closed* system wherein matter does not increase. Only very slow changes are considered, total energy E is conserved, potential energy seeks a minimum, and entropy (dS) seeks a maximum at equilibrium.

Open-system (irreversible) "thermodynamics" then extends this energy dissipation model to dynamical or flow systems that take in matter and retain and reorganize it. The decrease in entropy, or increase of internal organization $-d_i S$ is purchased by a greater overall external entropy increase $+d_e S$ (disorganization) of the region *outside* the system. Even in the recent thermodynamics, however, organization and disorganization still refer primarily to the randomness of thermal motion, making some allowance for spatial or *pattern* randomness (without increased kinetic randomness) as occurs in rubber or intermolecularly. Since such an open system may continually increase or lose matter, energy, or entropy, the total E within the system is not conserved. Even at steady state it can steadily grow; its total mass is not fixed—only its input, local, or output rates.

For these open systems, however, a new law of stability holds which resembles the minimum of potential energy characterizing stability in equilibrium systems. Whereas in thermostatics a system approaches maximum internal entropy, in irreversible thermodynamics the *rate of internal entropy production* is minimal at steady state (Prigogine). We learn that this "extremum" principle has very broad application, beyond such inert systems as simple mixtures or reactions in flux (e.g., solutions, electrons in solids) for which the new theory and formalism were originally devised.

Also emerging in open-system thermodynamics is the strong distinction, forces versus fluxes (intensive versus rates of extensive variables), as well as a formalism that seeks possible cross-coupling (entrainment) between such forces and flows. The formulation of the cross-coupling coefficients unifies and quantifies many well-known physicochemical phenomena (e.g., thermoelectric and electrokinetic effects, diffusion-coupled chemical reactions, membrane flows). Some new and useful concepts result, such as affinity and advancement of a reaction, and solute–solvent friction. The whole analysis, which is at a level of organization above that of molecular events, can be conducted without knowledge of them. The attempt to formulate macrodynamics statistically in terms of microdynamics has been fairly successful only with *un*coupled molecules or compounds suspended in dilute solution. The statistical model breaks down for biosystems, where reactants and processes are so constrained and geared together that entropy (disorganization) is produced at a minimal rate. Consonant with this (Prigogine's) principle, such systems have proved genetically persistent (i.e., are stable genetically also).

Outstanding in biosystems are the "natural" (degradative) processes associated with and linked energetically and entropically to active *synthetic* processes, a system product. This dual aspect of the living system, whereby a catabolic chain feeds an anabolic chain by transferring reduction-oxidation and other energy (derived from the system input) into a net increase and output of organization, has a curious origin: the rotation of the earth. Rothstein [2.7] showed quite generally that any *cyclic* open system, receiving external energy and negentropy *intermittently*, has the "natural" (self-acting) thermodynamic property of *increasing* its internal organization or negentropy. That is, it does not tend to disorganization but must use incoming energy and negentropy to create new linkages or complex stabilities locally.

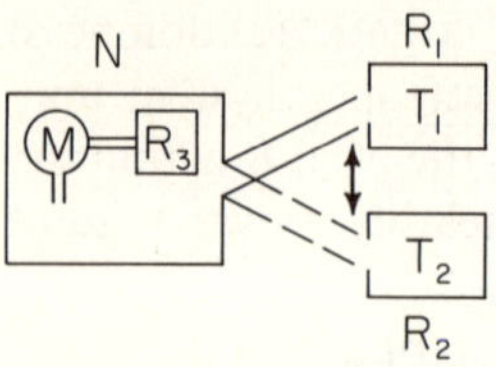

Fig. 2.7. Rothstein's open system N, in cyclic exchange with systems R_1 and R_2. If there is constant energy influx Q_2 from reservoir R_2 during one phase of the cycle, and if cycles repeat indefinitely, entropy can increase in N. The R_3 is an isolated store coupled to M, a heat engine, which interacts with R_3 mechanically only.

Consider a system N, which alternatingly is in contact and exchange with a reservoir R_1 (at temperature T_1) and a reservoir R_2 (at temperature $T_2 > T_1$) (Fig. 2.7). Assume that the exchange with R_1 consists of heat Q_1 flowing out of N, and that Q_2 flows into N when contact with R_2 is established. Assume further that steady state is reached, then the entropy change in N is

$$\Delta S_N = \frac{Q_1}{T_1} - \frac{Q_2}{T_2}$$

In any case, $d(\Delta S_N)/dt$ is minimal (Prigogine). However, ΔS_N is maximum, that is, $Q_2(1/T_1 - 1/T_2)$, when $Q_1 = Q_2$. It is minimum ($= 0$) if $Q_1/T_1 = Q_2/T_2$. In the latter case $Q_2 - Q_1 = W$ is the energy taken up by system N; W can be stored and it can do work. Rothstein [2.7] considered the case in which N comprises a heat engine M and a store R_3, which is isolated except that it can interact mechanically with M.

In this system (Fig. 2.7), Q_1 (heat lost to R_1) depends on the state of N during N's interchange with R_1. If Q_2 is fixed, N will tend toward a state, among those available after Q_2 flows in, corresponding to the smallest Q_1. Hence, if there is only one state where $Q_1 = (Q_2 - W) < Q_2$, ΔS_N is certainly less than maximum even if W is small. With W stored, more states become available to N, and it will tend to assume a state where Q_1 is smaller. That is, the trend of N to go from $\Delta S_{N,\max}$ toward $\Delta S_{N,\min}$ is possible if the cycling of W between R_3 and M goes on indefinitely. Then, however, during the phase when M works on R_3, the *organization* in R_3 can *increase*. Rothstein thus showed that evolution is not incompatible with thermodynamics and that cyclic, directed radiation from the sun presents a necessary and sufficient condition for increasing order.

This physical law (that cyclic open systems are self-organizing) obviously applies to all terrestial transformations, but especially to biosystems. Together with Prigogine's principle, it may be considered the "driver" of evolution. That is, "minimization of entropy production" helps determine which of several alternative synthesizing processes will be more stable in the face of perturbation and is, therefore, one aspect of defining the "fittest" in a noncircular logic. Conversely, "least rate of waste" or least effort per task is obviously of survival value at any level, if we define waste in a way much broader for higher systems than is used for the simple thermal and energetic phenomena of cellular chemistry.

If then we *confine* our "thermodynamics" to a terrestrial living system, and afford it adequately organized "bond-rich" materials from lower forms, we find need for a mathematical formalization in which the exchange of *organization* ($-dS$) or complexity per se becomes paramount. The inevitably associated disorganization is balanced out and may often be small enough, relatively, to omit from the equation. Free energy gradients are no longer the focus of interest, but rather the synthesis of various new degrees of freedom sustained by dynamic stability barriers.

Degradative thermodynamics recognized several kinds of "potentials" whose gradients were (intensive) *forces* driving the (*extensive*) fluxes (e.g., $p\,dV$ or $\mu\,dn$). Since in principle the number of these is unlimited, the resulting *kinds* of disorder are also unlimited. Similarly, in the associative or "syntactic" thermodynamics, novel kinds of order are constantly being devised. (How many kinds and levels of order has a chromosome?) We are dealing here with terms describing *kinds of order* at each level of organization, and the thermal and energetic terms are no longer of significance or interest; thus we might coin the name "taxodynamics" for the science of ordering (ταξον) and its time variation, much as (irreversible) thermodynamics—the time variation of thermostatics—is the science of what happens to energy rather than to order.

Taxodynamics is barely related to the ordering process of taxonomy, the simple stage of biological science (whether plants or diseases). It relates more closely to measures of the flux of *information* concerning goals and performance that characterize the stability of biosystems and the modulation or control of these for external tasks. From mitochondrial to emotional levels, hormonal or nervous operators monitor the processes and control the rates of organizing for a task. Thus sensory information "flow" is processed (reorganized) by a set of neuroanatomical and physiological formative "forces" (gradients) toward enhancement or optimization of informational features desired. Similarly, muscular action, in terms of information about position and speed, is driven by the user's intent at the higher levels of process organization (the lower levels are subconscious).

One of the main operators in information theory is the limited memory "window" or time-span of a filter (its weighting function). Such a filter also operates widely in the processes of animals. Thus hormones have a half-life, as does one type of conscious memory. This means that we are dealing continually with internal organization which, partly transient, is constantly tested and renewed, and cyclically disorganized (in sleep). Here the loosening of connections facilitates new combinations and recall (as in dreams and contemplation).

What kind of analytic format is suitable for these taxodynamic phenomena? Is a modified conservation equation, like the second law of thermodynamics, useful?

Information resembles entropy in that it is conserved or stored in a memory, possibly multidimensional. It is dissipated in a leaky memory or in a jumbling process. Redundancy opposes dissipation. A further similarity is that information has no zero. Each new pattern that is discovered in a situation is a new dimension of negentropy. Until its discovery, it does not exist, so that "zero" organization depends on what has been discovered.

Closed systems have no use for memory; open systems, with entropy throughput, can store both temporal and spatial information. With this, new linkages can be devised, new patterns perceived. These recalled, purposive behavior can result.

Man has a "metaperception" of patterns whereby his own behavior is actively perceived (self-consciousness) in real time and is compared with the behavior of others or with idealized patterns of behavior. Invention is self-recognized as such. Projected ideas are compared, criticized, and selected before action. Yet entering decisions (and so behavior) is far more than information. Perceptions and appreciation, emotions, prejudice, habit, and inertia, contribute from without and from within to the processing of increasing information. These other contributions can be formulated in terms of *patterns* and in some way be "measured" negentropically.

It is true that in *closed* systems at the simple level of linkages, exhibited by biophysics and biochemistry, "determinism" throws out the concept of "intent." But it is hardly possible to understand human organization or human communications (or even the behavior of a dog or a frog) without it. The "*force*," *intent*, or *meaning* of an idea is the essence of communication, not its choice of words (Shannon and Weaver [0.10]). The form and "flux" of language are ancillary vehicles, however primary they may be to the diplomat or telephone engineer.

Thus for every message there is a "driver"—a writer or speaker trying to express and transmit an idea, or a reader or listener trying hard to understand (as with poetry) the *intent* of the message. Moreover, attention is of the essence; and in dialogue a certain forcefulness, animation, or eagerness, a fire in the eye, a color and intonation of voice, are part of effective (and so affective) human communication. These considerations are indeed relevant in a book on human systems (as opposed to engineering).

History and applied psychology alike testify that arousal, "heat" as well as "light" in dialogue, are required to stimulate action or to change habitual ways. This phenomenon, which resembles the heat of activation in chemistry, is accompanied by positive entropy (waste energy incurred). Someone may (perhaps must) get hurt: war, coercion, and civic strife are forceful communications resulting in action, involving friction. The linkage of human knowledge as a social pattern becomes evident and contributions are realized, with division of intellectual and experimental labor. A *fabric* of society (economic, scientific, constructional, sociological, creative, artistic) becomes recognized, along with social responsibility. The complementary values of competition and cooperation become clear, and we recognize their impact on evolution, and so in turn on selection and on genes. *Genetic control* thus becomes a cyclic process, but open, especially in the noosphere where not only information, but also behavioral patterns of a culture—esthetics, pleasure, all major "taxodynamic" *forces*—are stored and transmitted from generation to generation.

Obviously a science of open-system taxodynamics, a force–flux–friction theory of organized information flow and the poesis of organization within and between humans (as foreshadowed earlier) is not yet developed. But if this open-system dynamic formalism can prove useful, a study of its simpler applications, powers, and limitations at the subcellular level, where only thermal entropy is involved, should be provocative to the scientific thinker. Some of these areas were considered in this chapter.

G. REFERENCES

2.1. Snell, F. M., S. Shulman, R. P. Spencer, and C. Moos. *Biophysical Principles of Structure and Function.* Addison-Wesley Publishing Co., Reading, Mass., 1965.

2.2. Ruch, T. C. and H. D. Patton (Eds.). *Physiology and Biophysics*, 19th ed. W. B. Saunders Co., Philadelphia, 1965.

2.3. (a) Mountcastle, V. B. (Ed.). *Medical Physiology*, 12th ed., C. V. Mosby Co., St. Louis, 1968. (b) Bard, P. (Ed.). *Medical Physiology*, 11th ed. 1961.

2.4. Finkelstein, A. and A. Mauro. Equivalent circuits as related to ionic systems. *Biophys J* **3**: 215–237, 1963.

2.5. Katchalsky, A. and P. F. Curran. *Nonequilibrium Thermodynamics in Biophysics.* Harvard University Press, Cambridge, Mass., 1965.

2.6. Bray, H. G. and K. White. *Kinetics and Thermodynamics in Biochemistry*, 2nd ed. Academic Press, New York, 1966. pp. 114–115.

2.7. Rothstein, J. *Communication, Organization and Science.* Falcon's Wing Press, Indian Hills, Colo., 1958, pp. 81–87.

2.8. Colby, K. M. Research in psychoanalytic information theory. *Amer Sci* **49**: 358–369, 1961.

2.9. Tribus, M. *Thermostatics and Thermodynamics.* D. Van Nostrand Co., Princeton, N.J., 1961.

2.10. Franck, U. F. Models for biological excitation processes. *Progr Biophys* **6**: 171–206, 1956.

2.11. Hempling, H. G. Application of irreversible thermodynamics to a functional description of the tumor cell membrane. *J Cell Physiol* **70**: 237–256, 1967.

2.12. Tosteson, D. C. Membrane transport of Na and K: a synopsis. *Physiologist* **9**: 89–96, 1966.

2.13. Katchalsky, A. and R. Spangler. Dynamics of membrane processes. *Quart Rev Biophys* **1**: 127–175, 1968.

CHAPTER THREE

Nerve Membrane: Dynamics

The time-varying, dynamic properties of nerve were not considered in Chapter 2. Although dS/dt, dn/dt, and $d\xi/dt$ were encountered in thermodynamics (as defining the dissipation function Φ and the flows J_i and J_{ch}) the fluxes were regarded as in steady state rather than dynamic.

In this chapter we consider the dynamic effect of applying subthreshold (small signal) and suprathreshold exciting currents to axonal membrane: the analyses of Hodgkin and Rushton (HR) and of Hodgkin and Huxley (HH), respectively. Next, to understand the generation of impulse trains and of the frequency modulation found in a single fiber carrying a portion of a message pattern, a deeper comprehension of the unstable repetitive operation of nerve is needed. For this, the phase plane representation best displays the nonlinear properties of excitable tissue. This formulation of nonlinear dynamics applies quite generally to instability in systems physiology. We consider it first in relation to Fitzhugh's exposition of the nonlinear properties of the HH model. These he has generalized to excitable tissues in the Bonhoeffer-van der Pol membrane model.

It turns out that the stability properties of linear open-system steady-state thermodynamics, treated in Chapter 2, apply directly to the nonlinear unstable case whenever a conductance becomes negative, which happens when large potential changes occur in the nerve membrane.

Finally, we consider transient, dynamic effects at the input end of a neuron: the dendrites. The approximate model needed to handle this mathematically was developed by Rall.

A. TRANSIENT LINEAR SPREAD OF NERVE POTENTIAL: HODGKIN–RUSHTON EQUATIONS

Electric double layers form across the membrane to balance the chemical potential of the ions. The membrane is thin (80 Å) and leaky (Fig. 3.1) ($R_m = 2000\ \Omega/\text{cm}^2$ at D.C.) but can still separate charges. If we check the order of magnitude (using, for the capacitance C_m, $1\mu F/cm^2$), the dielectric constant ε comes out

$$\varepsilon = \frac{dC}{\varepsilon_0 A} = \frac{8 \cdot 10^{-9} \cdot 10^{-6}}{8.8 \cdot 10^{-12} \cdot 10^{-4}} \simeq 10$$

where d = thickness, C = capacitance, and A = area. This is about right for

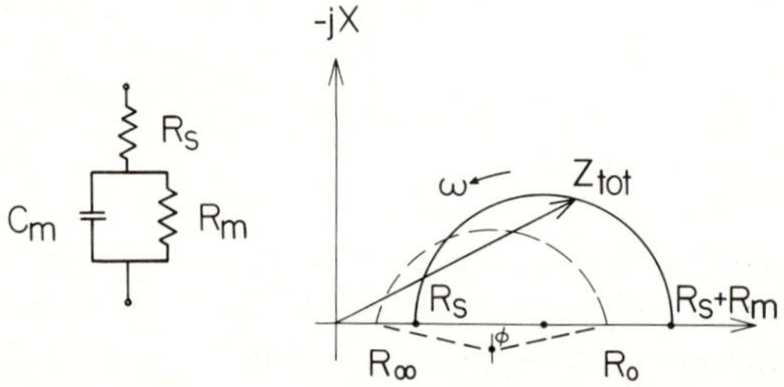

Fig. 3.1. Equivalent circuit (left) and impedance plot for unit area of nerve membrane (right). If the components of the network are constant, the locus on which the impedance Z_{tot} must lie is a semicircle centered on the real axis (solid curve in right panel). At D.C. impedance is $(R_s + R_m)$; at very high frequencies it approaches R_s. In biological membranes there are dielectric losses which change with frequency. It then is often possible to establish the impedance locus as a circle whose center is below the real axis in the Z-plane, so that ϕ, R_0, R_∞ characterize the locus (interrupted circle). (See Cole [3.6].)

a dielectric that is both aqueous and oily (i.e., dissolves both fat- and water-soluble materials). The capacitative charge is $Q = VC = (10^{-6}\ \mu F \cdot 0.25\ cm^2)/10 = 0.25 \cdot 10^{-7}$ C/cm of a 0.75-mm lobster nerve. This negative coating inside is only about 10^{-9} of the cationic (+) charge within the axoplasm.

This distributed capacitance makes of the axon a "leaky" RC cable (Fig. 3.2) along which the potential due to a local step current will spread with a certain time lag. In fact, it is modeled very well by an ideal RC line (i.e., R's and C are constants) calculable from its physical properties, as Hodgkin and Rushton [3.1] showed in a classical paper. We follow their argument (using mathematical derivations differing only slightly from the original treatment) and the experimental proof in abridged form. By deleting the batteries and adding the capacitance, as given in Fig. 3.1, the dynamical analysis of the membrane, previously considered statically, can be carried out. This capacitance is also an electric reservoir that makes possible a rapid reverse impulse of polarization, requiring diffusion of ions from the adjacent axoplasmic sol. However, it is interesting that the momentary high flux in the axoplasm stresses and aligns it anisotropically enough to polarize it optically during the impulse.

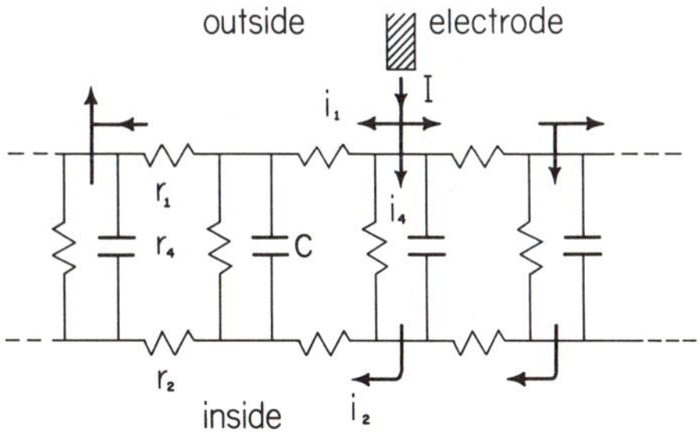

Fig. 3.2. Equivalent circuit for a section of the membrane of the distributed, leaky RC axon, together with resistance per centimeter of axoplasm (r_2) and of outside medium (r_1), and currents flowing in HR's experiment. The active electrode imposes total current I, part of which (i_4) enters the axon. On the outside, i_1 flows directly to the indifferent electrode (on the left; not shown); i_2 is current inside the axon; i_4 is current through the membrane comprising both ohmic and capacitive components (eq. 3.2). Cf. Fig. 3.4.

Using the symbols in Fig. 3.2, we can write the differential equations for the membrane potential. (See [0.16] for the theory of partial differential equations.) Using the following definitions:

$$V_m = V_1 - V_2 \qquad \frac{\partial V_1}{\partial x} = -r_1 i_1 \qquad \frac{\partial V_2}{\partial x} = -r_2 i_2$$

V_1 = outside potential; i_1 = current outside
V_2 = inside potential; i_2 = current in axoplasm
V_m = membrane potential; $I = i_1 + i_2$, the impressed current
r_1 = resistance/cm outside
r_2 = resistance/cm inside

we obtain

$$\frac{\partial V_m}{\partial x} = (r_1 + r_2)\, i_2 - r_1 I \tag{3.1}$$

The membrane current i_m is

$$i_m = \frac{V_m}{r_4} + C\,\frac{\partial V_m}{\partial t} \tag{3.2}$$

where r_4 and C are the membrane elements.

This equals
$$\frac{\partial i_2}{\partial x} = \frac{1}{r_1 + r_2}\,\frac{\partial^2 V_m}{\partial x^2} + \frac{r_1}{r_1 + r_2}\,\frac{\partial I}{\partial x} \tag{3.3}$$

(using eq. 3.1). Hence
$$-\lambda^2\,\frac{\partial^2 V_m}{\partial x^2} + \tau\,\frac{\partial V_m}{\partial t} + V_m$$

$$= r_1 \lambda^2\,\frac{\partial I}{\partial x} \quad \text{at electrode} \tag{3.4}$$

$$= 0 \quad \text{away from electrode}$$

where $\lambda = \sqrt{r_4/(r_1 + r_2)} \simeq \sqrt{r_4/r_2} \equiv$ characteristic length
$\tau = r_4 C =$ time constant

Besides representing, here, the voltage gradient around a current source, eq. 3.4 describes temperature gradients around an applied source of heat. It also models the density gradients due to diffusion of matter from a local source, as from a hole into a pipe. To solve it, first change the variables: $X = x/\lambda$, $T = t/\tau$, $U = V_m e^T$,

$$\frac{-\partial^2 U(X, T)}{\partial X^2} + \frac{\partial U(X, T)}{\partial T} = 0 \quad \text{(except at electrode)} \tag{3.5}$$

The time integration can be rendered algebraic by a Laplace transform, which avoids the partial derivatives:

$$\frac{d^2 U(X, s)}{dX^2} = sU(X, s) \tag{3.6}$$

This is satisfied by

$$U(X, s) = Ae^{-|X|\sqrt{s}} \tag{3.7}$$

where $|X|$ = absolute value.

There is only one constant A for this double integration because of symmetry and $V = 0$ at $|X| = \infty$. To evaluate A, consider the derivatives at the boundary region of the electrode, of width $\pm\delta$. The change of slope at the electrode is:

$$\Delta'' V_m = \left.\frac{\partial V_m}{\partial x}\right|_{x=\delta} - \left.\frac{\partial V_m}{\partial x}\right|_{x=-\delta}$$

$$= (r_1 + r_2)\left[\left. i_2\right|_{x=\delta} - \left. i_2\right|_{x=-\delta}\right] - r_1\left[\left. I\right|_{x=\delta} - \left. I\right|_{x=-\delta}\right]$$

$$= r_1 I_0 \quad \text{as} \quad \delta \to 0$$

Changing from V to U:

$$\Delta'' U = -r_1 I_0 \lambda e^T \quad \text{in normalized time domain}$$

$$= \frac{-r_1 I_0 \lambda}{s - 1} \quad \text{in the Laplace transform}$$

But from eq. 3.7 this difference of slopes is also

$$\Delta'' U = \left.\frac{\partial U}{\partial X}\right|_{X=\delta/\lambda} - \left.\frac{\partial U}{\partial X}\right|_{X=-\delta/\lambda} \simeq -2A\sqrt{s} \quad \text{in the limit as } \delta \to 0$$

Therefore,

$$A = \frac{r_1 I_0 \lambda}{2} \frac{1}{\sqrt{s}(s-1)} \tag{3.8}$$

When the solution of eq. 3.7 in the s domain is expanded in partial fractions, it becomes

$$U(X, s) = \frac{r_1 I_0 \lambda}{2} \frac{e^{-|X|\sqrt{s}}}{\sqrt{s}(s-1)} = \frac{r_1 I_0 \lambda}{2}\left[\frac{1}{2}\frac{e^{-|X|\sqrt{s}}}{\sqrt{s}(\sqrt{s}-1)} - \frac{1}{2}\frac{e^{-|X|\sqrt{s}}}{\sqrt{s}(\sqrt{s}+1)}\right] \tag{3.9}$$

The inverse Laplace transform is tabulated by Churchill [3.2] as follows:

$$L^{-1}\left\{\frac{e^{-K\sqrt{s}}}{\sqrt{s}(\sqrt{s}+a)}\right\} = e^{aK} e^{a^2 T} \operatorname{erfc}(a\sqrt{T} + K/2\sqrt{T}) \quad \text{in the time domain}$$

where $\operatorname{erfc} = 1 - \operatorname{erf}$ and $\operatorname{erf} z = \dfrac{2}{\sqrt{\pi}}\displaystyle\int_0^z e^{-w^2}\,dw$

Since $V_m(x, t) = U_m e^{-T}$, we obtain the membrane potential explicitly

$$V_m(x, t) = \frac{r_1 I_0 \lambda}{4}\left\{ e^{-|X|}\left[1 - \operatorname{erf}\left(\frac{|X|}{2\sqrt{T}} - \sqrt{T}\right)\right] - e^{|X|}\left[1 - \operatorname{erf}\left(\frac{|X|}{2\sqrt{T}} + \sqrt{T}\right)\right]\right\} \tag{3.10}$$

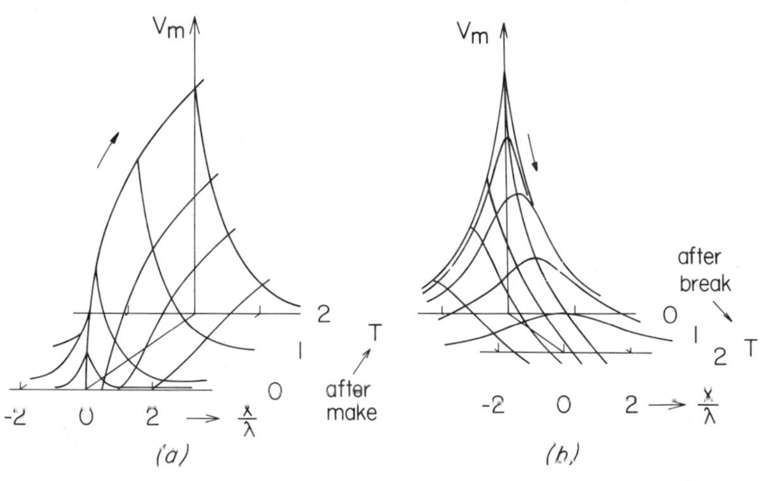

Fig. 3.3. Spatial and temporal voltage distributions (subthreshold) on HR axon in response to current pulse. (*a*) Conditions for *onset* of stimulus (leading edge of pulse or "make"). The growing of the spatial distribution of V_m, as a function of distance along the axon normalized by the space constant λ, and as a function of the normalized time. (*b*) Decay of V_m in space and time after stimulus is disconnected. Note the strong attenuation of signals over X and the apparent delay, in time, at sites remote from the origin $X = 0$.

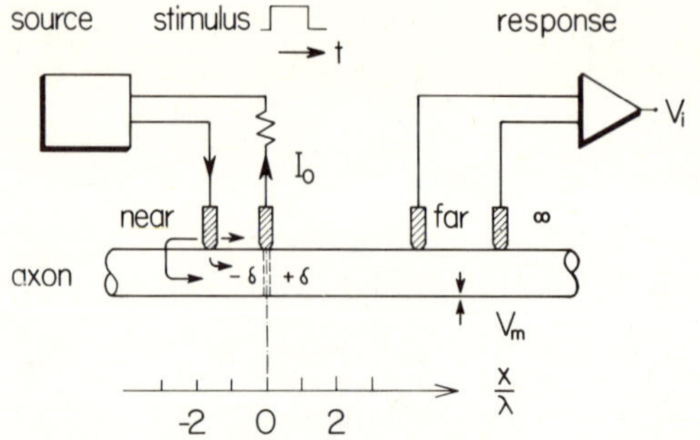

Fig. 3.4. Experimental setup (HR) for measuring spatial distribution of membrane potential. See text for relation between V_1, the only directly observable quantity, and V_m. The indifferent measuring electrode is marked ∞. The electrode picking up V_1 senses potentials far away from stimulating electrode; that is, where spatial distribution is exponential (characteristic length λ). The stimulating electrode has physical size 2δ; this is important when determining A in eq. 3.7.

The nature of this surface may be learned from Fig. 3.3, showing $V_m(X, T)$, the potential resulting from the make and break of I_0, as functions of $X = x/\lambda$ and $T = t/\tau$. The erf functions appear in more familiar form in Fig. 3.3*b*, at constant $|X|$; the exponential fall $e^{-|X|}$ is given in Fig. 3.3*a* at fixed times.

Note that in Fig. 3.3, at both make and break, the remote rise and fall are not only attenuated heavily but also delayed from that at the input to the line; a propagation effect in this noninductive transmission line.

Although this calculated distribution of the *transmembrane* potential V_m cannot be directly measured experimentally, it was verified by finding other properties which are observable (Fig. 3.4). Thus from the *external* potential alone, that is,

$$V_1 = \frac{r_1}{r_1 + r_2} V_m$$

we can obtain $\lambda, \tau_m, y \equiv \dfrac{r_1{}^2\lambda}{2(r_1 + r_2)}$ and $m \equiv \dfrac{r_1 r_2}{r_1 + r_2}$

From these, defining a as the radius of the axon, we can calculate

$$\text{internal resistivity } R_2 = \pi a^2 r_2 = \pi a^2 m \left(1 + \frac{m\lambda}{2y}\right)$$

$$\text{membrane resistivity } R_4 = 2\pi a r_4 = 2\pi a \lambda^2 m \left(2 + \frac{m\lambda}{2y} + \frac{2y}{m\lambda}\right)$$

$$\text{membrane capacitance } C_m = \frac{\tau_m}{r_4}$$

$$\text{characteristic length } \lambda^2 = \frac{r_4}{r_2} = \frac{aR_4}{2R_2}$$

$$\text{external resistance per length } r_1 = \frac{2yr_2}{m\lambda}$$

To do this, note from eq. 3.10 that the maximum potential at the electrode, in terms of V_1 at make, is

$$V_1\Big|_{\max} = V_1\Big|_{\substack{X=0\\T=\infty}} = \frac{r_1{}^2}{2(r_1 + r_2)} I_0 \equiv yI_0$$

We can easily check (using linearized plots) that at *make*

$$V_1(x)\big|_{t=\infty} = V_1\big|_{\max} e^{-|X|}$$

is a descending exponential, spatially, and

$$V_1(t)\big|_{x=0} = V_1\big|_{\max} \operatorname{erf} \sqrt{T}$$

is an ascending erf curve, temporally, whereas at *break*

$$V_1(t)\big|_{x=0} = V_1\big|_{\max} [1 - \operatorname{erf} \sqrt{T}]$$

Directly from these rectified slopes [or by comparing with $V_1(X, T)$] we find $\lambda = 1.6$ mm and $\tau = 2.3$ msec as average values from 13 experiments. Next, from the space derivative of eq. 3.10 or from Fig. 3.3*a*, it is evident that, at large distances β from the electrode ($\beta \gg \lambda$), we can use eq. 3.2 to obtain

$$\frac{\partial V_m}{\partial x}\Big|_{x=\beta} = 0 \qquad \text{so that} \qquad \frac{1}{I_0}\frac{\partial V_1}{\partial x}\Big|_{x=\beta} = \frac{r_1 r_2}{r_1 + r_2} \equiv m$$

Calculating the parameters as shown above, the average values are

$$R_2 = 60.5\ \Omega\text{-cm} \qquad \frac{r_2}{r_1} = 0.8 \qquad C_m = 1.33\ \mu\text{F/cm}^2$$

$$R_4 = 2290\ \Omega\text{-cm} \qquad y = 81{,}000\ \Omega \qquad 2a = 0.75\ \text{mm}$$

These are reasonable values for lobster nerve in the light of calculations by other methods.

The experimental arrangements for so measuring λ, T, y, and m are simple (Fig. 3.4). Optimal values of current density and I_0 must be used, however, keeping the current I_0 below half the threshold for excitation, to avoid eliciting the neural amplification (subthreshold response) which just precedes unstable depolarization. In this way, the transmembrane potential distribution can be found from external measurements alone.

Hodgkin and Rushton's work, published in 1946, is perhaps an ideal example of quantitative biological modeling. Not only does the mathematical function fit well the experimental observations with little adjustment, but there happens also to be a direct mechanistic homology between theory and biophysical structure and function. The fact that the analysis is one dimensional does not significantly alter the results, as Clark and Plonsey [3.3] have shown. This modeling is done at the organizational level above the electrochemical conductance mechanisms previously considered. Moreover, it uses the "pure electrical" interpretation of Finkelstein and Mauro [2.4] which, as we have seen, is a considerable abstraction from the ionic battery (electrochemical) model. The assumptions of pure capacitances and resistances arranged as in Fig. 3.2 therefore seem justified, although the uniqueness of the model does not logically follow.

B. NONLINEAR DYNAMICS OF THE NERVE MEMBRANE: HODGKIN–HUXLEY MODEL

We have just seen that for small ($< 5\%$) changes, membrane potential is proportional to an applied test current step, although delayed temporally and spread spatially by the capacitance. This output–input change is quite local ($\lambda = 1.6$ mm) and rapid ($\tau = 2.3$ msec) in an unmyelinated axon.

For stronger depolarizations the membrane becomes unstable, because conductance depends on the membrane potential V_m. As V_m decreases, the conductance increases even faster so that transmembrane current increases—by definition a negative variational conductance ($dI/dE < 0$), which we examine in detail in Section 3.D.

To study these instability effects experimentally, two new techniques were needed: (1) Cole and Marmont's methods [3.4–3.6] render the function stationary $[V_m(x, t) \rightarrow V_m(t)]$ by depolarizing only a limited segment of nerve. This enables us to study the time course of depolarization alone, without the propagation induced by adjacent local currents [2.2, p. 46]. (2) Hodgkin and Huxley's method [3.7] allows observing I_{Na} separately from I_{K} during depolarization. The potassium current I_{K} increases nonsynchronously, independently of I_{Na} and in the opposite direction (outward).

The temporal behavior of the corresponding ion conductances turns out to be quite complex. Some refinements were used on Cole and Marmont's method in the giant axon of the squid, as shown schematically in Fig. 3.5.

A feedback amplifier so drives current I that the axoplasm changes potential by a voltage step V. The voltage, which is sensed by a separate electrode, is fed back and compared with a reference, the difference controlling the current amplifier. The resulting membrane current is measured by its potential drop in external seawater, in a guarded

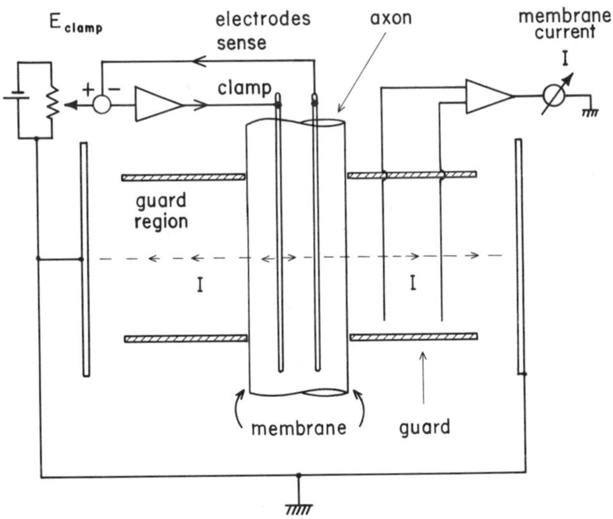

Fig. 3.5. Voltage-clamp method, combined with space-clamp technique, of Cole and Marmont [3.4, 3.5]. The current electrode marked "clamp" delivers a current such that potential within the axon, monitored by a special "sense" electrode, is equal to E_{clamp}, owing to the high gain amplifier. Current so imposed is indirectly measured (by way of voltage drop) by the two elecrodes in the outer compartment. Since both "clamp" and "sense" electrodes extend over the full length of the axon, no axial currents can flow and no suprathreshold signals can ever begin to propagate along the nerve. This technique is needed for determination of nonlinear, negative membrane resistances.

section of the test pool. Both current and voltage wires ran the full length of a 15-cm squid nerve, about 0.5 to 1.0 mm diameter.

The conductance for K^+ alone was studied by using seawater wherein choline$^+$ replaced Na^+. The clamped conductance of K^+ was found to depend nonlinearly on the impressed clamp voltage (not on current); the step response in time (Fig. 3.6) shows first a time lag, then a strong power law, and finally an exponential approach to steady state (g_∞). On "break" of the voltage clamp ($V = 0$), conductance g_K fell back exponentially. Sodium conductance g_{Na} is also activated with a more gradual power law (lagging) rise, but soon is inactivated.

Hodgkin and Huxley [3.7] devised an ingenious quantitative model of the nerve membrane which is credible physicochemically and could be fitted fairly well to the complex behavior represented in Fig. 3.6, as well as to a full range of clamping voltages. Yet the model is simple enough mathematically to predict, without *ad hoc* adjustments, other typical behavior of nerve: size and shape of action potential, refractory period, and impulse velocity. The

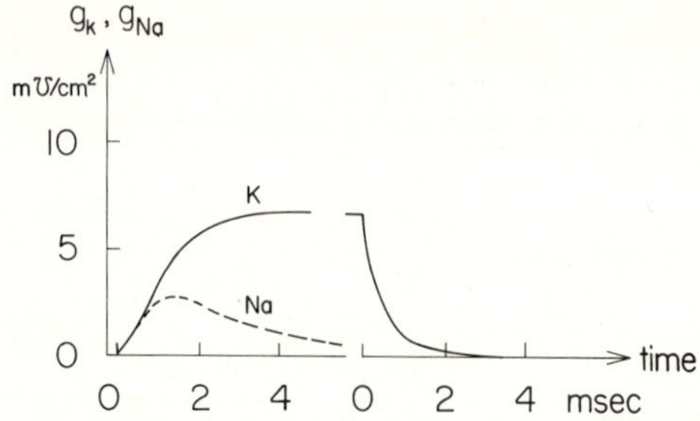

Fig. 3.6. Comparison of conductance changes, in time, for K^+ and Na^+, in response to a clamped, pulselike depolarization of 25 mV. Note change of time course at "break" for K^+ conductance. (℧ ≡ 1/Ω) After Hodgkin and Huxley [3.7].

methods, analysis, and application of these investigators earned them the Nobel prize, and Huxley's oration on that occasion gives a succinct summary [3.8].

The basic membrane circuit model (Fig. 3.7) is the static one used in electrochemistry (Fig. 2.1) but is explicit in V: the *departure* from normal resting voltage (E) for the membrane or for each ionic potential. In these terms, the total current density, including leakage (l) becomes:

$$I = C\frac{dV}{dt} + g_K(V - V_K) + g_{Na}(V - V_{Na}) + g_l(V - V_l) \qquad (3.11)$$

To account for the initial lag and rapid rise of g_K, fourth-order kinetics was assumed in some conductance–controller process "n," namely,

$$g_K = \bar{g}_K n^4 \qquad (3.12a)$$

where n varies from 0 to 1.

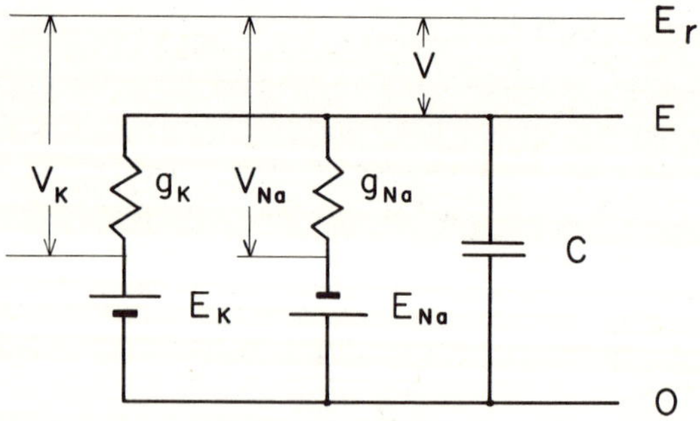

Fig. 3.7. Equivalent circuit of membrane showing notation used by Hodgkin and Huxley (eq. 3.11), but omitting leakage current path. The resistance in the K^+ path is g_K^{-1}, the one in the Na^+ path g_{Na}^{-1}.

At the same time, to obtain the simple exponential asymptote shown in Fig. 3.6, a first-order process was invoked:

$$\frac{dn}{dt} = \alpha_n(1 - n) - \beta_n n \tag{3.12b}$$

where α_n and β_n are complementary rate constants.

This more protracted first-order process might be diffusive migration of control particles N (for details see [3.7]) or their relaxation times as polarizable gates. The rate constants α, β depend on V only and change instantly with it. They thus provide for the asymptotic change of time course with clamp voltage. This simple formalism not only separates the variables but, by using separate time constants for forward and reverse processes, makes possible the rather different "break" behavior observed in Fig. 3.6.

This model resulted from the investigators' desire to stimulate mechanistic thinking about conductance-control processes in terms of membrane dynamics and field theory; hence, accuracy of fit to the data was somewhat sacrificed (n^{25} would have fitted better [3.9]). Clearly, the probabilistic mechanism outlined by Hodgkin and Huxley [3.7], developed in detail by others [3.10], and challenged by alternatives [3.11] has fulfilled this desire. The model is parsimonious in that fewer parameters would hardly cover the variables observed.

However, for g_{Na} a product of two similar elementary processes (simultaneous probability) was assumed: activation m and inactivation h. To fit the initial knee of g_{Na} (Fig. 3.5), a third-order process in m sufficed, whereas inactivation h was taken as monomolecular.

$$g_{Na} = \bar{g}_{Na} m^3 h \tag{3.13}$$

where m, h may vary from 0 to 1.

For each of these species of control particles, an independent set of voltage-dependent time constants was postulated, forward and reverse:

$$\frac{dm}{dt} = \alpha_m(1 - m) - \beta_m m \tag{3.14a}$$

$$\frac{dh}{dt} = \alpha_n(1 - h) - \beta_h h \tag{3.14b}$$

Being voltage-controlled, these sodium conductivity governors M and H clearly must be charged or polarized, yet sparse enough so that their motions contribute no measurable current [3.7]; probably, too, they must be arranged in mosaics [3.7] rather than randomly. These conductances have not yet been studied in relation to the L's of irreversible thermodynamics, although this is mechanistically eligible. Ordinarily the L's are assumed to be linear; this would have to be modified.

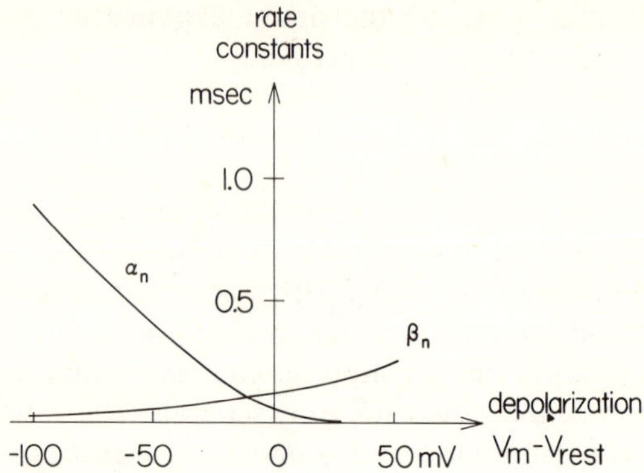

Fig. 3.8. Dependence of potassium rate constants α_n and β_n on depolarization (clamp) voltage. See eq. 3.12b. After Hodgkin and Huxley [3.7].

The evaluation of the six α's and β's versus voltage, and of n, m, and h versus time from voltage-clamp conductance data proved to be arduous. Using this model to separate the t and V variables, the α and β seemed to vary monotonically enough with the clamp voltage (Fig. 3.8) to attempt exponential fits. However, some 15 adjustable constants [zero (intercept) terms, scale, and slope factors] had to be evaluated empirically in order to express the six rate constants α, β as functions of V (see eq. 3.26). We can write at once the equations for n, m, and h and for g_K and g_{Na}. Thus, the step response for potassium

$$n = n_\infty - (n_\infty - n_0) \exp\left(-\frac{t}{\tau_n}\right) \tag{3.15}$$

where

$$n_\infty = \alpha_n(\alpha_n + \beta_n)^{-1}$$
$$\tau_n = (\alpha_n + \beta_n)^{-1}$$

makes the conductance:

$$g_K = \left\{ g_{K\infty}{}^{1/4} - [g_{K\infty}{}^{1/4} - g_{K0}{}^{1/4}] \exp\left(-\frac{t}{\tau_n}\right)\right\}^4$$

However, for sodium (m and h) the data (Fig. 3.9) show that

$$m_0 \simeq 0 \quad \text{if} \quad m > 30 \text{ mV}$$
$$h_\infty \simeq 0 \quad \text{if} \quad h < -30 \text{ mV}$$

Thus for sodium the step response of conductance reduces to

$$g_{Na} = \bar{g}_{Na} m_\infty^3 h_0 \left[1 - \exp\left(-\frac{t}{\tau_m}\right)\right]^3 \exp\left(-\frac{t}{\tau_h}\right) \tag{3.16}$$

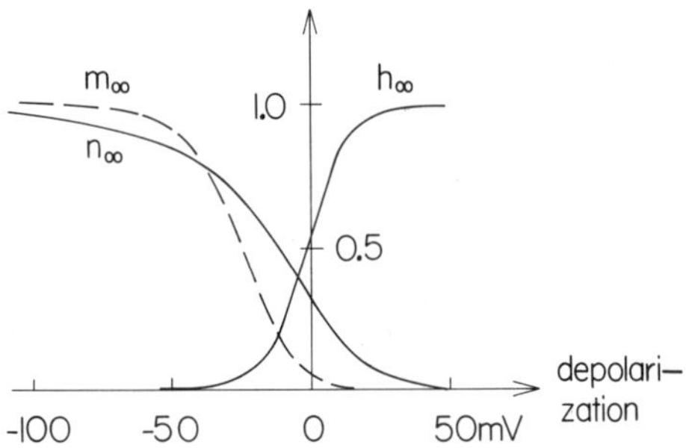

Fig. 3.9. Dependence of final n, m, and h values on depolarization voltage. These constants determine the membrane conductances in response to voltage-clamp step changes; see eqs. 3.15 and 3.16. Note symmetry of n_∞ and h_∞, compared with m_∞. After Hodgkin and Huxley [3.7].

It is interesting to note in detail how the steady-state n, m, and h vary with clamped membrane depolarization (Fig. 3.9). The sodium gating mechanism (m) operates most rapidly around a depolarization of 25 mV, whereas sodium inactivation (h) begins at -50 mV and increases fastest at $V - 0$ and toward hyperpolarization. The delayed potassium gate (n), however, vanishes with hyperpolarization and is nearly maximal beyond 50 mV of depolarization. The n and h curves are symmetric and complementary, and this is important to the nonlinear analysis of Fitzhugh (Section 3.D).

If we fit the Na behavior with a second-order differential equation d^2n/dt^2 [not bimolecular or second degree] rather than two first-order ones, somewhat fewer adjustable constants are needed, and an equally plausible physical chemistry for the gating can be invented [3.11]. As Hodgkin and Huxley remark, alternative molecular mechanisms are speculative until experimental discrimination is devised.

This stationary depolarization behavior can be related to the propagated potential by using cable theory (eq. 3.3 for I constant):

$$i_m = \frac{1}{r_1 + r_2} \frac{\partial^2 V_m}{\partial x^2} \simeq \frac{1}{r_2} \frac{\partial^2 V}{\partial x^2} \qquad \text{since } r_1 \text{ outside is relatively small} \tag{3.17}$$

where by definition $r_2 = R_2/\pi a^2$, $i_m = 2\pi a I$, a = radius, and R_2 = resistivity of axoplasm.

If now we assume that the waveform in space is geometrically similar to the one in time (i.e., the propagated one), which formally is

$$\frac{\partial^2 V}{\partial x^2} = \frac{1}{\theta^2} \frac{\partial^2 V}{\partial t^2} \tag{3.18}$$

where θ = conduction velocity, the differential equation for the waveform $V(t)$ at any point in space is no longer partial. We then have, using eq. 3.11,

$$\frac{a}{2R_2\theta^2}\frac{d^2V}{dt^2} = C_m\frac{dV}{dt} + \bar{g}_K n^4(V - V_K)$$
$$+ \bar{g}_{Na} m^3h(V - V_{Na}) + \bar{g}_l(V - V_l) \quad (3.19)$$

Since θ is unknown, the equation is solved for repeated values of θ until the computed value of $V(t)$ restores to zero after the action potential. The resulting waveform of $V(t)$ resembles satisfactorily the observed action potential $V(x, t)$, improving at higher temperatures. The conduction velocity depends on some of the passive-cable parameters:

$$\theta = \sqrt{Ka/2R_2 C_m} \quad (3.20)$$

where $K = 10.47$ m/sec experimentally but does *not* depend on $\tau = r_4 C_m$, the membrane time constant for passive propagation. This suggests axoplasmic conduction of ions to depolarize the capacitance ahead.

This waveform of the membrane potential $V(t)$ should be compared with the time course of the component conductances (Fig. 3.10*a*) and of the component currents (Fig. 3.10*b*). Note that the action-potential threshold is overcome by the effect of the capacitance discharge current ($I_C = -C_m dV/dt$) resulting from the incoming axoplasmic cathodal current of the approaching wave. This well precedes any change of membrane conductance or increased transmembrane flow to further discharge the capacity. The triggering is

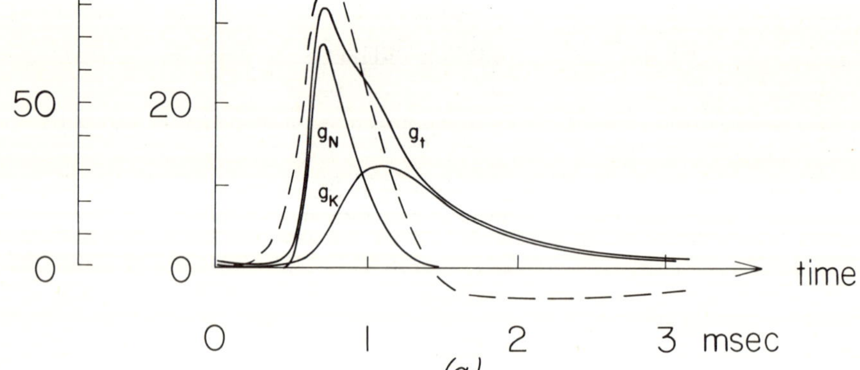

(*a*)

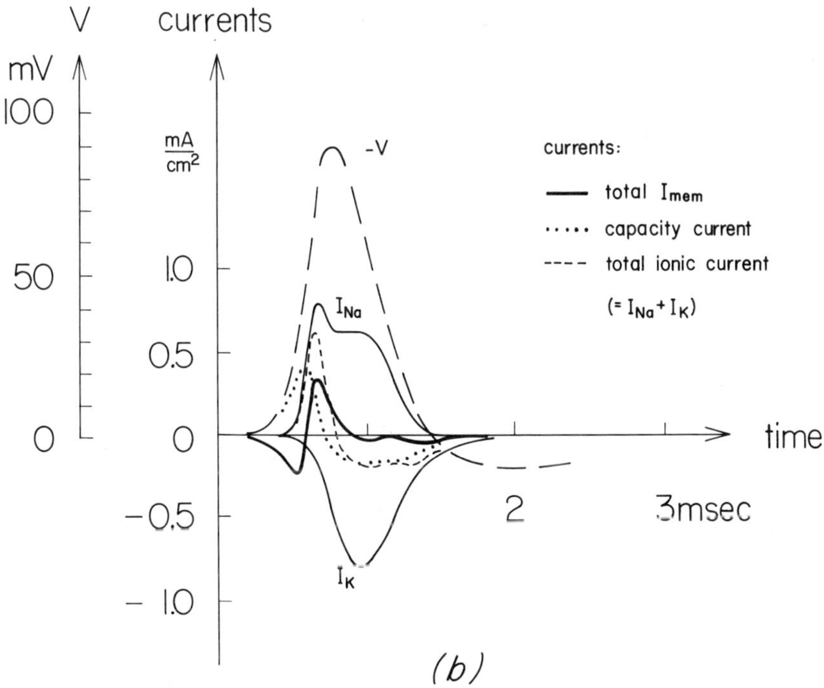

Fig. 3.10. (*a*) Conductance changes during an action potential in squid nerve. (*b*) Currents flowing during an action potential. In both (*a*) and (*b*) the membrane voltage ($-V$) is shown for time reference. Note in (*a*) that conductance changes are delayed, that sodium deactivation sets in early, and that potassium activation persists well after action potential, whereas (*b*) shows how capacitive current precedes transmembrane flow and that repolarizing I_k terminates with negative membrane potential. Total current (heavy line) exhibits rapid cessation, considerably before the end of ionic flows. Note also that currents stop well before conductance changes. After Hodgkin and Huxley [3.7].

therefore accomplished by current rather than by voltage; although the capacitative voltage then determines the onset of instability $dI_{Na}/dV \leq 0$ at threshold.

The original paper of Hodgkin and Huxley [3.7] contains quantitative explanations for the refractory period (in terms of inactivation h and potassium conductance g_K) both absolute and relative, for subthreshold behavior, for anode break excitation, and for apparent inductive (i.e., oscillatory) behavior.

The general analysis is clearly applicable to other excitable tissues (muscle and other nerves) with suitable changes in values of parameters. More detailed mechanistic models have been offered, and these give a better fit to the conductance–time $g(t)$ curves and avoid the high power law [3.10, 3.11]

However, such models involve rather complex expressions in $\dot{n}$, which have no detailed mechanistic basis and require simultaneous probabilities of high order in their kinetic theory. A completely different analysis [3.12], using a model from semiconductor theory, simplifies the postulated mechanism but is hard to justify physically. Its main contribution is to incorporate the local potential drops at the bounding surfaces of the membrane, which the various "gated-hole" theories neglect. Electron microscopy and artificial proteolipid bilayers are the current experimental approaches to improved models at this level.

C. ANALYTIC PROBLEMS OF INSTABILITY AND CYCLIC ACTION IN NONLINEAR SYSTEMS

Physiologists for years have studied the nerve impulse: its triggering, thresholds, component processes, and recovery cycle. Clearly there is still much to be learned about the membrane mechanisms at the molecular level. However, the main task of nerve is carrying information. As our study of sensory and control systems reveals, this is done mainly by *frequency modulation* of impulse trains in groups of nerve fibers, where information is coded by frequency (Δf) rather than by pulse interval (Δt). Hence to understand and formulate the function of the nervous sytem as such, we must study the phenomena, mechanisms, and analysis of nervous tissue as a *frequency generator*. This is not a large step from the foregoing "axonology" because the membrane mechanism just discussed does itself have an oscillatory or repetitive mode of operation. This modulation appears in many places: at the axon hillock, in the cardiac pacemaker, and in the proportionality of impulse frequency to voltage in sensory receptors. (Chapters 4 and 5).

The nerve oscillator is not of the class of linear oscillators (tuning forks, LC circuits, balance wheel) where energy oscillates between the potential and kinetic form; even if sustained by a quite nonlinear magnet drive, or escapement. Nor is it of that class of feedback oscillators (phaseshift oscillator), where by a phase delay and suitable gain, the signal reexcites itself linearly. Repetitive firing of nerve is in the class of nonlinear negative-resistance (conductance) phenomena which include neon flasher, carbon arc, dynatron, and tunnel diode. In these, where $R = \Delta X/\Delta J < 0$, a decrease of force produces an increased flow. We first consider electric and hydraulic examples and then apply this analysis to the nerve membrane.

The simplest example is charging a condenser to the firing threshold of a gas tube (Fig. 3.11). Avalanche conduction ensues, the conductivity becoming negative, and the capacitor discharges through a near short whose inductance must be taken into account in the analysis. If R is large enough, I_N will fall

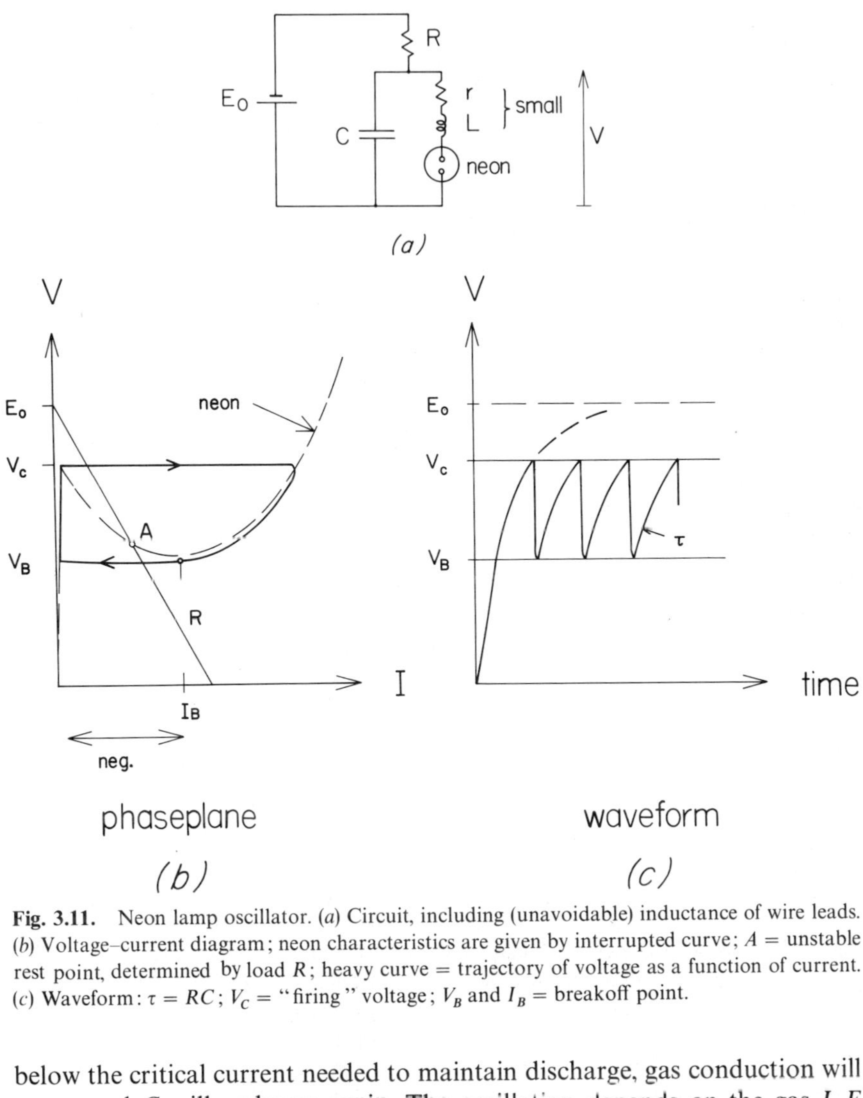

Fig. 3.11. Neon lamp oscillator. (*a*) Circuit, including (unavoidable) inductance of wire leads. (*b*) Voltage–current diagram; neon characteristics are given by interrupted curve; A = unstable rest point, determined by load R; heavy curve = trajectory of voltage as a function of current. (*c*) Waveform: $\tau = RC$; V_C = "firing" voltage; V_B and I_B = breakoff point.

below the critical current needed to maintain discharge, gas conduction will cease, and C will recharge again. The oscillation depends on the gas I–E curve having an N shape (Fig. 3.11*b*); that is, a section of negative slope. The condenser voltage V_c (Fig. 3.11*c*) executes a sawtooth with an exponential approach to the threshold or firing point. Time constant and threshold determine frequency. This process is called a relaxation oscillation.

Many models of nerve transmission [3.13] have been based on such elements, one of the closest analogies being the tunnel diode circuit of Nagumo et al. [3.14]. It was in analyzing the neon oscillator as a model of the heart's pacemaker that van der Pol (1926) first formulated the nonlinear differential equations and studied their topology in phase space. This furnishes to the study of nerve impulse trains a powerful analytic approach that displays the detailed conditions of firing and recovery as well as the limit cycle representation of nonlinear oscillators.

Some fundamental considerations of stability can be brought out by returning to the simple hydraulic model of Fig. 2.4. It was mentioned in Chapter 2 that if K_1 or K_2 were to be negative, the plot of entropy rate Φ would become parabolic downward, and the phase diagram $\dot{h}(h)$ would be reversed in slope. If increased potential h increases K_1 or decreases K_2 (Fig. 3.12), its effective conductance becomes negative and the system unstable.

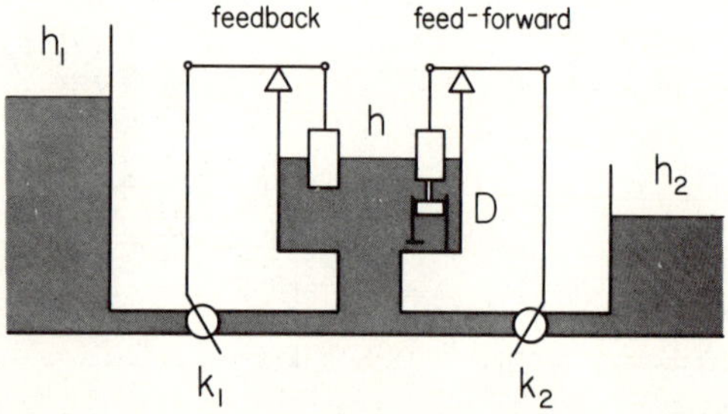

Fig. 3.12. Hill's hydrodynamic model of nerve membrane showing positive feedback (K_1 smaller if h increases), and negative feed-forward (K_2 larger if h increases), which also causes instability because it is delayed by dashpot D. After Franck [2.10].

It is a matter of definition whether this back-coupling in the model, or the dependence of g_{Na} on V in nerve, is called "feedback." It is what Milsum [3.15] calls "passive feedback," as opposed to "active feedback" wherein a comparator (subtractive element) is present to compare the output with some system purpose (input setpoint or guidance pattern). The term "feedback" is also becoming embedded in biochemical usage for any coupling or effect of a downstream value onto an upstream process, even if this does not fit the ordinary theory of control systems. But in this case, it is clear that g_{Na} is not "upstream" to V_m, so that "positive feedback" does not describe the effect of membrane potential on conductance, although "feed-forward" might apply. To call any coupling or linkage whatsoever "feedback" would be to rob the term of analytic value. With Milsum, we use the term only for the active type to which engineering feedback control theory applies.

As Hill [2.10] showed, it is possible to add other potential-controlled hydraulic gating mechanisms (Fig. 3.12). Such a model can represent several features of nerve behavior. The same negative total conductance was ob-

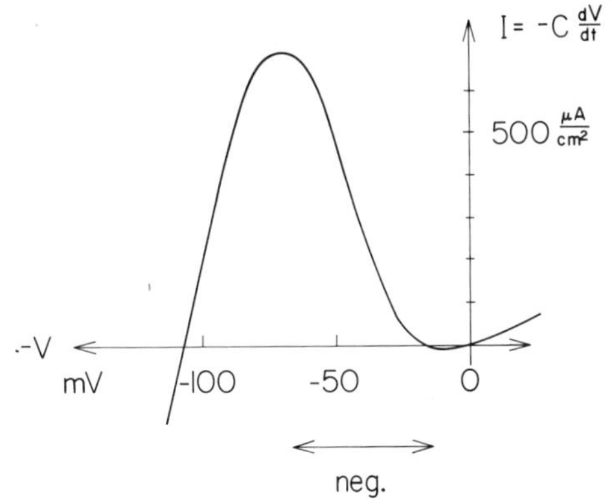

Fig. 3.13. Cubic relation of current versus membrane potential for squid nerve, showing negative dI/dV below 12-mV depolarization. Data from Hodgkin–Huxley. The negative conductance causes, for rapid voltage changes, regenerative depolarization of the nerve membrane.

served in squid nerve by Hodgkin et al. (Fig. 3.13). This happens also to exhibit the positive slope of the unstable phase diagram $d\dot{V}/dV > 0$ shown in the hydraulic model. The same analysis applies to the older metallo-chemical (iron wire) models of nerve [2.10] as well as to other examples of chemical instability (explosions).

Representation of Nonlinear and Unstable Systems in the Phase Plane. Mathematically, we see most clearly the transition from linear to nonlinear oscillations by studying them in the phase plane: defined physically by the velocity–displacement (momentum–position) diagram. Furthermore, the nature of all solutions of a nonlinear (autonomous) differential equation is evident in the phase plane by its topology. If the system defies analytic solutions in some regions of the phase plane, graphical techniques of obtaining trajectories can always be applied. (Note, e.g., the technique of using the isoclines in the following discussion.) For a more detailed introduction, the reader is referred to Minorsky [3.16, Chapter I]. For engineers, Ku [3.17] brings in more electrical examples.

Taking ω_0 to be the undamped natural frequency, one way to transform the familiar linear oscillator (damped pendulum, LC resonance circuit)

$$\ddot{x} + 2bx + \omega_0{}^2 x = 0 \tag{3.21a}$$

into a pair of first-order equations is by assuming

$$\dot{x} = y \tag{3.21b}$$

so that

$$\dot{y} = -2by - \omega_0{}^2 x \tag{3.21c}$$

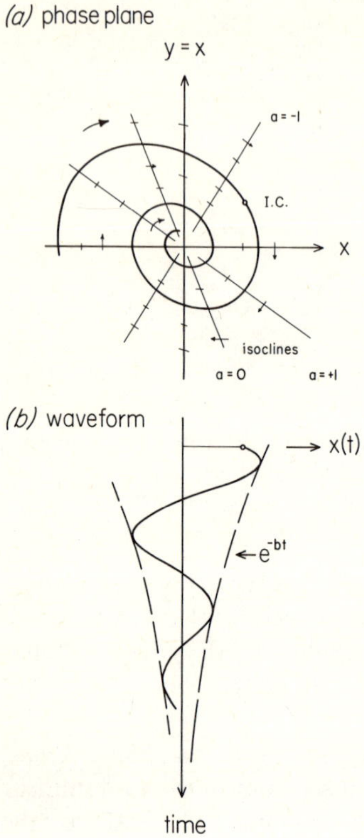

Fig. 3.14. (*a*) Phase plane representation of a linear damped oscillator ($\ddot{x} + 2b\dot{x} + \omega_0{}^2x = 0$), showing the isoclines $dy/dx = (-2by - \omega_0{}^2x)/y = a$, where $a = 0, \pm 1, \infty$. Trajectories can be constructed from their segments at the isoclines. Thus solutions for any nonlinear system can be obtained. Here, a logarithmic spiral into the origin results. The origin is a stable focal point. (*b*) Waveform $x(t)$, with initial condition (I.C.) as shown in (*a*).

Time can be eliminated from this system, which then is studied in the $\dot{x}(x)$ phase plane. Solutions are families of trajectories, for which the loci of equal slope, the isoclines (dy/dx = constant = a), are

$$\frac{\dot{y}}{\dot{x}} = \frac{dy}{dx} = \frac{-2by - \omega_0{}^2x}{y} = a \tag{3.21d}$$

If now we draw (Fig. 3.14) a few of the isoclines (e.g., $a = -1, +1, 0, \infty$) we can trace out the paths (trajectories) of points in phase space obeying eq. 3.21a. If damping $b = 0$, the trajectory becomes a set of stable concentric ellipses (not shown) about a stable "center" at the origin (0, 0). If the oscillator is slightly positively damped, the trajectories from any given initial condition (Fig. 3.14*a*) spiral into the stable center at 0, 0, with $x(t)$ a damped sine wave (Fig. 3.14*b*). Note that the original differential equation must be used when the time course $x(t)$ is obtained. If more than critically damped, $b^2 - \omega_0{}^2 > 0$, there are two stable roots $Ae^{-r_1t} + Be^{-r_2t}$ and a family of asymptotic solutions in $x(t)$. In the $\dot{x}(x)$ plane, the trajectories will then approach a stable "*nodal point*" at 0, 0 (not shown).

However, if the damping can be *either* negative *or* positive, a new kind of equilibrium path ensues. In van der Pol's simplified equation for the idealized nonlinear oscillator having both inertial and springy (capacitative) elements:

$$\ddot{x} - c(1 - x^2)\dot{x} + x = 0 \qquad \text{(excursion damping)} \tag{3.22}$$

Alternatively, we have Rayleigh's equation with velocity damping:

$$\ddot{x} - c(1 - \dot{x}^2)\dot{x} + x = 0$$

The damping coefficient $b \equiv -c(1 - x^2)$ is small and negative (i.e., introducing instability) for small excursions, but it is positive and large for greater amplitudes. This positive damping increases nonlinearly with amplitude, much as did the resistance of a gas tube (Fig. 3.11) or the conductance of nerve (Fig. 3.13). There must exist, therefore, an intermediate region of amplitudes in which a type of dynamic equilibrium is encountered which exists only in nonlinear systems. The system follows a limit cycle; $x(t)$ is periodic although not sinusoidal.

When transforming eq. 3.22 to the phase plane $\dot{x}(x)$, van der Pol's equation is

$$\dot{x} = y \tag{3.23a}$$

so that

$$\dot{y} = c(1 - x^2)y - x \tag{3.23b}$$

By plotting a few isoclines $dy/dx = [c(1 - x^2)y - x]/y = a$, we obtain a trajectory field (Fig. 3.15) wherein small oscillations spiral exponentially out from the unstable nodal point 0, 0, toward the *stable* limit cycle L, while those starting larger go inward to L. The oscillations continue at steady amplitude as long as energy is supplied to the negative resistance. Typical van der Pol oscillations are presented in Fig. 3.15*b* [3.16], where $c = 1.0$.

However, for analysis of the nerve impulse this transform of van der Pol's is ill suited because the $\dot{y}$-nullcline (N in Fig. 3.15) is multibranched and discontinuous. Moreover, repetitive oscillations are not always seen. Lienard's [3.20] transformation of van der Pol's equation turns out better. Substituting in eq. 3.22 yields

$$\frac{d}{dt}\left[\dot{x} - c\left(x - \frac{x^3}{3}\right)\right] + x = 0 \tag{3.24}$$

Hence $\dot{y} = -(1/c)x$ [since $cy \equiv \dot{x} - c(x - x^3/3)$] and $\dot{x} = c[y + (x - x^3/3)]$.

The isoclines $dy/dx = -x/[c^2(y + x - x^3/3)] = a$, a family of cubics, yield for the y-nullcline ($\dot{y} = 0$) the line $x = 0$, whereas the x-nullcline ($\dot{x} = 0$) is now the simple N-shaped cubic: $y = x^3/3 - x$.

In these coordinates, a small analog computer at once displays the field of trajectories, isoclines, and limit cycle (see Fig. 3.15*c*). Note that we no longer have a variable plotted against its time derivative as in Figs. 3.14 and 3.15*a*.

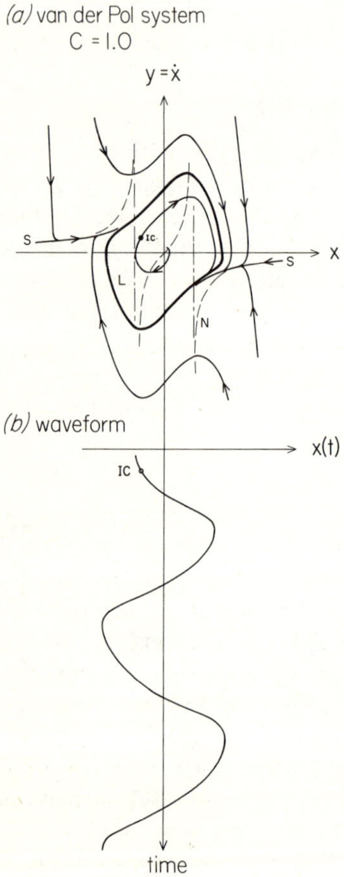

Fig. 3.15. (*a*) Phase plane representation of van der Pol's equation $\ddot{x} - c(1 - x^2)\dot{x} + x = 0$ with $c = 1.0$. Near the origin (small x), damping is negative and oscillations grow. Damping is positive for large x; oscillations decay. There exists, in the midrange, a stable limit cycle L, representing a new type of dynamic equilibrium. The $\dot{x}$-nullcline N (interrupted curves) is a three-branched system. Trajectory S is a "line sink" collecting many trajectories and leading directly into the limit cycle. (*b*) Waveform, with system beginning at initial condition IC [see (*a*)], growing rapidly into the limit cycle oscillation.

Clearly, the field topology and the shape of the limit cycle depend somewhat on the choice of first-order equations (variables and constants) on which are projected the second-order form. However, the character of the singular points does not show this dependence. We introduce Fitzhugh's [3.18] variation of this equation later in this chapter.

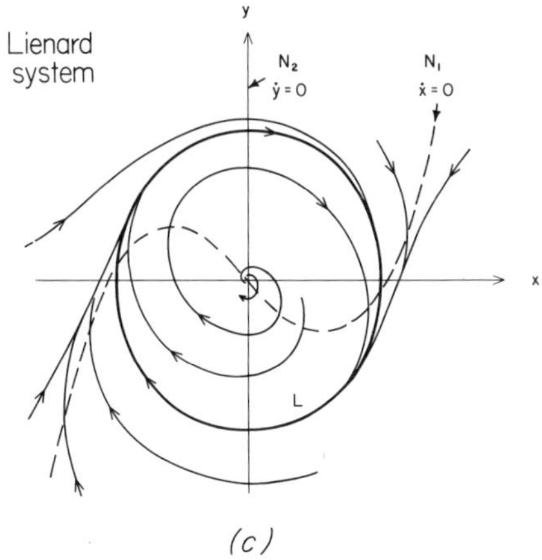

Fig. 3.15c. Phase plane of Lienard's transformation of van der Pol's equation: $\dot{x} = c(y + x - x^3/3)$, $\dot{y} = -x/c$, for which the $\dot{x}$-nullcline is a cubic (N_1) and the $\dot{y}$-nullcline (N_2) is coincident with y-axis. Similar to the van der Pol system, oscillations grow if x and y are small. The origin is an unstable focus. Large oscillations decay and approach limit cycle L.

D. STABILITY AND REPETITIVE PROPERTIES OF THE HH MEMBRANE

For comparison, we present the equations of HH describing nerve membrane behavior:

$$I = C\dot{V} + \bar{g}_{Na} m^3 h(V - V_{Na}) + \bar{g}_K n^4(V - V_K) + \bar{g}_l(V - V_l) \quad (3.25a)$$

$$\dot{m} = \phi[(1 - m)\alpha_m(V) - m\beta_m(V)] \quad (3.25b)$$

$$\dot{n} = \phi[(1 - n)\alpha_n(V) - n\beta_n(V)] \quad (3.25c)$$

$$\dot{h} = \phi[(1 - h)\alpha_h(V) - h\beta_h(V)]$$

where:

$$\alpha_m = 0.1(V + 25)\left[\exp\frac{V + 25}{10} - 1\right]^{-1}$$

$$\alpha_n = 0.01(V + 10)\left[\exp\frac{V + 10}{10} - 1\right]^{-1}$$

$$\alpha_h = 0.07\left[\exp\frac{V}{20}\right]$$

$$\beta_m = 4\left[\exp\frac{V}{18}\right]$$

$$\beta_n = 0.125\left[\exp\frac{V}{80}\right]$$

$$\beta_h = \left[\exp\frac{V+30}{10}+1\right]^{-1}$$

$$\phi = 3^{(T-6.3)/10} \tag{3.26}$$

Note that V is in mV, the α's and β's in msec^{-1}, T in °C.

Fitzhugh [3.18] observed that the conductivity factors $n(K^{act})$ and $h(Na^{inact})$ vary timewise but slowly (Fig. 3.16) and that their sum is almost constant (as though related); $m(Na^{act})$ and V_m vary rapidly but asynchronously. This suggested that the foregoing system of four first-order differential equations (leading to a formidable four-dimensional phase *space*) might be "reduced" to a more intelligible two-dimensional "V–m space" if n and h were regarded as constant.

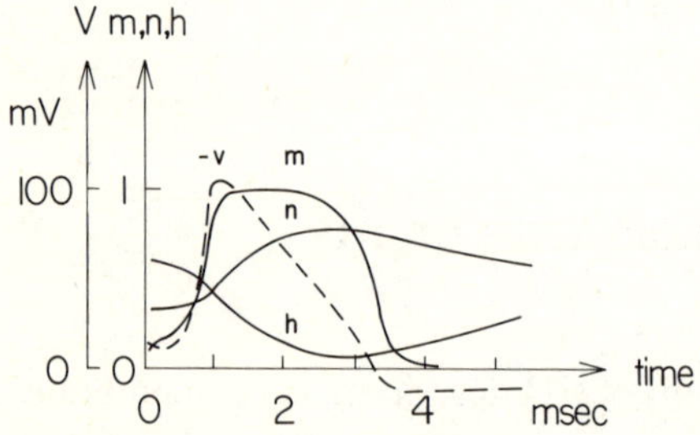

Fig. 3.16. Activation and deactivation functions, h, m, n, described by eqs. 3.12 to 3.14, as functions of time during action potential (interrupted curve $-V$). Note that $h(t) + n(t) \simeq$ constant. Action potential therefore can be studied, in a first approximation called the reduced HH, in a V–m plane. After Fitzhugh [3.20].

In the latter case, with $\dot{n} = \dot{h} = 0$ and letting $I = 0$ after stimulus has ceased to exist, the isoclines are obtained directly from eqs. 3.25a and 3.25b. (See [3.18] for mathematical details.) They are given by

$$\frac{dm}{dV} = \frac{\dot{m}}{\dot{V}} = \frac{C\phi[(m-1)\alpha_m(V) + m\beta_m(V)]}{\bar{g}_{Na}\, m^3 h(V+115) + \bar{g}_K\, n^4(V-12) + \bar{g}_l(V+10.6)} = a \tag{3.27}$$

Then we find, solving eq. 3.25b for m, that the $\dot{m} = 0$ nullcline ($a = 0 = dm/dV$) is,

$$m(V) = \frac{\alpha_m}{\alpha_m + \beta_m} = \frac{0.1(V+25)[\exp(V+25)/10 - 1]^{-1}}{0.1(V+25)[\exp(V+25)/10 - 1]^{-1} + 4[\exp V/18]} \tag{3.28a}$$

and with $I = 0$ the $\dot{V} = 0$-nullcline ($a = \infty = dm/dV$) is

$$m(V) = \left[\frac{\bar{g}_K\, n^4(12-V) - \bar{g}_l(V+10.6)}{\bar{g}_{Na}\, h(V+115)}\right]^{1/3} \tag{3.28b}$$

which is clearly a cubic in $m(V)$.

These nullclines are drawn in Fig. 3.17, slightly altered from [3.18] to clarify their intersection in the three singular points *A*, *B*, and *C*. The trajectory map shows that *A* and *C* are (stable) focal points or resting states, whereas *B* is an unstable "saddle point" through which passes the *threshold* separatrix *S–S* separating the stable fields around *A* and *C*.

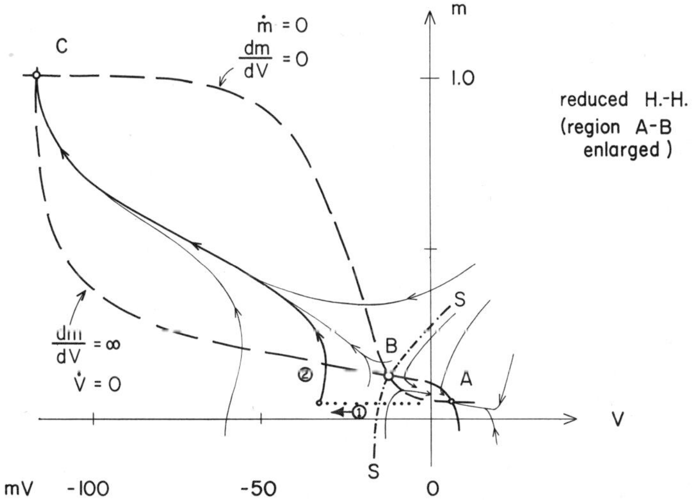

Fig. 3.17. Phase plane of the reduced $m(V)$ system, showing nullclines that intersect at three singular points. Only excitation (depolarization) can be represented here. *A* and *C* stable, *B* unstable saddle. Separatrix *S*——*S* divides the realms of *A* and *C*. A few trajectories for varied initial conditions are given by the fine lines. Dotted path labeled (1) represents a stimulus carrying the system from resting state *A* to the left of the separatrix (*threshold*). Then trajectory (2) ensues, which stops at point *C*, the excited or depolarized state. For repolarization, changes in *n* and *h* are important. These shift the $\dot{V}$-nullcline (see text) so that the system can return to *A*.

Thus in *single-pulse* operation of the HH equations, with an axon membrane resting at *A* (Fig. 3.17), a negative voltage pulse (1) carries the phase point beyond the separatrix *S–S* to trajectory (2), along which it moves to stable depolarization at *C*. But as time progresses the values of *n* and *h* also change, as in Fig. 3.16. This raises nullcline $\dot{V} = 0$ so that intersections *C* and *B* coalesce and vanish, leaving the potential to return by a path through the upper half-plane to its rest point at *A*. Several features of nerve excitation (accommodation, facilitation, refractoriness, subthreshold response) can be illustrated on this plot [3.18].

In *oscillatory* operation, the $\dot{x}$-nullcline of HH moves upward by changes in *n* or *h*, which are voltage-dependent. The nullcline intersection *A*, therefore, moves leftward into the negative-slope region of $\dot{x} = 0$, and the rest potential finds itself to the left of the separatrix. Oscillations ensue on a limit cycle. This means that the four HH equations on an analog computer will exhibit continuous oscillations in this complex 4-space, suggesting that these equations might also be used to model such frequency generators as the cardiac pacemaker or the axon hillock. However, the live squid nerve does not show continued oscillations, but only a limited burst or train. This suggests the existence in nerve of some additional slow accommodation process requiring still another differential equation.

The foregoing discussion shows that the HH equations are too cumbersome for a generalized nerve model to be used as a part of a larger system. However, their stability properties are of some interest to the biophysicist [3.21] dealing with membranes. In addition, other excitable tissues exhibit several variations from the values and behavior of squid nerve (Noble [3.19]). It seems better for analytic purposes, therefore, to have a simpler and more generalized mathematical model, more tractable and intelligible than the four-dimensional phase space of the HH model. In this spirit Fitzhugh proposed to use the very general two-dimensional van der Pol form (eq. 3.24), with such minor modifications as needed to show the characteristic behavior of the nerve impulse.

The reduced (V–m) HH differential equations (Fig. 3.17) are topologically similar to the van der Pol equations (Fig. 3.15) already. They belong to the same class, in that one nullcline ($\dot{V}$)is a cubic (N shape) with respect to the other nullcline ($\dot{m}$). With this in mind, Fitzhugh [3.20] shifted the nullclines of the van der Pol equations slightly by adding a few terms (Z and $a - by$) to $\dot{x}$ and $\dot{y}$:

van der Pol, Lienard
$$\begin{cases} \dot{x} = c\left(y + x - \dfrac{x^3}{3}\right) \\ \dot{y} = \dfrac{-1}{c}(x) \end{cases}$$

Fitzhugh, Bonhoeffer-van der Pol
$$\begin{cases} \dot{x} = c\left(y + x - \dfrac{x^3}{3} + Z\right) \\ \dot{y} = \dfrac{-1}{c}(x - a + by) \end{cases}$$

This adds a Z dimension to take care of I in eq. 3.25a, but it leaves $y(x)$ a cubic for $\dot{x} = 0$, $Z = 0$. It straightens and *slants* the HH nullcline $\dot{m} = 0$ to the linear form $x = a - by$ (or $m = a - bV$) as in Fig. 3.18*a*, yet preserving the relations of the singular points ABC of the original HH equations plotted in Fig. 3.17.

These Bonhoeffer-van der Pol (BVP) nullclines are plotted in Fig. 3.18*a*, with constants as shown. With $Z = 0$, the rest point P is stable and a stimulus pulse in Z causes the $\dot{x}$-nullcline to move upward. The state or phase point finds itself below the separatrix and follows path (1) to (2) through the active state, returning through the "absolutely refractory" region. Just after this, a strong stimulus will cause it to recross the separatrix to another (but smaller) active state. Subsequently, the threshold remains "depressed" till the state returns to P.

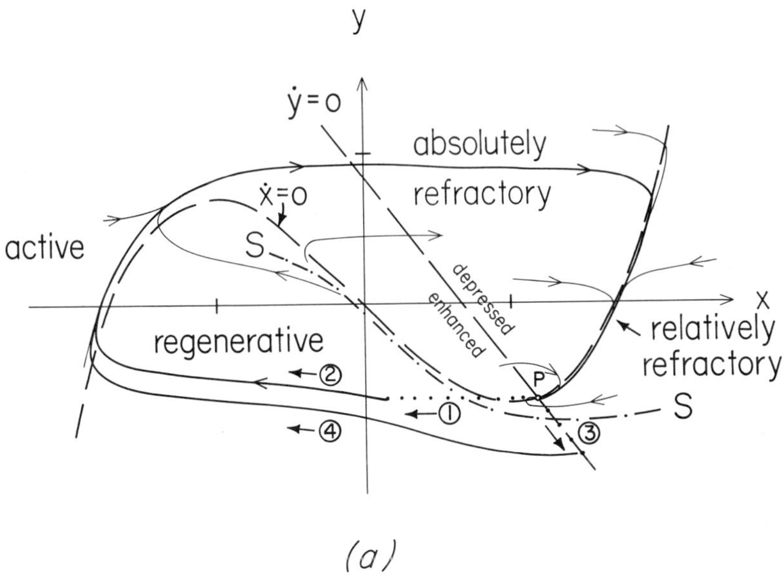

Fig. 3.18*a*. Fitzhugh (or Bonhoeffer-van der Pol) system, exhibiting (quasi) threshold properties (separatrix *S*——*S*), regenerative depolarization, an active or depolarized state, absolute and relative refractoriness, where $P =$ stable resting state. The two nullclines are given by interrupted curves. Path (1) represents the impulse stimulus after which action potential (2) follows. Paths (3) and (4) demonstrate how anode break excitation is achieved.

A large cathodal square wave [$Z = -0.4$] will raise the $\dot{x}$-nullcline, which is represented in Fig. 3.18*b* by *lowering* the abscissa to ($0'$–x'), carrying down with it the $\dot{y} = 0$ nullcline to intersect now at P'. This is now an unstable point on the negative slope of $\dot{x} = 0$, lying below the separatrix. The phase point therefore moves into limit cycle *LL* and continues to "fire" periodically until the cathodal stimulus ceases. Fitzhugh shows how anodal break, break reexcitation, and other complex phenomena can be represented on this plane diagram. See [3.20] for a thorough mathematical discussion.

This BVP model differs from real axons in that submaximal responses can be produced, although these can occur in other excitable tissue. An analog computer, connected this way, will trace out the time course of an impulse or train; this is simply the projection of the trajectory on the x-axis displayed versus time (Fig. 3.18*b*). But this simplified model will not generate a limited train in response to a single stimulus, as would a real nerve or an axon hillock. Nor does this model provide for frequency modulation proportional to voltage, as required for receptor cells. This property resides perhaps in the cell as a system (Chapter 4), rather than in the membrane alone.

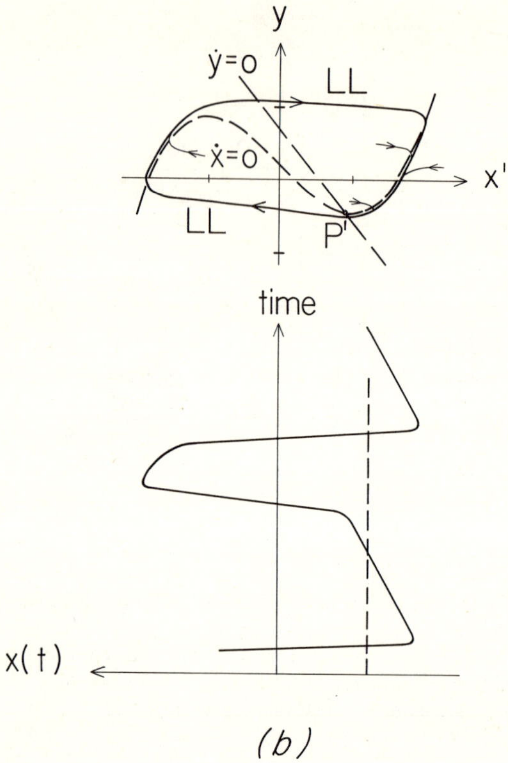

Fig. 3.18*b*. Continuous firing with steady stimulus creating unstable nullcline intersect P'. (Large Z in BVP model). A limit cycle (LL) ensues, producing the waveform in the lower panel. After Fitzhugh [3.20].

Although not intended to reproduce the shape of experimental action potentials, as Fitzhugh states [3.20], the BVP model is meant to exhibit the basic dynamic interrelationships between the variables of state that are responsible for the properties of threshold, refractoriness, and finite and infinite trains of impulses. The algebraic form was chosen as the simplest, and it could be changed without altering the properties of interest. Such a model not only represents certain facts of axon physiology in organized fashion, but relates nerve membrane models to those of physically quite different systems that behave similarly. This generalizing property of mathematical models, which has done so much to unify the physical sciences, should do the same for physiology. Such behavior in the phase plane would also be a test of future differential equations that purport to describe more detailed membrane mechanics.

We have passed from models of squid axon at the level of electrochemical systems, as proposed by Hodgkin and Huxley, to models of nerve tissue behavior, where electrochemistry is now a subsystem. To make the model tractable in terms of the new variables of nerve behavior, the dimensionality of the subsystems had to be reduced from four to two. Fitzhugh was able to do this by a mathematical transformation [3.20] that essentially collapsed or projected the four-dimensional system into a two-dimensional view.

It proves to be true in general, as systems become subsystems of the next higher one, that the dimensionality of subsystems must be reduced to make room for new dimensions of behavior needed to characterize the next higher system.

E. TRANSIENT PROPERTIES OF THE DENDRITIC TREE

The passive local currents ahead of a depolarized area play a major part in initiating the progressive change in axon voltage which actuates the conductivity change in active discharge. But in many cases at the receptive or dendritic end of a neuron, the whole transfer and integration of information is by way of this passive linear process. Rall has shown how the theory of such propagation can be applied to the complex dendritic geometry, in order to study the effects on the soma of transient synaptic excitation of the dendrites.

In general, the dendrites include 90 to 99% of the cell area and, with the soma, perform the most important function of the neuron. This is the spatiotemporal reorganization or "translation" of incoming information before it is retransmitted in frequency-modulated form as a small part of the next bundle of spatiotemporal information. Only the simplest steps in this processing are known. Rall [3.22–3.24] has demonstrated how the information can be weighted differently with respect to amplitude, delay, rate of onset, and averaging (smoothing), according to its entrance proximally or distally on the dendritic tree. Its subsequent integration at the soma to a voltage that controls outgoing spikes or spike frequency, and a model of its transform into a conductance variation at the axon hillock, are sketched in the next chapter. Because of capacitative considerations discussed there, there is doubt that the entire soma membrane itself (of motoneurons at least) undergoes the nonlinear conductance change ("fires"), even though an electrode tip can sense within it the large action potential spike being generated nearby at the outgoing hillock and axon.

The somatic and dendritic surface is thought to respond to incoming synaptic excitation (ε) and inhibition (j) by linear transient changes of the cell-surface conductance—G_ε (permitting flow of Na^+ down its gradient) quite locally depolarizing C to some degree short of the resting potential E_r

but not firing; or G_j permitting flow of K^+ down its chemical gradient (Eccles [3.25]) locally increasing the charge on C (hyperpolarization). This is done by a complex system of intermediate synaptic processes (electrical, chemical, and hormonal) described in modern physiology texts [2.2, see also 3.26] but not yet amenable to quantitative modeling.

For analysis, this control function (conductivity modulation) is modeled (Fig. 3.19) as in parallel with the resting potential E_r and with the lumped resting conductance G_r. This conductance is subject to nonlinear depolarization as well, but only at the axon hillock, where thresholds are relatively low.

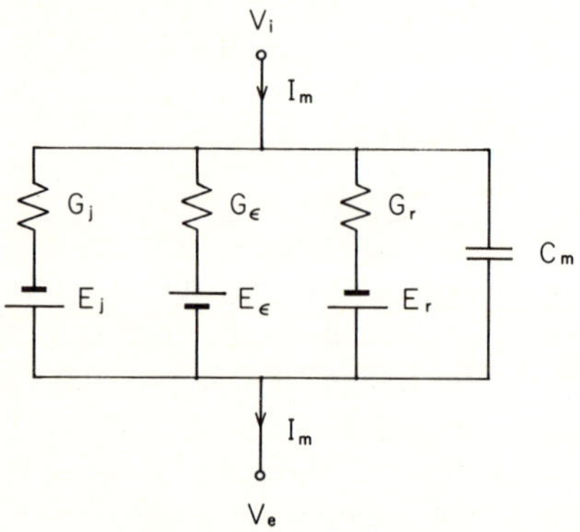

Fig. 3.19. Electric model of dendrite membrane, with resting potential and conductance (r) and synaptic control of conductance G, inhibitory (j), excitatory (ε). After Rall [3.22].

The somadendritic model (Fig. 3.19) may be described electrically and mathematically (for details see Rall [3.24]) as

$$I_m = C_m \frac{dV_m}{dt} + [G_r(V_m - E_r) + G_\varepsilon(V_m - E_\varepsilon) + G_j(V_m - E_j)]$$

or briefly

$$I_m R_m = \tau \frac{dV}{dt} + K^2(V - V^*) \tag{3.29}$$

where

$$\tau = R_m C_m \qquad R_m = 1/G_r$$
$$V = (V_m - E_r) \qquad K^2 = (G_r + G_\varepsilon + G_j)/G_r$$
$$V^* = [G_\varepsilon(E_\varepsilon - E_r) + G_j(E_j - E_r)]/K^2 G_r$$

This simple model has some interesting physiological properties. If we set $I_m = 0$, eq. 3.29 becomes the differential equation of the somatic membrane. Its transient solution for the initial condition of a *step increase* in G_ε or G_j is

$$V = V^*(1 - e^{-K^2 t/\tau}) + V_0 e^{-K^2 t/\tau} \tag{3.30}$$

where $V_0 = V|_{t=0}$.

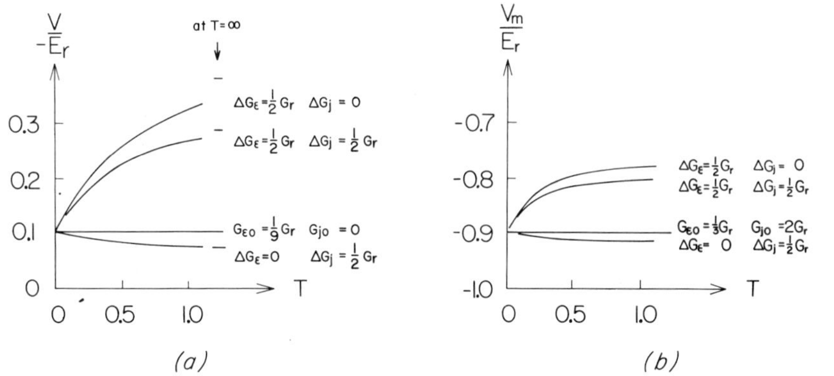

Fig. 3.20. Local transients of dendrite membrane potential, for isolated patch, given $E_\varepsilon = 0$, $E_r = E_j$ initially, and the two types (*a* and *b*) of initial conductance distribution. Transient responses to three different types of conductance change patterns (see labels) are presented. After Rall [3.22].

The initial slope of this is

$$\left.\frac{dV}{dt}\right|_{t=0} = \frac{\Delta G_\varepsilon(E_\varepsilon - V_{m0}) + \Delta G_j(E_j - V_{m0})}{G_r}$$

This behavior is shown in Fig. 3.20 for

1. a uniform excitatory step $\Delta G_\varepsilon = \frac{1}{2} G_r$ at $\Delta G_j = 0$
2. a uniform inhibitory step $\Delta G_j = \frac{1}{2} G_r$ at $\Delta G_\varepsilon = 0$
3. both steps superimposed

under the following typical *preexisting* steady states of excitation and inhibition:

$$\text{(a)}\quad G_\varepsilon = \frac{1}{9} G_r \qquad \text{(b)}\quad G_\varepsilon = \frac{1}{3} G_r$$

$$G_j = 0 \qquad\qquad G_j = 2G_r$$

In both (a) and (b) the membrane responses proves to be nine times faster initially to excitation than to inhibition, but its amplitude depends strongly on the preexisting steady excitatory and inhibitory conductances. Furthermore, the effect of inhibition on *transient* excitation is much less than is its final or *background* effect. These properties are independent of the effects of dendritic branching, discussed next.

Since the soma in general is connected to between 10 and 50 times its area in dendrites, the core conductor model of passive propagation on an "infinite" line (Section 3.A) can be used. To a local step excitation, this "dendritic dominance" gives the "erf" response (eq. 3.10) in the soma. This is significantly steeper than an exponential (Fig. 3.21). Looking backward from

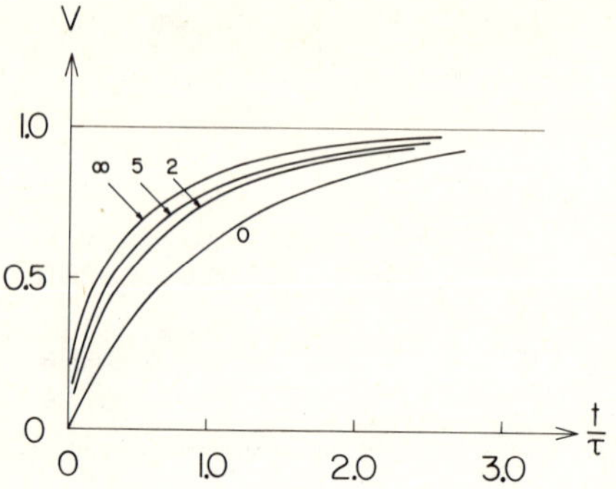

Fig. 3.21. Effect of dendrite area on transients in soma potential. Membrane voltage V, after current step at $t = 0$ at the soma, as a function of dendritic dominance. The V rises exponentially if no dendritic structure is involved (curve labeled 0); it rises according to an erf function (i.e., much quicker) if the dendrite structure is very large (curve labeled ∞). The parameter on these curves represents the ratio of combined dendritic conductances and the soma membrane conductance. After Rall [3.23].

the axon hillock, the soma and the tree present such a large internal damper or current sink of parallel capacitance, that we can expect summated soma changes in conductance G_ε amounting to $10G_r$ without exceeding its firing threshold. To assess next the relative contribution of proximal versus distal synaptic excitation and inhibition, we must calculate the change of potential *at* the soma, caused by positive and negative conductance changes at various distances in the dendritic tree.

The differential equation at any *point* representing this potential distribution $V(x, t)$ resembles that previously solved (eq. 3.4). But because boundary conditions change at each branching of the dendritic trunk, a simpler analytic form is needed. Rall [3.24] has proposed a model wherein length Z is *scaled* according to the total surface area A at a given order of ramification (Fig. 3.22): $dZ/dx \propto dA/dx$. The argument does not require bifurcations at equal areas, and it has been extended to the point at which Z changes exponentially with area; nevertheless, we hold to these restrictions. The analytic problem then is to reduce a stylized dendritic tree (Fig. 3.22) to an equivalent cylinder of length Z. (Rall discusses further the adequacy of this approximation.)

This interesting result is approached as follows: at branching level i from the soma of n equal branches of radius r, (e.g., $n = 8$ at level 3) the total *axial* dendritic current can be written

$$I_i(r, x) = -\frac{\pi r^2 n}{R_i}\frac{\partial V_i}{\partial x} \tag{3.31}$$

where R_i = specific intracellular resistance.

The *trans*membrane current, differentiating eq. 3.31 partially with respect to x, is

$$I_m = \frac{\partial I_i/\partial x}{dA/dx} = -\frac{\pi r^2 n}{R_i}\left[\frac{\partial^2 V_i}{\partial x^2} + \frac{\partial V_i}{\partial x}\frac{d}{dx}\ln(r^2 n)\right]\left[\frac{dA}{dx}\right]^{-1} \tag{3.32}$$

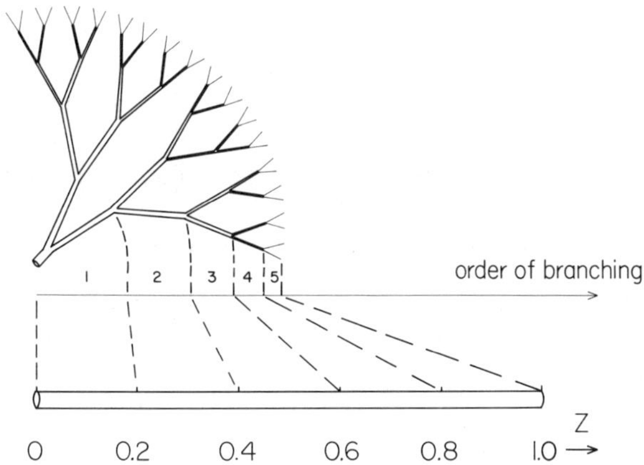

Fig. 3.22. Uniformly branched dendritic tree and its electrotonic equivalent cylinder. Branching of the tree must be such that $\sum_j d_j^{3/2}$ = constant; that is, the sum, over all branches of (diameter)$^{3/2}$ at a level of ramification, is a *constant* and the same constant at all levels of branching. Then the tree can be replaced by a cylinder whose length is measured in space constants. After Rall [3.22].

Assuming n cylindrical branches at level i, we have

$$\frac{dA}{dx} = 2\pi rn \frac{ds}{dx} \tag{3.33}$$

since $\quad A = \pi r^2 ns$ and $\left(\dfrac{ds}{dx}\right)^2 = 1 + \left(\dfrac{dr}{dx}\right)^2$ in general

ds = surface element

Therefore,

$$I_m R_m = \frac{rR_m}{2R_i}\left[\frac{ds}{dx}\right]^{-1}\left[\frac{\partial^2 V_i}{\partial x^2} + \frac{\partial V_i}{\partial x}\frac{d}{dx}\ln(r^2 n)\right] \tag{3.34}$$

Changing variables from x to Z, with $\partial V_i/\partial x = (\partial V_i/\partial Z)(dZ/dx)$, the partial derivative is

$$\frac{\partial^2 V_i}{\partial x^2} = \left[\frac{dZ}{dx}\right]^2\left[\frac{\partial^2 V_i}{\partial Z^2} + \frac{\partial V_i}{\partial Z}\left(\frac{dZ}{dx}\right)^{-1}\frac{d}{dx}\ln\frac{dZ}{dx}\right]$$

Substituting these in eq. 3.34, we have

$$I_m R_m = \frac{rR_m}{2R_i}\frac{(dZ/dx)^2}{ds/dx}\left[\frac{\partial^2 V_i}{\partial Z^2} + \frac{\partial V_i}{\partial Z}\left(\frac{dZ}{dx}\right)^{-1}\frac{d}{dx}\ln\left(r^2 n\frac{dZ}{dx}\right)\right] \tag{3.35}$$

This goes into the form of the cable equation by setting

$$\left[\frac{dZ}{dx}\right]^2 = \frac{2R_i}{rR_m}\left[\frac{ds}{dx}\right] \text{ and } r^{3/2}n\left[\frac{ds}{dx}\right]^{1/2} = \text{constant}$$

These assumptions mean that the dendritic branches proliferate according to $n = n_0\,(r/r_0)^{-3/2}(ds/dx)^{-1/2}$. Moreover, Z becomes a generalized "electrotonic distance" such that $dZ = dx/\lambda(x)$, since $\lambda^2 \equiv rR_m/(2R_i)$ and since $ds/dx = 1$ in a cylinder.

Equation 3.29 therefore becomes, at the ith level of a tree with cylindrical branches

$$I_m R_m = \frac{\partial^2 V_i}{\partial Z^2} = \frac{\partial V_i}{\partial T} + K^2(V - V^*) \tag{3.36}$$

where $T = t/\tau = t/R_m C_m$.
(We omit the generalization [3.22] to trees that ramify at an exponential rate.) Along such a "cylinder," conduction velocity varies with dZ/dx, that is, decreases with ramification.

Because of its piecewise boundary conditions, the equation was integrated [3.22] by a Fourier series expansion rather than explicitly (as HR did), for initial values of step input uniform over a portion Δi of the whole tree, and bounded by zero voltage slope ($dV/dZ = 0$) at the ends.

Two solutions (boundary-value problems) are examined:

1. The transient distribution of depolarization when all the peripheral half receives excitatory synaptic input (stepwise or pulse).
2. The proximal and distal transients of depolarization due to strong excitation or inhibition at the soma only (central tenth).

We consider each solution in turn.

1. It is of considerable interest to seek the effect of a continued background of synaptic excitation at the remote dendrites: How is this weighted at the soma? In Fig. 3.23, at $T = 0$, G_ε steps from 0 to $2G_r$ over the peripheral half of the tree. We notice that centrally ($Z = 0$) the resulting excitation is delayed but eventually is little less than its value peripherally. That is, a state of arousal, adaptation, or "center frequency" of neural discharge can be maintained by continuous bombardment of the peripheral synapses. Conversely, a steady inhibitory background causes inhibition of spontaneous discharge. Also, if the central end is the "hot" end (upper curve), the *rate* of rise is much faster (has less accommodation); that is,

$$\left.\frac{dV}{dt}\right|_{Z=0} = 2\left.\frac{dV}{dt}\right|_{Z=1/2}$$

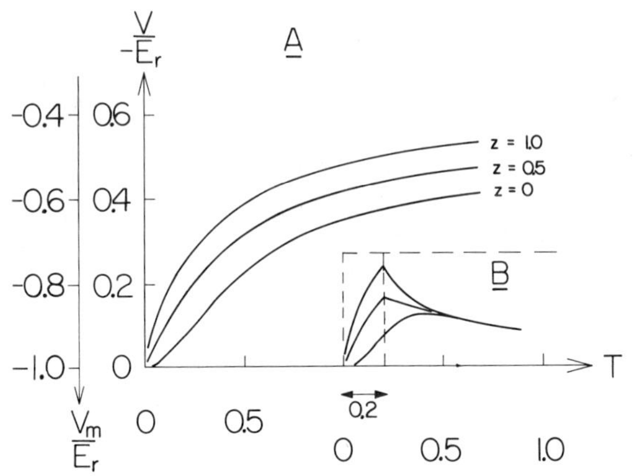

Fig. 3.23. Effects of step-excitation (A) and of pulse-excitation (B) of peripheral half of dendritic tree, observed at three locations (soma, $Z = 0$; midpoint, $Z = 0.5$; periphery, $Z = 1.0$). These curves also apply to the inverse case—central half excited, periphery at $Z = 0$, soma at $Z = 1$. After Rall [3.22].

A short test volley lasting 0.2τ ($T \sim 6$ msec) delivered to the proximal half of the dendritic tree, (Fig. 3.23B) produces twice the conductance change of the same volley delivered peripherally. Thus synapses in the immediate vicinity of the soma are better suited to rapid control.

2. The foregoing differential effects are enhanced by confining a step of synaptic excitation (ε) or inhibition (j) to the central or peripheral tenth of the tree (e.g., body of soma). This is illustrated by Fig. 3.24. There the initial

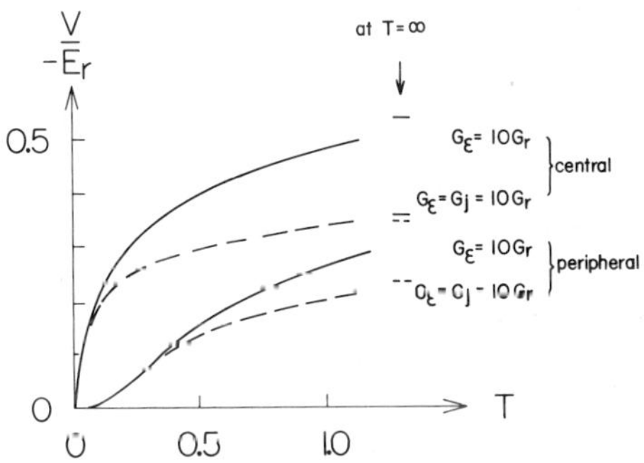

Fig. 3.24. Effect of step excitation, or combined excitation-inhibition, according to labels, of one tenth of the dendritic tree. If the central part of the dendrite is stimulated, then potentials vary as shown (quickly and strongly at soma; delayed at periphery). Curves can be applied to the case where the peripheral portion (0.1) of the dendrite is stimulated. Then labels "central" and "peripheral" must be interchanged. For further interpretation see text. After Rall [3.22].

slopes of ε and j are both great centrally but remain low peripherally. As before, the effect of incremental excitation is considerably more than of inhibition, and the asymptotic effect on central control by distal depolarization is not negligible.

Enough progress has been made to demonstrate the accessibility to analysis of temporal and spatial aspects of synaptic control on a dendritic neuron. Although in principle these results are verifiable experimentally (e.g., in the spinal motoneuron), this has not yet been reported.

F. REFERENCES

3.1. Hodgkin, A. L. and W. A. H. Rushton. The electrical constants of a crustacean nerve fibre. *Proc Roy Soc* [*Biol*] **133**: 444–479, 1946.

3.2. Churchill, R. V. *Operational Mathematics*, 2nd ed. McGraw-Hill Book Co., New York, 1958. p. 329.

3.3. Clark, J. and R. Plonsey. A mathematical evaluation of the core conductor model. *Biophys J* **6**: 95–112, 1966.

3.4. Marmont, G. Studies on the axon membrane. *J Cell Comp Physiol* **34**: 351–382, 1949.

3.5. Cole, K. S. Dynamic electrical characteristics of the squid axon membrane. *Arch Sci Physiol* **3**: 253–258, 1949.

3.6. Cole, K. S. *Membranes, Ions and Impulses*. University of California Press, Berkeley, 1968.

3.7. Hodgkin, A. L. and A. F. Huxley. A quantitative description of membrane current and its application to conduction and excitation in nerve. *J Physiol (London)* **117**: 500–544, 1952.

3.8. Huxley, A. F. Excitation and conduction in nerve: quantitative analysis (Nobel Lecture). *Science* **145**: 1154–1159, 1964.

3.9. Cole, K. S. and J. W. Moore. Potassium ion current in the squid giant axon: Dynamic characteristic. *Biophys J* **1**: 1–14, 1960.

3.10. Tille, J. A new interpretation of the dynamic changes of the potassium conductance in the squid giant axon. *Biophys J* **5**: 163–171, 1965.

3.11. Hoyt, R. The squid giant axon: mathematical models. *Biophys J* **3**: 399–431, 1963.

3.12. Wei, L. Y. A new theory of nerve conduction. *IEEE Spectrum* **3** (no. 9): 123–127, 1966.

3.13. Roberge, F. A. and J. H. Milsum. A feedback of the neuron for the study of small neuron networks. *Med Biol Engin* **4**: 357–366, 1966.

3.14. Nagumo, J., S. Arimoto, and S. Yoshizawa. An active pulse transmission line simulating nerve axon. *Proc IRE* **50**: 2061–2070, 1962.

3.15. Milsum, J. H. *Biological Control Systems Analysis*. McGraw-Hill Book Co., New York, 1966.

3.16. Minorsky, N. *Nonlinear Oscillations*. D. Van Nostrand Co., Princeton, N.J., 1962. Chapter 1.

3.17. Ku, Y. H. *Analysis and Control of Nonlinear Systems*. Ronald Press Co., New York, 1958.

3.18. Fitzhugh, R. Thresholds and plateaus in the Hodgkin-Huxley nerve equations. *J Gen Physiol* **43**: 867–896, 1960.

3.19. Noble, D. Applications of Hodgkin-Huxley equations to excitable tissues. *Physiol Rev* **46**: 1–50, 1966.

3.20. Fitzhugh, R. Impulses and physiological states in theoretical models of nerve membrane. *Biophys J* **1**: 445–466, 1961.

3.21. Chandler, W. K., R. Fitzhugh, and K. S. Cole. Theoretical stability properties of a space-clamped axon. *Biophys J* **2**: 105–128, 1962.

3.22. Rall, W. Theory of physiological properties of dendrites. *Ann NY Acad Sci* **96**: 1071–1092, 1962.

3.23. Rall, W. Electrophysiology of a dendritic neuron model. *Proc Intl Biophys Cong, Biophys J* **2/2 part 2**: 145–167, 1962.

3.24. Rall, W. Theoretical significance of dendritic trees for neuronal input–output relations, pp. 73–97 in *Neural Theory and Modeling*. R. F. Reiss (Ed.). Stanford University Press, Stanford, Calif., 1964.

3.25. Eccles, J. C. Ionic mechanism of postsynaptic inhibition. *Science* **145**: 1140–1147, 1964.

3.26. Comroe, J. H., Jr. The mechanism of action of some drugs on the sympathetic nervous system. *Physiol Physicians* **1** (no. 10): 5–8, 1963.

CHAPTER FOUR

Systems Behavior of Sensory Input in General

A. THE TRANSFORMATION OF NEURAL INFORMATION AT THE CELLULAR LEVEL

At the neuronal level, temporal and spatial patterns of changes in membrane conductance at the synapses are transformed into temporal patterns of action potentials, as seen, for instance, at the axon hillock. No complete mathematical treatment of the neuron as a signal processing system exists. But the models presented in Sections 4.A.1 and 4.A.2, based on simplifying and restrictive assumptions, demonstrate the various analytic tools needed to describe some of the mechanisms involved and some of the phenomena observed. These techniques should be developed further, since the processing of information by one nerve cell is important for the behavior of neuron groups, a notch higher in the systems hierarchy.

Most neuron models assume that each presynaptic pulse initiates, at one or two synapses only, release of a "quantum" of synaptic chemical transmitter that subsequently is rapidly destroyed. It is not proven that all synapses engage chemical transmitters, although acetylcholine and noradrenalin have been identified at some synapses. Electrical transmission also is possible—for example, if presynaptic and postsynaptic structures are in contact over relatively large areas. Most model builders studying neuronal firing patterns as functions of postsynaptic voltage fluctuations have disregarded the fact that the dendritic–somatic structure, with its many synapses, forms a very complex signal processor. Rall [4.1] has investigated the temporal and spatial voltage distributions on such structures (see also Chapter 3) but has not treated the active firing mechanism.

Difficult mathematical problems are posed by the dependence of the postsynaptic potentials (PSPs) on membrane potential, essentially a dependence on previous membrane history. "Elegance" may be lost when numerical techniques are employed, yet they are needed to establish the structure–function relationships for a system as complicated and nonlinear as a nerve cell. Generality of applicability of numerical results can, in some cases, be obtained by systematic search. The work by Segundo et al. [4.2] points in this direction, although their particular model represents a relatively simple structure. Stein [4.3] chose a different approach, cataloging performances of many models, incorporating different types of synaptic actions, and involving various degrees of nonlinearities. One set of his calculations is based on the Hodgkin-Huxley equations (see Chapter 3).

Such investigations, restricting attention to sequences of neuronal spikes of one cell, were reviewed by Moore et al. [4.4], who also discussed critically the applicability of statistical techniques. These studies may yield limited—but not unimportant—insight into mechanisms involved in the genesis of action potentials. Although in most models the distributed neuron elements were lumped, it is important to gain knowledge about the statistics of impulse trains, inasmuch as they influence the coupling between neurons. The timing of a spike may determine whether firing occurs in the cell on which it impinges [4.2].

Purple [4.5, 4.6] explained the integrative action on excitatory and inhibitory inputs by a neuron of simple structure: the eccentric cell in the eye of *Limulus*. He carried out his analysis in terms of distributed structures and their properties. It is known, however, that many other neurons offer an amazing range of mechanisms in processing different types of input. Some synapses cause facilitory effects that are not detected as membrane potential changes; thus excitability of a neuron is not necessarily expressed in terms of threshold voltage in every case. Some neurons show synapses near the axon hillock, and we must accept the possibility that electric fields from neighboring cells may influence the spatial and temporal patterns on the dendritic trees mentioned earlier. Nor do we have a model of a neuron that is affected, in addition, by a pacemaker region. Only a few of these are discussed in detail here. Closest to a systems analysis of the neuron, probably, is the model by Roberge and Milsum [4.7a], who took into account the phenomena linking membrane voltages on the soma with active depolarization of the axon hillock—a possible model of a motoneuron—while neglecting the partly nonlinear summing effects that may occur on dendritic trees [4.1].

1. Statistical Analyses of Temporal Discharge Patterns

In this section we consider statistical tools and measures used for characterizing (spontaneous) sequences of action potentials obtained experimentally or from models. On the one hand these techniques open avenues toward defining possible measures of information conveyed; on the other hand they allow the model builder to compare models' performances with observed spike data. We concentrate on neuron models based on deterministic mechanisms. The synaptic inputs and the neurons' outputs, however, are considered to be random processes. Moore et al. [4.4] surveyed also the models that involve randomness of threshold, noisy PSPs, and so on.

It is hoped that available designs, notably by Rodieck et al. [4.8], Stein [4.3], and Segundo et al. [4.2], will yield insight into spike-generating processes and thus, with certain restrictions, into the nature of data transformation at the neuronal level. Unfortunately all models include only one level (of amplitude) of postsynaptic potentials, a rather poor picture of most neurons in nature! Furthermore, most of the statistical analyses were applied to spontaneous nerve activity. A good counterexample treating stimulus–response transformation was worked out by Werner and Mountcastle [4.9]. In addition, there is great need for statistical tools for describing nerve impulse trains that are highly structured in time.

For analysis of *trains of action potentials* (being, in principle, either synaptic input or output of a neuron) a semilogarithmic plot of occurrence of interspike intervals versus length of interval can suggest a variety of further steps. Rodieck et al. [4.8] found the following examples of (semilogarithmic) histograms in single units of the cochlear nucleus of anesthetized cats: (*a*) data lying essentially on a straight line, suggesting an exponential distribution, which indicates an underlying Poisson process; (*b*) a bell-shaped graph, calling for replotting of the data on Gaussian density paper on which a normal process has to fall on a straight line; (*c*) a bimodal distribution that stemmed from neural activity occurring in bursts, so that the second "mode," at large intervals, was a measure of interburst times. The last case clearly is not suited for a simple interval analysis.

In other analyses of spike trains [4.3] the following methods of attack are useful. Let $f(t)$ be the interval density function (e.g., giving the probability that membrane potential is at threshold), whose mean is μ. Then

$$F(t) = \int_0^t f(t)\, dt \qquad (4.1)$$

is the interval distribution function, and $[1 - F(t)]$ is the probability that an interval is longer than t. In a graph where $\log [1 - F(t)]$ is plotted versus t, an exponential distribution appears as a straight line ($-$slope $=$ mean frequency). Furthermore, the negative slope of log $[1 - F(t)]$ is the *conditional probability* that a discharge occurs between t and $(t + dt)$ (event A), given that an impulse has not occurred by t, that is, that the interval is longer than t (event B):

$$p(\mathrm{A} \mid \mathrm{B}) = \frac{p(\mathrm{AB})}{p(\mathrm{B})} = \frac{f(t)}{1 - F(t)} \equiv -\frac{d}{dt} \ln [1 - F(t)] \tag{4.2}$$

where $p(\mathrm{A} \mid \mathrm{B})$ = probability of A occurring, given B, since $f(t)$ by definition is the probability that the potential is at threshold at time t *and* that the impulse has not occurred previously, that is, $p(\mathrm{AB})$. Stein [4.3] discussed a simple neuron model exhibiting (*a*) EPSPs (excitatory PSPs) of long durations (e.g., caused by membrane pores opening during time spans randomly distributed) and constant amplitude, (*b*) an exponential decay of the membrane potential with time constant τ, and (*c*) a membrane potential that would be reset after each active discharge (no refractoriness). For such a "neuron," the spike intervals at the output are distributed according to one exponential law for small intervals between the input signals and according to another at large intervals. The latter set of data scatters, in the representation of Fig. 4.1, along lines of relatively small slopes. It would cause a long tail in a simple histogram [4.8], which is easily discerned in the plot of log $[1 - F(t)]$ versus t/τ. In a more complete computer study, Stein [4.3] considered the cell to follow the Hodgkin-Huxley model for the action potential. This introduces such features as accommodation, dependence of magnitudes of PSPs on membrane voltage, and refractoriness. Without variability in the input, accommodation would limit the lowermost frequency at the output (i.e., no intervals beyond a certain maximum value would be found). Variability removes this limitation, and in a semilog plot of $[1 - F(t)]$ (Fig. 4.1), *slope changes* are readily distinguished at this maximum interval.

Temporally structured, frequency-modulated spike trains require analysis in more dimensions. Stein [4.3] computed correlation coefficients for successive spikes, finding them to be positive for the simple pore model and negative and small for the cell obeying the Hodgkin–Huxley equations. Rodieck et al. [4.8] plotted *joint* histograms for the kth and $(k + 1)$st intervals, which are symmetric about the 45° line *if* the two intervals are distributed independently. Temporally structured pulse trains create joint histograms asymmetric with respect to the 45° line. These authors developed a more elegant test for independence by computing, at each value of the kth interval, the mean of all $(k + 1)$st intervals, and vice versa.

These averages of all the rows and columns of the joint histogram are then plotted against interval durations. If the means are independent of the succeeding (or previous) beat, their values are the same for all k's [or $(k + 1)$'s].

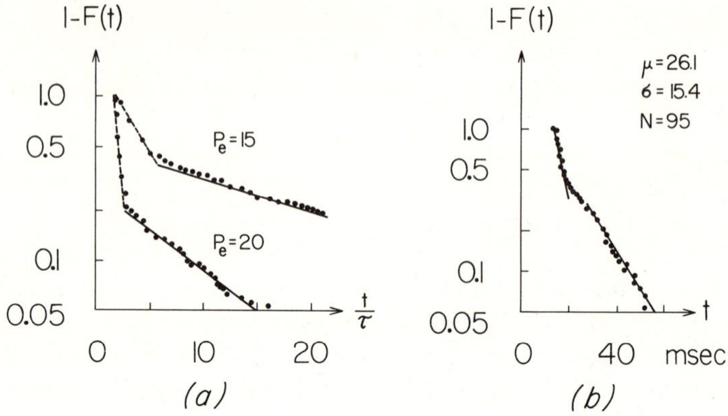

Fig. 4.1. Statistical analysis of intervals between spikes of a nerve impulse train. The logarithm of the conditional probability that a spike occurs between t and $(t + dt)$, given that no spike has occurred by time t, is $\ln [1 - F(t)]$, which is plotted against the duration of interval. (*a*) Stein's neuron model stimulated by EPSPs of constant amplitude, of random duration, and occurring at fixed rates (15 and 20/sec, respectively). The membrane potential is assumed to decay with time constant τ, and not to exhibit refractoriness. Here spikes distributed according to an exponential distribution would lie on one line (with negative slope). The model clearly shows two such regions, one for short and the other for long spike intervals. The circles are results of computer simulations. (*b*) Stein's model incorporating the Hodgkin-Huxley equations (PSPs a function of membrane potential, refractoriness, etc.). The "break" between the two regions of exponential interval distributions occurs near the "critical" frequency below which periodic, deterministic EPSPs would not elicit spikes owing to accommodation. The randomness of the input is responsible for the wide range of intervals occurring. Here μ is the mean rate of EPSPs, σ is the standard error of $f(t)$ distribution, and N is the number of events analyzed by computer simulation. After Stein [4.3].

When plotted, they then fall on horizontal and vertical straight lines, indicating independence. Dependence of successive spike intervals, positive or negative, will cause the averages to be scattered about lines of positive or negative slopes [4.8]. In Fig. 4.2 one of the cases reported by Rodieck is reproduced. Both semilogarithmic histograms and means from the joint histogram of successive beats are shown. The deviation from an exponential distribution, as we indicate in the first-order histogram, must be due to interdependence of successive intervals, seen to exist strongly for intervals shorter than 30 msec.

For statistical analysis of the *transformation between synaptic inputs and neuronal outputs*, the relations explained by the following example are useful. Stein [4.3] worked out the performance of the single-pore neuron model (assumption given previously) as a function of mean duration of pore opening. For this, the coefficient of variation is a convenient measure: Let μ be the mean interval of the output spike train [the mean of $f(t)$], and σ^2 the

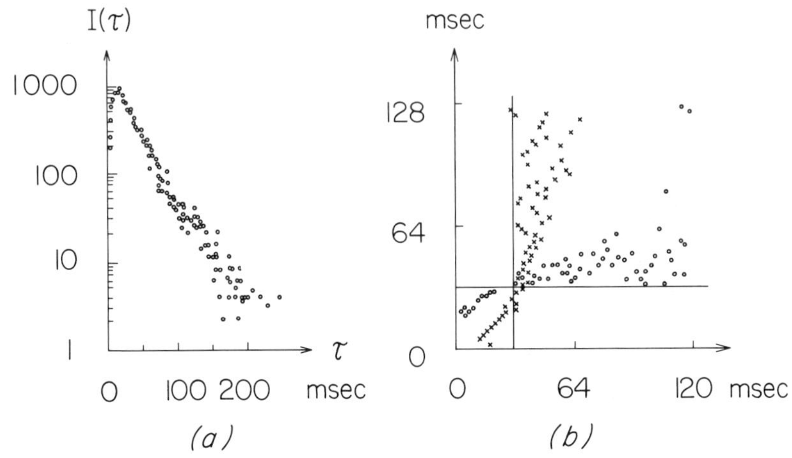

Fig. 4.2. Statistical analyses of a spike train observed in a single unit in the cat's cochlear nucleus. (*a*) Semilogarithmic interval histogram, where $I(\tau)$ is the number of intervals of duration τ. These data are from a " complex " cell showing, for most intervals, nonexponential distribution. Some temporal structure and dependence of succeeding intervals on each other is suspected. (*b*) The same spike train shown in (*a*) is analyzed as follows: From a joint interval histogram [kth versus $(k + 1)$st interval] the mean of all $(k + 1)$st intervals is computed and plotted at each kth interval, and *vice versa.* Since the data lie scattered about oblique lines, we know that there is interdependence of successive intervals. This shows clearly for short intervals. For longer intervals the data may be scattered about vertical (and horizontal) lines, indicating independence. From Rodieck et al. [4.8].

variance of $f(t)$ at any given average input rate ν_{in}. Then the coefficient of variation is given by σ/μ. In this model, the variability of the input, following a Poisson law, causes σ/μ to deviate from the square-root law predicted for nonvariable inputs [4.3]. This is shown in Fig. 4.3, as well as the results found from Stein's calculation on the Hodgkin-Huxley model.

A more revealing technique of relating temporal structure of synaptic activity with active depolarization of the neuron may involve multidimensional joint scatter diagrams (of successive intervals) for the synaptic input. If a spike at the output was observed, the corresponding data point is distinguished by the special marks in Fig. 4.4. Segundo et al. [4.2] lucidly discussed this approach, analyzing data obtained from cells of visceral ganglion in *Aplysia*. Assume that three successive EPSPs determine significantly the probability of an action potential being elicited. They may be numbered $(k - 2)$, $(k - 1)$, and k. In the scatter diagram all observed intervals between $(k - 2)$ and $(k - 1)$ are plotted versus the intervals between $(k - 1)$ and k. Lines of $-45°$ slope connect in this graph all triplets of EPSPs of constant overall duration, as in Fig. 4.4. When going along such a line, from

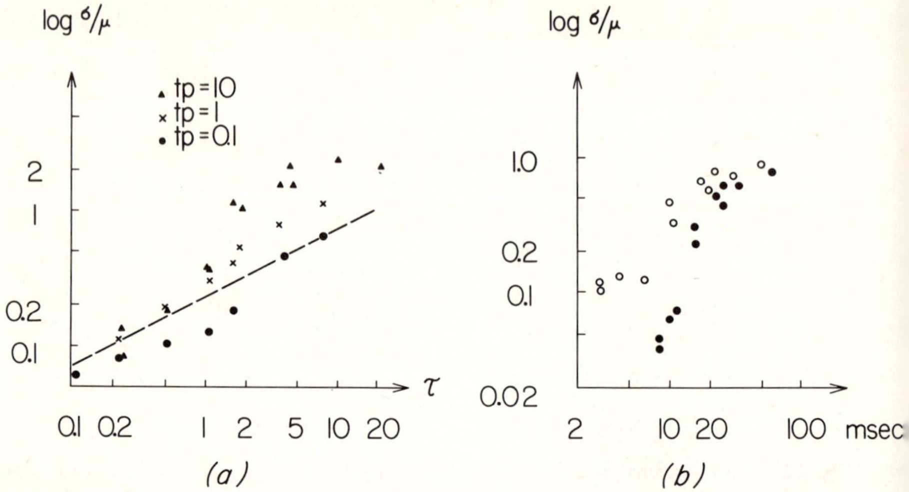

Fig. 4.3. Coefficient of variation, a statistical measure of spike trains introduced by Stein [4.3]. The standard deviation (σ) divided by the mean (μ) of the interval density function, given a fixed repetition rate of EPSPs of random duration, is plotted against the mean firing interval on a log-log scale. (*a*) Same model as in Fig. 4.1*a*, with three different "pore"-opening durations. The data, obtained by computer simulation, lie near the theoretical (dashed) line of slope 0.5 only for rapid impulse sequences (short τ). (*b*) Coefficient of variation for model including Hodgkin-Huxley equations (solid circles), as shown in Fig. 4.1*b*, with fluctuating currents. Refractoriness reduces variability at short intervals, adaptation increases at long ones. The open circles are from a model including a correction for refractory period; see Stein [4.3].

ordinate to abscissa, we find triplets of different temporal structure. Near the ordinate the first two spikes are widely separated and the second and third impulses are close to each other. Thus all straight lines with positive slope, passing through the origin, connect triplets of variable total duration but of constant (relative) temporal structure. Furthermore, certain regions in the scatter diagram of successive intervals encompass triplets belonging to a certain class of a (mean) duration and of a certain temporal pattern. We now can define (the estimates of) the following probabilities.

Let N = total number of observed spike triplets

N_r = number of observed triplets in region r, belonging to class R

M = total number of triplets after which action potential was observed

M_r = number of triplets triggering an output spike in region r, belonging to class R

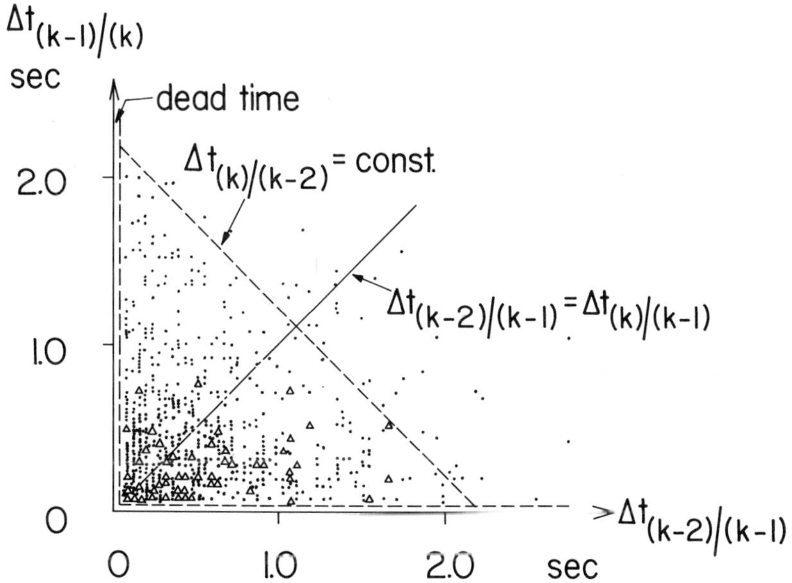

Fig. 4.4. Joint interval scatter diagram, giving the intervals between $(k-1)$st and kth EPSPs versus interval between $(k-2)$nd and $(k-1)$st events, *including* special marks (open triangles) for triplets that cause a spike at the *output* of the cell. From this plot we can estimate the probability of an action potential occurring, as dependent on temporal sequence and patterning of EPSPs (see text). The data, obtained from cells in the visceral ganglion of *Aplysia*, indicate that triplets for which $\Delta t_{(k-1)/k} < \Delta t_{(k-2)/(k-1)}$ are much more likely to trigger the cells than triplets lying above the 45° line. Furthermore, the probability of a spike occurring is high if the total duration of a triplet is below 1 sec. From Segundo et al. [4.2].

Then N_r/N = probability (i.e., estimate thereof) that a spike triplet belongs to class R

M_r/N_r = probability that an action potential is elicited, given that a triplet of class R has occurred, called the "prospective" probability [4.2]

M_r/M = probability that a triplet belongs to class R, given that an output spike occurred

M/N = spike-triggering probability (for entire experiment)

Note that any of these four estimates can be obtained if the other three are known, since

$$\frac{N_r}{N}\frac{M_r}{N_r} = \frac{M_r}{M}\frac{M}{N} \tag{4.3}$$

These tools allow us to define the probabilities with which certain sequences of EPSPs trigger an action potential. These sequences are not only defined by particular (average) temporal pulse patterns, but also by their (average) mean duration. Segundo et al. [4.2] developed additional techniques of estimating the number of (previous) EPSPs which influence the probability of spike initiation. They also applied concepts and tools of information theory to the problems raised. These, however, may be useful only when entire neuron pools and pulse sequences (at synapses as well as at axon hillock) are considered. Werner and Mountcastle [4.9] found that the information transmission of single, first-order mechanoreceptor fibers, when expressed in measures used by information theorists, was quite limited. They concluded that the fine discrimination achieved by an observer, which requires channels of much higher "capacities," probably relies on finely graded spatial and temporal patterns over whole neural populations. This question doubtless also involves the problems of identification of neurophysiological observables with the concepts used in information theory, such as source alphabet and channel.

2. Input–Output Relations for Deterministic Signals

Jenik and Hoehne [4.10, 4.11] investigated many aspects of the performance of a model incorporating the Hodgkin–Huxley (HH) equations for depolarization (see Chapter 3) and several directly coupled synapses obeying linear first-order differential equations. The spatial separation and the "processing" of PSPs by the dendritic structure [4.1] were neglected. The equations governing synaptic transmission yield fast-rising PSPs with exponential decay; the time constant was assumed to be 1 msec as observed in some structures (e.g., giant axon of *Loligo*). The synaptic conductance changes and the HH system were coupled through the currents elicited. Before whole firing patterns were computed, Jenik tested the model extensively. The excitatory and inhibitory PSPs (EPSPs and IPSPs), at subthreshold level, were found to depend realistically on instantaneous membrane voltage (i.e., on the differences between resting potential and equilibrium potentials for Na^+ and K^+, respectively). Furthermore the peak amplitudes of the EPSPs increased strongly as soon as the threshold (of firing) amplitudes were approached. Takeuchi et al. [4.12] obtained essentially the same pattern experimentally. During the decay, EPSPs tended to undershoot the baseline, as observed in nature, especially when the membrane initially was slightly depolarized. Jenik could also investigate in detail the nonlinear superposition of two subthreshold PSPs (conditioning and testing pulses) spaced at various intervals.

Superthreshold responses, of main interest here, showed the following characteristics. When steady (D.C.) excitation was increased by a factor of 4 (presynaptic), the firing frequency increased steadily but not exactly linearly from 250 to 400 Hz (exact values would depend on assumed component values of the model). With ramplike excitation, the typically observed dependence of the neuron's response on rate of rise of the membrane voltage was also determined. As expected, with *periodic* (superthreshold) excitatory inputs of short duration, the firing frequency was found to depend not only on input frequency but also on strength of stimulus. Jenik [4.10] therefore graphed his results in the so-called ν-diagram, where the stimulus frequency is plotted along the abscissa and its amplitude along the ordinate. Each point in this plane is associated with an output firing frequency, which is divided by the input frequency. Points of equal frequency ratios can be connected, or regions of constant ratios delineated. Figure 4.5 indicates that below a certain stimulus intensity, which is a function of frequency, ν vanishes; this is because the output rate is zero (i.e., threshold is never reached). At low input frequencies and high input amplitudes, each PSP elicits an action potential (i.e., $\nu = 1$). At higher frequencies, however, some of the stimuli fall into the refractory periods and it is not possible that all the stimuli cause a firing. These are regions of frequency division, seen in Fig. 4.5, where the output spike train is no longer periodic but patterned. Jenik expressed ν in these situations in terms of the mean output firing rate divided by the input frequency. The statistics of the outgoing pulse train have not been investigated.

In earlier studies, Kuepfmueller and Jenik [4.13] worked with a simpler pulse processor not incorporating all nonlinearities of the HH systems. When stimulating this model with two periodic stimuli, through two "synapses," they found ranges of stimulus amplitudes and frequencies in which the output

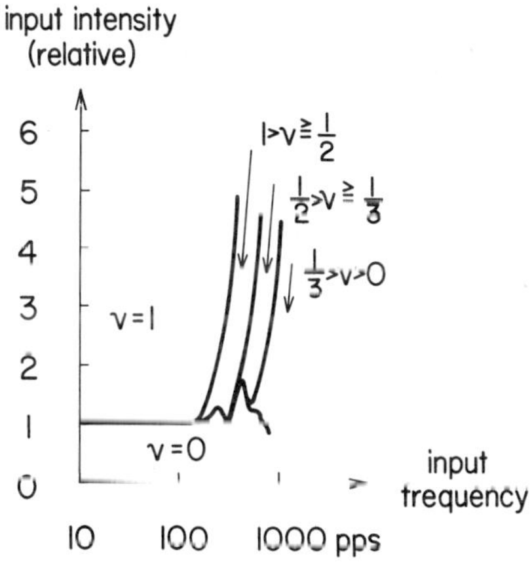

Fig. 4.5. Jenik's ν-diagram showing regions in which a theoretical neuron model, including simple RC synapse models, shows the same firing frequency at the output as the input EPSPs train ($\nu = 1$), and in which the frequency at output is less than that at the input ($\nu < 1$). The input–output frequency relation is determined by input frequency (abscissa) and amplitude of EPSPs. The ν diagram only applies to deterministic input train impulses. The model represented here exhibits most of the characteristics of the Hodgkin-Huxley model. Note that a spike train at the output, for which $\nu < 1$, may not be a regular impulse train. In these cases ν is found from the ratio of *mean* output rate to input rate.

firing rate would be the product of the two incoming frequencies. This was repeated [4.11] with a much more carefully designed model, including accommodation and refractory periods, which essentially was governed by the HH equations. The finding here was that the output firing rate could indeed be proportional to the product of the two stimulating pulse rates, provided the two input frequencies were noncoherent and subthreshold.

The great number of observations carried out on the giant axon and giant synapse of squid allow the model builder to construct analogs capable of simulating and explaining many aspects of the measured behavior. In the case of other neural structures whose performance may be of equal interest, this is not always the case. It then becomes important not only to obtain pertinent experimental results but also to establish collaboration between neurophysiologist and neuron model builder.

In the model of the motoneuron, Jenik [4.11] lumped the spatially distributed elements. The spike-generating mechanism, incorporating one positive feedback loop to simulate the negative Na^+ conductance and one positive loop for the polarizing effects of changes in K^+ conductances, and exhibiting PSPs without oscillatory tails, led Jenik to investigate some basic properties that should be expected and looked for in motoneurons.

The ν-diagram (Fig. 4.5) strongly depends on the particular nature of refractoriness that exists in the neuron [4.11]. If it is possible to elicit two action potentials with one synaptic impulse of large intensity, the motoneuron can act as a frequency multiplier. Frequency divisions occur when some of the input stimuli fall into the refractory periods of previous discharges. In these situations, the output pulse train will appear to be specially patterned (e.g., two closely spaced output pulses for each input pulse). As Jenik emphasizes, it would be desirable to obtain a match between the ν-diagrams of the model and those of the motoneuron. On the other hand, it should be emphasized that these graphs can represent the performance of the neuron as given by *one* synaptic input only, which is an unrealistic situation.

Jenik also investigated the transformation of two subthreshold synaptic pulse trains into one output sequence of action potentials. Two excitatory synapses, causing EPSPs of equal intensities, were stimulated by two noncoherent, periodic trains of impulses. Whenever incoming pulses occurred within a short time of each other, or simultaneously, their EPSPs, assumed to add linearly, elicited an action potential. In the earlier study [4.13], it was found that any threshold device incorporating an absolute refractory period (and exponentially decaying PSPs) can act as an adder or as a multiplier of the incoming frequencies, depending on the intensities. In the multiplying range, found when both PSPs were small and equal, the performance of the model could be (approximately) given by

$$\bar{f}_{\text{out}} = \frac{1}{K} f_{\text{in}1} \cdot f_{\text{in}2} \tag{4.4}$$

where $\bar{f}_{out}$ = average output firing rate
$f_{in1,2}$ = incoherent input frequencies of excitatory channels with subthreshold PSPs

At larger input intensities of two equal excitatory inputs, it was found that

$$\bar{f}_{out} = f_{in1} + f_{in2} \tag{4.5}$$

The same relations for the average frequencies were observed to hold in the motoneuron model. Equally important is the finding that the output pulse train is *patterned* in occasional dense pulse packages separated by periods of less frequent occurrence. These patterns depend strongly on the amplitudes of the inputs (i.e., EPSPs), as do their mean frequencies. In a particular study [4.11], $\bar{f}_{out}$ decreased by 60% for a 20% decrease in input intensity. The statistical characteristics were not worked out by Jenik, yet they may be of interest when, subsequently, another cell is stimulated (or inhibited) by such pulse patterns.

So far we have mentioned only special and simple cases of neuron stimulation. As Rall [4.1] pointed out, not only are the waveforms of the membrane potentials at the soma dependent on temporal structure of the synaptic inputs, but they are functions of the spatial separation of the synapses from the soma and of the electric properties of the dendritic tree. The latter, in turn, depend on synaptic activity at sites between the synapse under consideration and the soma (parametric driving).

3. Spatiotemporal Integration of Neural Information in the Neuron

Although the tools for an elaborate treatment and description of the behavior of the components of a nerve cell have been prepared (HH, with low threshold at axon hillock; Rall, dendritic structure; see Chapter 3 and [4.1]), no integrative analysis has been worked out for neurons as complex as motoneurons. A large amount of information is available, however, on secondary receptor cells of the faceted eye of *Limulus*, the horseshoe crab. This cell is a neuron; it has one dendrite of simple geometry where the cell is excited, in addition to quite well-defined regions where purely inhibitory synapses are located. The lateral eye of *Limulus* is a two-dimensional array of uniform ommatidia, the photoreceptors, each of which is a unit eye. Each ommatidium incorporates a simple dioptric system (corneal lens and crystalline cone) and the retinular cells with rhabdomes, into which the (only) dendrite of the neuron protrudes (see Fig. 4.6). The eccentric cell received this name because its dendrite originates on one side of the soma (i.e., off the main axis). The soma is a well-defined, ovoid structure of approximately 60-μ diameter. The axon of the eccentric cell gives off collaterals to neighboring cells and receives many synapses that are innervated by axonal side branches from other cells in the vicinity. These synapses, situated 50 to 80 μ proximal of the soma forming the neuropile, are purely inhibitory. The actively depolarizing, action-potential-producing region is just proximal of the neuropile. More centrally, the optic nerve is formed by the axon, together with all others from the other ommatidia.

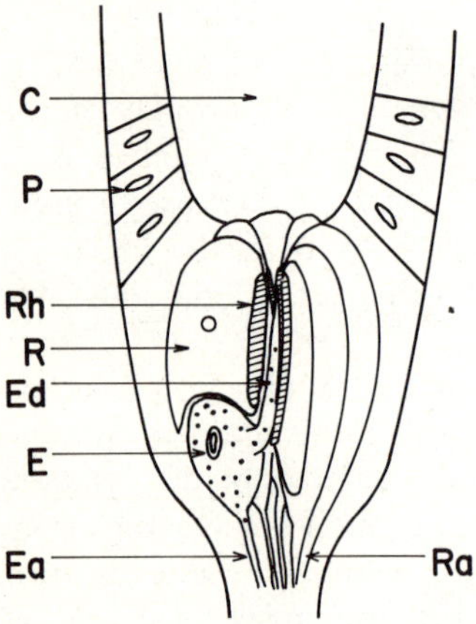

Fig. 4.6. Schematic view of an ommatidium, the "unit eye" of facetted eyes of horseshoe crabs. Light enters from the top, and passes through the crystalline cone (*C*) which is a simple dioptric apparatus; *R* represents retinular cells with rhabdomes (*Rh*), which contain pigments; *P* are pigmented epithelial cells; *E* is the eccentric cell, whose dendrite (*Ed*) is embedded among the (12) retinular cells. The axons of the eccentric cells (*Ea*) lead to the neuropile, where interaction with neighboring cells takes place. The function of the axons of the retinular cells (*Ra*) is not understood.

When light is causing the pigments in the rhabdomes to decompose, the membrane of the dendrite changes its conductivity, in steady state, as a function of the logarithm of the light intensity. At the dendrite and at the soma, therefore, the membrane voltage decreases. These changes, the generator potentials, spread electrotonically along the axonal structure to the axon hillock and elicit spike frequencies proportional to the generator potentials. The mechanisms involved, here only of indirect interest and treated from the phenomenological point of view, were investigated by Fuortes [4.14] and analyzed by Rushton [4.15] (see [4.16] for a review). For our treatment, on the next higher, integrative level, we need to note the logarithmic (D.C.) relationship and the capacity of the generator potential to exhibit a considerable phasic response. Upon a stepwise change in light intensity, the generator potential increases very rapidly to a high level from which it decays exponentially to settle at a new static level.

At the neuropile, inhibitory signals impinge from neighboring cells. The effects on information processing by the entire eye of this mutual lateral inhibition form the subject of the next section; here we simply consider the neuropile to be the site at which inhibitory signals strongly interact with excitatory signals and thus affect the response of the eccentric cell.

Experimental evidence suggested that the section of the membrane that is actively depolarizing must be remote from the cell body. It was observed that the membrane resistance of the soma, as measured through an intracellular microelectrode, does not change much during an action potential. On the other hand, spike amplitudes observed in the soma decrease when the ommatidium is illuminated.

Owing to the simple structure of this special case of a neuron it is possible, as Purple [4.5] and Dodge [4.6] have shown, to explain in almost complete detail the integration of excitatory and inhibitory information as processed by this nerve cell. The explanation is based on the particular geometry (inferred from electron micrographic findings) and on properties (membrane resistance and capacitance) of the structures involved. (Later in this chapter

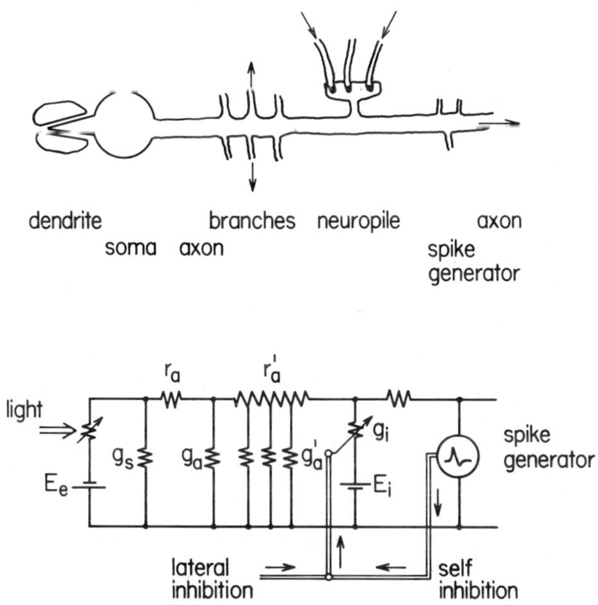

Fig. 4.7. Simple schema of an eccentric cell, its axon with collaterals, the neuropile, and the actively depolarizing region. The electric network is an equivalent circuit which represents main functional parts satisfactorily. Here, the variable resistor exposed to light represents the conductivity of the lumped dendrite membrane (its change causes excitation); E_e is the excitation equilibrium potential; g_s is the lumped membrane conductance of the soma; r_a and g_a are the series and shunt resistances of the first section of the axon and its branches, and g_i is the conductance of the neuropile membrane. The changes of g_i cause inhibition by bringing the local membrane potential closer to E_i, the inhibitory equilibrium potential. Also, g_i is affected by the activity of neighboring ommatidia and by the discharge rate of its own eccentric cell. The circuit illustrated pertains only to deviations from resting state. See text for values of g's. After Purple [4.5].

we discuss another model of a motoneuron relying almost entirely on a phenomenological description, enabling us to discuss systems aspects rather than relations between structure and function.)

The electric analog that can be given for the eccentric cell is also shown in Fig. 4.7. We have not drawn the analog circuits responsible for the resting potentials. Hence the batteries (E_e and E_i) shown represent the difference between resting potential and equilibrium potentials for excitation and inhibition, respectively. With the eye exposed to light, g_e, the conductance in series with E_e, deviates from zero, and excitation currents start to flow through passive RC structure. The linear relation between changes in g_e and (steady) membrane voltage was established experimentally. By extrapolation of resistance versus voltage curves, the value of E_e was found to be 50 mV. The conductance (g_i) allowing inhibitory currents to flow, depends not only on synaptic activity but also on a factor called self-inhibition. Self-inhibition is seen, for example, when the neuron is stimulated by a current pulse through an intracellular micropipette. The discharge frequencies measured are initially high and then decrease exponentially to about 50% of their first, highest value. This change is independent of mean, initial membrane potential and therefore is unlike accommodation. It is due, in all probability, to a temporal summation of synaptic inhibitory potentials [4.5]; or, it may be attributable to an unrecognized, uncharacterized mechanism similar to refractoriness.

Values were assigned to g_s (Fig. 4.7) from measurements of the resting value of conductance of somatic membrane. With a microelectrode in the soma, Purple measured typically 8 MΩ across the membrane structure. Part of this resistance must be due to the shunts (g_a, r_a) provided by the axon and its branches. Theoretical calculations, based on reasonable assumptions about the geometry of soma, dendrite, and axons, showed that the axonal shunt resistances should amount to approximately 20% of that of the soma [4.5]. At the same time, this means that the distributed nature of the axon and its collateral cannot be neglected. It also was determined that the properties of the dendrite can be lumped into concentrated elements (as in Fig. 4.7) since the dendrite's largest dimension is smaller than its space constant. The membrane resistivity thus was calculated to be of the order of 9000 Ωcm^2.

Although the spatial extent of the axonal structure can be estimated from electron micrographs, its electrical properties can be found only indirectly. It can be considered a "leaky cable" for which the space constant (λ) must be established (see eq. 3.4 for definition of λ). Purple measured relative resistance changes in the soma versus membrane potential, with varying hyperpolarization caused by antidromic stimulation (of neighboring cells) which, in turn, increased g_i. The (D.C.) voltage changes seen at the soma are, however, smaller than those created at the neuropile because of attenuation in the cablelike axon segment, whose load is provided by the soma. The voltage that would appear, if g_i could be brought to infinity, can be estimated from observed (changes in) firing frequencies. These tend to approach asymptotically a saturation level which is estimated by extrapolation. The voltage at which saturation is reached must be the equilibrium voltage (E_i) as it is seen through the axonal segment. These data can be compared with predictions by models, using leaky cables of different lengths. Purple [4.5] found λ/x to range from 0.4 to 0.6 and E_i from 10 to 17 mV (below resting potential).

The site of the axon hillock can be estimated from the previously mentioned decrease in spike amplitude, as measured within the cell body, when the soma–dendritic membrane resistance is decreased by illuminating the eye. Purple considered this resistance and the (soma's) membrane capacity to be the variable load of a cable whose electrical length can be computed from data on spike attenuation at several levels of light intensity. He found that the spike-producing region must be separated by 0.5 to 0.8 λ from the site of the inhibitory neuropile. The rate of discharge of the axon hillock (as a function of depolarization) can be determined only indirectly. At the soma, two impulses/sec occur per millivolt of depolarization. Considering the electrical separation, as given previously, the active region must therefore fire at a rate of 3.2 impulses/(sec mV).

We now have enough information to compare the model's performance with the data from eccentric cells. The steady-state firing rates from cell and model, as functions of depolarizing voltage at the soma, can be made to match perfectly if the separation of the neuropile from the soma is assumed to be 0.5 λ. In these plots [4.6], inhibitory signals apparently dominate the control over the cell's firing range. This is not so, however, since inhibitory generator potentials are larger near the axon hillock than they are at the soma where they are measured (i.e., equal voltage changes per frequency decrement are required for both excitatory and inhibitory inputs).

The firing frequency of the eccentric cell is decreased, owing to lateral inhibition, by an amount proportional to the inhibitory presynaptic activity and independent of the initial value for the cell's discharge rate. This is the principle underlying the Hartline–Ratliff equations (see below) and it is explained by the model when the relative component values are considered (Fig. 4.7). The considerable shunting action of the axon segment between soma and neuropile causes any change in g_i (proportional to afferent inhibitory firing rate) to produce a current variation essentially given by the (change in) g_i and the input resistances of both the peripherally and centrally lying cable structures. The one going to the soma, however, is almost independent of g_e, since the shunt resistors in the cable are much smaller than $1/g_e$. Hence the membrane voltage at the axon hillock, due to Δg_i, is proportional to the activity on the inhibitory tracts from the neighboring cells. The predictions of the "constant decrement effect" by the model are compared with data obtained in the *Limulus* eye by Hartline et al. [4.17], who exposed neighboring cells to different levels of illumination.

The temporal interaction in the model of a pulselike excitatory light stimulus (simulated by changing g_e following a phasic *and* a static time course) with the cell's self-inhibitory properties was in accord with observations, at least qualitatively. The initial overshoot in instantaneous output *frequency* reaches approximately seven times the steady-state level in both eccentric eye and model. Upon decay of this transient, phasic response, the frequency undershoots by almost 50% the ultimate level reached by the frequency after 1.3 sec in both model and cell. The generator *potential*, at the soma, only exhibits an overshoot and an exponential, but not an oscillatory, decay. In the model, the phasic response peaks at about 2.8 times the new steady-state value. In the cell, the observed amount of peaking is 2 to 3 times the steady level.

Thus the basic characteristics of the cell are explained by the model that was developed from electrophysiological and electromicrographic observations, as well as from some theoretical considerations. The constant decrement effect of lateral inhibition is largely explained, too. This result is employed later, when the optical image processing of the eye of *Limulus* is discussed. Clearly, so far we have not considered fine details (temporal structures, etc.) of the interaction at the level of the ommatidia between light stimuli and inhibitory influx. This question is also relevant when the entire faceted eye and its performance are treated (Section 4.B).

Attempting to describe analytically the most important *interactions* between compartments and mechanisms determining the behavior of a neuron, we move up one level in the hierarchy of systems. We would like to understand the effects of a (given) synaptic distribution over a dendritic tree—not in terms of local conductance changes, resistances, and capacitances of dendritic segments, but rather in regard to their importance to the behavior of the soma and the spike-initiating region. Roberge et al. [4.7] developed such a model from the systems point of view, and it can be compared with many neuron models, as Fitzhugh's nonlinear system is compared with the Hodgkin–Huxley equations (Chapter 3). As in the latter case, the concepts needed for description of the performance of the entire system differ from those used at the lower levels; the questions asked now concern stability and general relations between systems characteristics and output waveform.

Roberge considers the soma–dendritic region and the axon hillock as the two functionally important compartments. (Pacemakers, dendrites sensitive to local electric fields, etc., are disregarded.) These subsystems differ in their excitability (later we actually assume the soma and the dendrites to be inexcitable) as well as in their subthreshold voltage–current relationship. The latter, which is given by the net membrane resistance and capacitance, is considered to be linear in this model. Generally, the soma–dendritic region exhibits a total capacity much larger than that of the axon hillock; its net resistance across the membrane is much smaller than that of the spike-initiating region.

The active discharge mechanism involves a voltage threshold and refractoriness after an action potential. Thus each of the compartments can be considered as a feedback system, as in Fig. 4.8*a*. The current i_1, given by synaptic inputs below threshold, through the soma–dendritic membranes is transformed into the potential e_1 by the net membrane impedances (G_1 for soma–dendritic region). If the membrane potential passes the high threshold, the nonlinear feedback path renders the subsystem unstable, and active discharge occurs. For the axon hillock region the same type of system can be assumed, except that there is no current input. The membrane voltage e_2, however, is affected by electrotonic spread of the potential e_1 as transformed by an operator T_1. As soon as $T_1 e_1$ reaches the threshold of the nonlinear element NL_2, the axon hillock fires. The coupling element from axon hillock to the first subsystem (soma + dendrites) is similar and is called T_2. As Roberge [4.7] points out, T_1 and T_2 are not equal.

The compartmental method, suggested by Rall [4.1] to treat complex dendritic systems, can be used to compute the voltage transfer from axon hillock to soma and dendrites, or phenomenological equations that match direct observations can be given. Since the dendrites have a very large net capacitance, their membrane voltage cannot change abruptly. High frequencies of a signal transmitted from axon hillock to soma are attenuated.

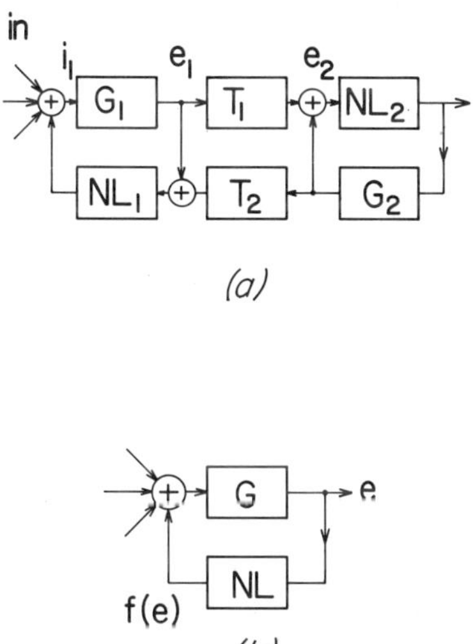

Fig. 4.8. Evolution of Roberge's model of a whole (moto)neuron. (*a*) The main functional elements are defined as follows: G_1 and NL_1 for the soma–dendritic portion and G_2 and NL_2 for the axon hillock region, where G is a linear transfer function and NL is a nonlinear impulse generator; T_1 and T_2 are (linear) coupling elements. Operator T_1 represents electrotonic spread from soma to axon hillock. Operator T_2 differs from T_1 mainly because the soma has a large capacity making rapid transfer from axon to soma impossible. This may not apply in the reverse direction. (*b*) If the threshold of the soma is much higher than that of the axon hillock, the system can be simplified. Here G essentially represents the soma–dendritic section; NL, the nonlinear operator, is the axon hillock. From Roberge [4.7b].

In the other direction, from soma to axon hillock, the high frequencies may not be attenuated at all. Roberge [4.7b] quotes the findings of Coombs et al. as indicating an emphasis of high frequencies. Hence the following transfer functions suggest themselves (in the Laplace domain):

$$T_1 = \frac{k}{1 + s\tau_1} \tag{4.6}$$

$$T_2 = \frac{\tau_2 s}{1 + s\tau_3} \tag{4.7}$$

where $\tau_{1,2,3}$ are time constants, k is a constant, and s is the Laplace operator.

This system was further simplified by Roberge. Since the soma's threshold of firing is much higher than that of the axon hillock, the effective coupling within the system is almost always through path T_1. It is therefore possible to reduce the model to that of Fig. 4.8*b*. The nonlinear element (NL) now has to incorporate possible time delays; its spike-producing mechanism includes NL_2 and is also influenced by the properties of NL_1. The linear element, G_1, essentially is given by the properties of the soma–dendritic compartment. At the summing point we again introduce the synaptic inputs (which should, however, be weighted according to the electric distance from the soma). Temporal summation is emphasized by this scheme. Approximating measurements in motoneurons, Roberge let

$$G = K\frac{1 + s\tau_4}{(1 + s\tau_5)(1 + s\tau_6)} \tag{4.8}$$

where
$$\tau_4 = \text{of the order of 50 msec}$$
$$\tau_5 = \text{of the order of 22 msec}$$
$$\tau_6 = \text{of the order of 7 msec}$$

Roberge showed that the nature of NL, the spike-initiating characteristics of this "neuron," does not have to be specified completely. Certain requirements for $f(e)$, the transfer function of NL, evolve from stability analysis. From analysis of the system in Fig. 4.8*b* we find, in terms of the operator G, that

$$e = G[p + f(e)]$$

where p = synaptic input.

With eq. 4.8 we obtain in the time domain

$$\tau_5\tau_6\ddot{e} + (\tau_5 + \tau_6)\dot{e} + e = K(p + \tau_4\dot{p}) + K\left[f(e) + \tau_4\frac{\partial f}{\partial e}\dot{e}\right] \tag{4.9}$$

This can be given as a system of first-order, nonlinear differential equations.

Let

$X_1 = e$

then

$$\dot{X}_1 = X_2 \tag{4.10}$$

$$\dot{X}_2 = -AX_1 + Bf(X_1) - \left[A(\tau_5 + \tau_6) - B\tau_4 \frac{\partial f}{\partial X_1}\right]X_2 + B(p + \tau_4 \dot{p}) \tag{4.11}$$

where
$$A = 1/\tau_5 \tau_6$$
$$B = KA$$

Roberge obtained stability criteria for this system (with $p = 0$), using Liapunov's second method. He found that, for positive and negative e and $\dot{e}$, stability can be obtained if

$$f(X_1) < 0 \qquad \text{and} \qquad \frac{\partial f}{\partial X_1} < \frac{\tau_5 + \tau_6}{K\tau_4} \tag{4.12}$$

Instability is encountered whenever $f(X_1) > 0$. These conditions can be established without introducing a specific $f(X_1)$ and without integrating the differential equations, thanks to the method of Liapunov. Let us rewrite eqs. 4.10 and 4.11 in a more general form ($p = 0$):

$$\dot{X}_1 = F_1(X_1, X_2) \tag{4.13a}$$

$$\dot{X}_2 = F_2(X_1, X_2) \tag{4.13b}$$

If $f(X_1)$ is zero for $X_1 = 0$, the origin ($X_1 = X_2 = 0$) is a singular point. Liapunov's first theorem states that, in this case, the equilibrium is stable if a function $V(X_1, X_2)$ can be determined so that

1. The $V(X_1, X_2)$ is a definite function (V_d) in a certain domain D: $|X_1| < h$, (h constant, >0); that is, it has values of only one sign and vanishes at $X_1 = X_2 = 0$.
2. The Eulerian derivative $W(X_1, X_2)$ of $V_d(X_1, X_2)$ either vanishes in D or is of opposite sign to V_d

where
$$W(X_1, X_2) = \frac{dV}{dt} = \sum_{i=1,2} \frac{\partial V}{\partial X_i}\frac{dX_i}{dt} = \sum_{i=1,2} \frac{\partial V}{\partial X_i} F_i \tag{4.14}$$

To interpret this theorem geometrically, let $V_d > 0$ and $dV_d/dt < 0$, and let $V_d = C_1$ define a closed curve in the X_1, X_2 plane; $V_d = C_2 < C_1$ defines another curve that lies within $V_d = C_1$. The assumption of a negative dV_d/dt means, however, that any trajectory defined by eq. 4.13, originating at a point in the vicinity of the origin, will tend toward smaller values of V_d. The trajectory never crosses a curve $V_d = C$ from inside to outside, that is, the system is stable. We refer the reader to the book by Minorsky [4.18] for theorems on asymptotic stability and instability.

In our application to Roberge's neuron model it is interesting to see that the conditions of eq. 4.12 can be used to determine $f(X_1) \equiv f(e)$, since $X_1(t) \equiv e(t)$ is known (as the action potential). Generally, eq. 4.12 has to be fulfilled below threshold and after the absolute refractory period, and $f(e)$ must be positive near and above threshold and during the absolute refractory period. Since $f(e)$ is determined from $e(t)$ and de/dt by these conditions, the refractoriness and its time course can be inferred from a given discharge waveform. As an example, consider the action potential and the corresponding $f(e)$, with stable and unstable domains, in Fig. 4.9. Roberge pointed out that during the spike of the action potential, when $\dot{e}$ is large, we can put

$$f(X_1) \simeq \tau_6 \dot{X}_1 \tag{4.15}$$

since the smallest time constant (τ_6) governs the behavior of G. Hence Fig. 4.9*b* represents the phase plane of X_1, the membrane voltage, during this (short) time interval.

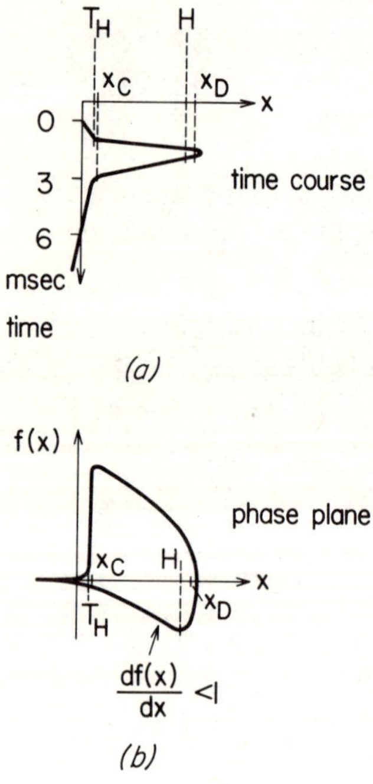

Fig. 4.9. (*a*) Schematic time course of membrane potential $x(t)$. (*b*) General nature of NL; that is, $f(x)$, obtained from values in (*a*) and stability criteria. After Roberge [4.7b].

Note that the waveform of net ionic current during the action potential through the membrane, which is in the model equal to $f(X_1)$, must therefore resemble the waveform of the time derivative of the membrane potential. This is indeed what Hodgkin and Huxley calculated [3.7].

Thus it is possible to study the behavior of a neuron on a level where general regions of stability, instability, and refractoriness are defined in terms of the nonlinear interplay of the compartments introduced here, such as the soma and dendritic tree and axon hillock. Roberge [4.7] also indicated that the model in this form may be used as one of the elements in a neuron pool. On this next higher level we investigate the processing of information rather than the transformation of one spike train into another. As we have seen, the Hodgkin–Huxley system of equations for the membrane potential, the "electrotonic" spatiotemporal PSP patterns, and the properties of synaptic (chemical) transmission can be taken into account when dealing with neuron descriptions, although all have not received simultaneous consideration. Model builders have neglected synapses of other kinds that may facilitate a cell without influencing membrane potentials, pacemaker regions, or effects of extrinsic fields. Harmon and Lewis [4.19] discuss some of the neuron networks built up from nerve cell simulators in their excellent review of neural modeling. In the next section we attempt to define concepts needed on this next higher level.

B. LAYERED PROCESSES

In the previous section we tackled the problem of information processing at the neuronal level and found that application of conventional engineering principles (e.g., transfer function, signal-to-noise ratio, stability) does not cover as many dimensions as we know exist. Several of the phenomena described by physiologists were not incorporated in the models. This is due not only to the limited amount of quantitative data available but also to the restricted nature of physical analyses, which emphasize linear systems, or nonlinear but autonomous systems. This is felt even more strongly when intermediary neuronal systems, such as second- and third-order sensory systems, are described. Thus the notion of the transfer function as a complex ratio of output and input is inadequate. The coding of spatiotemporal patterns of inputs into those of outputs, from the point of view of the purpose and task performance of the system, is of prime importance. Generally, the central nervous system is informed about the state of peripheral affairs, their history, and ongoing changes in space and time, and informed in such a way that the organism can react suitably for its effective operation and preservation. There exist various levels of importance for the organism.

Furthermore, as Schmitt [4.20] points out, the codes cannot necessarily be separated from the messages to be transmitted, as is possible in man-made systems. Time, which most often can be assumed to be an independent variable, becomes one of the bases of signal patterns. Distance may result in temporal separation; it may also be a measure of coupling between or within systems (as in the retina), or a factor introducing informational differences (as in the cochlea). Threshold or excitability must be defined for entire neuron pools.

Rall and Hunt [4.21] showed for motoneuron pools that one of the factors determining pool thresholds is given by the interaction within the population. Distribution of thresholds within a pool also is of physiological significance. The dynamic range of a system can be thus increased (as in baroreceptors), although in some cases this is achieved by inhibitory interaction of subsystems (cochlea). Classification of " codes " (i.e., transformative exchange of information) is difficult unless approached from a task-oriented viewpoint. In the baroreceptor reflex loop, the directional rate sensitivity, the differentiating action of the pressure-sensing organs, has been recognized as carrying the information about the current state of affairs (pressures caused by current ejection). The static response of the baroreceptors depends on averages of flows and peripheral resistance during the (immediate) past. In analyses of other systems, such as the mechanoreceptive organs, the phasic response has so far been largely ignored [4.9].

Similarly, in the following discussion of lateral inhibition in the compound eye, where detailed analyses of *steady*-state phenomena have been carried out, we can offer only the general statement that *transient* responses of a tightly intercoupled system enhance changing or moving input patterns even more than stationary ones. Owing to the highly organized structure of the vertebrate retina, transients are detected to a much finer degree. Most of the ganglion cells (third-order cells) respond to changes in illumination only, or in some retinas to motion of an optical pattern in certain preferred directions. Contours of stationary images blur; " fixation blindness " occurs.

The varied collection of coding is achieved not only by lateral inhibition within a pool. Other examples are inhibition by specialized Renshaw cells, or a delayed negative feedback. In their theory of pain Melzack et al. [4.22] postulated a delayed negative feedback, by way of the *substantia gelatinosa.* onto the dorsal horn cells by *pre*synaptic inhibition of the first-order afferent fibers. The functional significances range from enhancing of (moving) contours in the visual system to finely graded control of " action systems " in the perception of pain. In the following we present first, as an example, the work of Reichardt and his collaborators quantitating the effects of lateral inhibition in the faceted eye of *Limulus*. We then take up Mueller's work on special algorithms that can deal with many of the phenomena mentioned.

1. Horizontal Coupling in the Eye of Limulus

The anatomy of an individual ommatidium, the "unit" of the compound eye, is shown in Fig. 4.6 for *Limulus*. A schematic section through the mosaic-like structure of the entire eye, containing up to 1000 ommatidia, appears in Fig. 4.10 where the directional sensitivity of each ommatidium is indicated, as well as the geometrical relation of one unit to the other. Figure 4.11 is a

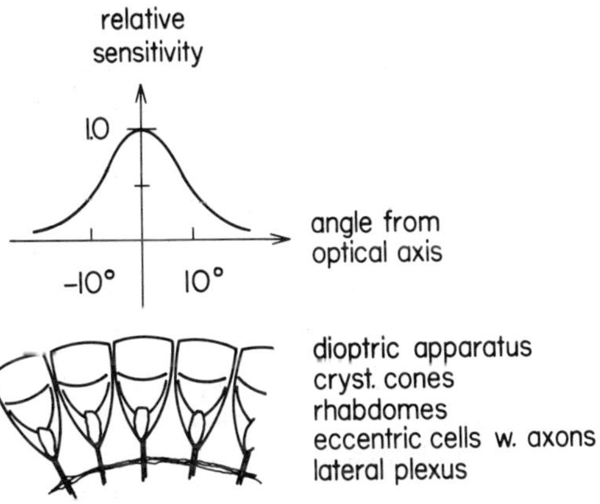

Fig. 4.10. Spatial, directional sensitivity of one ommatidium, the "unit-eye" of the faceted eye (*Limulus*), whose anatomy is given in Fig. 4.6, together with a schematic cross-sectional view of the faceted eye. The lateral plexus is formed by the branches of the axons of the eccentric cells, which synapse with neighboring units. After Reichardt [4.26].

block diagram indicating the paths of signal flow in a small area of the eye. This type of (apposition) photoreceptor is also found in locusts and bees. A slightly more complicated variance with superposition is found in nocturnal insects and lobster, [4.23]. Light through the crystalline cone of one cell can also, depending on background illumination, strike the (proximal) pigment layer of neighbouring units.

The directional sensitivity of the ommatidia (see representative example plotted in Fig. 4.10) was determined by Waterman [4.24] and verified by Kirschfeld and Reichardt [4.25], who also obtained indices of refraction of the crystalline cone ($\simeq 1.5 \pm 0.1$). The "half-width" of sensitivity, between 6° and 8°, is considerably more than would be expected from the diffraction patterns of a circular aperture of the same size (160 μ). Near the center of the eye the angle between the optical axes of neighbouring units is also approximately 6°. Half of the light intensity of a point source on the axis of one ommatidium, therefore, is received by the adjacent unit; 10 to 20% is seen by the second next ommatidia. Hence any sharp discontinuity in an optical image, such as a gray–black intensity step, is blurred at the level of the

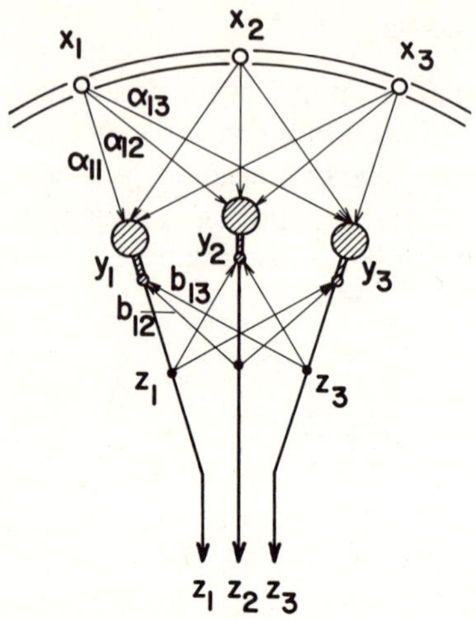

Fig. 4.11. Signal-flow diagram within the faceted eye. The directional sensitivity of the ommatidium is such that a light source at point x_i is not only seen by ommatidium i, but also by ommatidia $i \pm 1$, $i \pm 2$, ... The α's, determined by the sensitivity curve in Fig. 4.10, determine the degree of optical overlap. Here y_i is the generator potential of ith unit and z_i is the firing frequency of ith unit. The b's determine the degree of lateral inhibition. See text. After Reichardt and MacGinitie [4.28].

receptor cells because of overlap of the optical fields of corneal lenses and crystalline cones. The degree of overlap varies from center to edge of the eye, where the angles between optical axes are larger [4.25]. The signal flow from optical image (spatial distribution of X) to light intensity at the ommatidia therefore can be given in principle, as in Fig. 4.11. We can write the following expression for the generator potential (y_j) of one of the eccentric cells (the jth), which is proportional to the logarithm of the light intensity at its rhabdom, which in turn is the weighted sum of intensities (x_i) of the point sources:

$$c_j y_j = \ln \sum_{i=1}^{m} \alpha_{ij} x_i \tag{4.16}$$

where
- c_j = a constant of proportionality, accounting for light absorption along the optical path and for the relation between the logarithm of light intensity and generator potential
- y_j = generator potential at jth eccentric cell, proportional to firing frequency if no lateral inhibition exists
- α_{ij} = fraction of light intensity of ith source reaching jth ommatidium
- x_i = intensity of ith (point) light source
- m = number of point sources whose light reaches jth cell

As pointed out earlier, the generator potentials are observed at the peripheral end (dendrite and soma) of the bipolar eccentric cells. The axons originate centrally and are uninterrupted (no synapses in series). Thus there exists a one-to-one correspondence between ommatidia and fibers in the optic nerve. The axons, however, give off branches, forming the lateral plexus, just proximally of the eccentric cell bodies. (Axons also emanate from the retinular cells—their function is not thought to be involved in the light detection process.) The axonal branches innervate the neuropiles of neighboring units, as in Fig. 4.10. The corresponding signal flow is seen in Fig. 4.11. Hartline and Ratliff [4.17] established that these side connections are purely inhibitory. Their action is described by the "constant decrement law" stated earlier (inhibition proportional to activity on axonal branches and independent of activity of cells they innervate). This was formulated by Hartline and Ratliff as follows, valid for the steady state, above the threshold case:

$$z_j = z_j(y) - \sum_{i=1}^{n} b_{ji}(z_i - z_{ji}^*) \tag{4.17}$$

where z_j, z_i = activity (firing frequency) of jth, of ith unit
$z_j(y)$ = firing frequency of jth unit if illuminated alone (without lateral inhibition)
y = generator potential (of jth cell)
z_{ji}^* = threshold of inhibitory action
b_{ji} = coefficient of lateral inhibitory action of ith receptor on jth cell; often dependent on distance between these two units (b_{ji} smaller for larger distance)
$b_{ji} = 0 \quad \text{if} \quad z_i < z_{ji}^*$
n = number of units contributing to inhibition of jth cell

The (static) relation of eq. 4.17 can be formulated more rigorously when noting that the discharge frequency due to excitatory generator potential y is proportional to y, above its (excitation) threshold y^*:

$$b_{jj} z_j = y_j - y_j^* - \sum_{i=1,\, \neq j}^{n} b_{ji}(z_i - z_{ji}^*) \tag{4.18}$$

where b_{jj} = coefficient, giving the ratio of generator potential and firing frequency, and including the spatial and electric separation of soma and axon hillock [4.5]
$b_{ji} = 0 \quad \text{if} \quad z_i < z_{ji}^*$
y_j^* = threshold below which generator potentials do not elicit any action potential
$(y_j - y_j^*)$ = zero or positive only

Thus the overall transformation of optical information (a spatial pattern of x_j's) into a pattern of firing frequencies in the optical nerve (distribution of z_j's, also spatial within nerve) is established by eqs. 4.16 and 4.18. These are nonlinear relations owing to the logarithmic nature of the generator potentials and because, near threshold of firing, terms in eq. 4.18 may disappear. Reichardt [4.26] suggested that eq. 4.16 can be linearized, over the range (1–10) of physiological light intensities encountered by a *Limulus*, by considering deviations (Δx and Δy) from the minimum values ($x_{\min}$, a constant for all ommatidia, and $y_{\min}$) so that

$$\Delta y_j \simeq \frac{1}{c_j} \frac{\sum_{i=1}^{m} \alpha_{ij} \Delta x_i}{x_{\min} \sum_{i=1}^{m} \alpha_{ij}} \equiv \sum_{i=1}^{m} a_{ij} \Delta x_i \tag{4.19}$$

where
$$a_{ij} = \frac{\alpha_{ij}}{c_j x_{\min} \sum_{i=1}^{m} \alpha_{ij}}$$

Or, in matrix notation we can put

$$\mathbf{Y} = \mathbf{AX} \tag{4.20}$$

where $\mathbf{Y}$ = column vector of all generator potential differences Δy
$\mathbf{A}$ = matrix of coefficients a_{ij}
$\mathbf{X}$ = column vector of all discrete light differences Δx

This equation expresses the existence of a linear transformation between optical image in space and the spatial distribution of generator potentials, at least for the static case and for relatively small intensity variations within a pattern. Linearity still applies if $x_{\min}$ is changed to a different value, since the Δx's then change by the same factor. For the special case of high light intensities, we would expect both generator potentials and firing rates to be above their thresholds by significant amounts. We then can also linearize eq. 4.18:

if $y_j > y_j^*$ and $z_i > z_{ji}^*$ for a certain subset of j, i

and

$(y_j - y_j^*) = (z - z_{ji}^*) = 0$ for complementary subset

then

$$y_j = b_{jj} z_j + \sum_{i=1,\, \neq j}^{n} b_{ji} z_i = \sum_{i=1}^{n} b_{ji} z_i \tag{4.21}$$

or, in matrix notation:

$$\mathbf{Y} = \mathbf{BZ} \tag{4.22}$$

where $\mathbf{B}$ = matrix of coefficients b_{ji}
$\mathbf{Z}$ = column matrix of firing rates

This equation relates the (steady-state) spatial distribution of generator potentials to that of neural activities in the optic nerve. The overall transform therefore can be given, since we have a unique one-to-one correspondence between the Δy's and the Δz's (and vice versa, i.e., determinant of $\mathbf{B}$ is non-zero). From eqs. 4.20 and 4.22, we have

$$\mathbf{Z} = \mathbf{B}^{-1}\mathbf{AX} \tag{4.23}$$

It is obvious from this formulation that there exists the possibility of perfect pattern transfer: if $\mathbf{B}^{-1}\mathbf{A} = \mathbf{1}$, we get $\mathbf{Z} = \mathbf{X}$. If this were possible, the coefficients (b_{ji}) of lateral inhibition would have to assume such values that the blurred image at the Y-plane, which is due to overlap of fields of dioptric systems, would be "sharpened" (i.e., restored completely).

This immediately raises several questions. First, it should be noted that the overlap coefficients (α) depend on the distance between the optical image and the plane of the corneal layer of the eye. Contours in patterns in the vicinity of the eye are blurred only slightly by the dioptric system. Hence they may be greatly enhanced at the next level, the optic nerve, and Mach bands may appear. Conversely, the existing lateral coupling between the eccentric cells is not sufficient to enhance intensity steps in remote images by an equal amount, although the curvature of the surface of the ommatidia lets the overlap be less than would be expected from the individual angular sensitivities. These effects were discussed lucidly by Kirschfeld and Reichardt [4.25]. In their careful study they went from the discrete (matrix) treatment of all transformations to the continuous description by integral equations. (For a discussion of this transition see Whittaker [4.27]). When using such a continuous description eq. 4.21 becomes

$$y(u, v) = bz(u, v) + \iint_{\text{eye}} B(\xi, \eta) z(u - \xi, v - \eta)\, d\xi\, dy \tag{4.24}$$

where u, v — coordinates of (an) ommatidium
ξ, η — coordinates of neighboring ommatidia
$B(\xi, \eta)$ = continuous, spatial distribution of inhibition

We can introduce $B(\xi, \eta)$ in this form because it can be shown that the mutual inhibitory effects do not depend on u, v, the position on the eye. Kirschfeld was able to approximate the observed coupling by a Gaussian function, $B_0 \exp[-c^2(\xi^2 + \eta^2)]$, with which the neighboring neural activities are to be convolved; B_0 and c were adjusted to match direct experimental observations. The half-width of B is between six and nine ommatidia.

Some results of this particular form of lateral inhibition appear in Fig. 4.12. The impulse responses (response to a *line* of high intensity across the eye) of the y–z transform and of the overall x–z transform are given. The Gaussian interaction is seen to be relatively weak, but it spreads over 10 to 15 ommatidia. Its effect is mainly to sharpen the blurred impulse at the y level and to enhance the contrast (by undershooting the zero-intensity level). The existing lateral inhibition is not sufficient to compensate for the "defects" of the dioptric systems. This is also seen in Fig. 4.13, representing a response to a (remote) light step in which a "Mach band" is evident, however.

We have so far treated the special case of large mean light intensities, and thus have ignored nonlinearities introduced by the thresholds. If parts of the

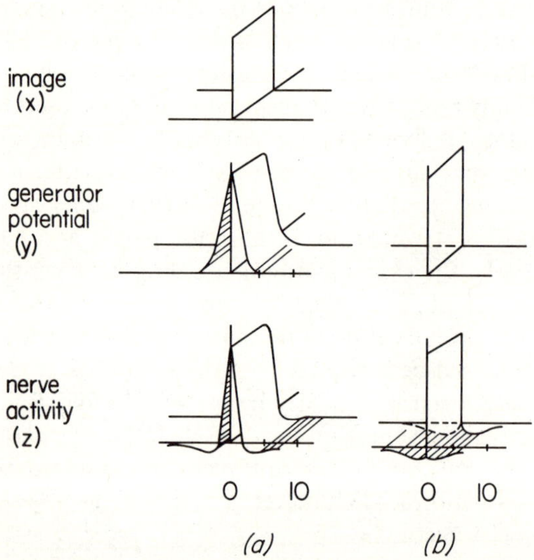

Fig. 4.12. Effects of lateral inhibition as observed in the faceted eye of *Limulus*. (*a*) The transformation of the optical image x, here a narrow bright line on a dark background (i.e., a "one-dimensional impulse"), into the spatial distribution of generator potentials of the primary receptor cells involves blurring due to optical overlap. The blurred line extends over five ommatidia. Lateral inhibition in the Kirschfeld-Reichardt model sharpens the pattern. The responses of the eccentric cells create a pattern z, which is narrower than y and shows contrast enhancement. (*b*) The transformation due to mutual lateral inhibition is seen best if the response of the eccentric cells is evaluated for (spatial) impulse function in the y-domain. Experimentally this can be achieved by imaging the stimulus onto the (physical) plane of the ommatidia. Then optical overlap does not occur. Contrast enhancement is clearly seen. The values of x, y, and z are plotted as functions of coordinates in space, measured in terms of ommatidia. After Kirschfeld and Reichardt [4.25].

eye are only dimly illuminated, their inhibitory effects on the surrounding regions may be vanishing. The pattern of firing frequencies then is not a faithful representation of the optical image. Indeed, it should not be expected a priori that the principal task of this eye is to present the central nervous system with a pattern of neural activity that is a replica of the optical image.

Reichardt and MacGinitie [4.28] investigated the extent to which the *sum* of all firing frequencies of the optical fibers may be an index of static shapes of patterns when part of the eye works in the nonlinear domain. With relatively small and coarse arrays of analog circuits mutually inhibiting each other and with numerical values of b_{ij}'s slightly different from those established experimentally by Kirschfeld [4.25], it was found that the total neural activity in fact distinguishes between more or less dense clusters of light sources. On the other hand, there is no unique differentiation between triangular and rectangular shapes. An image of the form of a triangle may cause the same amount of net discharge activity as a rectangular stimulus, with the rectangle slightly higher and narrower than the base of the triangle. But a long, narrow rectangle induces much more overall neural activity than a wide one. However, whether the sum of the frequencies on all fibers is the physiologically important measure has not been investigated.

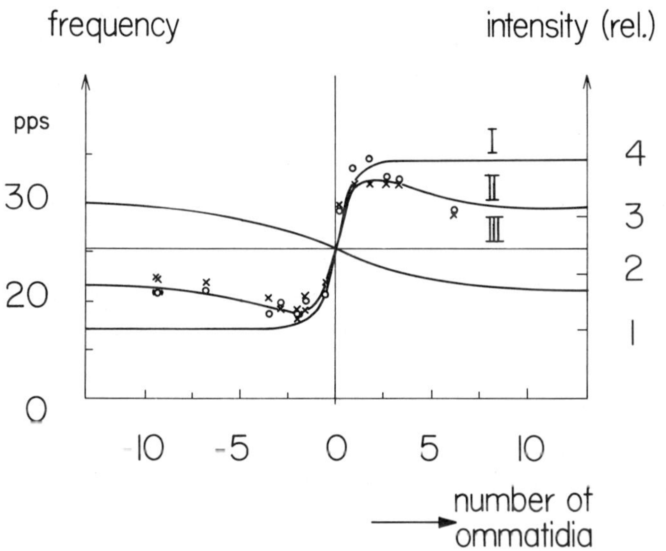

Fig. 4.13. Theoretical and experimental responses to a step in brightness (distance to image is 3 cm). Curve I: distribution of intensity at ommatidia; curve III: distribution of lateral inhibition according to $B(\xi, \eta)$ of eq. 4.24; curve II: the sum of I and III, yielding predicted pattern of firing rate. Crosses and circles are experimental points. Abscissa in terms of numbers of ommatidia, taking into account curvature of the surface of the eye. After Kirschfeld and Reichardt [4.25].

Furthermore, it should be noted that we have concentrated, in the foregoing discussion, on stationary patterns. The detection of moving objects is much more important. Two physiological mechanisms cause moving contours to be enhanced much more than steady steps in illumination: the phasic response of the individual receptor cell (whose amplitude may be 3 to 5 times the steady-state value), and the delayed onset of lateral inhibition after approximately 0.1 sec. In addition, we realize from what has just been said that there must exist cases of nonstationary, small contrast steps, for which the darker side causes signals below or near threshold and where the moving boundary is detected very easily owing to the powerful enhancement occurring in this situation. Existence of a unique one-to-one correspondence between optical information at the ommatidia and neural activity at the axonal fibers should not lead us to conclude that the compound eye contributes little toward information processing. The strong enhancement of moving contours and borders of contrast (much less analyzed than the static transformations) is the physiologically important feature of this receptor system.

2. Spatiotemporal Processing of Neural Information

We have not yet developed any analytic formulation of the functionally significant operations a neuron (or a neuron pool) can perform, nor of the different types of couplings between them. The only example was the space function $B(\xi, \eta)$ of lateral inhibition at the level of the secondary receptor cell in the compound eye. The value of $B(\xi, \eta)$ is constant in time. In addition to position in *space* and *time*, the membrane *voltage* is a variable that also determines the behavior, or the nature, of the function of a neuronal cell. It is the voltage rather than the rate of discharge, since the incoming action potentials are integrated at the synapse. In other words, the pulselike nature of the action potential is important for the transmission of signals over distances in space; it is, however, unimportant for the signal processing by the individual neuron. In the spatial domain we can define ranges of influences, of intercoupling between units, and also particular locations. Events have certain epochs—that is, they occur at certain instants in time and are of particular durations. Thus by letting time t, space s, and voltage e (sometimes also called intensity when dealing with signals at pools of cells) be the three most important variables, we can postulate certain operations following the suggestions by Mueller et al. [4.29].

The basic operations on t, s, and e are those of addition and subtraction in space or time, or integration. Integration with respect to *time* can be seen at individual synapses bombarded with high-frequency pulse trains. It is achieved at one stage in the vestibulo-ocular reflex loop, as discussed in Chapter 8, to mention just one example. At the third-order relay of the joint receptor system, at the thalamic cells, we see a running *spatial* integral of all signals coming from the peripheral sensors (Mountcastle et al. [4.30]). Furthermore, the nature of the coupling mechanisms between neurons introduces "sentential" connections: *if* there exists steady excitatory bombardment at synapses scattered over a dendritic tree, *then* one EPSP originating from a synapse close to the axon hillock brings forth a change in the output. Discontinuities in space, such as sharp definitions of regions of influence, as well as thresholds, make such logical operations possible.

Operators defining points in space or instants in time are called $P(s)$ and $P(t)$, respectively. We should expect ordered sets of such operators according to the sequence of the P's in space or time, for which, however, points of references have to be recognized. Other operators directly involving time and position in space are, for example, derivatives such as de/ds, de/dt, or dQ/dt, where Q is an interval of t or s. Already we have encountered the operation de/dt when the phasic response of the eccentric cell was discussed. These are indeed important in many other organ systems; see Chapter 7. The (static) contrast enhancement achieved by lateral inhibition is equivalent to obtaining, at the level of the optical nerve, operators of the form d^2e/ds^2 (valid at high light intensities) [4.31].

Additional important operators are necessary in the spatial and temporal domains: those operating on time intervals, or durations of events, called $Q(t)$, and those operating on fixed distances, ranges of influence, or spatial extensions, called $Q(s)$. The intensity e can be considered to be available in quantized form. Thus we can introduce transformations $TQ_e(s)$, defined as the transformation of a $Q(s)$ *into* a $Q(e)$. A dendrite should be assigned the $TQ_e(s)$ established by Rall [4.1] for simultaneous PSPs. As Mueller pointed out, Q represents a scaling or a projection of extrinsic quantities of the variables into the space limited by the characteristics of the neurons, or neuron pools. When only two (or one) of the three domains are considered at the input of a system, the remaining domain has to be normalized. If e and s are input domains, $Q(e)$ and $Q(s)$ must be made independent of time. Mueller gave examples of neuron nets capable of carrying out such normalizations.

Generally, transformations can be carried out algebraically or logically. In one neuronal cell these operations may be mixed. For example, $TQ_t(s, t)$, the transformation of the spatiotemporal pattern of postsynaptic potentials by the soma–dendritic structure of a cell, must reach the threshold at axon hillock before an output occurs (a logical operation). The output, however, may be proportional to the generator potential (i.e., it is given by an algebraic operation).

Neuron pools also perform analog together with logical operations. The work of Rall et al. [4.32] offers an instructive example of analog processing of spatiotemporal activity patterns in the olfactory bulb. The system is shown

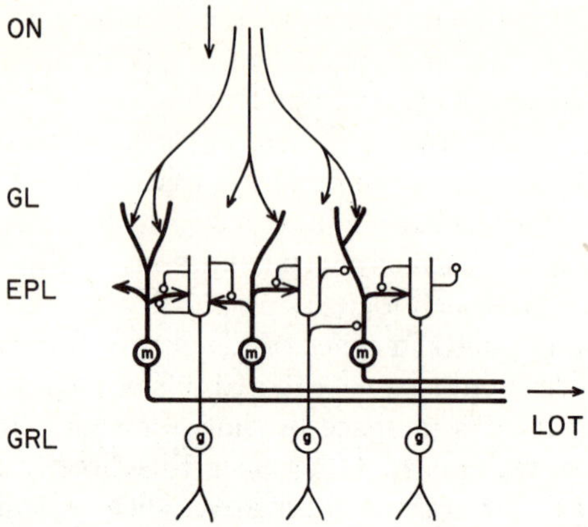

Fig. 4.14. Schematic diagram of the neural network in the olfactory bulb, a network whose nature is verified by electrophysiological and electron microscopic observations. Here *ON* is the olfactory nerve, *GL* the glomular layer, *EPL* the external plexiform layer, *GRL* the granular layer, *LOT* the lateral olfactory tract, *m* the mitral cells, and *g* the granulae cells. Synapses with arrows are excitatory, the ones with circles are inhibitory. After Rall [4.32].

schematically in Fig. 4.14. The active depolarization spikes, as triggered by the input on the olfactory nerve, spread (electrotonically, passively) along secondary dendrites in the external plexiform layer. These in turn are coupled to the radial (centrifugal) dendrites of the multipolar granule cells through excitatory dendro-dendritic synapses. These radial dendrites may depolarize actively. The potentials distribute passive toward the soma of the granule cells situated below, centrally of the mitral cells, thus setting up strong electric potentials in the surrounding.

There also exist, within the region of the external plexus, inhibitory dendro-dendritic synapses through which the radial dendrites exert an inhibitory influence on the mitral cells. Due to the centrally flowing currents, the external fields are negative near the somas of the mitral cells, thus hyperpolarizing and inhibiting the mitral cells further. Here we have a case, hitherto unmentioned, of interneuronal coupling: two types of dendro-dendritic synapses (discerned in micrographs [4.32]). This type of coupling can be made variable because the granule cells also are under influence of signals from more central structures.

The input to this system can be considered the two-dimensional spatial pattern (in x, y) of activities at the mitral cells, that is

$$\text{input} = T_{\text{in}}\, Q_e(x, y) \tag{4.25}$$

The thresholds (θ) of firing at the radial dendrites of the granule cells enable a logical operation in s and e to be carried out. If the e pattern of eq.

4.25 reaches θ at (x_i, y) belonging to a region R, then a second pattern $TQ_{e2\Delta s}$ is set up within the region $R^+ : (x_i \pm \Delta x, y_i \pm \Delta y, \Delta z)$. Let

$$(x_i, y_i) \text{ belong to region } R$$

If

$$T_{\text{in}} Q_e(R) \geqslant \theta$$

then

$$T_{\text{in}} Q_e(R) \to T_{\text{in}} Q_e(R) + TQ_{e2}[TQ_{\Delta s}(R)] \tag{4.26}$$

The sign of TQ_{e2} is positive in deep layers of the olfactory bulb; more peripherally, however, it is negative, causing inhibitory effects. We then further observe that within the region R^+ (and possibly a fringe added to R^+) the compound voltage pattern (eq. 4.26) inhibits the mitral cells synaptically, by a third voltage pattern called TQ_{e3}:

$$T_{\text{in}} Q_e(R) + TQ_{e2}[TQ_{\Delta s}(R)] \to T_{\text{in}} Q_e(R) + TQ_{e2}[TQ_{\Delta s}(R)] - TQ_{e3}[TQ_{e2}] \tag{4.27}$$

The output of the neuron pool considered is determined by the overall $e(x, y)$ resulting from eq. 4.27. Particular forms could be chosen for the transformations indicated. Temporal structures, neglected earlier, are partly available from experiments involving antidromic stimulation [4.32] of the olfactory bulb. The ranges Δs also could be predicted (as Rall did for the radial field distributions), and the computed predictions then should be compared with experimental data. See Chapter 5 for similar neural networks in the vertebrate retina.

In this example we mentioned only quantization in the spatial dimension. Time can be quantized, as Mueller [4.29] suggested, by a ring of neurons. The beginning of the event is given by the input, its end when the pulse has traveled around the ring. In a transformation $TQ_e(t)$ both measurement and transformation can be done in one step. Mueller also gave examples of nets performing logical operations and pattern recognition. The elegant formalism introduced here deserves further development and application in connection with experimental work, as do many of the more recently described theories of neuronlike systems, automata, and pattern recognition (Caianiello [4.33]; Leibovic [4.34]). Knowledge about mechanisms known to exist in the nervous system, about behavior of entire nets of cells, and about theoretical–analytical devices then can be brought to bear, each upon the others.

C. PRECOGNITIVE AND COGNITIVE LEVELS OF COMMUNICATIVE BEHAVIOR

In the previous sections of this chapter, we developed concepts as they were required in the description and mathematical treatment of successively more complex transformations of information by single nerve cells, by neuron pools, and by entire (intermediate) processors of sensory information *within* an organism. Here we present a hierarchical classification scheme of higher neural processing systems, on the precognitive and cognitive level, allowing systematic analysis of *communicative input–output behavior*. For this, as in all other systems analyses, it is clear that the understanding of the operations performed by the subsystems (i.e., the neuron pools) is necessary but not sufficient. Modes of interaction resulting in compound operations of many subsystems, defined in new terms, must be described if the overall performance at the next higher systems level is to be analyzed.

Mark [4.35] postulated a hierarchical structure of such compound mechanisms (brain factors) which are logically necessary when an organism carries out tasks, solves problems, answers questions, and reacts "intelligently" to its environment by progressively more complex systems. He demonstrated how brain factors can be defined, from the operational point of view, so that the postulates can be tested at each level within the hierarchy. The categorized brain factors are those which are likely to be involved when carrying out the following operations, which are listed in an order that indicates the nature of the hierarchy mentioned:

1. When information about the nature and structure of the environment and the "inputs" (the questions asked, dilemmas presented, problems posed) is received, accumulated, and classified. "Problems" may also arise within the organism. The nature and structure of the available information may also be categorized and associated with others.
2. When such information, in the form of (temporal and spatial) patterns of neuronal activity within the brain, brings forth behavioral responses or reactions.
3. When errors in the responses are detected, classified, and reduced.
4. When motives for answering, for reacting, and for producing responses are *generated*, or when motive-generating operations are remembered.
5. When error-minimizing and answer-finding *operations* are learned and new combinations of available information are made, creating new answers *and* new questions to be recognized at an even higher level of cognitive activity.

These operations and transformations are functions of the brain's outstanding ability and capacity in relating *nerve activity patterns*, transforming them, and so on. They therefore depend on the *probabilities of occurrence* of such patterns. Central patterns causing activities appearing with very high or very low probability elicit "preprogrammed" action or suppression of responses, respectively, without necessarily using information on the effects of the output regarding efficiency or accuracy. Such responses may be strongly dependent on input stimuli and internal regulatory phenomena; they are dominant in animals on lower levels of the evolutionary scale. If the probability of availability of brain patterns can be changed (increased or decreased) as in more complex organisms, we can speak of *learning*. Memory, the altered appearance probability, allows for error-reducing operations; it also permits the *efficiency* of responding to be enhanced by retention of entire error-reducing schemes and answer-finding operations.

The *versatility* of the brain can be described in terms of the capacity to learn at all five levels listed previously. In particular, nerve activities are grouped and classified according to their similarities as well as their differences from the operational point of view. This means that one (input) information structure may be substituted for another, allowing for abstractions. Input patterns of one subgroup may or may not cause the same response. Activity patterns yielding answers are also categorized according to the similarities and differences in their effects (i.e., their error-reducing power). Substitution of one activity pattern for another then makes it possible to predict both the questions that are to be answered (input) and the responses to be chosen (output). Clearly, feedback from the external structure and logical decisions play important roles in this, and we can talk of flexible *motivators* to react to a given problem (i.e., to reduce the errors in the response).

In summary, the motivation lies in the task—given by environmental situation or generated internally—of reducing errors, answering a question correctly, or adapting to the environment in an efficient way designed to minimize hunger, pain, embarrassment, or anger, on the one hand, and of achieving such goals as experiencing pleasure, establishing rapport, or coming to positive or negative agreement, on the other. The efficiency in carrying out this task is determined by memory for transformations, categories, relations, and implications or associations of neural activity patterns. As we have seen, these depend on versatility in categorizing or grouping of stimuli and outputs. Since all these patterns have temporal structures, predictions can be made by associations with and within classes. Furthermore, man has learned most about his environment, and his efficient behavior in reacting to it, by developing abstracted quantitative measures for relations between activity patterns *and* observable structures and phenomena.

On the basis of the foregoing loosely defined hypothesis, we are in a position to define the brain factors together with quantitative behavioral tests for their successful operation. As an example, we can estimate a person's efficiency in relating various patterns previously acquired and questions previously raised. Mark [4.35] proposed this "behavioral hierarchy" of increasingly complex information processing systems, each being a cumulative function of the subsystems. A system at one particular level includes a combination of all systems at the next lower level. Failure of one system causes failure of the systems on all subsequent (higher) levels. When referring to Fig. 4.15, which shows a two-dimensional array of postulated brain factors and behavioral

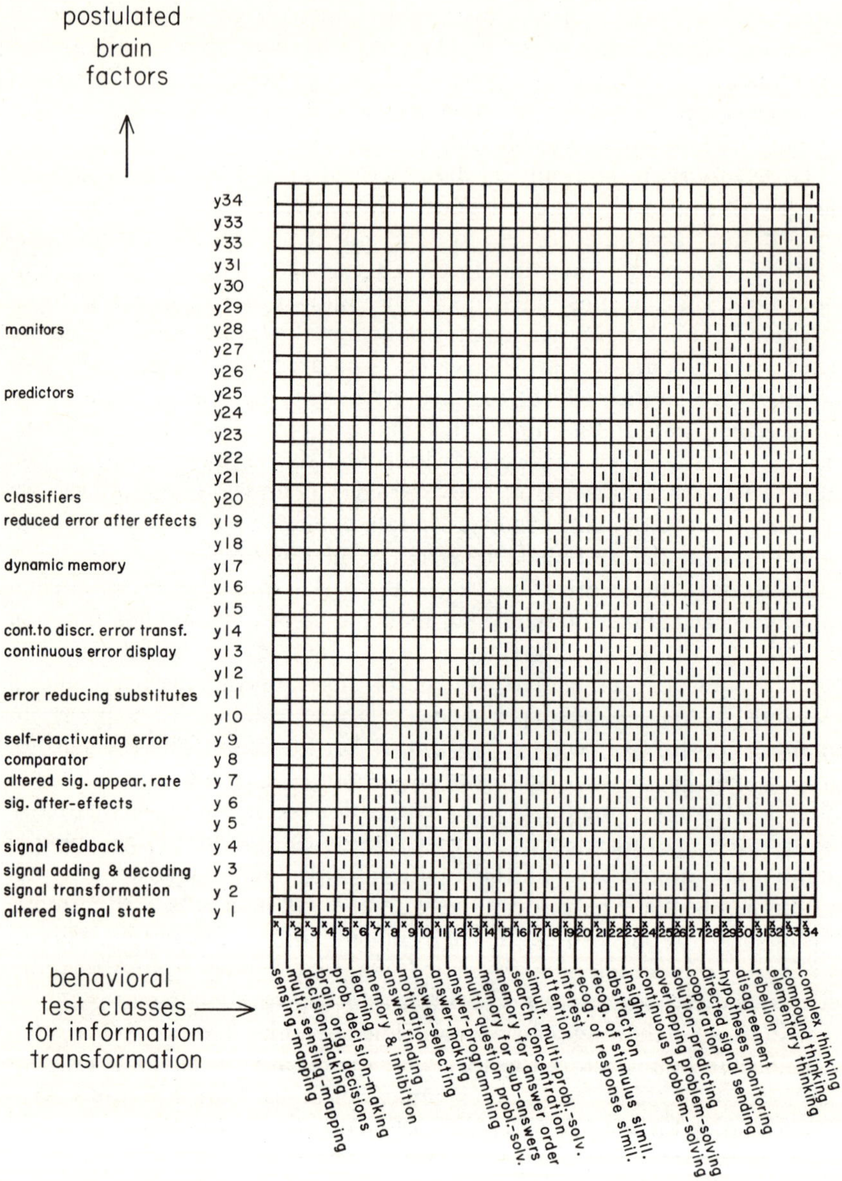

Fig. 4.15. Postulated brain factors (y_1–y_{34}), at the precognitive and the cognitive levels, together with their test classes (x_1–x_{34}). Note the hierarchical structure of the system and also that temporal changes and developments introduce a third dimension. After Mark [4.35].

test classes, we describe the brain's activities at any one time. Time is a third dimension, along which brain factors vary and along which changes that accompany maturing may be plotted.

At the bottom of the hierarchy lies the ability of the system to receive information from the environment, to relate activity patterns, and to produce simple, stimulus-triggered outputs determined only by information from the environment (brain factor y_1). The behavioral test class called x_1 includes experiments investigating sensing ability and (rigid) mapping. When multiple inputs and outputs are related, perhaps by internal chains of transformation, we talk of information *transformations* (y_2), and multiple sensing and mapping tasks (x_2) can be used for testing. For unique testing in classes x_1 and x_2 it is sufficient to establish criteria for probability of occurrence "close to unity" (for the responses) and to demonstrate that the probability does not change with time.

At a next higher level, the nervous system becomes more adaptive by using internally stored information when stimuli occur. Patterns are compared. Mark called this operation "*adding and coding*" of signals (information) (y_3), which obviously depend on y_1 and y_2, and can be tested by requiring "primitive" decisions (x_3). Closely associated is the case of such activity comparisons and decisions depending on physiological homeostats and tending to cause reactions best for preservation of the system. *Information feedback* then is involved (y_4, x_4) and the decisions originate within the brain, in many cases by coupling between various parts of the brain. A slightly more complex situation arises when the internally available activity patterns appear only with a certain probability (y_5). This requires corresponding evaluatory tests (x_5).

More general adaptiveness requires the existence of *aftereffects*, unidirectional changes in pattern appearance probability. The "answers" are formed on the basis of decisions made using the (transiently) altered configurations. Again, we see that the brain factor (y_6) depends on all lower levels (y_1–y_5). It is tested by measuring the ability for elementary learning (x_6). If the changes (positive or negative) in probability of occurrence of the patterns are retained, the altered internal information appearance *rate* (y_7) is tested in terms of memory and inhibition, as a function of learning (x_7), by observing changes in the probability of appearance of error-reducing action from an initial to a final value. The levels 1 through 7 thus comprise systems that receive, transform, and compare activity patterns. They further are able to make simple decisions and can adapt, in an elementary manner, to reach decisions more efficiently.

The next three brain factors are involved when individual, discrete problems are resolved. A comparator (y_8) must be engaged when the system tries to match, in one trial, the activity pattern elicited by the external stimulus with a *memory*. No further search is begun if the correct answer is not found. Tests can be devised (x_8) that measure the "one-try" answer-finding capability. Factor (y_9) allows for repetitive search for an answer to an individual question, and we can test for "motivation" (x_9), defined by the repetitive error-minimizing attempt. We can prove that motivation exists by demonstrating the involvement of memory. Because of motivation, the problem-solving capability continues to exist after the removal of the input stimulus upon which the answers were found and correctly maintained. Furthermore, it should be shown that, before the test, the maintained answer-finding operation could never be carried out as it is after the test. If the search involves more complex decision making by searching for response patterns of different classes, we also have to test the answer-*selecting* ability (y_{10}, x_{10}). On a next level, where we speak of "original" trials, error-reducing "substitutes" (y_{11}, y_{12}) enable *internal* answer-making, internal trial and error, by searching for different combinations. The answer-shaping ability has to be tested (x_{11}, and x_{12}) by verifying that some answers were newly considered but rejected, and that the answer patterns occur only if the particular problem is encountered.

When the brain receives continuously changing patterns and the observed errors thus also have to be observed continuously (y_{13}, x_{13}), the system must be able to react in a much more complex manner, which we can call multi-question problem solving. Memories of subanswers that must be tested (x_{14}) can greatly alter the original decision thresholds. The continuous problems and errors may be split into discrete patterns (y_{14}). Internal interest influences the search, and efficiency is increased by memory for *order* or appearance of output patterns (y_{15}, x_{15}). The search concentration for internal memory patterns can be tested (y_{16}, x_{16}); several subpatterns may be combined for obtaining one output. On the next level we can postulate a dynamic memory (y_{17}) enabling simultaneous solving of multiple problems (x_{17}). In these multiple-answer search situations, attention, defined as "multipattern search concentration," should be tested (x_{18}).

Classes x_{13} to x_{18} involve careful statistical analyses of frequencies of test results to distinguish correct answers initiated by the stimuli and worked out by the system, from coincidentally successful error reductions. Mark [4.35] pointed out that these are needed at each level (in x_{13}–x_{18}) and also at different stages in the development with age. Furthermore, the system arrives at guesses much faster than at correct answers involving all factors y_{13} to y_{18}. This provides for another method of discriminating test results.

In higher animals we would expect to find *aftereffects* of error-reducing patterns (y_{19}). Interest (x_{19}) is aroused; the organism "enjoys" continuing the search. The quest for better answers may go on after a solution has been presented, perhaps further lowering the decision thresholds. Tests belonging to x_{19} need to last for some time. Beyond this level we can test for recognition of the response (x_{20}), asking: are similarities or equivalance detected among the different answer classes that allow for compound substitution? Likewise, x_{21} represents the class of tests for recognition of similarity or equivalence in problem patterns allowing for substitutions. The next test (x_{22}) inquires about the ability of double, complex, or abstract substitution. Conversely, conditioned discrimination is also tested by this class.

When all subsystems are functioning successfully, insight can be obtained on the next level. Class x_{23} tests for internalized trial-and-error operations. Mark [4.35] observes that this level, and higher ones, probably exist only in primates. In these we also expect the abilities to solve problems continually (x_{24}) and to solve overlapping complex problems (x_{25}) continually. At the next level we then can postulate predictors (y_{26}), tested by asking for anticipated solutions (x_{26}). This is the basis of simple forms of cooperation (x_{27}) (problems can be shared) which can be initiated internally only in the next higher form (x_{28}). The same ability enables the system to test simple hypotheses and to communicate in elementary form. Errors are reduced by directed input–output and output–input transformations. This requires a new brain factor, namely, *monitors* (y_{28}). We next must inquire whether communication can be maintained (x_{29}). Continuous functioning on all lower levels then allows for continuing hypothesis monitoring and improvement.

Further test classes should be developed at the next higher levels. To probe interest in internal and external answer pattern differences, questioning, or disagreement, there is x_{30}. When the environment—the external stimuli—is restructured to achieve changes in question patterns, we can talk of rebellion (x_{31}). If, on the other hand, question patterns, originating *within* the system, are manipulated so that internal problems are solved continuously, the system thinks on an elementary level (x_{32}). *Compound* thinking (x_{33}) also involves internally initiated problems and problem structures, but the errors are experienced by the system itself.

Complex thinking (x_{34}) involves storage of information, problems, and so on outside the system. Questions are structured by other brains, errors are known to exist. It is hoped that the new solutions, answers, or inventions will enrich the bulk of knowledge.

So far, the *modality* of input stimuli and of the outputs has not been specified. The model indeed is based on the assumption that this system of brain factors (y_1–y_{34}) is engaged whether auditory, visual, tactile, or kinesthetic information is processed; whether gross or fine movements, or vocalizing movements are tested. Furthermore, Mark [4.35] suggested that his model also serves in the combined modes of speech based on audition and in writing based on vision or on tactile information. The hierarchical system may, however, be fully functioning when gross movement is tested while it fails at a particular level in other modalities. Modality-specific handicaps thus can be defined in terms of the brain factors stipulated, by carrying out tests over all sensor–effector systems in search for failing levels.

In addition, it should be realized that the (sub-) systems may not be stationary during short time intervals. A correct answer may be found at one time but not at another, or correct response patterns may be repetitively elicited without allowing for time to correct the error completely. Tests should be planned accordingly and carefully evaluated [4.35]. During the development over *long* time intervals, the organism becomes progressively more complex. Each brain factor evolves only when all the subsystems have become operational. It should also be noted that the number of neural activity patterns that can be processed increases.

D. REFERENCES

4.1. Rall, W. Theoretical significance of dendritic trees for neuronal input–output relations, pp. 73–97 in *Neural Theory and Modeling*. R. F. Reiss (Ed.). Stanford University Press, Stanford, Calif., 1964.

4.2. Segundo, J. P., D. H. Perkel, and G. P. Moore. Spike probability in neurones: Influence of temporal structure in the train of synaptic events. *Kybernetik* **3**: 67–82, 1966.

4.3. Stein, R. B. Some models of neuronal variability. *Biophys J* **7**: 37–68, 1967.

4.4. Moore, G. P., D. H. Perkel, and J. P. Segundo. Statistical analysis and functional interpretation of neuronal spike data. *Ann Rev Physiol* **28**: 493–522, 1966.

4.5. Purple, R. L. The integration of excitatory and inhibitory influences in the eccentric cell in the eye of *Limulus*. Ph.D. thesis, Rockefeller University, New York, 1964.

4.6. Purple, R. L. and F. A. Dodge. Interaction of excitation and inhibition in the eccentric cell in the eye of *Limulus*, in *Sensory Receptors. Symp Quant Biol* **30**: 529–537, 1965.

4.7. (a) Roberge, F. A. and J. H. Milsum. A feedback model of the neuron for the study of small neuron networks. *Med Biol Engin* **4**: 357–366, 1966. See also: (b) Roberge, F. A. Stability analysis of a mathematical neuron model. *Bull Math Biophys* **29**: 217–226, 1967.

4.8. Rodieck, R. W., N. Y.-S. Kiang, and G. L. Gerstein. Some quantitative methods for the study of spontaneous activity of single neurons. *Biophys J* **2**: 351–368, 1962.

4.9. Werner, G. and V. B. Mountcastle. Neural activity in mechanoreceptive cutaneous afferents: Stimulus–response relations, Weber functions, and information transmission. *J Neurophysiol* **28**: 359–397, 1965.

4.10. Jenik, F. and H. Hoehne. Ueber die Impulsverarbeitung eines mathematischen Neuronenmodells. *Kybernetik* **3**: 109–128, 1966.

4.11. Jenik, F. Pulse processing by neuron models, pp. 190–212 in *Neural Theory and Modeling*. R. F. Reiss (Ed.). Stanford University Press, Stanford, Calif., 1964.

4.12. Takeuchi, A. and N. Takeuchi. Electrical changes in pre- and postsynaptic axons of the giant synapse of *Loligo*. *J Gen Physiol* **45**: 1181–1193, 1962.

4.13. Kuepfmueller, K. and F. Jenik. Ueber die Nachrichtenverarbeitung in der Nervenzelle. *Kybernetik* **1**: 1–6, 1961.

4.14. Fuortes, M. G. F. Initiation of impulses in visual cells of *Limulus*. *J Physiol (London)* **148**: 14–28, 1959.

4.15. Rushton, W. A. H. A theoretical treatment of Fuortes's observations upon eccentric cell activity in *Limulus*. *J Physiol (London)* **148**: 29–38, 1959.

4.16. Wolbarsht, M. L. and S. S. Yeandle. Visual processes in the *Limulus* eye. *Ann Rev Physiol* **29**: 513–542, 1967.

4.17. Hartline, H. K. and F. Ratliff. Inhibitory interaction of receptor units in the eye of *Limulus*. *J Gen Physiol* **40**: 357–376, 1957.

4.18. Minorsky, N. *Nonlinear Oscillations*. D. Van Nostrand Co., Princeton, N.J., 1962.

4.19. Harmon, L. D. and E. R. Lewis. Neural modeling. *Physiol Rev* **46**: 513–591, 1966.

4.20. Schmitt, O. H. Biological transducers and coding. *Rev Mod Phys* **31**: 492–503, 1959 or in *Biophysical Science—A Study Program*. John Wiley and Sons, New York, 1958.

4.21. Rall, W. and C. C. Hunt. Analysis of reflex variability in terms of partially correlated excitability fluctuation in a population of motoneurons. *J Gen Physiol* **39**: 397–422, 1956.

4.22. Melzack, R. and P. D. Wall. Pain mechanisms: A new theory. *Science* **150**: 971–979, 1965.

4.23. Kuiper, J. W. The optics of the compound eye, pp. 58–71 in *Biological Receptor Mechanisms*. Symposia of the Society for Experimental Biology, No. 16, Cambridge University Press, Cambridge, England, 1962.

4.24. Waterman, T. H. Directional sensitivity of single ommatidia in the compound eye of *Limulus*. *Proc Nat Acad Sci USA* **40**: 252–262, 1954.

4.25. Kirschfeld, K. and W. Reichardt. Die Verarbeitung stationaerer optischer Nachrichten im Komplexauge von *Limulus*. *Kybernetik* **2**: 43–61, 1964.

4.26. Reichardt, W. Ueber das optische Aufloesungsvermoegen der Facettenaugen von *Limulus*. *Kybernetik* **1**: 57–69, 1961.

4.27. Whittaker, E. T. *A Course of Modern Analysis*. Cambridge University Press, Cambridge, England, 1958.

4.28. Reichardt, W. and G. MacGinitie. Zur Theorie der lateralen Inhibition. *Kybernetik* **1**: 155–165, 1962.

4.29. Mueller, P., T. Martin, and F. Putzrath. General principles of operations in neuron nets with application to acoustical pattern recognition, pp. 192–212 in *Biological Prototypes and Synthetic Systems* (Proc. 2nd Bionics Symp.). E. E. Bernard and M. R. Kare (Eds.). Plenum Press, New York, 1962.

4.30. Mountcastle, V. B., G. F. Poggio, and G. Werner. The relation of thalamic cell response to peripheral stimuli varied over an intensive continuum. *J Neurophysiol* **26**: 807–834, 1963.

4.31. Ratliff, F. *Mach Bands: Quantitative Studies on Neural Networks in the Retina*. Holden-Day, San Francisco, 1965.

4.32. Rall, W., G. M. Shepherd, T. S. Reese, and M. W. Brightman. Dendrodendritic synaptic pathway for inhibition in the olfactory bulb. *Exp Neurol* **14**: 44–56, 1966.

4.33. Caianiello, E. R. (Ed.). *Neural Networks*. Springer-Verlag, New York, 1968.

4.34. Leibovic, K. N. (Ed.). *Information Processing in the Nervous System*. Springer-Verlag, New York, 1969.

4.35. Mark, H. J. Elementary thinking and the classification of behavior. *Science* **135**: 75–87, 1962.

Part II

EXTEROSENSORY SYSTEMS

CHAPTER FIVE

The Visual System: Neural Information Extraction and Psychometrics

A. INTRODUCTION

It is impossible in one chapter to summarize the great body of knowledge about the visual system. Important basic data and functional descriptions are presented in physiology books [2.2]; deeper insight is gained from more specialized surveys, for example, those in the *Handbook of Experimental Psychology* [5.1] or in the *Handbook of Physiology* [5.2]. For this chapter we have selected certain topics which especially contribute to synthesis of these findings and may therefore promote understanding of the integrative action of the different "subsystems." Thus discussion of single-cell behavior at the various levels of the retina allows us to define more accurately mechanisms involved in the coding of light patterns into (static) patterns of ganglion cell activity. In consequence, theories of acuity suggest themselves if we study dynamic modes of operation of the greatly diverging projection systems (proximally of the retina), with their mutual overlap.

Generally, at each level of organization, we have tried to describe those properties and mechanisms which are important for performance of the system at this level and which, in addition, allow correlation with phenomena at the next higher level. Hence we restrict attention to those functions of the visual system which are most important to the organism in receiving information about the environment—that is, information about the topology of visual targets and their three-dimensional relationships, and information required for detection, recognition, and comparison of stationary and moving patterns—all involving high acuity and contrast detection, under many different conditions of background light intensity.

Many topics are omitted, such as physiological optics even though this plays an important role as first link in the "chain of events" [5.2]. Only passing mention can be made of processes taking place in the photoreceptor cells, noting that, among others, Pinter and Jones [5.3] have developed a mathematical model of photoreception. In addition, Levine [5.4] proposed enzyme amplifier systems, possibly involved when many pigment molecules are activated after one or a few molecules are excited by light. A great wealth of knowledge on pigment chemistry is available, as lucidly summarized by Wald [5.2], who received the Nobel prize for work in this field. However, little is known about the mechanisms connecting photochemical change and membrane (generator) potential of the receptor cells (i.e. the "output"). Theories of color vision have attracted new interest since direct evidence has been obtained by Marks et al. [5.5] for the existence of three cone pigments, but these cannot be pursued further here.

Thus the functional descriptions given below of the successive levels, from neural organization in the retina to psychophysical performance, must remain incomplete. Many questions remain open, and many experimental findings cannot be systematized yet by mathematical relations.

B. NEURAL ORGANIZATION AND MECHANISMS

1. Retina

The eye fulfills its function of converting optical into neural information in several stages. In vertebrates, the eye is moved and the lens accommodated so that image formation on the retina is optimal. Optimality here implies relatively stable imaging (focusing) of the visual target on the fovea (or on the area centralis). Image formation of course is never ideal because of such factors as optical defects, deviation from sphericity of refracting surfaces, scattering, and wavelength-dependent properties [5.2]. Many details of the structure of the primate retina were worked out and summarized by Polyak [5.6], and from these it is clear that neural processes taking place at this level must be complex and that some coding is achieved here. In fact, we might expect exceedingly complex "preprocessing" mechanisms, since convergence is great (125 million rods, 4–7 million cones converging onto approximately 1 million fibers in the optic nerve), and since a large amount of information must be conveyed by this system. Distribution of rods and cones varies markedly over the retina, and thus preprocessing mechanisms can also be expected to be a function of location within the retina. Indeed, synaptic relations, recently elucidated by Dowling [5.7], change from fovea (one cone to one ganglion cell; high acuity) to periphery (ca. 250 rods and a few cones to one ganglion cell; static summation for low absolute threshold of vision and dynamic summation for high sensitivity to changes in illumination).

As a rule, each ganglion cell carries information (types are specified later) from a group of receptor cells lying within its so-called receptive field. Kuffler [5.8], exploring the fine structure of the receptive field in cats with stimuli of small size (0.2 mm, 1° on retina), discovered the antagonistic center-surround organization of the receptive field, but stressed that the field is not constant. It depends on size of the stimulating light spot, intensity of the "exploring" spot, and the state of dark adaptation. Receptive field size may shrink from 10° to 5° when the retina adapts to high background intensities. Firing activity of the ganglion cell either increases when the center of its receptive field is stimulated and decreases when the "antagonistic surround" is illuminated, or vice versa. Generally, the center needs some stimulation if the surround is to exert its antagonistic influence. More complex types of coding have been found in the frog [5.9], the pigeon [5.10], and the rabbit [5.11]. Information about directionality and motion of stimuli is extracted at higher neural levels in the primates.

In the primate retina it is difficult to explore the activity of individual cells *within* the retina and their contributions to the center-surround organization. The cells are small and densely packed. It is a fortunate finding that the retina of *Necturus*, a type of salamander, contains the same type of cells as the human retina (see Fig. 5.1). These cells are large, and most synapses have been identified by Dowling and Werblin [5.12]. Werblin was able to determine the response of individual cells, thus shedding light on the distribution of functions among various units. His findings also clarify earlier observations, particularly the S(vaetichin)-potentials of the distal retina. The *Necturus* retina appears to be an excellent "model" of the primate's, although a few of the synapses are unique to *Necturus*.

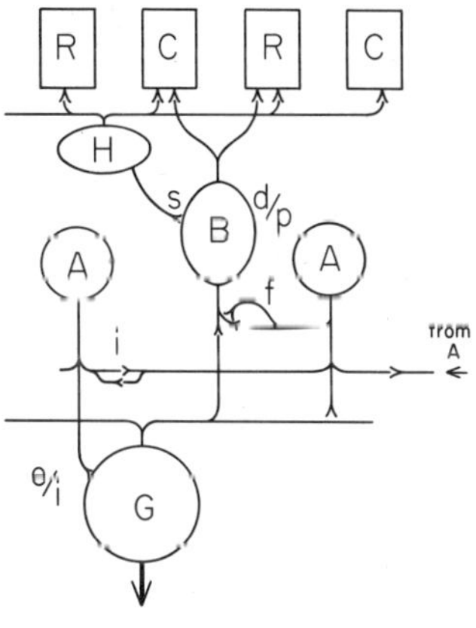

Fig. 5.1. Schematic view of the functional organization of the retina of *Necturus* (mud puppy): R and C are receptor cells, H = horizontal cells, B = bipolar cells, A = amacrine cells, G = ganglion cell. The same types of cells are found in the primate retina, although the connectivity varies slightly. On the one hand, synapse s is not found in primates, and the connection between receptor and horizontal cell is more complex. (There exists, in primates, an inhibitory feedback synapse from H onto R.) On the other hand, *Necturus* has ganglion cells (not shown) driven mostly by amacrines, which may not exist in the primate. Synapse f forms an (excitatory) feedback path to the bipolar cell. The i's are mutually inhibitory synapses between amacrines; e/i may be excitatory or inhibitory synapses; d/p indicates that two types of bipolars exist (cf. text and Fig. 5.4*b*). After Dowling and Werblin [5.12a]

It is now known that receptor, horizontal, and bipolar cells respond with slow, tonic, and graded potentials. Both receptor cells and horizontals always hyperpolarize, as is shown in Figs. 5.2 and 5.3. Both horizontal and bipolar show extended receptive fields, to be expected from their morphology (Fig. 5.1). Bipolar cells divide into two classes. Some hyperpolarize, some depolarize (Fig. 5.4), and their receptive fields (of the order of 1–1.2 mm radius) are slightly narrower than those of the horizontal. Bipolar cells appear to carry (static) contrast information, and this over a dynamic range larger than at the receptor cell level.

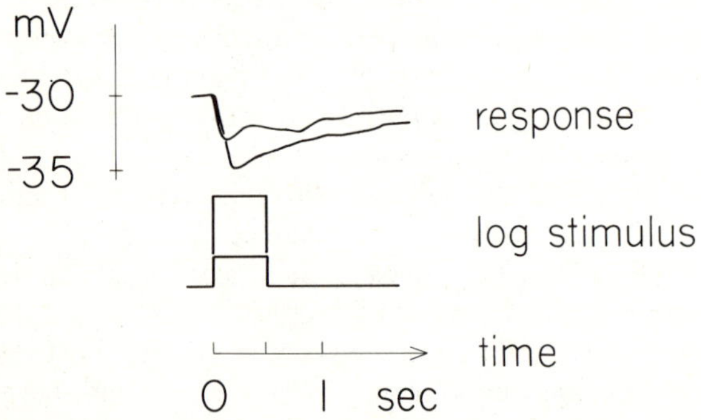

Fig. 5.2. Typical response of a receptor cell in the *Necturus* retina to a small spot stimulus, centered on the receptor, of 0.5-sec duration. This intracellular response, plotted against time, is independent of illumination of the surround. Increase in light stimulus causes hyperpolarization beyond the resting (inside) potential of approximately −30 mV. The response is "graded" with light intensity but only over a relatively narrow dynamic range. At very large stimulus intensities, the response is a constant. The "on" response is always quite distinct. From Werblin [5.12b].

Amacrine and retinal ganglion cells both respond by depolarizing and by active (spike) discharge above their thresholds. Amacrine cells, mostly responding with one single spike, can be divided into two classes, as summarized in Fig. 5.5. Half are the on-center–off-surround type; the others do not have center-surround-type receptive fields and respond at onset or at end of a stimulus or at both. All are very sensitive to changes in illumination of their receptive fields. It is interesting that they interact with each other (synapse *i* in Fig. 5.1). In fact, the idea that they probably inhibit each other led Werblin [5.12] to hypothesize a simple mechanism of motion detection in retinas or in retinal areas with high amacrine cell density (e.g., in the frog retina).

Some ganglion cells have narrow receptive fields (ca. 0.5 mm), with an on-center and an off-periphery. The center, however, must be stimulated before the inhibitory effect of the surround can act. Their responses are

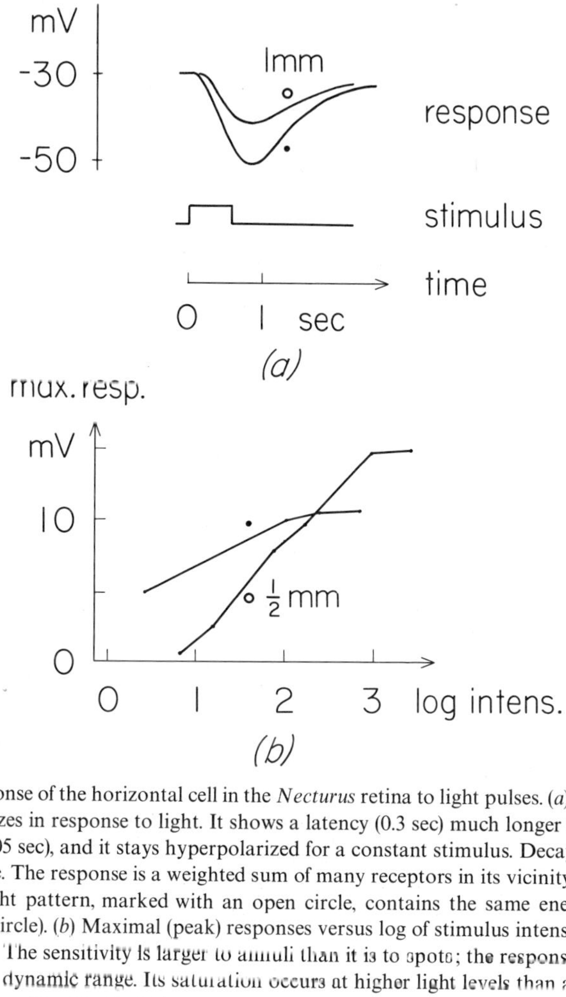

Fig. 5.3. Response of the horizontal cell in the *Necturus* retina to light pulses. (*a*) The horizontal cell hyperpolarizes in response to light. It shows a latency (0.3 sec) much longer than that of the receptor cell (0.05 sec), and it stays hyperpolarized for a constant stimulus. Decay time is similar to response time. The response is a weighted sum of many receptors in its vicinity (0.25–0.5 mm). The annular light pattern, marked with an open circle, contains the same energy as the spot stimulus (solid circle). (*b*) Maximal (peak) responses versus log of stimulus intensity for spot and annulus stimuli. The sensitivity is larger to annuli than it is to spots; the response to annuli also exhibits a larger dynamic range. Its saturation occurs at higher light levels than are recorded for spot stimuli. From Werblin [5.12b].

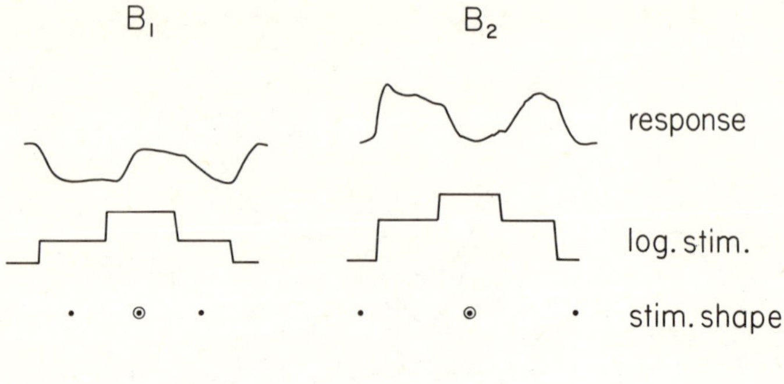

(a)

Fig. 5.4*a*. Two types of bipolar cells in the *Necturus* retina. All bipolar cells have receptive fields of the antagonistic center-surround type. The response on the left shows the case in which a light spot, centered in the middle of its receptive field, causes maintained (and graded) hyperpolarization. If a light annulus is added (middle of record), the " center response " is suppressed. The sequence of stimuli is indicated by the symbols at the bottom. The response on the right shows depolarization due to a spot at the center. The antagonistic surround, if illuminated by an annular stimulus, causes hyperpolarization. Latency is about 0.1 sec. The diameter of the center region and of the surround are about 0.1 and 0.25 mm, respectively. The influence of small annuli is difficult to measure owing to light scattering. From Werblin [5.12b].

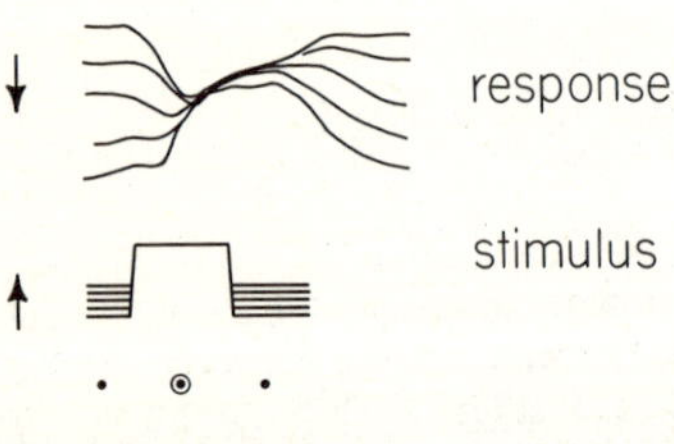

(b)

Fig. 5.4*b*. One type of bipolar response to the same type of stimulus as is used in the experiment shown in Fig. 5.4*a*. After a light spot stimulus is centered on the receptive field of the unit, a concentric annulus is projected to stimulate the (antagonistic) surround as indicated by the symbols at the bottom. This is done for different illumination intensities of the center spot. The arrows indicate the directions of increasing center illumination and the corresponding responses. The scale for the responses (upper figure) is such that the total range, at the onset of the curves, covers approximately 20 mV. From Werblin [5.12b].

maintained " on " or " off," which suggests direct, strong influence by the bipolar cells. When the whole receptive field is illuminated, there may be short bursts both at beginning and end of stimulus, suggesting that the antagonistic center can act only after a latency period. Other ganglion cells show transient changes (positive or negative) in firing rate either at onset or end of stimulus, or both. These have small receptive fields. It is probable that these are dominated by amacrine cells and receive excitatory or inhibitory inputs from them (synapse *e*/*i* in Fig. 5.1).

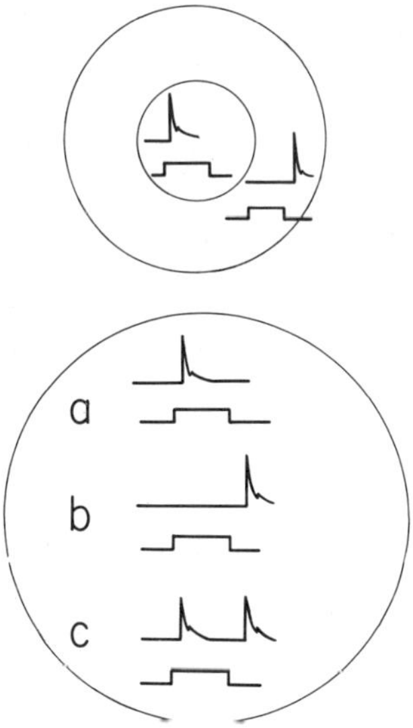

Fig. 5.5. The amacrine cell, the most peripheral unit responding with action potentials in the *Necturus* retina. The resting potential is of the order of -35 mV. In response to light pulses, one or two impulses occur at the onset and/or at the end of the stimulus, followed by a slow depolarizing potential (5–10 mV) lasting approximately 0.25 sec. The five different types of amacrine cell responses are shown in this figure. Only some amacrines have the center-surround type receptive field (top). Stimuli at the center cause an on-response; stimuli at the periphery cause an off-response. Widespread stimuli induce responses both at the beginning and the end of the light pulse. Other amacrine cells exhibit very wide receptive fields and may show on-responses (*a*), off-responses (*b*), or both (*c*). None of the responses includes more than one or two spikes. After Werblin [5.12b].

Ganglion cell responses with concentric opponent receptive fields have been studied in a great variety of animals (see the review by Bishop [5.13]). Very detailed quantitative studies have been carried out in the cat by Stone and Rodieck [5.14, 5.15], at least for stimuli of relatively low intensity and short duration (<1 sec), with weak background luminance. These authors verified findings by Kuffler [5.8] and demonstrated, furthermore, that no motion-sensitive units per se exist in the cat. A typical receptive field (center-on

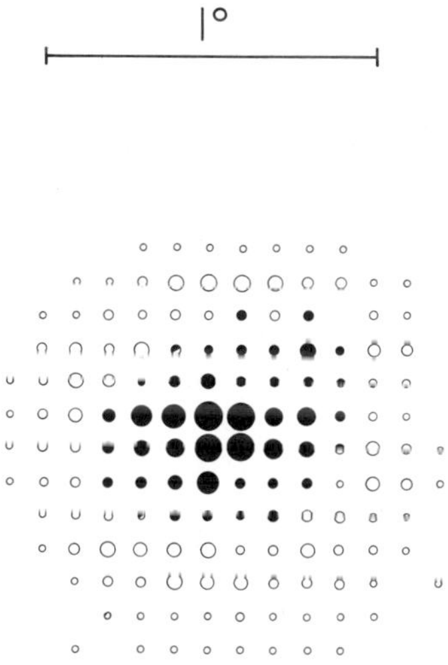

Fig. 5.6. Details of the receptive field of a ganglion cell of the cat. This unit has an off-center–on-surround response; that is, small flashing spots projected anywhere on the center area suppress, on the average, the spontaneous firing rate (solid circles), whereas small spotlike stimuli at the periphery of the receptive field cause an increase (open circles). The grid indicates the detailed geometry of the center-surround organization. The size of the circles indicates the *average* change in ganglion cell firing rate caused by the flashing spot (2′–4′ in diameter, 1-sec duration) projected on the location shown. The temporal behavior of the cell during such stimuli generally contains information too. See Fig. 5.7. From Rodieck and Stone [5.14].

–surround-off) type is graphed in Fig. 5.6. Solid circles indicate locations that elicit, on the average, an increased discharge. The diameter of the circle is qualitatively proportional to the magnitude of response. Quantitative studies are presented in Fig. 5.7, where histograms of 40 responses are shown for points lying on a straight line across the receptive field of a ganglion cell. The temporal structure of the responses, which is lost when plotting receptive fields as in Fig. 5.6, is indeed of importance as will be seen in the following.

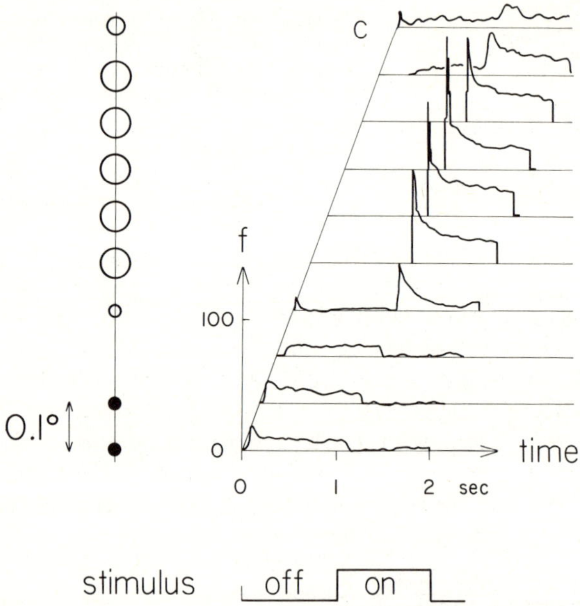

Fig. 5.7. Temporal structure of the response of a ganglion cell (cat retina) stimulated by a small flashing spot, projected at different places along a line passing through the center of the receptive field. Left: The type of receptive field of this cell is defined by the row of circles. Open circles indicate enhancement, solid circles indicate depression of (spontaneous) response. The size of the circles corresponds to the average degree of change in firing rate. Thus this cell is of the on-center–off-surround type. Right: The time course of cell activity is given for all stimulus conditions. Both off-responses (stimulus in surround) and on-responses (stimulus in center area) contain much more information than the row of circles indicates. Both onset and offset of light pulse cause "phasic" responses, or overshoots. From Rodieck and Stone [5.14].

Rodieck and Stone [5.14] concluded from their findings that, as a first approximation, the typical center-surround organization is the result of summation of two spatially overlapping and antagonistic fields: one of small extent and strong influence on the cell under investigation, the other of wider spread over the retina, of lesser influence, and acting with a larger delay than the first region. Responses of a ganglion cell due to step stimulus in any field show a quickly adapting, phasic transient (time constant of the order of 0.05 sec). This model, developed mathematically below, explains the basic nature of most responses seen. Particularly, we note not only that a ganglion cell can show different responses due to stimuli moving into or out of its receptive field (although no true motion sensing network is introduced), but also that a single unit can be very sensitive to small targets moving about within its field.

Following Rodieck [5.15], we introduce the following expressions for the step responses of a ganglion cell, centered at $x = y = 0$, due to a stimulus of spatial extent $\Delta x\, \Delta y$, at (x, y), and of unit intensity:

due to signal from center field:

$$f_1(x, y, t) = \frac{F_1}{\sigma_1{}^2 \pi^{1/2}} e^{-(x^2+y^2)/\sigma_1{}^2} (M + Te^{-\alpha t}) \Delta x\, \Delta y \tag{5.1}$$

due to signal from "surround field":

$$f_2(x, y, t) = \frac{F_2}{\sigma_2{}^2 \pi^{1/2}} e^{-(x^2+y^2)/\sigma_2{}^2} (M + Te^{-\alpha t}) \Delta x\, \Delta y \tag{5.2}$$

where $F_{1,2}, \sigma_{1,2}$ = constants determining amplitude and spatial spread of Gaussian fields
α = time constant of decay of phasic response

Note that in this formulation the two fields are made to overlap. Furthermore, we have assumed a temporal response with an overshoot T and a maintained activity M, independent of x and y. Thus the total response of the unit with the typical antagonistic surround (see Fig. 5.8) is

$$\begin{aligned} \Delta R(x, y, t) &= f(x, y)\, h(t)\, \Delta x\, \Delta y \\ &= [f'_1(x, y) - f'_2(x, y)] h(t)\, \Delta x\, \Delta y \end{aligned} \tag{5.3}$$

where $f'(x, y)$ = Gaussian distributions of eqs. 5.1 and 5.2
$h(t) = M(1 + T/Me^{-\alpha t})$

Clearly, if the influence of the center region (f_1) strong, we let $F_1 > F_2$. The spreads of the fields differ in that $\sigma_2 > \sigma_1$. In a numerical simulation, Rodieck [5.15] chose

$$\frac{F_1}{F_2} = 1.2 \qquad \sigma_1 = 0.59^\circ \qquad \frac{\sigma_2}{\sigma_1} = 3 \qquad \frac{T}{M} = 4 \qquad \alpha = 25\ \text{sec}^{-1}$$

such that center and surround regions have diameters of approximately 2° and 5°, respectively.

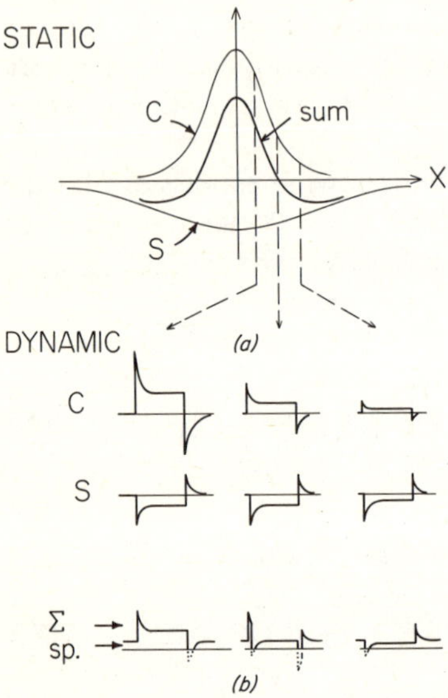

Fig. 5.8. Model of receptive field organization of ganglion cells. (*a*) The static response explains the "classic" center-surround response by a superposition of a narrow excitatory center area (*C*) plus a wider inhibitory area (*S*). In the periphery, where inhibition is larger than excitation, the off-surround response results. The curves in this model are assumed to be Gaussian. (*b*) The dynamic response depends on the position of a spot stimulus. Three different cases are shown. In the first row, the contribution of curve *C* becomes smaller the farther away the stimulus is placed from the center. The same holds for the inhibitory contribution, shown in the second row (*S*). Note that only deviations from a spontaneous firing rate are graphed. Component *S* probably contributes to the total response with a slight delay. This becomes important when *C* and *S* are about equal in magnitude (see second column). The bottom row pictures the sum ($\sum$) of *C* and *S* plus a "spontaneous" component (*sp.*). This model fits quite well what is found in the retina of cats. Extension of the model to other than small flashing spot stimuli is possible (see text and Fig. 5.10). From Rodieck and Stone [5.14].

The stimulus area, which is assumed to create uniform illumination of the receptor cells within its boundaries, can be defined (see Fig. 5.9) by its leading edge $g_1(x, y)$ and its trailing edge $g_2(x, y)$. If it is moving across the receptive field at a velocity v, parallel to the x-axis, the response $R(t)$ is

$$\begin{aligned} R(t) &= \iint_{-\infty}^{+\infty} \left[f\left(x, y, t - \frac{x - g_1(y)}{v}\right) - f\left(x, y, t - \frac{x - g_2(y)}{v}\right) \right] dx\, dy \\ &= \iint_{-\infty}^{+\infty} \left[h\left(t - \frac{x}{v} + \frac{g_1(y)}{v}\right) - h\left(t - \frac{x}{v} + \frac{g_2(y)}{v}\right) \right] f(x, y)\, dx\, dy \end{aligned} \tag{5.4}$$

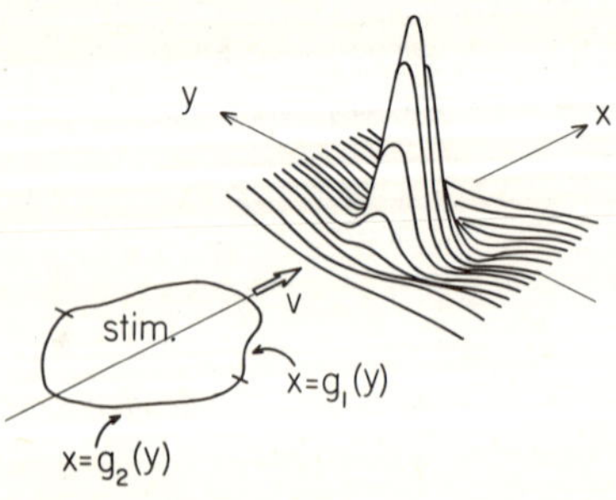

Fig. 5.9. View of the spatial distribution of the static response of the model of an on-center ganglion cell, together with symbols used to describe a stimulus. The stimulus (*stim*) projected on the retina may have any shape. Its "leading edge" is defined by the function $g_1(y)$, its trailing edge by $g_2(y)$. The hypothesized properties of the receptive field exhibit an on-center (high "peaks") and an off-surround, symbolized by a "valley" surrounding the peak. Here v is the velocity of stimulus on the retina. After Rodieck [5.15].

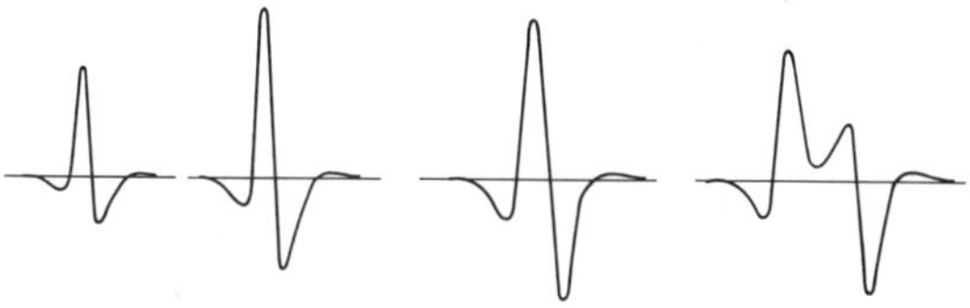

Fig. 5.10. Temporal structure of responses of the receptive field model, with on-center response, to long, bright bars passing over the receptive field. The bars are assumed to have different widths (0.5, 1.0, 2.0, and 5.0°); the velocity is 10°/sec. The curves give only changes in spontaneous activity of the ganglion cell. Deep negative peaks may silence the cell. The contributions of center and surround areas, and of leading and trailing edge of the 5° bar, are given in detail in Fig. 5.11. From Rodieck [5.15].

This double integral seldom can be solved in closed form. Rodieck [5.15], however, pointed out that in principle the method of Green's function can be applied. Indeed, he computed the "impulse response" [i.e., the $R(x, y)$ for a point stimulus], which can be used when complex stimuli are studied. Responses to stimuli of simple shapes, such as bars moving across the field, are quite easily found, however. They compare satisfactorily with Rodieck and Stone's [5.14, 5.15] experimental findings. Figure 5.10 presents responses to long bars of different widths, which are moved at 10°/sec across the field. The initial peak of activation is mostly due to the transient center response of the leading edge. The second peak, seen in the 5° bar, is due to the transient surround response of the trailing edge.

The genesis of the curves of Fig. 5.10 can be derived from Fig. 5.11, where all components of the widest bar are given individually. It is interesting to note that peaks for the 1° and 2° bars are larger than those for narrower or wider ones. All the curves indicate changes in firing rate from spontaneous frequencies. Thus negative going responses may silence the unit (cf. Fig. 5.8). Calibration of the curves is difficult. Experimentally, the firing rate histograms of one unit (Fig. 5.7) must be obtained from many responses. The retina, however, may change its property and adapt during the course of the experiment [5.15].

In preparation for the next section, it is useful to report Rodieck's model studies on responses due to a small stimulus moving back and forth over a small region within the receptive field. Figure 5.12 reveals that it is right at the boundary between center and surround that such small oscillating stimuli may elicit large output signals at the level of the ganglion cell. Interestingly, the signals essentially depend on the slope or gradient of the shape of the receptive field distribution. The model predicts the experimental findings [5.14] quite well. This, together with the calculation on bar responses,

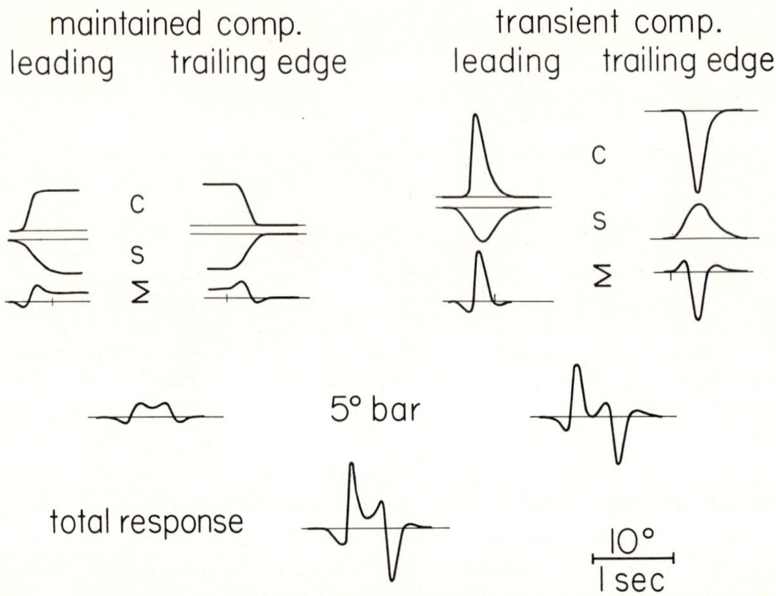

Fig. 5.11. Synthesis of the response of the ganglion cell model (on-center type) to a bright bar passing over its receptive field. The top row (*C*) gives the steady activity changes (left) and the transients (right) of leading and trailing edges due to the excitatory influence of the center. The second row (*S*) shows the equivalent signals from the inhibitory area. The third row gives the sum ($\sum$) of areas *C* and *S*. The fourth row combines leading and trailing edge responses for a 5° bar passing at a velocity of 10°/sec. The total response (bottom), as already shown in Fig. 5.10, indicates the strong " on " and " off " peaks. Note that signals generally are added onto a (steady) spontaneous activity level. After Rodieck [5.15].

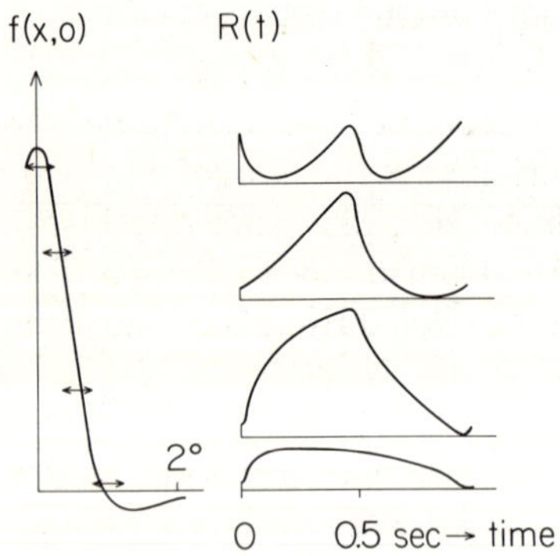

Fig. 5.12. Response of a receptive field model of a ganglion cell to a small bright spot oscillating, as indicated by the double arrows, at different locations within the field. The relatively good comparison of the response curves $R(t)$ with experimental curves gives further support to the hypothesis of receptive field organization (Fig. 5.8), assuming superposition. Note that the ganglion cell essentially detects the slope of the overall sensitivity curve $f(x, y = 0)$. The first half of $R(t)$ in time is due to motion from center toward the periphery. The spot stimuli are assumed to move $\pm 0.25°$. After Rodieck [5.15].

indicates clearly that no true motion-sensitive units exist in the cat. However, as Fig. 5.10 makes clear, the information on velocity of the bar is not lost. The leading edge of the response due to large objects is larger than the trailing edge and, with increasing velocity, the "wake" is also increased.

The principal limitation of the model is that only stimuli lasting less than one second can be satisfactorily processed. Furthermore, the ratio F_1/F_2—that is, the relative strengths of the two antagonistic fields (at $x = y = 0$)—would be predicted to be smaller from probings with very small stimuli. Thus flashing spot experiments, probably dependent on local adaptation processes, yield information on the general nature of the center-surround fields but do not allow prediction of the effective F_1/F_2 for stimuli of any shape.

A later discussion indicates that analysis of the receptive fields of retinal ganglion cells, although indispensable in probing into the neural organization of the retina, does not account for many other phenomena observed at higher levels. For example, we have not explained so far how the visual system achieves its high acuity. Likewise, when trying to describe in detail the interactions and distribution of function of cells within the retina, we are again at a loss, although Werblin's findings undoubtedly provide an important foundation to investigations in this area.

2. Lateral Geniculate Body and Visual Cortex

Other questions of importance arise when we consider functional performance of the system at the higher neural levels. Besides questions pertaining to the types of receptive fields of single units, importance attaches to areas of projection of one afferent fiber and the overlap of such areas, coding of brightness information, facilitation due to temporal or spatial summation, color discrimination, flicker fusion frequencies of single units, and couplings from neural centers not primarily involved in vision. (We restrict attention to visual and omit oculomotor control function.) Particularly, phenomena significant for acuity and pattern coding are discussed. When Talbot and Marshall [5.16] studied retino-cortical projection in the (rhesus) monkey with needle electrodes on the surface of the optical cortex, they observed extraordinary expansion of projection. The fovea (within a 1° circle) is projected at a ratio of 2′/mm on the cortex surface. The entire perimacular area (from a 1° 8° circle) is widened at the cortex at a ratio of 18′/mm. Thus the ratio of cells involved may be as large as 100:1. Furthermore, the diverging paths greatly overlap (see below).

The first station proximal to the retina, the lateral geniculate nuclei of the thalamus, appears to be mostly a relay station. However, there is coupling between the reticular formation and the "relay station," and fibers that emanate from this area project to other centers important for sensory–motor reflexes. It is known that the center-surround organization of receptive fields of single units in the cat's lateral geniculate body is preserved, and that the associated antagonistic action is enhanced as well. Hubel and Wiesel [5.17] found in the monkey that the lateral geniculate cells have receptive fields almost identical to the fields of the optic nerve fibers, although it is known that some divergence of projection takes place at this level.

Three large classes are found: two of the center-surround type (type I: red-on center with green-off surround, or other color combinations, and type III: on-center essentially independent of color) and one without center-surround organization (type II: green-on and blue-off). Type I occurs frequently, and under white light it is equivalent to the retinal ganglion cells mentioned earlier. Antagonistic action of the surround is only possible when the center is stimulated. These cells can mediate dark–bright contrast information well. Center fields are small in many cells, suggesting that only a few cones project onto this area. Type II chiefly conveys asymmetries in the spectral distribution of the stimulus, independent of stimulus size. Type III is sensitive to nonhomogeneity of illumination of its receptive field. Hubel and Wiesel point out that it therefore is not clear whether color contrasts are actually already coded at this level. Black–white information is processed, however, as indicated.

DeValois et al. [5.18] reported similar but not entirely equivalent findings in the macaque monkey. They also investigated (among a great number of other phenomena) the types of codes for light intensity changes encountered at this level in the visual system. In steady state the maintained firing frequencies are nearly independent of degree of adaptation. The activity may depend also on activity in other brain centers; but changes in activity, which can be either positive or negative, are approximately proportional to the logarithm of the intensity steps.

At the level of the *striate cortex*, Hubel and Wiesel [5.19] confirmed, in their many admirable investigations, the topographical retino-cortical projection first seen by Marshall and Talbot [5.16]. Furthermore, they investigated the receptive fields of many units at all depths and were able to propose, in some cases, schemes of neural organization among cortical cells. Cells that respond to small stimuli (as opposed to diffuse light) react most intensely to patterned stimuli, particularly to bars, stripes, or edges. It appears as though a large number of lateral geniculate cells, whose "on" centers are arranged on straight lines (across the retina), feed these cortical units, called simple cells (see Fig. 5.13). Within the columnar structure of the striate cortex, a great number of simple cells respond to bar-shaped stimuli of certain orientation and location on the retina. Conversely, for each stimulus of rectilinear nature, for each position on the retina, and for each orientation, there apparently exists a set of "simple" cortical cells with the corresponding trigger features.

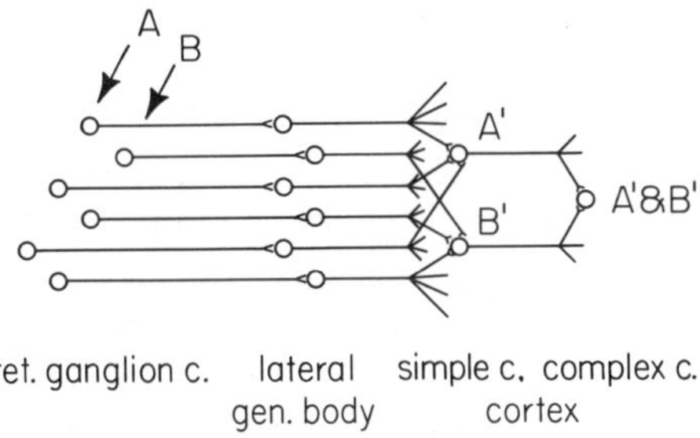

Fig. 5.13. Possible organization of neural pathways between retinal ganglion cells and cortical cells as proposed by Hubel and Wiesel. If all ganglion cells, which lie along a straight line on the retina (*A* or *B*), influence dominantly a particular cortical cell (*A′* or *B′*), the latter responds to stimuli in the form of lines, projected on the retina at the particular location of line *A* or *B* (arrows). Such cortical cells have been named " simple " ones. If many simple cells that respond to line stimuli of a particular tilt (i.e., angle with respect to horizontal) dominantly excite a cell at the next higher level, this second cell then responds to lines projected at *any* location on the retina, provided the angle of the line is the same as that of line *A*. Such cells have been called complex cells. Note that simple and complex cells might have receptive fields also in the form of bars or linear dark–bright edges. If many different complex cells are influencing more central cells in a concerted fashion, shape-detecting units may result. This scheme represents a static " wiring " diagram; see Fig. 5.14 for dynamic effects.

The complex cells, also defined by Hubel and Wiesel, cannot be associated with " on " or " off " regions. They seem to receive inputs from many simple cells, which are all sensitive to the same pattern orientation. Thus the trigger feature of a complex cell is a barlike stimulus of particular orientation, more or less independent of its exact position. In some cases, the stimulus must be moved in a particular direction in order to elicit response. The complex cells therefore convey pattern and motion information to higher levels.

Wurtz [5.20] studied cortical cells in awake monkeys which he had trained to fixate gaze direction for a few seconds, long enough to establish (in consecutive runs) their trigger features. He essentially confirmed that the same basic organization that Hubel and Wiesel had found in anesthetized animals exists when the animals are conscious. He also found the cells to fall into two large categories: first, nonadapting cells whose trigger features are of the simple rectilinear, the complex or the circular type, assuming steady stimulation; and second, rapidly adapting cells most sensitive to moving stimuli. Directionally selective units were not often observed. Although Wurtz maintained that activity of the first class may be due to small eye movements (physiological nystagmus), it need not be. Minute eye movements become important, however, for small foveal receptive fields. It is not known to what extent attention would influence these findings, because the animals in Wurtz's study paid attention to the fixation stimulus.

3. Mechanisms of Attention, Pattern Vision, and Acuity

Jung and his co-workers [5.21] have categorized cortical units in very general fashion, indicating that the visual cortex serves not only as a processing center of visual information but even more as a coordination center important in visual stabilization *and* attention. Some cells (type A), which they observed responding to electrical stimulation of the (intralaminar) thalamus, *change* their activity when light is flashed upon the retina. Others, type B, essentially carry information on brightness and contrast within their (large) receptive fields; these cells also change the level of their activity during and after thalamic stimulation. For example, their "flicker-fusion" frequency is raised by nonspecific thalamic stimuli. Conversely, stimulation in the mesencephalic reticular formation decreases neuronal flicker-fusion frequency. In behavioral studies it was found that the same inhibitory stimuli cause a cat to cease to search visual targets. Thus facilitation and inhibition of cortical units by such inputs from other neural centers may well correspond to the influence of attention and fatigue on visual perception. Types D and E, which essentially carry information on darkness and on contrast within their wide receptive fields, are therefore the counterparts of type B. Type C exhibits inhibitory phases during both "on" and "off" of a stimulus, which means that it extracts information on temporal changes.

In order to strengthen the above-mentioned clues concerning mechanisms involved in pattern vision, contour sharpening, and acuity, we need to introduce more information on processing of temporal sequences of stimuli, local adaptation, and physiological nystagmus. Note that these are dynamic phenomena; they play a significant role functionally. Structurally, diverging projection fibers between retina and cortex would seem to prohibit high acuity if it were not for lateral interaction and temporal structures in most natural stimuli.

On the basis of Lorente de No's observations on overlap (lateral interaction) of neighboring projection fields at the higher centers and on facilitation by volleys of firing, Talbot and Marshall [5.22] hypothesized peaking and sharpening mechanisms that may be of great functional significance. Statically, the reciprocal overlap causes peaking of the spatial distribution of neural activity. This is shown schematically in Fig. 5.14. If individual activity fields in fact are submaximal, spatial integration in the center area may cause maximal response, in a localized area, of a few cells only. This lateral integration causing peaking is useful in pattern vision; contour gradients thus can be sharpened if receptive fields of peripheral units are small (fovea). In the periphery, it serves to integrate intensity from larger numbers of ganglion cells. There is additional sharpening of excitation distributions due to (temporal) facilitation. The central units in the middle region (Fig. 5.14) receive stimuli more frequently. Talbot and Marshall [5.16], for example,

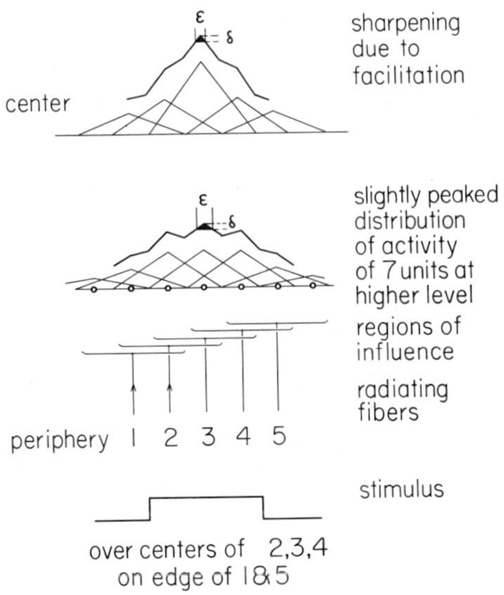

Fig. 5.14. Mutual lateral overlap of projection regions of peripheral ganglion cells (bottom) causes sharpening of wide or blurred patterns. At higher levels, particularly if temporal firing sequences (e.g., due to physiological nystagmus) are considered, sharpening is considerable. For a given threshold δ of central units (top of figure), the number (proportional to range ε) of units stimulated can be very small, although the image spreads over relatively large ranges at the retina. The scales used are arbitrary; however, the relative degree of peaking (due to facilitation by rapid sequences of stimuli) assumed in the top figure approximately corresponds to observations. After Marshall and Talbot [5.16].

showed that successive bombardment of lateral geniculate units by weak signals within 5 to 10 msec increases excitability by a factor as large as 2.5. This represents a *vertical* factor in peaking (Fig. 5.14) with the effect that functionally, and only in a dynamic sense, there may exist a cortical mosaic much finer than that at the retina.

It is known that the vertical amplification also results in faster response (shorter latency) of central units. Thus if acuity is lost due to convergence within the retina, it is regained by reciprocal overlap multiplication of pathways and the vertical component of peaking at higher levels. We therefore speak of a neural lens refocusing the "smeared" neural pattern of activity at the optic nerve. Its function depends greatly on temporal fluctuations, partly due to statistical variations in discharge rate and partly due to physiological nystagmus.

As we have seen, variations near the edge of a receptive field may cause large signals (Fig. 5.12). Facilitation of neighboring units may therefore really be strong. Interestingly, although thresholds of higher level units are lowered by rapid successions of inputs, there later exists a supranormal threshold. This compresses the dynamic range. Weak signals are amplified; strong ones less so (cf. De Valois' [5.18] observations, quoted previously). Talbot and Marshall used very similar arguments for explaining the extremely high foveal acuity such as is experienced when viewing a star. Visual targets which are point sources first are "smeared" owing to nonideal optics and overlapping receptive fields of retinal ganglion cells. Second, their images on the retina are constantly shifted because of physiological nystagmus. This not only prevents image fading (adaptation, bleaching) but, by its dynamic nature, elicits both vertical and lateral amplification and the sharpening discussed earlier. Thus fine spatial discrimination is possible, although the sensory organ per se would never allow it.

C. VISUAL PERCEPTION: INFORMATION PROCESSING AND PSYCHOPHYSICS

It has been demonstrated several times in previous chapters that all system analyses require not only knowledge of the properties of subunits of a system, their connectivity, and their interactions, but also understanding of couplings between the system and its surroundings. Thus when analyzing behavior and performance of the whole organism, we must consider as "input channels" the sensory systems that establish rapport with the outside world by extracting information important to the organism, and then cause sensations ultimately leading to reactions. In fact, sensations are no longer considered to be "private mentalistic entities" [5.1, 5.23], not amenable to objective analysis. Psychometric studies have allowed organization of sensations into orderly, quantitative schemes, playing an important role in psychology. Furthermore, basic studies in the theory of measurements have been stimulated, perhaps because "things" have been quantitated in psychophysics which rarely were measured before [5.23, 5.24]. Here, the operational approach discussed in general respects in Chapter 1 contributed a great deal to understanding possibilities and limitations in psychophysics. In this light, sensations are considered to be "constructs that derive their meaning from (verbal) reactions of the organism to stimuli" [5.23].

Many psychophysical laws have been established since investigators learned that a test person can respond not only to "just noticeable differences" in stimuli but also, and more important, to intervals and to ratios of sensory magnitudes [5.24]. A test person also can respond by assigning numbers to sensory magnitudes. In short, what S. S. Stevens calls the "manageable" measurements are: absolute threshold, differential threshold, equality, order, equality of intervals or of ratios, and stimulus rating [5.1]. This pertains to the so-called prothetic continua [5.24] involved when sensory intensities are judged and measured (e.g. brightness or loudness). In addition, metathetic continua, involving other attributes of sensory response such as color or pitch, also can be quantitated. One of the problems often encountered in the latter case concerns the search for independent attributes of sensations (see Chapter 6). Generally, the definition of stimulus is *the* problem in psychophysics. From the viewpoint of information processing in the nervous system, we would search for the characterization of all transformations of the input data causing particular, constant responses.

In the previous section we presented examples of neurophysiological phenomena that explain, in part, types of information extraction and processing found in the visual system: the dynamics of overlapping, diverging fields of projection between retina and cortex, the color-contrast or spectral-imbalance sensing units in lateral geniculate bodies, and the power law of frequency modulation by which brightness is coded at the same level. These show that it is sometimes possible to point to the differences in processing between metathetic and prothetic continua. In the former, substitution of neural mechanisms or of excitation is important; in the latter, additive mechanisms may be engaged [5.24]. As we learn in the following section, however, analysis of the overall performance can be carried out at the level of the entire visual system, regardless of neurophysiological processes. This must be done when the behavior of the system as a whole is of prime concern: as a pattern recognizer or as a detector of brightness variations, for example. Such system aspects are important when we study man–machine interactions or behavior within natural surroundings.

1. Absolute Threshold and Some Remarks on Adaptation

It is well known that the visual system retains its astonishing pattern-recognition ability over a tremendous range of background intensities [ca. 90–100 dB above 1 mμ L (lambert)]. Thus phenomena of light- and dark-adaptation, brightness discrimination, and so on always have attracted the interest of physiologists [5.1]. The absolute threshold of vision is extremely low and close to the physical limit of one light quantum. Obviously, then, for studies near threshold, at very low light intensities, the quantal nature of light must be taken into account [5.25]. The energy of light E at absolute threshold, in the (blueish-green) band of maximal sensitivity, is of the order of 6×10^{-10} erg. This equals $nh\nu$ (n – number of quanta, h = Planck's constant = 6.6×10^{-27} erg-sec, ν = frequency = 10^{15} Hz), hence

$$n = \frac{E}{h\nu} \simeq 100 \qquad \text{number of quanta, at cornea, at threshold}$$

Since approximately 50% of the light energy is absorbed in the eye before reaching the retina, and since pigment cells cover only approximately 20% of the retina, the sensation of vision may be elicited by as few as 10 light quanta. From careful tests, with completely dark-adapted eyes and small and rapid light stimuli, it has become clear that the absolute threshold in fact is lower. (In these tests, the stimulus was a monochromatic light pulse, ca. 10′ wide and 0.02–0.1 sec in duration, projected to a spot approximately 7° off the fovea.)

The quantal nature of light forces us, however, to consider statistical fluctuations in stimuli of this type. Given that m quanta are emitted on the average, we can predict the probability that r quanta exist in any one light pulse of duration Δt from the Poisson probability distribution:

$$p(r) = e^{-m}\frac{m^r}{r!} \tag{5.5}$$

where m = average number of quanta per light pulse at retina

$p(r)$ = probability that r quanta are present in any one light pulse at retina

This probability mass function applies to many other random processes; it also determines the number of gas molecules in a subvolume, of particles emitted by a radioactive source, or of cars going through a tollgate, per unit time (see Fig. 5.15). As a rule it applies to populations of independent phenomena.

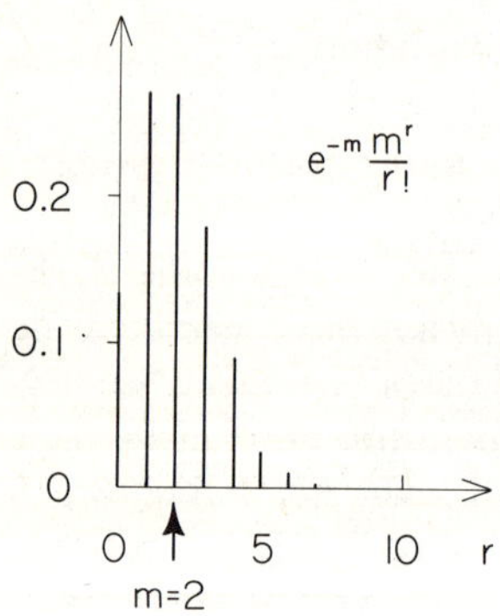

Fig. 5.15. The Poisson distribution. The sample space for the Poisson distribution is the set of nonnegative integer numbers. The distribution function is defined by the formula given. It may apply to phenomena such as the number of counts by a Geiger counter within a given period of time, or to the number of white cells on a microscope slide within a certain small area on the slide. The Poisson process plotted has, on the average, two occurrences (per time). Its standard deviation is $m^{1/2} = 2^{1/2}$.

If at least k quanta are needed to elicit light sensation, the probability that threshold is reached by any one stimulus is

$$p(r \geqslant k) = 1 - p(r < k) = 1 - \sum_{s=0}^{s=k-1} e^{-m}\frac{m^s}{s!} \tag{5.6}$$

Or, if the average number of quanta in the stimulus entering the eye is n, and if the fraction f thereof reaches the retina, we have,

$$p(r \geqslant k) = 1 - e^{-(fn)} \sum_{s=0}^{s=k-1} \frac{(fn)^s}{s!} \tag{5.7}$$

Figure 5.16 shows these (accumulated) probabilities as a function of fn or m for different k's. It also shows Bouman and van der Velden's [5.25] experimental frequency-of-seeing curve. Its shape compares quite well with the theoretical curve of $k = 2$, and f indeed is close to 0.2.

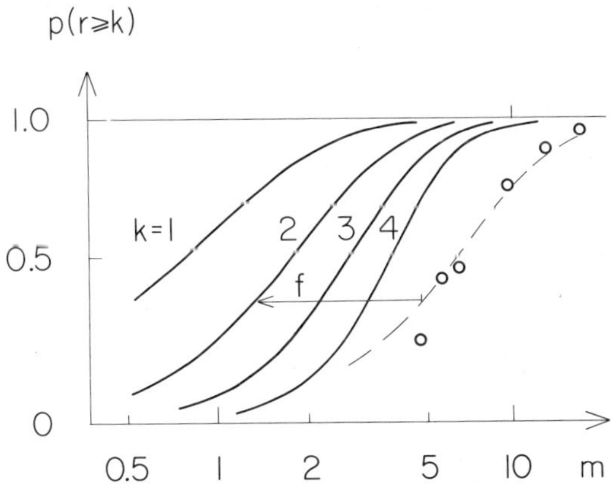

Fig. 5.16. The probability that in a short flashing spot of light, with an average number of quanta m, at least k quanta exist, is drawn for $k = 1, 2, 3$, and 4. The frequency-of-seeing curve, approximating the experimental observations (circles) is fitted quite well by the curve $k = 2$ when shifted by the factor $f \simeq 0.2$. In fact, this f corresponds well to the expected loss of light energy within the eye. After Bouman [5.25].

The outcome of experiments involved in establishing the frequency-of-seeing curve depends on the training and motivation of the test person. Probably for this reason, the analysis of frequency-of-seeing curves is still controversial. It has, however, stimulated a great amount of work in the theory of detection of signals in noise [5.26]. In this particular case it is very unlikely that spontaneous decomposition of rhodopsin creates any noise [5.25]. In fact, when the absolute threshold is investigated by different methods, as described below, we not only can corroborate the two-quanta threshold for light sensation, but also explain two "classical" psycho-physical laws on the basis of the particular threshold condition and the statistics of quanta impinging on the retina.

The curves in Fig. 5.16 base on very narrow stimuli of very short duration and variable intensity. Threshold properties also can be obtained by determining sensation as a function of light pulse duration while spatial spread and intensity are kept constant, as van der Velden pointed out. Assume again that light is sensed when k quanta reach the retina within a time τ. If stimulus duration T is smaller than τ, k quanta are received according to the Poisson distribution. If $T > \tau$, the probability of k quanta received, at least once, within τ must be computed. This of course is a function of m. It can be shown that the number of quanta for 60% frequency (m_{60}) of vision shows the following dependence on pulse duration:

$$m_{60} \propto \left(\frac{T}{\tau}\right)^{(k-1)/k} \qquad T > \tau = 0.1 \text{ sec} \tag{5.8}$$

This function is shown, for different k's, in Fig. 5.17. Clearly, when $k = 2$,

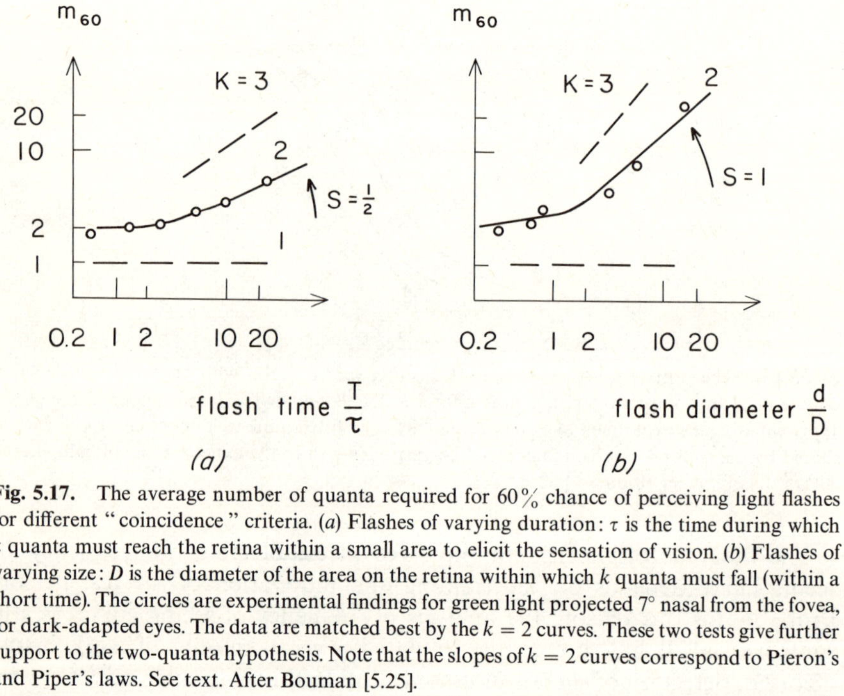

Fig. 5.17. The average number of quanta required for 60% chance of perceiving light flashes for different "coincidence" criteria. (*a*) Flashes of varying duration: τ is the time during which k quanta must reach the retina within a small area to elicit the sensation of vision. (*b*) Flashes of varying size: D is the diameter of the area on the retina within which k quanta must fall (within a short time). The circles are experimental findings for green light projected 7° nasal from the fovea, for dark-adapted eyes. The data are matched best by the $k = 2$ curves. These two tests give further support to the two-quanta hypothesis. Note that the slopes of $k = 2$ curves correspond to Pieron's and Piper's laws. See text. After Bouman [5.25].

the threshold energy for relatively long stimuli (of small size—10′) varies with the square root of exposure duration. This is also found experimentally, as originally shown by Pieron, after whom this law is named.

Analogous considerations hold for short stimuli, spread over larger retinal areas. In this case we find

$$m_{60} \propto \left(\frac{d^2}{D^2}\right)^{(k-1)/k} \qquad d > D = 10' \tag{5.9}$$

where d = diameter of stimulus area
D = diameter of stimulus area causing certain light sensation

From this it follows that for stimuli of short duration, threshold energy increases proportionately to the diameter of the area exposed on the retina. This is Piper's law.

On the basis of the foregoing relations, Bouman [5.25] separated quantitatively the different mechanisms involved in *adaptation*. The threshold of vision immediately after prolonged exposure to bright light is much higher than the value mentioned earlier, particularly if the area exposed on the retina is large (ca. 3×10^4 quanta for a 5°30′ size stimulus versus 10^2 quanta for an 11′ stimulus). The theory introduced here allows us to conclude that changes must occur in the size of the receptive fields, in addition to pigment concentration and quantum effects. Assume that we deal with short stimuli of approximately 20° in size on the retina and that in the light-adapted eye the minimal number of quanta increases from 2 to k', then, with $d/D = 100$, the ratio of threshold energies becomes

$$\rho = \frac{\Delta E_{\text{adapted}}}{\Delta E_{\text{abs}}} = \frac{k'}{2} \frac{(10^4)^{(k'-1)/k'}}{(10^4)^{1/2}} \tag{5.10}$$

where $\Delta E_{\text{adapted}}$ and ΔE_{abs} are threshold energies of the light-adapted and of the dark-adapted eye (absolute threshold), respectively.

Assume that $k' = 10$, then $\rho \simeq 200$ (for large target sizes), due to quantum effects. Bouman estimates that adaptation due to "bleaching" or regeneration of photopigment contributes a factor of approximately 30. In fact, it may be lower. A factor of 6000, however, is still less than is seen experimentally. Thus neural effects changing the effective size of receptive fields by a factor 2 to 3 (4–9 in area) suffice to bring about adaptation factors of the order of 3 to 4×10^4. The spatial neural reorganization, together with the "k-adaptation," therefore provides the visual system with a mechanism that can adjust field size and threshold per recipient unit such that energies eliciting sensation are minimal for maximal D and T (stimulus size and duration). The above-mentioned change in k must be caused by neural mechanisms, too. We refer the reader to Dowling's paper [5.27] for a discussion of possible mechanisms.

2. Thresholds in Pattern Vision

The *static* transfer characteristics for a great number of stimuli on the prothetic continuum have been found to be governed by Stevens's law:

$$\psi = k(\phi - \phi_0)^n \tag{5.11}$$

where ψ = psychological magnitude
ϕ = (physical) magnitude of stimulus
ϕ_0 = threshold stimulus

The exponent n is smaller than unity, especially in the visual and auditory systems: $n = 0.33$ for brightness of 5° target seen by dark-adapted eye; $n = 0.5$ for brightness of point source in dark adaptation. This obviously means compression of the dynamic range. In the chemical senses, especially for taste, and for pain sensation due to electric shock in the finger, n's larger than unity are exhibited. This clearly can represent a protective device.

In most normal behavior, however, the *dynamic* characteristics are much more important. In the following we concentrate on the elegant studies by Kelly [5.28] concerning temporal brightness variation discrimination. This work summarizes several "classical" psychophysical laws and affords a beautiful example of psychophysical methodology and interpretation.

The visual response to time-varying, wide-field stimuli is best measured by modulating sinusoidally the radiance of a target in the form of

$$B(t) = \hat{B}(1 + m \cos \omega t) \tag{5.12}$$

where B = luminance
$\hat{B}$ = average luminance to which eye is adapted
m = modulation factor = $(B_{max} - B_{min})/(B_{max} + B_{min})$
ω = (angular) frequency

Kelly chose a target size of 50° of visual angle. Peripheral to the target, B was made to fall off gradually to avoid spatial transients (edge enhancement). In the experiment the threshold value of m is determined as a function of $\hat{B}$ and ω. This is what might be called a "sine-wave flicker threshold determination." Of course m is a measure of temporal contrast. In the following, B is measured in trolands where 1 troland (td) = retinal illuminance produced when viewing a surface whose luminance is 1 candle/m^2 through an artificial pupil of 1 mm^2 area (centered on natural pupil).

Kelly's observations may be shown in the form of Fig. 5.18 where $(m\hat{B})^{-1}$, the absolute amplitude sensitivity for temporal contrast vision, is plotted against frequency, with adaptation ($\hat{B}$) as a parameter. At low frequencies, $(m\hat{B})^{-1}$ increases as strongly as adaptation level decreases, as is generally observed with sensory sensitivities. It is interesting that sensitivity decreases precipitously above a cutoff frequency (4 decades/decade in frequency). All curves join into a common one along which sensitivity is independent of adaptation level. In fact, above the cutoff frequency, any periodic waveform seems to have the same brightness as a constant stimulus of the same $\hat{B}$. This also holds for stimuli modulated by square waves (i.e., all higher harmonics are filtered out). However, this statement is the Talbot-Plateau law, which

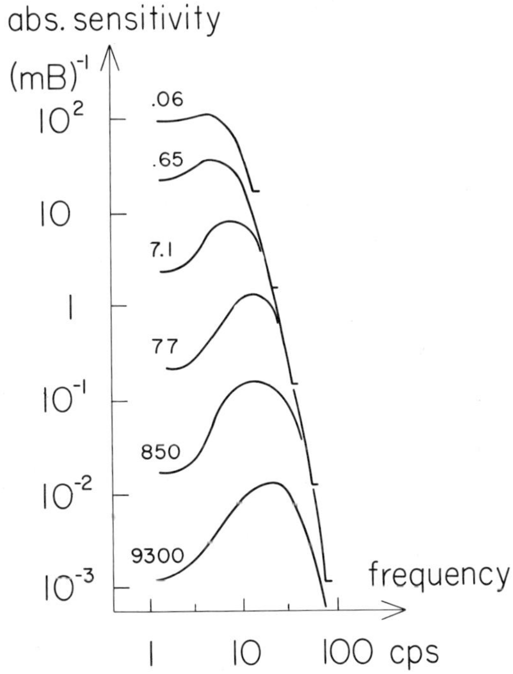

Fig. 5.18. Absolute sensitivity $(m\hat{B})^{-1}$ for detection of temporal fluctuation of light intensity (wide-angle target, with "smooth" edges), as a function of frequency of the fluctuation, for different background intensities (0.06, 0.65, 7.1 td, etc.). The short horizontal lines indicate where 100% modulation is reached. Note that it is impossible to define one unique cutoff frequency. For a particular $(m\hat{B})^{-1}$ there exists a "cutoff" frequency; for example, if $(m\hat{B})^{-1} = 10^{-2}$, $f_c \simeq 50$ Hz. If f is raised just above 50 Hz, a *steady* brightness is perceived corresponding to the time average of the stimulus (provided it is below 850 td, in this case). This is the Talbot-Plateau region, which is seen more clearly in Fig. 5.19. After Kelly [5.28].

usually is encountered in experiments with square-wave stimuli that are matched, above "flicker-fusion frequency" (CFF), by a steady light. From Kelly's data it is clear that the CFF is difficult to define. The transfer function, at a given adaptation level, may be written [5.29] in the Laplace domain as

$$F(s) \propto \frac{as^2 + bs + 1}{(s + k)^4} \tag{5.13}$$

where s = Laplace operator
a, b, k = constants

This function is characterized by poles and zeros rather than by one cutoff frequency. Kelly [5.29] added to the model of eq. 5.13 a second stage introducing adaptation. In essence it incorporates a low-pass filter whose "3-dB point" is progressively shifted to higher frequencies when mean brightness increases.

It is revealing to consider Kelly's data three dimensionally in the m^{-1}, $\hat{B}$, f space (Fig. 5.19). Then it is seen that m^{-1} is nearly a constant at *low* fre-

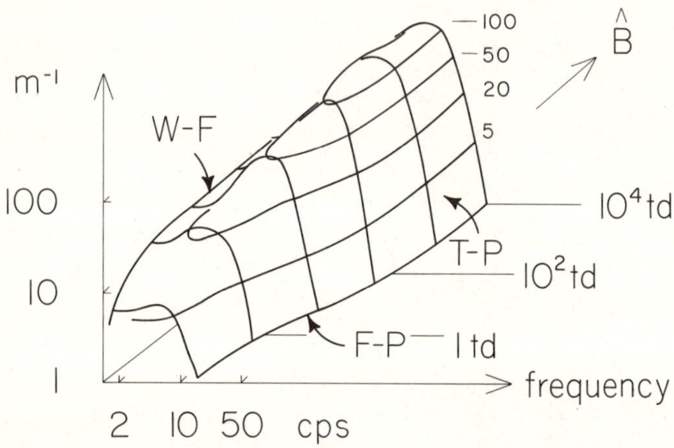

Fig. 5.19. The inverse temporal modulation ratio (m^{-1}) or the relative modulation sensitivity at threshold of perception of brightness fluctuation, as a function of frequency and mean background intensity ($\hat{B}$). The modulation ratio is a measure of temporal contrast. The regions of applicability of three psychophysical "laws" are indicated. (1) At low frequencies, m^{-1} is independent of $\hat{B}$. At frequencies below 2 Hz, m^{-1} becomes independent of $\hat{B}$ if $\hat{B} > 30$ td. This is the Weber-Fechner region, marked *W-F*. (2) In the range between 20 and 60 Hz, the high-frequency threshold is proportional to log $\hat{B}$, for a given m. This corresponds to the Ferry-Porter law for periodic waveforms. This region is marked *F-P*. (3) At high frequencies ($f > 30$), the threshold frequency is independent of $\hat{B}$, the background intensity. This is the Talbot-Plateau (*T-P*) region, as discussed in the text and in the legend of Fig. 5.18. Data are from Kelly [5.28].

quencies (and in the photopic range of operation, i.e. $\hat{B} > 0.1$ td). In other words, the *relative* sensitivity to brightness changes is independent of the adaptation level. This corresponds to the Weber-Fechner law, which also applies to spatial contrast thresholds. At higher frequencies ($20 < f < 60$ Hz) and in the midrange of retinal illuminance ($5 < \hat{B} < 400$ td) we find that frequencies (f_c) at threshold are linearly related to the logarithm of adaptation level. Formally,

$$f_c = k + \log \hat{B} \tag{5.14}$$

where k is a constant.

This holds in the range of stimuli delineated earlier, assuming that m is held constant and that $1 < m^{-1} < 50$. This corresponds exactly to the law of Ferry and Porter, originally determined by quite different methods. Thus

Kelly found his data conforming to three "classical" laws of visual psychophysics. Moreover, he was able to define the limits of applicability of these laws.

The *spatial* modulation threshold properties must be determined if acuity, or resolution, in pattern vision is to be defined. Acuity usually depends on the pattern observed [5.1]. However, in the following we summarize the findings by Van Nes and Bouman [5.30] on the spatial modulation transfer function of the eye, which they determined using a "striped" pattern whose intensity varied sinusoidally in one direction only, that is,

$$B(x, y) = \hat{B}(1 + m \cos \omega x) \tag{5.15}$$

The field of vision was 4.5° × 8.25°, three different colors were used, and threshold modulation was determined as a function of frequency for different $\hat{B}$'s. When data are corrected for chromatic aberration and for effects of λ-dependent diffraction on modulation transfer, contrast sensitivity functions are the same for all colors [5.30]. The findings are summarized in Fig. 5.20, where the spatial modulation sensitivity m^{-1} is plotted, as in Fig. 5.19, against frequency and retinal illumination (i.e., adaptation) level. Frequency

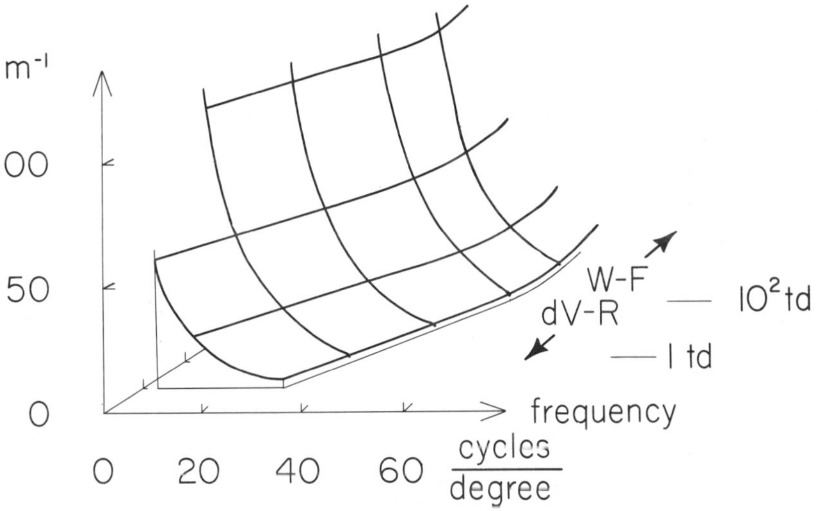

Fig. 5.20. The inverse spatial modulation ratio (m^{-1}), a measure of contrast sensitivity, plotted as a function of spatial frequency in cycles per degree, and background illumination (i.e., adaptation) level, in trolands. Two main regions of performance are seen. At low $\hat{B}$ (i.e., below 300 td), m^{-1} is proportional to $B^{1/2}$. This is the deVries-Rose law. At higher mean illumination (i.e., for $\hat{B}$ > 300 td), m^{-1} is a constant, dependent on frequency only. This is the Weber-Fechner law. The value of m^{-1} decreases rapidly with increasing frequency; between 20 and 40 cycles/degree it drops from 180 to 25, and reaches 2 at 55 cycles/degree. After Van Nes and Bouman [5.30].

now is given in cycles (or "stripes") per degree of visual angle. There are two main regions in this diagram. First, at $\hat{B}$ below approximately 300 td, m decreases proportionally to $\hat{B}^{1/2}$ at a fixed frequency. Most of the surface in Fig. 5.20 belongs to this domain. (Lines of constant f are not shown; the straight lines of constant m^{-1}, in the horizontal planes, indicate a linear dependence of $f_{\text{threshold}}$ with $\hat{B}$.) This behavior corresponds to the de Vries-Rose law. At intensities above 300 td, m remains constant independently of $\hat{B}$. This situation was encountered before: it is the region where the Weber-Fechner law applies.

In summary, this chapter presents a few illustrative examples of processes and mechanisms in the visual system, beginning with retinal organization and its role in transforming optical into neural patterns, and ending with a few aspects of the overall response of the system tested at the psychometric level. Many other important areas of study were omitted: obviously, the visual system conveys *directional* information about targets. The relative orientation between viewer and target and its change are important (e.g., when reaching for or avoiding objects). *Topological* information on relative orientation of different targets in space and on structure of the environment also must be transferred to the organism. *Learning* and plasticity or task adaptation are other topics not discussed here. Although considerable knowledge [5.1] is available on some of these tasks fulfilled by vision, little is known on mechanisms. Thus we concentrated the discussion on some aspects of *pattern feature* extraction, *contrast coding* and enhancement, *moving target* sensing, and color contents processing. We are far from being able to present a complete "theory" of the visual system, but the few examples selected from the tremendous amount of vision research should indicate how quantitative methods can contribute to our understanding.

D. REFERENCES

5.1. Stevens, S. S. (Ed.). *Handbook of Experimental Psychology*. John Wiley and Sons, New York, 1951.

5.2. Field, J. and H. W. Magoun (Eds.). *Handbook of Physiology, Section 1: Neurophysiology*, Vol. 1. American Physiological Society, Washington, D.C., 1959.

5.3. Jones, R. W., D. G. Green, and R. B. Pinter. Mathematical simulation of certain receptor and effector organs. *Fed Proc* **21**: 97–102, 1962.

5.4. Levine, S. N. Enzyme amplifier kinetics. *Science* **152**: 651–653, 1966.

5.5. Marks, W. B., W. H. Dobelle, and E. F. MacNichol, Jr. Visual pigments of single primate cones. *Science* **143**: 1181–1183, 1964.

5.6. Polyak, S. *The Retina*. University of Chicago Press, Chicago, 1941.

5.7. Dowling, J. E. and B. B. Boycott. Organization of the primate retina: electron microscopy. *Proc Roy Soc* [*Biol*] **166**: 80–111, 1966.

5.8. Kuffler, S. W. Discharge patterns and functional organization of mammalian retina. *J Neurophysiol* **16**: 37–68, 1953.

5.9. Lettvin, J. Y., H. R. Maturana, W. S. McCulloch, and W H. Pitts. What the frog's eye tells the frog's brain. *Proc IRE* **47**: 1940–1951, 1959.

5.10. Maturana, H. R. and S. Frenk. Directional movement and horizontal edge detectors in the pigeon retina. *Science* **142**: 977–979, 1963.

5.11. Barlow, H. B., R. M. Hill and W. R. Levick. Retinal ganglion cells responding selectively to direction and speed of image motion in the rabbit. *J Physiol (London)* **173**: 377–407, 1964.

5.12. (a) Dowling, J. E. and F. S. Werblin. Organization of retina of the mudpuppy, *Necturus Maculosus*. I. Synaptic structure. *J Neurophysiol* **32**: 315–318, 1969. (b) Werblin, F. S. and J. E. Dowling. Organization of the retina of the mudpuppy, *Necturus Maculosus*. II. Intracellular recording. *J Neurophysiol* **32**: 339–355 1969.

5.13. Bishop, P O. Central nervous system: Afferent mechanisms and perception. *Ann Rev Physiol* **29**: 427–484, 1967.

5.14. (a) Rodieck, R. W. and J. Stone. Response of cat retinal ganglion cells to moving visual patterns. *J Neurophysiol* **28**: 819–832, 1965. (b) Analysis of receptive fields of cat retinal ganglion cells. *J Neurophysiol* **28**: 833 849, 1965.

5.15. Rodieck, R W Quantitative analysis of cat retinal ganglion cell response to visual stimuli. *Vision Res* **5**: 583–601, 1965.

5.16. Marshall, W. H. and S. A. Talbot. Recent evidence for neural mechanisms in vision leading to a general theory of sensory acuity. *Biol Symp* **7**: 117–161, 1942. Also pp. 195–211 in *Pattern Recognition*. L. Uhr (Ed.). John Wiley and Sons, New York, 1966.

5.17. Wiesel, T. N. and D. H. Hubel. Spatial and chromatic interactions in the lateral geniculate body of the rhesus monkey. *J Neurophysiol* **29**: 1115–1156, 1966.

5.18. De Valois, R. L., G. H. Jacobs, and A. E. Jones. Effects of increments and decrements of light on neural discharge rate. *Science* **136**: 986–987, 1962.

5.19. (a) Hubel, D. H. and T. N. Wiesel. Receptive fields, binocular interaction and functional architecture in the cat's visual cortex. *J Physiol (London)* **160**: 106–154, 1962. (b) Receptive fields and functional architecture of monkey striate cortex. *J Physiol (London)* **195**: 215–243, 1968.

5.20. Wurtz, R. H. Visual receptive fields of striate cortex neurons in awake monkeys. *J Neurophysiol* **32**: 727–742, 1969.

5.21. Jung, R. Neuronal integration in the visual cortex and its significance for visual information, pp. 627–674 in *Sensory Communication*. W. A. Rosenblith (Ed.). John Wiley and Sons, New York, 1961.

5.22. Talbot, S. A. and W. H. Marshall. Physiological studies on neural mechanisms of visual localization and discrimination. *Amer J Ophthalmol* **24**. 1255–1263, 1941.

5.23. Stevens, S. S. Measurement and man. *Science* **127**: 383–389, 1958.

5.24. Stevens, S. S. The psychophysics of sensory function, pp. 1–34 in *Sensory Communication*. W. A. Rosenblith (Ed.). John Wiley and Sons, New York, 1961.

5.25. Bouman, M. A. History and present status of quantum theory in vision, pp. 377–402 in *Sensory Communication*. W. A. Rosenblith (Ed.). John Wiley and Sons, New York, 1961.

5.26. Green, D. M. and J. A. Swets. *Signal Detection Theory and Psychophysics*. John Wiley and Sons, New York, 1966.

5.27. Dowling, J. E. The site of visual adaptation. *Science* **155**: 273–279, 1967.

5.28. Kelly, D. H. Visual responses to time-dependent stimuli. I. Amplitude sensitivity measurements. *J Opt Soc Amer* **51**: 422–429, 1961.

5.29. Kelly, D. H. Visual responses to time-dependent stimuli. II. Single-channel model of the photopic visual system. *J Opt Soc Amer* **51**: 747–754, 1961.

5.30. Van Nes, F. L. and M. A. Bouman. Spatial modulation transfer in the human eye. *J Opt Soc Amer* **57**: 401–406, 1967.

CHAPTER SIX

Auditory Information: Receptor, Intermediate, and Psychophysical Systems

A. TRANSMISSION OF SOUND THROUGH THE MIDDLE EAR

Analysis of the performance of the middle ear is based in part on the geometry of the conelike tympanic membrane, the malleus–incus complex, and the stapes. However, it often is difficult, if not impossible, to infer dynamic performance and function from structure. Information pertaining to structure is inherently static and should be supplemented by direct observations of sound transmission and impedances. These dynamic measurements, when obtained from abnormal ears, allow elegantly for validation of parts of the model. In spite of this, the model discussed in this chapter must remain of limited applicability (i.e., useful at frequencies below 1500 Hz only). This is partly due to representation of the eardrum by a lumped parameter system —that is, an elastically suspended piston. The motion of the tympanic membrane, in fact, contains a rotational component, as well as complex modes of oscillation at higher frequencies [6.1]. These are not fully taken into account. Moreover, the properties of the middle ear cavities are considered at low frequencies only. In this frequency range, details in distribution of pressures and velocities within the cavities are not thought to influence transmission of vibratory energy to the oval window of the cochlea. Concepts of "lumped-parameter" network theory can therefore be applied. Input impedance, transfer function, and impedance transformation, all functions of frequency, are *functionally* meaningful.

An electroacoustic analogy is employed through the following. The analogous quantities are listed in column 4 of Table 6.1., which summarizes some of the electromechanical correspondences. In this section pressure and

ELECTRICAL Variables and Elements	MECHANICAL ANALOG System 1	MECHANICAL ANALOG System 2	ACOUSTIC ANALOG
voltage e	force f	velocity v	pressure p
current i	velocity v	force f	volume flow $\dot{q}$
capacitor $C = \frac{\int i\,dt}{e}$	$\frac{\int v\,dt}{f} = k$ spring const.	$\frac{\int f\,dt}{v} = m$ mass	$\frac{\int \dot{q}\,dt}{p} = \frac{\Delta Vol}{\Delta p}$ = compliance
inductance $L = \frac{e}{di/dt}$	$\frac{f}{dv/dt} = m$ mass	$\frac{v}{df/dt} = \frac{1}{k}$ spring const. (inverse)	$\frac{p}{d\dot{q}/dt} = \frac{\text{mass}}{\text{Vol.} \times \text{length}} \rightarrow L' = \frac{\rho}{A} X$ inertance of mass in conduit of length X and lumen area A
resistance $R = \frac{e}{i}$	$\frac{f}{v} = \mu$ dashpot const.	$\frac{v}{f} = \frac{1}{\mu}$ dashpot const. (inverse)	$\frac{p}{\dot{q}} = \frac{\text{mass}}{\text{time} \times (\text{length})^4} \rightarrow R' = \frac{8\pi\eta}{A^2} X$ resistance to flow of conduit of X and A; η = viscosity

TABLE 6.1. SUMMARY OF ELECTROMECHANICAL CORRESPONDENCES

volume flow velocity are made analogous to voltage and current, respectively. Hence an inductance represents a mass divided by the square of the area across which the forces are coupled to it. A capacitor may represent the distensibility of a structure such as the tensor tympani muscle, or the compliance of an air chamber. In the latter case, compliance is defined as $\Delta V/\Delta p$ (V = volume, p = pressure). Its value is obtained from the velocity of sound c (at atmospheric pressure and body temperature) which is

$$c = \left(\frac{\Delta p V}{\Delta V \rho}\right)^{1/2} \tag{6.1}$$

where ρ = density of air; hence

$$\frac{\Delta V}{\Delta p} = \frac{V}{\rho c^2} \tag{6.2}$$

where V = volume of cavity.

The unit of resistance is 1 (dyne/cm^2)/(cm^3/sec) and is analogous to 1 Ω in the electric domain. Generally, topology of the analogous electric network is different from that of the mechanical system. For example, rates of change in length of a viscoelastic structure, such as the tensor tympani muscle, are the same for both its elastic and viscous elements, since these are in parallel mechanically. In the electric analog, however, the corresponding capacitor and resistor must be in series, and therefore the current passing through one equals the current through the other.

The input signal to the middle ear structure can be considered the same as the auditory stimulus at the inlet to the external auditory canal. Wiener and Ross [6.2] obtained the transfer characteristics of the canal, which strictly should be treated as a section of an acoustic wave guide. Quarter-wavelength resonance effects, possibly enhancing pressure amplitude at the eardrum by approximately 10 dB, show up at frequencies near and above 4000 Hz, which is beyond the band considered here. Definition of the absolute level of sound intensity at the input is more difficult. The intensity at a location in the sound field of a loudspeaker, measured in the undistorted field, changes when a listener is brought into it; this is because of diffraction of sound at the listener's head. The sound level produced by an earphone is equally difficult to obtain, since its load—the input impedance of the external canal—cannot easily be simulated in a calibration apparatus. We therefore restrict our attention to relative transmission characteristics.

The acoustic load of the middle ear structure (i.e., the input impedance of the cochlea) affects the transfer function and the input impedance of the ossicular chain. Strictly speaking, then, we should develop the theory of sound transmission in the cochlea first, as Zwislocki [6.3] has done. On the other hand, the segmental development of the network representing the middle ear system can be carried to the level of the incudo-stapedial joint without knowledge of the load. Performance of the entire system can be optimized by adjusting the load until the input impedance equals that found experimentally. This optimal value of the load can then be compared with the value predicted by the cochlear theory, the subject of the next section in this chapter.

Before we analyze the transmission of sound through the ossicular chain, as affected by the compliances of the cavities and the properties of the various viscoelastic structures, we should note that direct transfer of sound through the air within the middle ear cavity is not considered, since it is minimal. Bone conduction around the middle ear is not discussed here either, since it is not of importance in normal hearing of air vibrations at low intensities. Von Békésy discussed his measurements on bone conduction in several places (e.g., [6.4]).

1. Dynamic Properties of the Middle Ear Structures

Pressures acting on the tympanic membrane are balanced by viscoelastic forces within this structure, inertial forces due to the mass of the malleus, and the pressures within the middle ear cavities. The latter include the tympanic cavity behind the eardrum, connected by a narrow passage to the epitympanum and the antrum with the pneumatic cells. The rate of change in volume in these cells is given by the sum of the rates of change in displacements of the two regions of the eardrum: one coupled to the malleus, the other directly compressing the air in the cavities. Thus we can schematize the analogous electrical network (see Fig. 6.1) in which the total input current is divided into components flowing into both the impedances representing these sections of the eardrum.

The network elements of the *middle ear cavities* can be assigned numerical values using two approaches. The values of the capacitors C_a and C_t can be computed from eq. 6.2 if the volumes of the respective air pockets are known. Zwislocki [6.5] measured the volumes either directly by filling the cavities with water or indirectly by determining resonance frequencies. He concluded that it is possible, with reasonable accuracy, to put $C_a = 5.1\ \mu\text{F}$, $C_t = 0.35\ \mu\text{F}$, where $1\ \text{F} = 1\ \text{cm}^3/(\text{dyne/cm}^2)$. Indirect determinations yield values for the inductance L_a, representing the mass of air within the passage between tympanic cavity and the antrum, R_a, the resistance to flow of air through the same passage, and R_m, the resistance accounting for absorption of sound at the walls of all cavities. Onchi [6.6] directly measured impedance functions of the cavities, so that L_a can be computed from the known, fundamental resonance frequency, and the R's can be adjusted to fit the observed phase behavior of the measured impedance. We find, for example, that $L_a = 14\ \text{mH}$.

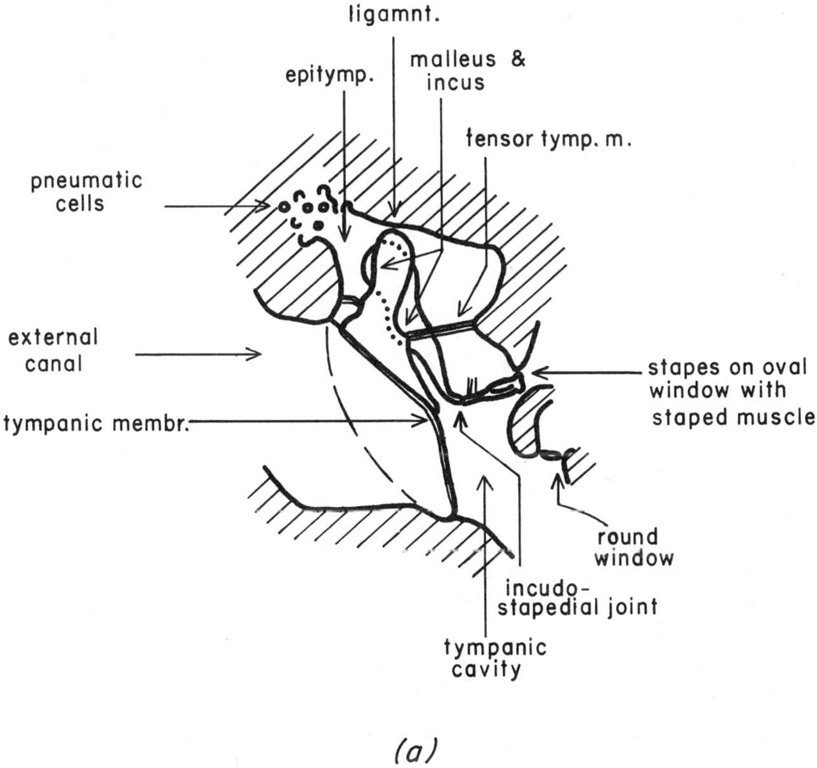

Fig.6.1*a*. Schematic view of middle ear structures. At low frequencies the tympanic membrane moves like a piston, thus turning the malleus–incus complex about an axis that is approximately perpendicular to the plane of the paper. The stapes then is moved very slightly in and out of the oval window, or it is slightly rocked about its position. This description probably is very much simplified, since the axes of rotation of both incus and stapes are not necessarily fixed in space. The tonus of the tensor and stapedius muscles, among others, influences the motions. (For a detailed description of these motions see, e.g., Stuhlmann [6.8].) Note that viscosity and compressibility of the air within the middle ear spaces represent dynamic loads on the system.

As mentioned earlier, the *tympanic membrane* is treated in two sections. Approximately 30% of the total area of the eardrum is moving relatively freely (total area − 0.85 cm^2). The remainder is intimately coupled to the malleus. The smaller area is considered, at low frequencies, as a viscoelastic structure represented by C_{d1} and R_{d1} (Fig. 6.1*b*). At higher frequencies, near and above 900 Hz, inertial forces must be taken into account; L_d represents the effective mass. Von Békésy [6.7, also in 6.4] observed more complex modes of vibration above 2400 Hz. Thus the range of frequency in which simple motion can be assumed is close to that in which a distributed-element model would be applicable. Zwislocki [6.5] therefore compromised by intro-

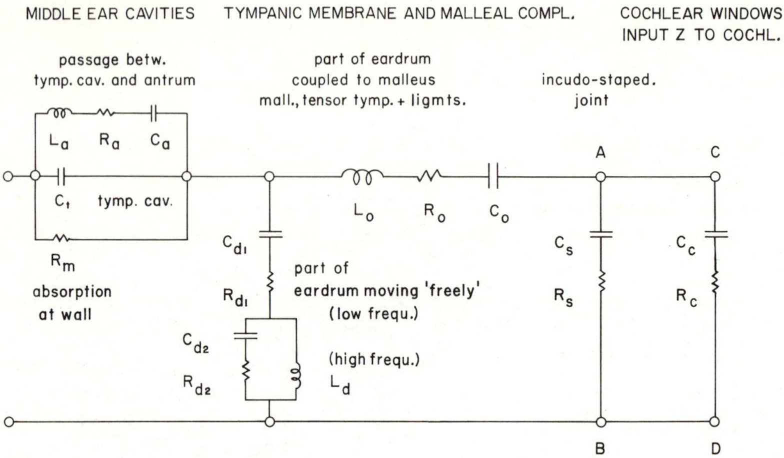

Fig. 6.1*b*. Electroacoustic analog of middle ear structures, after Zwislocki [6.5]. Note the different topology of this network compared with the physical structures; see text for explanation. Some of the elements can be determined in abnormal ears. For example, there exists a short circuit between points *A* and *B* in ears without incus. Immobilized stapes cause an open circuit to appear between points *C* and *D*. Numerical values are given in the text.

ducing the R_{d2}–L_d–C_{d2} circuit of Fig. 6.1*b*. The other part of the eardrum drives the malleus which, in turn, is constrained by the three ligaments holding it in position and the tensor tympani. These structures can be represented by the following three mechanical elements arranged in parallel: the combined effective masses of the malleus and approximately half of the eardrum, a spring for the elasticity of parts of the eardrum and the muscular systems, and a dashpot accounting for viscous elements in the suspension. In the electric network, we therefore need L_0, C_0, and R_0 in series connection as in Fig. 6.1*b*. Numerical values of the elements may be achieved by methods similar to those chosen for the cavities. Measurements of the input impedance of ears without incus yield resonance frequencies determined, essentially, by all *L*'s and *C*'s in the circuit. Furthermore, the impedance at low frequencies is given by C_{d1} and C_a in parallel. These two sets of information, together with the phase of the impedance at low frequencies, enabled Zwislocki [6.5] to obtain the following values. Note that there exists an electric short circuit (zero voltage) between points *A* and *B* in Fig. 6.1*b* if the malleus is moving freely (no force exerted):

$$
\begin{array}{lll}
C_{d1} = 0.23\ \mu\text{F} & R_{d1} = 40\ \Omega & \\
C_0 = 1.4\ \mu\text{F} & R_0 = 70\ \Omega & L_0 = 40\ \text{mH} \\
C_{d2} = 0.40\ \mu\text{F} & R_{d2} = 220\ \Omega & L_d = 15\ \text{mH}
\end{array}
$$

The input impedance of the entire electric network given so far predicts that of an ear without incus with satisfactory accuracy. Its imaginary part is within 20 Ω (i.e., ca. 2% at 100 Hz and 10% at 500 Hz) of the measured values. The real part also is well matched up to 500 Hz. Near 1000 Hz the results are less satisfactory, but this is partly due to experimental difficulties.

It is interesting to compare the network parameter values for the inductances with what is known about the masses of the structures concerned. On the one hand, we know that the mass of the malleus is approximately 23 mg, including the portion of the eardrum to which it is coupled. On the other hand, the indirectly determined value of L_0 is equivalent to $L_0 \times S^2$ (where $S \simeq 0.55$ cm^2; 70% of the eardrum's surface area), or 12 mg. It is not surprising that this "effective" mass is smaller than that found directly. The malleus is not moving linearly but is rotated about an axis near which most of its mass is concentrated. Since it is feasible to obtain an effective mass that is assumed to move linearly, it is not necessary to compute the rotational moment of inertia or to express the impedance in terms of torque per angular velocity. (Otto Frank, quoted by Zwislocki [6.5], computed the "effective" mass in this latter way and also obtained 12 mg.)

The values of C_s and R_s, providing a path for energy loss at the *incudo-stapedial joint*, are not easily found, although it appears that measurements on otosclerotic ears would yield results directly applicable. In otosclerosis the stapes are fixed, so that the electric network is open-circuited at points *C–D* of Fig. 6.1*b*, and the incudo-stapedial impedance is the only component affecting the network discussed so far. On the other hand, the mass of the incus and its two ligaments, probably rigidly coupled to the malleus, should be added. Furthermore, the action of the stapedius may modify considerably the values of the coupling elements between incus and stapedius. Stuhlmann [6.8] carried out studies with mechanical models of the ossicles and described linear and nonlinear regions of operation. We are trying, however, to obtain average parameter values, useful for a small signal analysis. Zwislocki [6.5] verified that R_s is quite large (3000 Ω), approximately five times the load offered at the oval window. Likewise, C_s was found to be quite large (0.25 μF). Errors entering into these values therefore do not perturb significantly the overall performance.

The remaining network to be discussed includes, strictly, the mass of the *stapes*, the elastance of the ligaments attaching the footplate to the *oval window*, the mass of the fluid in the *vestibule* (between oval window and beginning of cochlea), and the input *impedance of the cochlea*. These could be represented by a series R_c–L_c–C_c circuit. When matching the input impedance of the entire network with that measured at the tympanic membrane of normal ears, however, we realize that the inductance can be neglected. First, we would expect L_c to be quite large because of the small area of the oval window (0.032 cm^2 [6.5]). But it should be remembered as well that we now must introduce the "transformer ratio" of the middle ear, given by the square of the ratio of the "lever arms" of malleus and incus and of the areas of eardrum and oval window. The relatively high inertial forces at the oval window are decoupled from the eardrum, since the leverage of the malleus is about 1.35 times that of the incus and since the effective eardrum area is approximately 18 times that of the oval window. Taking this into account, C_c is found to be of the order of 0.6 μF and R_c is optimized to 600 Ω.

It is worth emphasizing that only the impedance at the oval window is transformed by the "transformer" often mentioned in physiology books. The compliances of both eardrum and middle ear cavities contribute directly to the input impedance (at the tympanic membrane), whose imaginary part therefore is large at low frequencies. Hence it is impossible to describe the entire function of the middle ear in terms of the "impedance transformer."

With the given values for the "load," the *input impedance* of the whole analog circuit is obtained (Fig. 6.2). Satisfactory correspondence is found between the experimental data from the normal human ears and that of the model. It is interesting that the impedance is minimal near 1000 Hz. The

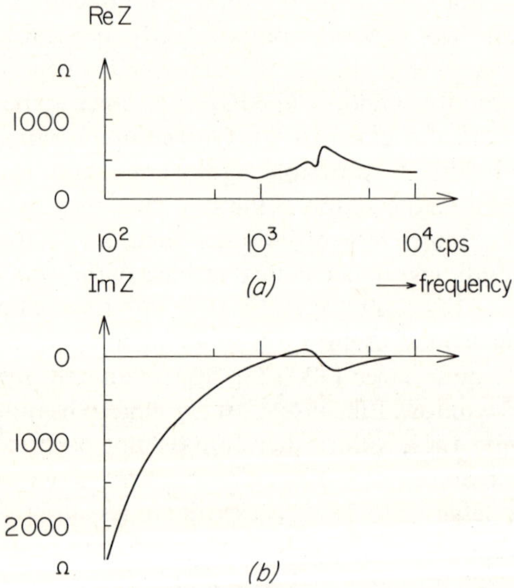

Fig. 6.2. Acoustic input impedance at the eardrum. (*a*) Real part or resistance; (*b*) imaginary part or reactance, both plotted against frequency. At low frequency the reactance is negative, or elastic. The data reported by most investigators lie close to the lines shown: the scatter is ±150 Ω for resistance; ±200 Ω for reactance. The curves were obtained by Zwislocki [6.3] from the electric analog in Fig. 6.1 using the values given in the text. Also see text for a discussion of pressure wave reflections at the tympanic membrane. Here 1 Ω = 1 dyne/cm^2 per cm^3/sec.

reactive part becomes smaller than its real part, which is of the order of 350 Ω. In this frequency range, reflection of sound from the tympanic membrane back into the external canal is also minimal. This effect does not explain fully, but it contributes to, the maximal sensitivity of the entire auditory system at these frequencies. On the other hand, at 100 Hz the input impedance is approximately $(350 - j2500)\Omega$. The characteristic impedance of the external canal is of the order of 100 Ω. Hence the reflection coefficient R', the ratio of retrograde and forward pressure waves, may be as large as

$$R'_{100} = \frac{Z_{\text{in}} - Z_0}{Z_{\text{in}} + Z_0} \simeq 0.985 - j0.08$$

where Z_{in} is input impedance, Z_0 is characteristic impedance, and j is $\sqrt{-1}$, as opposed to the optimum value near 1000 Hz, $R'_{1000} \simeq 0.55$. (See Chapter 14 for a thorough discussion of reflective transmission systems.) It should be realized that knowledge of R' at the eardrum is necessary but not sufficient when defining the stimulus. The geometry of the pinna and of the external canal also must be considered as determinants of the coupling between an external sound field and the auditory system.

2. Transmission of Vibratory Energy

Transmission of sound through the middle ear now can be easily predicted by measuring the transfer function of the model of Fig. 6.1. However, this is not easy to verify in the living ear. Several workers, quoted in [6.5], obtained such responses post mortem. Others used the size of the microphonic potentials in the inner ear as a measure of the transmission. It is qualitatively similar to the model's transfer function. Figure 6.3 shows that relation between volume *displacement* at the stapes and pressure at the eardrum, as

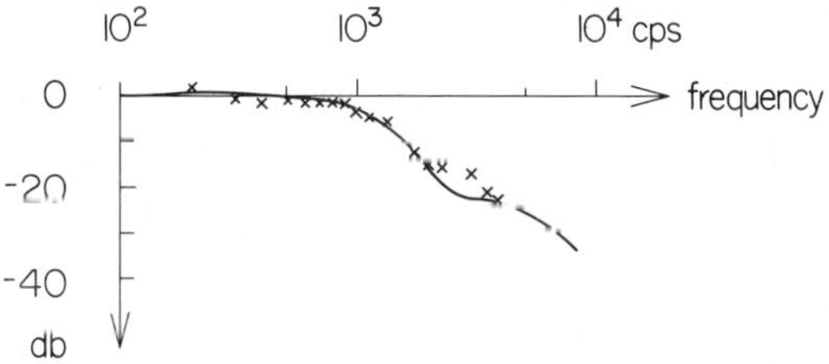

Fig. 6.3. Transmission characteristic of the middle ear. The ratio of the amplitude of volume displacement of stapes to the amplitude of (constant) sound pressure at the tympanic membrane is plotted against frequency. Note that the reference of 0 dB is given by the low-frequency transmission ratio. The crosses are measurements on ear preparations; the solid line is derived from the electric analog. From Zwislocki [6.3].

a function of frequency. The input is assumed to be constant. The amplitude of the output also stays constant up to 1000 Hz, beyond which it falls off by approximately 40 dB/decade. At very low frequencies the transmission is determined by the elastic structures acting on the malleus and at the incudo-stapedial joint. The transformer ratio mentioned earlier, which depends on the lever arms (of malleus and incus) and on the areas (of oval window and eardrum), also has an influence at these low frequencies.

No accurate description has been developed for the action on middle ear function of the *tensor tympani and the stapedius* muscles. These muscles attach to the malleus and the neck of the stapes, respectively, and can rotate these ossicles toward the center of the middle ear cavity. Thus the coupling at the incudo-stapedial joint is decreased. They probably do not exert significant tension, normally, but they reflexly and transiently contract, after a delay of approximately 10 msec, in response to strong acoustic stimuli and to strong stimuli to the vestibular apparatus.

Furthermore, it has been observed that differences in tonus in the middle ear muscles between the left and the right ear affect localization capability. Normally both ears are affected symmetrically by the auditory reflex. This can be used to induce reflex action at one ear and to test perception of loudness at the other. Changes observed psychophysically are of the order of 10 dB. Dallos [6.9] analyzed at 500 Hz the effect of the auditory reflex, characterized by a nonlinear servo system, on the impedance at the tympanic membrane. Changes up to approximately 20% may occur but only if the average sound intensity is very high (e.g., 110 dB above the absolute threshold of hearing at 1000 Hz). Thus a more important role in the *protection* of the ear against high sound intensities is played by the ossicles themselves which assume, for loud sounds, modes of oscillation different from the one considered in this chapter [6.4].

B. COCHLEAR DISCRIMINATION

To a large extent we owe to von Békésy our ability to develop linear mathematical models of the hydrodynamics of the inner ear. Some of the results in his outstanding collection of direct observations of static properties and of vibratory patterns of the cochlear partition are reported in [6.4], and from these we can conclude that the entire cochlear partition can be considered as one "membrane" whose main characteristics are highly nonuniform geometry and elasticity. The amplitude of its oscillation is linearly dependent on that of the stapes. Phase delays along the cochlear partition indicate that we must think in terms of propagated waves, perhaps superimposed on some reflected ones, whose spatial amplitude distribution changes with frequency.

Furthermore, the wavelength of most frequencies encountered is much larger than the diameters of the cochlear canals.

These are some of the experimental data forming the basis of the simplified, one-dimensional, hydrodynamic analysis with which we further attempt to separate transducer actions of the inner ear from the functions of the nervous system in the auditory system. Thus the following development largely neglects fine points such as flow velocity patterns in the cochlear ducts and eddies such as observed by von Békésy. The energies absorbed by the latter probably are too small to affect the displacement pattern of the basilar membrane. (It requires extreme acoustic stimuli, e.g., 120 dB above threshold, to create similar eddies within the vestibule that can stimulate the vestibular organs.)

The results of the hydrodynamic analysis must be related to theories of the peripheral processing of auditory stimuli.

1. *Sound Transmission in the Cochlea*

Only that fraction of the available information on geometry and anatomy of the inner ear which is of functional significance is given here. We refer the reader to von Békésy's work (e.g., [6.4]) and also to that of Zwislocki [6.3] for more details. The cochlea, forming a $2\frac{1}{2}$-turn spiral within the rigid bone structure of the skull, contains three compartments. First is the scala vestibuli, at whose base pressure waves are set up by the footplate of the stapes. Second is the scala media or cochlear partition, bounded by (*a*) Reissner's membrane, which is a very thin but quite elastic (little deformable) sheet; (*b*) the stria vascularis covering the bone; (*c*) the basilar membrane; and (*d*) the osseous spiral lamina. Von Békésy could not detect differences between the movements of Reissner's membrane and the basilar membrane. We therefore can consider the entire cochlear partition as one mechanical "unit." Third is the scala tympani, which is connected at the apex to the scala vestibuli through the helicotrema (an acoustic "short circuit"); at the base, the scale tympani ends at the round window. The cochlear spiral winds around the modiolus through which the nerve fibers reach the transducer organs [2.2].

Approximate dimensions of the basilar membrane in the human ear are given in Fig. 6.4, where a stretched-out cochlea is represented diagrammatically. Von Békésy also ascertained that the stiffness, opposing lateral movements of the partition, is essentially determined by that of the basilar membrane. The distensibility of the partition, plotted in Fig. 6.5 in terms of change in area of the scala vestibuli per unit pressure, is quite accurately approximated by an exponential function. Near the apex of the cochlea, area changes are one hundred times larger than those near the base. Additional significant information pertains to the geometry of the scalae vestibuli and tympani, for which only a few data are available. Figure 6.6 presents

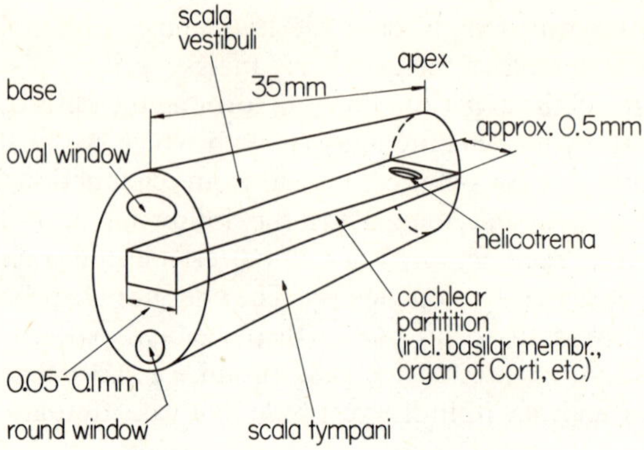

Fig. 6.4. Schematic view of stretched human cochlea, indicating the main geometric features and relationships. Normally, excitation occurs through the stapes at the oval window. The cochlear partition, whose mechanical property probably is given by the basilar membrane, changes its property drastically from base (input) to apex. See Fig. 6.5.

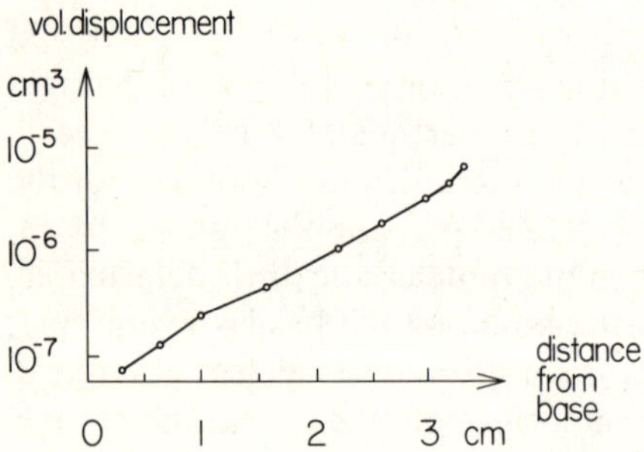

Fig. 6.5. Overall elastic property of cochlear partition. The area change (ΔS_1), or volume change of scala vestibuli per millimeter along the cochlea, and per cm H_2O pressure difference across the partition, is plotted logarithmically against distance from base. This "lateral" elasticity is one of the main determinants of wave transmission in the cochlea. The circles represent measurements by von Békésy; the data are matched well by a straight line in this semilogarithmic graph, justifying eq. 6.16. From Zwislocki [6.5].

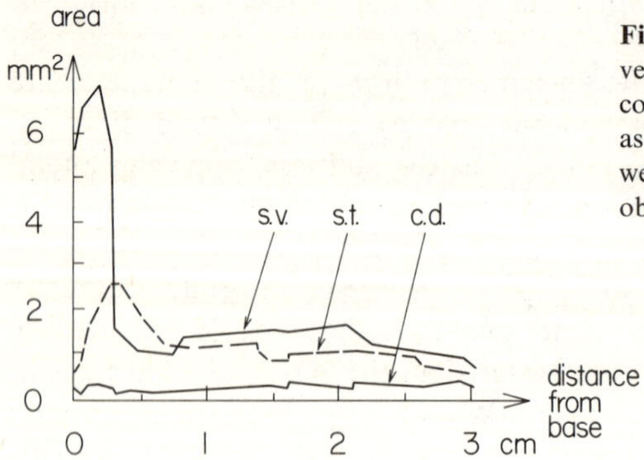

Fig. 6.6. Cross-sectional areas of scala vestibuli (*s.v.*), scala tympani (*s.t.*), and cochlear duct (*c.d.*) in one human cochlea, as published by Wever [6.11]. The lines were found by connecting Wever's observations.

one published case, as evaluated by Zwislocki [6.3], who approximated the spatial dependence of these data by an exponential function. These dimensions have bearing on the resistance to axial fluid flow within the scalae.

The hydrodynamic relations governing the propagation of pressure waves in a nonuniform structure such as the cochlea are very much simplified by assuming that the tapering is relatively small. Then we can treat the "transmission line" as being *locally* uniform; convective accelerations need not be taken into account and, similarly, the condition of continuity is essentially given by the distensibility of the partition and is not influenced by the axial change in cross-sectional area. Zwislocki, who summarized his work in [6.3], showed that this approach is satisfactory in that several of von Békésy's observations could be explained or predicted. In the following development we therefore can assume sinusoidal signals; we can also discuss the different impedances involved and define mechanic, cochlear discrimination.

a. Derivation of Differential Equations; Long-Wave Approximation. Formally, the relations between pressure gradient and axial flow, together with continuity equations, must be solved simultaneously for both canals. Some pertinent features of the geometry and the definitions of flow directions are indicated in Figs. 6.4 and 6.7, respectively. Note that we now express the

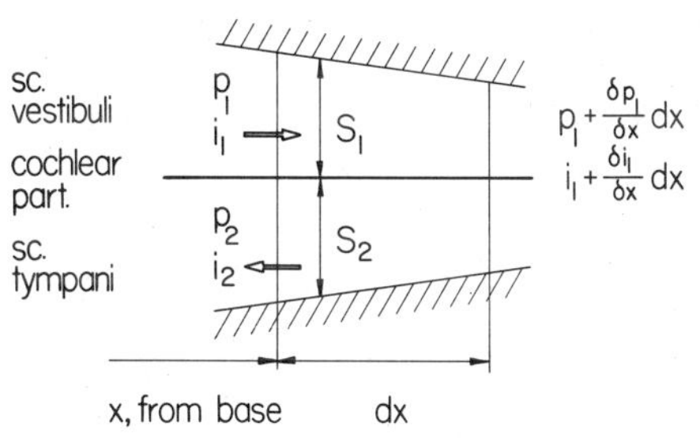

Fig. 6.7. Longitudinal view of a section of the cochlea defining symbols used in eqs. 6.3 to 6.6.

flow velocity in terms of particle velocity averaged over the cross-sectional area. Thus the axial pressure changes are

$$-\frac{\partial p_1}{\partial x} = (R_1 + j\omega\rho)i_1 \qquad \text{in scala tympani} \qquad \left(\rho i_1 \frac{\partial i_1}{\partial x} \text{ neglected}\right) \qquad (6.3)$$

$$\frac{\partial p_2}{\partial x} = (R_2 + j\omega\rho)i_2 \qquad \text{in scala vestibuli} \qquad (6.4)$$

where p = pressure, sinusoidal in time
i = flow velocity, sinusoidal in time
x = distance from base
R = resistance to axial flow per unit length; a function of x and frequency (introduced later)
ρ = density of perilymph

$j\omega = \sqrt{-1} \cdot 2\pi \cdot$ frequency of sound
$\rho i\, \partial i/\partial x$ = pressure gradient due to convective flow; neglected here

Furthermore, the axial flow changes are

$$-S_1 \frac{\partial i_1}{\partial x} = \frac{\partial S_1}{\partial t} \quad \text{in scala tympani} \quad \left(i_1 \frac{\partial S_1}{\partial x} \text{ neglected}\right) \tag{6.5}$$

$$S_2 \frac{\partial i_2}{\partial x} = \frac{\partial S_2}{\partial t} = -\frac{\partial S_1}{\partial t} \quad \text{in scala vestibuli} \tag{6.6}$$

where S_1, S_2 = cross-sectional areas of tympanic and vestibular canals, respectively, functions of x
$-i\, \partial S_1/\partial x$ = flow change due to geometric taper; neglected here

When differentiating eqs. 6.3 and 6.4 with respect to x and combining with eqs. 6.5 and 6.6, we obtain

$$-\frac{\partial^2 p_1}{\partial x^2} = Z_1(x) \frac{\partial i_1}{\partial x} = -\frac{Z_1(x)}{S_1(x)} \frac{\partial S_1}{\partial t} \quad \left(\frac{\partial Z_1}{\partial x} i_1 \text{ neglected}\right) \tag{6.7}$$

$$\frac{\partial^2 p_2}{\partial x^2} = Z_2(x) \frac{\partial i_2}{\partial x} = -\frac{Z_2(x)}{S_1(x)} \frac{\partial S_1}{\partial t} \tag{6.8}$$

where

$$Z(x) = R + j\omega\rho \tag{6.9}$$

The temporal change in S_1 can be related to the net force on the cochlear partition $(p_1 - p_2)$, through the (lateral) distensibility and viscosity of the basilar membrane. Zwislocki [6.3] found that the mass of the partition can be neglected. Hence for the long-wave approximation we have

$$\frac{\partial S_1}{\partial t} = j\omega S_1 = \frac{p_1 - p_2}{r + (1/j\omega C)} \tag{6.10}$$

where r = resistive element of impedance to lateral displacements, per length, due to viscosity of cochlear partition

C = distensibility of partition, in terms of change in S_1 per pressure difference $(p_1 - p_2)$; strongly dependent on x

Thus by adding eqs. 6.7 and 6.8 and by introducing eq. 6.10, we find

$$\frac{\partial^2 p}{\partial x^2} = \left[\frac{Z_1(x)}{S_1(x)} + \frac{Z_2(x)}{S_2(x)}\right] Y(x)\, p \tag{6.11}$$

where

$$p = p_1 - p_2$$
$$Z(x) = R(x) + j\omega\rho$$
$$Y(x) = [r + (1/j\omega C)]^{-1}$$

Equation 6.11 has the general appearance of a wave equation. Propagation is determined, essentially, by inertia of the perilymph and elasticity of the cochlear partition. Once the solution $p(x)$ of eq. 6.11 is known, the displacement patterns of the cochlear partition can easily be found and compared with von Békésy's observations. The solution, however, is not a simple sinusoidal function in space, since S and Y are functions of distance along the cochlea. These functions are formulated next. The term in brackets on the right side of eq. 6.11 is simplified considerably if it is assumed that $S_1 = S_2$ (cf. Fig. 6.6). Approximations like this should not influence the outcome of the calculations. The behavior of a highly nonuniform system, with some energy losses, is of main interest. In addition, von Békésy observed in model studies that the type of wave propagation does not depend strongly on the depth of the "channels." We follow the suggestion by Zwislocki [6.3] and approximate $S_1(x)$ by an exponentially decaying function. Thus

$$\frac{Z_1(x)}{S_1(x)} + \frac{Z_2(x)}{S_2(x)} = \frac{2\exp(\alpha x)}{S_0}(R + j\omega\rho) \tag{6.12}$$

where $S_0 = 2.5 \times 10^{-2}\ \mathrm{cm}^2$; $\alpha = 0.5\ \mathrm{cm}^{-1}$

The longitudinal resistance R in one of the scalae depends inversely on the cross sectional area of the channel, and for this reason we can also assume that $R_1 - R_2 \equiv R$. Furthermore, R depends on frequency. As we show in detail in Chapter 14, the oscillatory flow profiles in a cylindrical vessel become flatter, at high frequencies, than the parabolic profile existing in steady flow. The parameter α', which defines a particular flow profile, depends on frequency, viscosity (μ), density, and vessel radius (a) as follows:

$$\alpha' = a\sqrt{\omega\rho/\mu} \tag{6.13}$$

In the two scalae, α' varies from 10 to 30 for frequencies between 100 and 1000 Hz. But for large α' (> 10) the longitudinal resistance per length R can be given in a very good approximation:

$$R = \sqrt{2}\,\frac{\pi\mu}{S}\,\alpha' \tag{6.14}$$

This becomes, if the scalae are thought to be of nearly cylindrical cross section,

$$R = \left[\frac{2\pi\mu\rho\omega}{S_0}\right]^{1/2} e^{(\alpha x)/2} \tag{6.15}$$

It is seen that R not only depends on x but also on the square root of the frequency, which is explained by the change in shape of the velocity profiles (similar to the electrical "skin effect").

The impedance to lateral motion of the cochlear partition can be given as a function of x, using von Békésy's [6.10] observations. The change in cross-sectional area of the scala tympani, for a given pressure across the partition, depends exponentially on the distance from the oval window; that is,

$$C = C_0\, e^{\beta x} \tag{6.16}$$

where $C_0 = 4 \times 10^{-10}$ cm^3/[(dyne/cm^2)(cm)]
$\beta = 1.5$ cm^{-1}

The cochlear partition is one hundred times more compliant near the apex than near the oval window. (Note that $j\omega C/S$ yields the shunt admittance in the system of units chosen.) Zwislocki [6.3] let r, accounting for viscous losses in the cochlear partition, be a constant. Huggins [6.12] showed how this can be refined by using von Békésy's data [6.4] on the logarithmic decrement of transient vibrations of the cochlear partition, which is almost constant along the cochlea (i.e., of the order of 0.5). In most cases ($f < 1000$ Hz), $r < (\omega C)^{-1}$ and hence it is satisfactory to use first the simplification r = a constant (of the order of 100 Ω/cm) and introduce refinements later in order to see their effects. (Note that the observation of undercritical damping of transients really requires a second-order system for the cochlear partition.)

Thus by letting

$$\frac{1}{r + (j\omega C)^{-1}} \simeq j\omega C_0\, e^{\beta x}\,[1 - j\omega r C_0\, e^{\beta x}] \tag{6.17}$$

the differential equation for the spatial pressure distribution becomes, in Zwislocki's form,

$$\frac{\partial^2 p}{\partial x^2} = \frac{2}{S_0}(R_0\, e^{(\alpha x)/2} + j\omega\rho)(\omega^2 r C_0{}^2 e^{2\beta x} + j\omega C_0\, e^{\beta x})e^{\alpha x} p \tag{6.18a}$$

The type of differential equation to be dealt with is recognized when eq. 6.18a is written as follows:

$$\frac{\partial^2 p}{\partial x^2} = -\frac{k(\omega)\rho C_0}{S_0'}\,\omega^2 e^{(\alpha+\beta)x} p \tag{6.18b}$$

where

$$k(\omega) = 1 - j\left[\omega r C_0\, e^{\beta x} + \frac{R_0'}{\rho}\,(\omega)^{-1/2} e^{(\alpha x)/2}\right] \tag{6.19}$$

$$rR_0\, C/\rho \ll 1$$

$$R_0' = (2\pi\mu\rho/S_0)^{1/2};\ \text{numerically } 2.25 \text{ dynes sec}^{3/2}/\text{cm}^4$$

$$S_0' = S_0/2 \simeq 1.25 \times 10^{-2}\ \text{cm}^2$$

b. Spatiotemporal Distributions of Pressure and Displacement of the Cochlear Partition. Equation 6.18, of the form $\partial^2 p/\partial x^2 + M \exp(\gamma x)p = 0$, is transformed into a Bessel differential equation by letting

$$z = e^{\gamma x} \tag{6.20}$$

that is,

$$\frac{\partial^2 p}{\partial z^2} + \frac{1}{z}\frac{\partial p}{\partial z} + \frac{M}{\gamma^2}\frac{1}{z}\, p = 0 \tag{6.21}$$

where $$M = k(\omega)\rho C_0\, \omega^2/S_0'$$

The complete solution is either a linear combination of Bessel functions of first and second kind of the zero order, or one involving the Hankel functions $H_0^{(1)}$ and $H_0^{(2)}$. (For the theory see [0.16].) Their argument is, in any case,

$$\arg = \frac{2}{\gamma}\, M^{1/2} e^{\gamma x/2} \tag{6.22}$$

whose imaginary part is negative (see eq. 6.19) due to the elasticity of the cochlear partition. When x increases (i.e., when going from base to apex), the magnitude of the imaginary part (as well as the real part) increases, $H_0^{(1)}$ increases and $H_0^{(2)}$ decreases. However, near the apex where the partition is more easily deformed than near the base, displacement of the partition was observed to decrease [6.4] (at least for frequencies above 50 Hz). Then the pressure across the partition must be decreasing; the solution therefore can only include $H_0^{(2)}$. Thus as a function of time and space we obtain

$$p(x, t) \propto H_0^{(2)}\left\{\frac{2}{\alpha+\beta}\sqrt{k(\omega)\rho C_0/S_0'}\ \omega e^{(\alpha+\beta)x/2}\right\} e^{j\omega t} \tag{6.23}$$

The Hankel function is a complex number (also true when its argument is real) yielding the amplitude and phase information of the spatial distribution at any one time for a given input pressure. Equation 6.23 indicates that the magnitude of a pressure wave decays very rapidly when it propagates into the cochlea, with the result that very little of the input signal reaches the helicotrema. Hence we do not have to introduce possible wave reflections from the apical end of the cochlea. However, the decay of pressure amplitude is much more rapid than could be explained by viscous losses. On a nonuniform transmission system such as the cochlea, distributed reflection sites can be thought to cause the amplitude changes. At each point along the path of travel, there exists a local impedance mismatch. The progressive softening of the partition causes the distributed reflections to cancel part of the incident wave. In Chapter 14 we discuss wave propagation through the system of systemic arteries. There the system widens and decreases in distensibility from source to load. Pressure amplitudes tend to increase toward the load. This is just the opposite from the phenomenon seen in the cochlea. Distributed reflections also play a functionally important role when impedance changes along the cochlea are considered, as will be seen later.

In order to relate the result from eq. 6.23 to the observable distribution of displacement of the cochlear partition, the following steps are required. At each point x, the velocity $(\partial y/\partial t)$ of the partition is given by

$$\frac{\partial y}{\partial t} = Y(x)p(x)\frac{1}{b(x)} \tag{6.24}$$

where $y = S_1(x, t)/b(x)$ the displacement of the partition, see eq. 6.10

$b(x) =$ width of basilar membrane

Hence the displacement $y(x, t)$, with sinusoidal excitation, becomes

$$y(x, t) = \frac{Y(x)}{j\omega b(x)}p(x, t) \tag{6.25}$$

Zwislocki [6.13] considered the effective width of the cochlear partition to be independent of x. Indeed, for the cochleae reported [6.14, 6.15], $b(x)$ may be approximated by exp $(0.3x)$, which has a much smaller range than exp $(1.5x)$ in the expression of the compliance (eq. 6.16). From eq. 6.25 it becomes clear that the displacement is approximately in phase with the pressure, at least toward the apex, where the imaginary part of $Y(x)$ is larger than its real part. The envelope of the spatial amplitude distribution is given essentially by the product of a strongly decaying function and an (exponentially) increasing function. This product in turn produces the typical peaking of vibration amplitude at a location that depends on frequency.

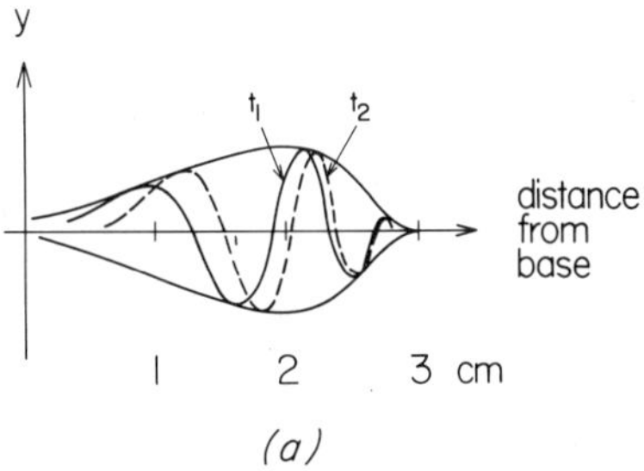

Fig. 6.8*a*. Instantaneous displacement distribution of a traveling wave, as calculated by Zwislocki [6.13], at a frequency of 1000 Hz: solid line, $y(x)$ at a fixed time t_1; dashed line, displacement distribution at $t_2 > t_1$. Local velocity is high near the stapes (i.e., at the base).

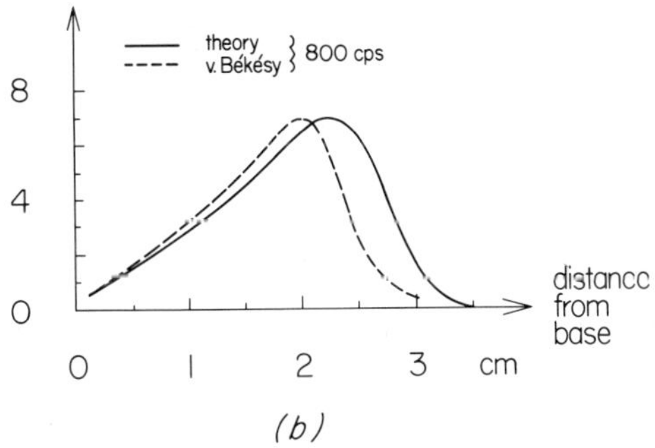

Fig. 6.8*b*. Envelope of $y(x, t)$ at 800 Hz. Zwislocki [6.3] computed the envelope of an 800-Hz wave (i.e., its spatial amplitude distribution) as given by the solid line. Theory predicts quite well what is observed (dashed line, measured by von Békésy). Both curves represent displacements of the cochlear partition. Note that the 800-Hz wave peaks slightly closer to the apex. From Zwislocki [6.3].

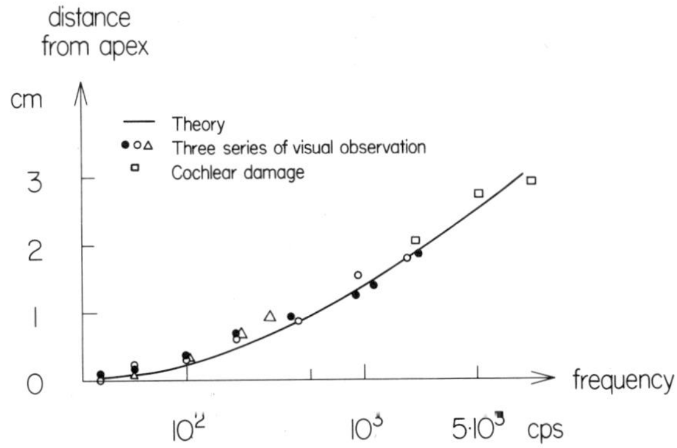

Fig. 6.9. Comparison of the theoretical location of amplitude maxima, as a function of frequency, with data from four types of observations. Circles and triangles are direct visual determinations of oscillation peaks; the squares are derived from frequency-selective hearing deficits due to localized cochlear damage. From Zwislocki [6.3].

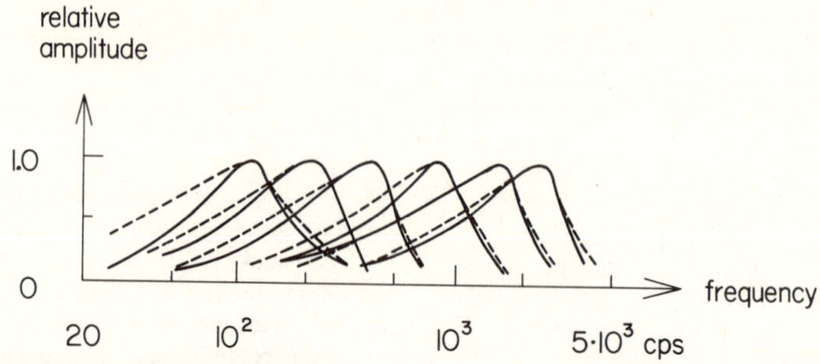

Fig. 6.10. Resonance or tuning curves for six points on the cochlear partition: solid curves, observations by von Békésy; dashed curves (only shown when different from the measured curves), theoretical predictions by Zwislocki [6.3].

Zwislocki's calculations on the spatial distribution of displacement, at fixed frequencies, [6.3, 6.13] are reproduced in Fig. 6.8. Figure 6.9 shows how the simple theory presented here also predicts satisfactorily the locus of the maximum vibration ($\partial y/\partial x = 0$) over quite a wide range of frequencies. The frequency dependency of the displacement at a given point within the cochlea is obtained directly from eq. 6.25. The theoretical approach described here also is remarkably successful in predicting these "resonance" curves, or *tuning curves*, as in Fig. 6.10.

Furthermore, it is of interest to compute the distribution of *velocity* of wave propagation along the cochlea. Zwislocki showed that the phase information (ϕ) in eq. 6.23 can be given, at relatively high frequencies and near the apex, as

$$\phi(x) = \omega \sqrt{k(\omega)\rho C_0/S'_0}\, \frac{2}{\alpha + \beta}\, e^{(\alpha+\beta)\, x/2} \tag{6.26a}$$

But the phase velocity c_0 is defined by

$$c_0 = \frac{\omega}{\partial\phi/\partial x}$$

hence

$$c_0 = \sqrt{S'_0/[k(\omega)\rho C_0]}\; e^{-(\alpha+\beta)x/2} \simeq \sqrt{S'_0/(\rho C_0)}\; e^{-(\alpha+\beta)x/2} \tag{6.26b}$$

Figure 6.11 plots this function against x. At the entrance of the cochlea pressure waves travel very fast, and c_0 is of the order of 50 m/sec. (Note that $k \simeq 1.0$). Near the apex, due to the large compliance of the partition, c_0 may be below 5 m/sec.

Before we discuss possible discriminatory mechanisms within the cochlea, we should investigate the *input impedance* of the inner ear. This has bearing on the transmission of sound from the outer ear to the oval window, as was explained in the previous section. In a highly nonuniform system, the local impedance is nearly independent of wave reflections from distributed sites

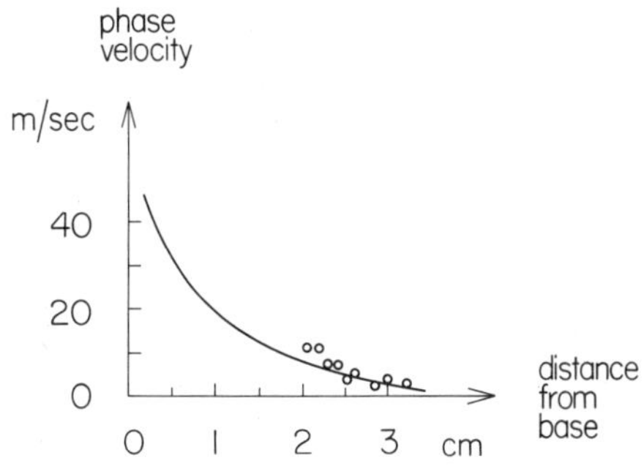

Fig. 6.11. Phase velocity of waves in the cochlea, plotted as a function of position on the cochlear partition. The curve is theoretical; the circles are derived from von Békésy's experiments. The elasticity of the basilar membrane, being much larger near the base (cf. Fig. 6.5), together with the geometrical taper of the lumen area of the scalae, essentially determines the spatial distribution of phase velocity. From Zwislocki [6.3].

(see discussion of the arterial vascular system, Chapter 14). The input impedance therefore is approximately equal to the (local) characteristic impedance at the base. By definition, the characteristic impedance is given by the following real number:

$$Z_0(x = 0) \equiv \rho c_0 = \sqrt{\rho S'_0/C_0} \tag{6.27a}$$

In this form, however, Z_0 is in terms of pressure per particle velocity. In the previous section we expressed impedances as pressure per volume flow velocity; hence we must equate

$$Z_{\text{in,cochlea}} = \frac{\rho c_0}{S'_0} = \sqrt{\rho/(S'_0 C_0)} \simeq 2 \times 10^5 \, \frac{\text{dynes/cm}^2}{\text{cm}^3/\text{sec}} \; (= \Omega) \tag{6.27b}$$

If the transformer ratio (T) of the middle ear (slightly over 20) is introduced, we find indeed that the load of the middle ear network (Z_0/T^2) must be of the order of 500 Ω, which is close to the value obtained in Section 6.A (p. 184).

2. Frequency Discrimination

The curves in Fig. 6.10 are of interest with respect to the functional description of the cochlea. At any one location, the response to a variable frequency input is similar to that of a very broadly tuned band-pass filter. The frequency of the stimulus has to be changed by 30 to 50 % to decrease the displacement to 90 % of its maximum value. The following calculation shows how equally little selectivity can be expected in the spatial domain. At a moderate sound intensity of 40 dB above threshold and at almost any one frequency, approximately 20 % of the length of the basilar membrane displaces sufficiently to stimulate hair cells. If the frequency is changed by 1 %, the range over which cells are possibly stimulated is shifted along the cochlea by approximately 0.05 mm. Hence 98 % of the hair cells that may be stimulated by one frequency may also be excited by another one, which differs only by 1 % in frequency. But a frequency change of 1 % is well above the just noticeable difference (at moderate intensities and above 250 Hz;

see below). It therefore appears impossible to explain our ability for fine frequency discrimination assuming neural responses depending on displacement curves only. In fact, it is known [6.16] that the neural tuning curves of single fibers of the eighth nerve are much narrower than those of Fig. 6.10. This is shown, for one case, in Fig. 6.12*b*.

It is remarkable how successfully Zwislocki's linear model predicts the lateral motion of the entire cochlear partition. However, it apparently fails to yield direct insight into the forces stimulating the hair cells. It is possible that the outer hair cells are stimulated only by relative motion between the tectorial membrane and the organ of Corti in the direction of the spiral ligament. This may not always occur when the pressure in the scala tympani is raised above that in the scala vestibuli. The fluid within the internal spiral sulcus is incompressible. Hence the sulcus cells of the limbus may be deformed such that the tectorial membrane essentially moves together with the organ of Corti. Such action could, however, put the inner hair cells under longitudinal stress, which might cause these receptors to be stimulated. In any case, it should be recognized that stress distributions within the partition, both in lateral and longitudinal directions, may be important for the transduction of the signals.

Huggins [6.12, 6.18] and Licklider [6.17] therefore investigated various hypotheses that could explain sharpening of the tuning curves on the basis of mechanical events, in particular at higher frequencies. For all signals above 50 Hz, not all points along the partition move in phase with each other (i.e., there exist spatiotemporal patterns of displacement and bending of the cochlear partition). It is likely that the densely packed system of receptor cells extracts information which, in a statistical sense, depends on displacement, slopes, bending curvatures, and so on, of the basilar membrane. Mathematically, then, it becomes interesting to examine not only the behavior of $y(x, t)$ of eq. 6.25 but its spatial derivatives. In particular, Huggins [6.17] showed that bending forces proportional to the fourth derivative along the axis of the scalae assume numerical values that are of detectable magnitude. Near the threshold of hearing, all lower-order derivatives are exceedingly small. It is well known, from the analysis of bending of structural beams, that the term $\partial^4 y/\partial x^4$ gives the external force necessary to bend an element of the length dx of the beam.

Consider an element of the "beam" such as a segment of the length dx of the tectorial membrane. At a particular time and at a point x it assumes a curvature $\partial^2 y/\partial x^2$. Thus, in order to compress (or elongate) this element, an (internal) torque M is needed:

$$M = -EI \frac{\partial^2 y}{\partial x^2} \tag{6.28a}$$

where E = Young's modulus of elasticity
I = bending moment

The bending torque must be balanced by external forces $F = \partial M/\partial x \propto \partial^3 y/\partial x^3$. The differential of the force that actually bends the element, therefore, is given by

$$dF = \frac{\partial F}{\partial x}\,dx = EI\,\frac{\partial^4 y}{\partial x^4}\,dx \tag{6.28b}$$

Thus the force distribution along the beam, or the cochlear partition, is proportional to the fourth spatial derivative of $y(x, t)$. Since y is a wave, $\partial^4 y/\partial x^4$ also is a wave, as graphed in Fig. 6.12*a*.

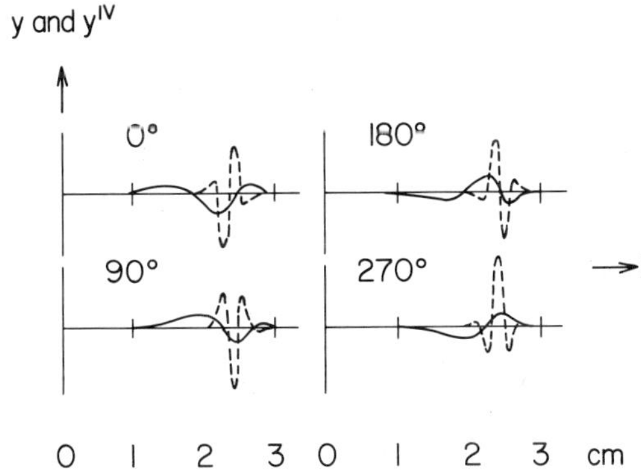

Fig. 6.12*a*. Instantaneous distributions of displacement (solid curve) and fourth space derivative of y (i.e., force) along the cochlea. Both curves are in relative units for a wave at 1000 Hz. From Huggins [6.17]. Note that force distributions exhibit much sharper localization.

The most interesting features of Huggins' result concern the relatively narrow envelope and the large spatial phase change of the force wave. As Fig. 6.12 indicates, the range over which the *amplitudes* of the force waves are large is much smaller than the range of large displacements. There exists no *direct* evidence that the activity of the first-order neural units is given by only this force distribution within the cochlear partition. But it is interesting that the spatial bandwidth, as indicated by the envelope in Fig. 6.12, is indeed very similar to that of the tuning curves of eighth nerve fibers. Moreover, it compares well with that obtained in psychoacoustical experiments on the effective bandwidth as determined by masking by narrow-band

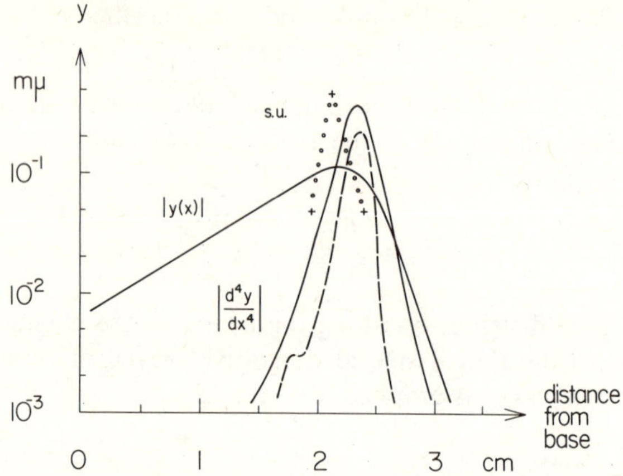

Fig. 6.12*b*. Envelopes (i.e., amplitude distributions) of displacement and force, for a tone of 1000 Hz, 55 dB above absolute threshold: solid curves, theoretical; dashed curve, derived from Fletcher's measurement of masking [6.34] by transforming the tuning curves into spatial distributions. From Huggins [6.17]. The dotted curve represents the neural tuning curve found by Kiang [6.16] at 800 Hz, after transformation into the spatial domain. The dotted curve is scaled by letting 20 dB = 1 decade on the ordinate.

noise. (See section on critical masking band, below.) The *phase change* along the cochlea of a tone of 1000 Hz, near the force peak, is of the order of 200°/mm, as opposed to 40°/mm near the displacement peak. Thus hair cells separated by approximately 0.9 mm are acted upon by forces that are out of phase with each other. Furthermore, there is only one point on the partition where the *y*-wave is in phase with the force wave, and this site is sharply defined. This may offer one very sharp "place mechanism" for frequency discrimination, assuming combined effects of y and $\partial^4 y/\partial x^4$ waves.

We have just emphasized possible mechanical events contributing to frequency discrimination. Analysis of the activity of single fibers of the *eighth nerve* reveals several additional aspects of the "coding" of acoustic stimuli at the level of the first-order units. It is known, mostly from experiments in cats, that temporal structure and spatial distribution of a stimulus essentially are preserved within the eighth nerve. In response to a sinusoidal stimulus at *one single* frequency, discharges are observed either with each half-wave (at low frequencies) or at intervals which are multiples of the period (at high frequencies) [6.19]. When obtaining the so-called tuning curves for high-frequency units, we establish the dependence on frequency of the threshold at which such a unit begins to respond in the manner described. Kiang et al. [6.16] demonstrated experimentally that knowledge of the response of a unit to sinusoidal stimuli suffices to predict its response to a click, although several nonlinear phenomena are part of the chain of events (latency, threshold, accommodation, and refractory effects of receptor cell). A unit whose threshold is lowest at a particular frequency responds to a click with a series of impulses whose probability of occurrence is highest at those times which are multiples of the period of the characteristic frequency. Weiss [6.20] put forward a model, including a noise source that introduces a probabilistic threshold and spontaneous activity, which successfully explains the statistics of the responses of a unit of the eighth nerve. The nature of these responses does not change significantly if the intensity of the stimulus is varied. However, the model does not explain unequivocally the sharpening of the neural tuning curves with respect to the mechanical ones.

Thus the task of elucidating the nature of the peripheral neural frequency discrimination remains. Distribution of forces within the cochlear partition, the particular mechanism of hair cell stimulation, electric interaction among unmyelinated fibers innervating the hair cells [6.17], or a combination of all these may be important.

Furthermore, the responses of first-order units to *two* or more sinusoidal stimuli are poorly understood. The activity of a unit driven by its characteristic frequency may be reduced or suppressed by a tone burst at a neighboring frequency. It is highly improbable that efferent neural signals are involved

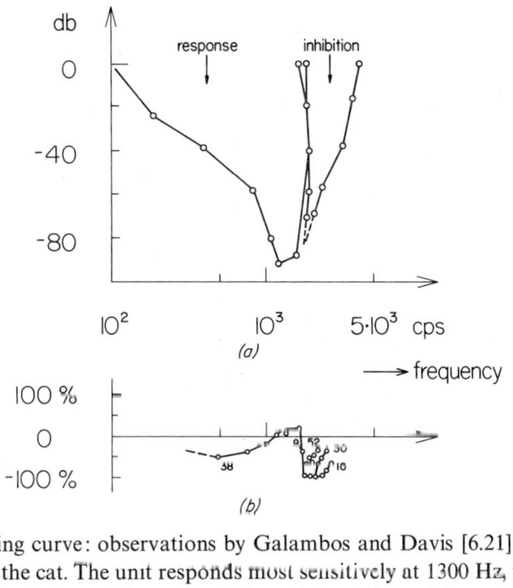

Fig. 6.13. (*a*) Tuning curve: observations by Galambos and Davis [6.21] in single units of the cochlear nucleus of the cat. The unit responds most sensitively at 1300 Hz, where its threshold is −94 dB below 2 V from the oscillator used. The activity of the unit is suppressed if, for example, a tone of 3000 Hz at −60 dB is received. (*b*) Plot of the percentage of change induced by a second tone, at the frequency and intensity indicated.

in this. The tuning curves, or the isorate curves of an eighth nerve fiber stimulated by two tones, exhibit characteristics very different from the ones mentioned so far. Inhibitory areas both above and below the characteristic frequency appear, depending on the energy of the second stimulus. The degree of inhibition by the second tone is a function of the ratio of the intensities of the two stimuli. Interactions such as these obviously increase significantly the ability of the system to discriminate among frequency components of complex stimuli. It has been known for a long time that this is observed in units of the *cochlear nucleus*. Galambos' [6.21] original results obtained in the cat are presented in Fig. 6.13.

It is worth pursuing, at this point, the previously developed idea of combined effects of various characteristics of an acoustic stimulus within the cochlea, which may determine the responses of both first- and second-order units. The model discussed in the following was developed by Huggins [6.12] to explain afferent inhibition among either first- or second-order units. We therefore relate the actions of " receptor units of two different kinds " rather than that of inner and outer hair cells. The model assumes that (*a*) certain receptor units are excited primarily by forces in phase with the displacement of the cochlear partition, and others primarily by the bending forces; (*b*)

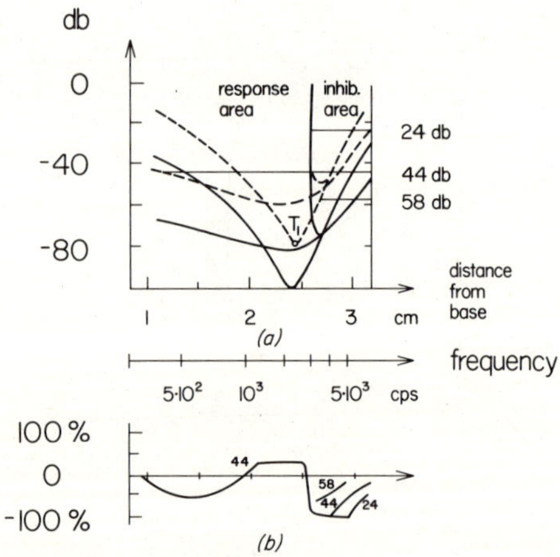

Fig. 6.14. (*a*) A model proposed by Huggins [6.12] whose performance is similar to units found by Galambos (cf. Fig. 6.13). Data of Fig. 6.12 are replotted in terms of stimuli required to excite two different types of receptor cells (solid and dashed curves, respectively). See text for details of assumptions. (*b*) Effects of a second tone, in terms of percentage of change in activity. The model works with a unit most sensitive to 1000 Hz. The " second " stimulus is fixed at 1300 Hz; the response of the unit to a variable " primary " tone is changed as shown, depending on its intensity (24–58 dB). Figures 6.13 and 6.14 can be compared, since the frequency–location relation is known.

units of the first kind may inhibit those of the second category, which lie within a small region (1.5 mm) toward the stapes; and (*c*) the threshold of response of the possible inhibitors is above that of the inhibited units. (In the following a difference of 20 dB is assumed.)

Thus at relatively high intensities, possible inhibitors excited maximally by a certain frequency may influence a unit of the other kind responding optimally to a slightly lower frequency (causing a peak force at a site closer to the helicotrema), provided the extreme effects of the two stimuli occur at sites spaced by 1.5 mm or less. This is due to the crossing of the phase distributions of $y(x)$ and $\partial^4 y/\partial x^4$ at the point of maximum force, beyond which the phase of $y(x)$ is leading. For weak signals, the so-called inhibitors may be stimulated before the other units, yet their discharges may occur later because the second kind fires at lower levels of intensity. In this case no or little inhibition would occur. It thus is possible to predict the performance of the model, as in Fig. 6.14, and to compare it with the observations (e.g., those of Fig. 6.13).

It thus becomes quite clear that the cochlear receptor systems cannot be simply compared with a frequency analyzer. We have mentioned some complicating factors inherent in the organ of the cochlea. Threshold and latency properties of single units, interaction among several units, adaptation of the receptors, multiple innervation of hair cells, and probably additional, unrecognized mechanisms require further investigation before their functional roles can be fully explained.

C. INTERMEDIATE STATIONS AND AUDITORY CORTEX

Questions of great interest to the systems physiologist concern the types of information defining acoustic stimuli which the auditory system conveys to the highest centers of the central nervous system (CNS), the nature of the neural "codes" employed by the system, and the explanation of how the system achieves the coding. Only very few of these questions can be answered today, even though the neural connections between the various nuclei identified as belonging to the auditory system are quite well known, and in spite of the many observations on the responses of single units at the different stations. In this section we limit the discussion to a few of the electrophysiological findings which are of particular interest to the systems analyst.

At the level of the *cochlear nucleus*, the units of the eighth nerve are represented threefold. Within these three areas, populations of neurons exist which exhibit behavior similar to that of the first-order units. For example, the tuning curves reveal that the neurons are inhibited by lower as well as by higher frequencies. Their excitatory frequency bandwidth is sharpened considerably over quite large ranges of intensity. Figure 6.15 reproduces an example reported by Greenwood et al. [6.22] showing an approximate bandwidth of excitatory performance of 10% (of f_{center}) for changes in intensity of almost 40 dB. Note that fine gradation of activity within the excitatory region (which is not seen in Fig. 6.14) can be obtained. When considering the activity of an entire population of this kind, it becomes clear that the units most sensitive to component tones of a complex stimulus are efficiently segregated since each unit is partly masked by all others. In other populations inhibitory regions are found over wide bands of frequency and intensities up *to* a certain amplitude. Such units may show decreasing activity when stimulus level is increased; above a certain intensity, however, activity increases again. This behavior is called nonmonotonic.

In any case it is important to know that it is difficult to gain insight into the nature of the coding at this level solely by defining frequency-intensity graphs, from which no information is available on temporal sequence of

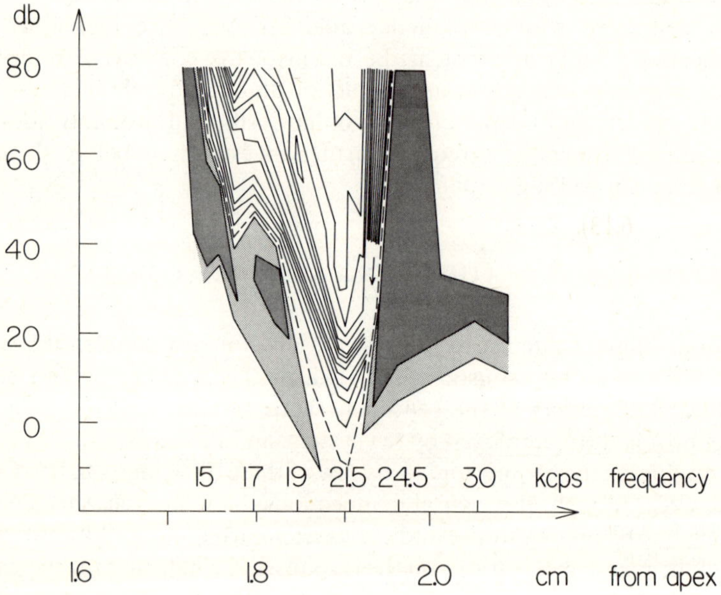

Fig. 6.15. Isoactivity map of a cell in the cochlear nucleus of the *cat*, as reported by Greenwood et al. [6.22]. An isoactivity contour line connects all those frequency–intensity combinations (of the stimulus) which elicit the same discharge rate (determined from 20 presentations of 0.2-sec tones). Dashed lines indicate the spike rates exceeding spontaneous activity by one standard deviation. "Distance" between solid lines is 100 spikes in 0.2 sec. Black areas cover those stimuli which suppress spontaneous activity by at least 3 standard deviations (below the mean). Within the gray areas, suppression is by 1 to 3 standard deviations. Both space and frequency scales are given along abscissa (cm from apex). Ordinate: dB *re* 0.0002 dyne/cm^2.

responses to such stimuli as bursts. Greenwood [6.22] found that some units within the cochlear nucleus would be temporarily inhibited; that is, at the onset of a complex stimulus containing more than one frequency, a period of no activity would be followed by one of intense activity. Units with similar responses occur at higher levels in the auditory system, but it appears that information, such as that pertaining to onsets of stimuli, is already available in simple form at the more peripheral levels. Note that these structured responses also may serve to inform the auditory cortex about small and sudden changes in frequency or amplitude of one of the component tones. Greenwood [6.22] determined that the initial silent period and the consecutive onset of activity vary from unit to unit. When considering a population of this kind, the information content, given by a spatiotemporal pattern of activity, may entail such additional parameters as rate of onset of stimulus and changes in frequency and amplitude.

At the level of the *medulla*, binaural interaction occurs for the first time in the auditory system (when going from periphery to center). Pathways originating within the cochlear nuclei connect to units of both ipsi- and contralateral superior olivary complexes. In response to maintained ipsilateral stimuli, steady-state activity of cells within the latter nuclei increases rapidly when stimuli are made more intense. Over a limited range, the rate of discharge increases approximately proportionally to the logarithm of the sound amplitude. It reaches a maximal level at a firing rate 5 to 10 times the spontaneous one. If sound intensity is further increased, some units maintain and others decrease their activity. Therefore we cannot say to what extent these particular cells at this level contribute to the intensity sensitivity of the entire system. The loudness sensation, at the psychophysical level, is a power function of sound intensity. But there are other subsystems whose properties apparently do determine to some extent the performance of the whole.

Transmission characteristics of mechanical structures of the ear influence quite strongly the frequency function of the threshold of audition (see below). Generally, however, it is very difficult to establish such correlations because the topology of pathways between subsystems, being quite complex in the auditory system, determines overall performance at least as strongly as do the properties of units at intermediate levels. Nevertheless, the time course at the response of these cells is of interest. At onset of a long stimulus, the typical unit exhibits a high initial firing rate which decays to settle at a low one. A minority of cases display a burst of increased activity when the stimulus ends [6.23]. At higher levels in the auditory systems the on–off type of response is seen more frequently. This indicates aspects of the temporal structure of a stimulus which are important to the system (e.g., beginning and end of a sound; rapid increase and decrease in intensity).

Binaural stimuli are processed differently by different populations within the nuclei of the superior olivary complex. For example, some units of the accessory nucleus are activated by simultaneous bilateral stimuli much more than they would be by stimulation of either side alone. Other cells respond less to simultaneous bilateral acoustic signals than to the individual stimuli. Furthermore, their activity depends on the difference in arrival times of the two stimuli. These exhibit, in a statistical sense, the ability to trade time differences against intensity differences. It appears very likely that these cells contribute to the system's outstanding ability to localize sources of sound. Hall [6.24] investigated such units in the cat, using clicks, and proposed one of the more recent models for binaural localization. His experimental results indicate that, in the accessory nucleus of the medial superior olivary complex, more cells on one side are excited by a stimulus to the contralateral ear which increases in intensity or precedes the one to the ipsilateral ear. If the signal to the ipsilateral side is increased or advanced in time, fewer cells are active. Thus the information relayed from this level to the higher center is contained in the sizes of the active areas on both sides. Whether the difference between the sizes or the ratio of the size on one side to the sum of both active areas is utilized at the higher levels is not known.

At the level of the medulla, the *efferent pathways* of the olivo-cochlear bundle must be considered, too. Little is known about the central control of the descending signals, but one aspect of their action on the periphery is of interest. Dewson [6.25] studied the eighth nerve activity in response to rapidly repeating clicks in the presence of noise while he stimulated the olivo-cochlear bundle. He determined that *masking* by noise is considerably reduced by activity of the efferent fibers. In other words, the signal-to-noise ratio is enhanced by information that must be carried (at least) by the fibers in the olivo-cochlear bundle. Dewson was also able to verify that this particular peripheral mechanism is essential at the behavorial level. Monkeys which had been trained to discriminate between two different acoustic stimuli within noise were less successful in doing so, at corresponding signal-to-noise ratios, after their olivo-cochlear bundles had been severed, although the animals discriminated perfectly the stimuli in the absence of noise. We refer the reader to the literature [6.26] describing additional studies of this system.

A great deal of data on the behavior of neurons at *higher levels* and the *auditory cortex* has been collected. On the basis of responses to a tone burst of smooth onset and decay, Erulkar [6.27], Goldstein [6.28], and others classified units in nonanesthetized cats into those which increase or decrease their activity, those which temporarily increase or decrease their firing rate at the end of the stimulus, and those showing a short burst of increased activity at the onset of the tone. Some show combined "on," "through," and "off" responses; that is, these cells may fire more rapidly (compared with their "spontaneous" activity) at the onset of, be quieter during, and show another period of enhancement at the end of, a tone burst. All these units are frequency-band limited, and their responses are so complex that it has not yet been possible to predict the response of any of the cortex units from the catalog established for some of them. In any case, information about the beginning and end of a burst of noise or tone indeed is carried by some of the "channels" involved. Among these, however, different channels are assigned different frequencies. Completely different units may be responsible in responding according to intensity.

Clearly, it is difficult to define codes and modes of information processing in the *high auditory* centers because entire temporal–spatial patterns of activity cannot easily be inferred from the knowledge of single unit responses. Furthermore, not all neurons in the auditory cortex can be shown to respond to (conventional) acoustic stimuli, whereas some cells that do respond may also be innervated by fibers not belonging to the classical auditory pathways. (See [6.29] and Schwartzkopff [6.30] for a review of recent results.) By ablation or lesions in certain areas, their importance can be established in behavioral responses to differences in duration of tones, or changes in frequency or intensity. However, the mechanisms causing these responses are unlikely to be revealed completely by such studies.

D. PSYCHOPHYSICAL METRICS

We discussed in Chapter 5 some of the difficulties that arise when the performance of sensory systems at the psychophysical level is investigated quantitatively. The main problem is posed by the need to define independent attributes of sensations belonging to the class of metathetic continua.

This applies particularly to the auditory system. Many techniques are available to specify a stimulus in physical terms (e.g., by its Fourier spectrum). But the subjective experience of an input such as continued sound involves "volume," "brightness," "density," and "pitch," and all these must be demonstrated to be independent attributes. Loudness, a prothetic continuum, seems clearly to be an independent feature of sensations. Still, complications arise when, for example, the pitch function is determined for a sound comprising several tonal (sinusoidal) signals. Pitch sensation of a sound changes less when frequencies are varied than it does for a pure tone. Complications are worse with tonal stimuli of short duration (i.e., below 0.25 sec). Tone bursts of equal energy but of different lengths in time appear to differ in loudness. The absolute threshold of perception of short tone bursts also depends on their duration. It increases by 3 dB if their duration is cut in half, as soon as fewer than 200 cycles (of 1000 Hz) are presented to the listener [6.3].

Neurophysiological observations may assist in the search for independent (orthogonal) features or dimensions of subjective experience. From work with complex stimuli of several sinusoidal components, we can conclude that the sensation of loudness probably depends on the intensity of neural activity in the brain, whereas "volume" may depend more on the spread of such activity patterns over larger cortical areas. The corresponding sensations differ, too. A stimulus may sound "big" even if not very loud.

As mentioned in Chapter 5, we need further information about the characteristics of the stimuli that elicit these independent ("basis") sensations. Considering the input, Huggins [6.18] suggested that all *natural* stimuli, brought forth by excitation of physical systems that are satisfactorily described by linear differential equations, can be decomposed into decaying (complex) exponential time functions. These may be an important class of signals which, however, have not been used extensively as stimuli.

Nevertheless, using sinusoidal stimuli, a wealth of information on significant features of the auditory system has been collected. Over a limited range, the sensation of complex sounds can be predicted therefrom. Indeed, part of the data can be partially explained by the dynamic properties of some of the "subsystems." Doubtless the interplays are complex between the different centers of the brain, as are the functions of the efferent pathways. It should not be surprising, therefore, if some sensations must be described by factors other than those used in defining performance of the peripheral organs.

We summarize, in the following, a small selection from the great amount of quantitative data available. The primary goal of this section is to discuss performance of the whole system, even though we are unable to present a complete systematic theory of hearing because of the system's adaptiveness, its capability for learning (and forgetting), the influence of attention, and, above all, lack of identification of truly independent attributes.

1. Thresholds and Psychophysical Dimensions of the Sensations of Pure Tones

It obviously is convenient experimentally to test the performance of a listener with sinusoidal stimuli. However, as emphasized previously, it is unlikely that the auditory system has evolved into an optimal detector and discriminator of pure tones. Nevertheless, some of its basic features can be described in terms of tonal sensation, as summarized by Licklider [6.31].

The *absolute threshold* of tonal audibility, often defined as the intensity of stimulus allowing for 50% chance of detection, is a function of frequency. In Fig. 6.16, the lowermost curve represents the average or typical minimal audible (pressure) field, called MAF, for persons under 20 years of age.

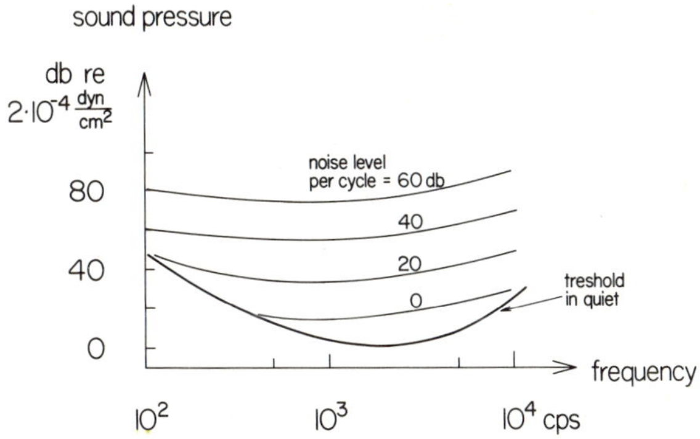

Fig. 6.16. Threshold of hearing as influenced by wide band noise ("masked threshold"). The ordinate gives the sound intensity which a tone stimulus must reach to be perceived against the noise of the spectral density level shown as parameter. After Licklider [6.31].

Individual differences as well as those due to differences in experimental technique may raise or lower the MAF by 5 to 10 dB. The value of the MAF at 1000 Hz is 0.0002 dyne/cm^2, which corresponds to an energy flux of 10^{-16} W/cm^2. This intensity level is often used as a reference for any stimulus. Tones at frequencies below 20 Hz are detected if their intensity is appropriately raised, and the MAF function exhibits quantum jumps at those frequencies.

Zwislocki [6.3] compared the MAF function with the (mechanical) transmission function of the ear. The shapes of these two frequency functions are quite similar, although the audibility curve moves progressively away from the ear's transfer function. The threshold of hearing does not increase as much when frequency increases. The two curves deviate from each other by approximately 3 dB/octave. The uppermost curve in Fig. 6.16 indicates the threshold of "feeling" the sound, or tickle. Tone intensities obviously do not have to be raised to that level before becoming annoying.

All curves below the "pricking" threshold indicate the shift of MAF if white, or broad-band, *noise* is added to the stimulus. These curves are labeled according to the power density of the noise. Fletcher (see [6.31]) found that this type of masking does not require white noise. The threshold shift is a constant, provided the spectrum of the noise spreads beyond a certain bandwidth, the *critical band*. The threshold shift becomes, when expressed in terms of noise power within the critical band, a linear function of this effective power which is independent of frequency. The width of the critical band, however, depends on its center frequency. Figure 6.17 plots the critical

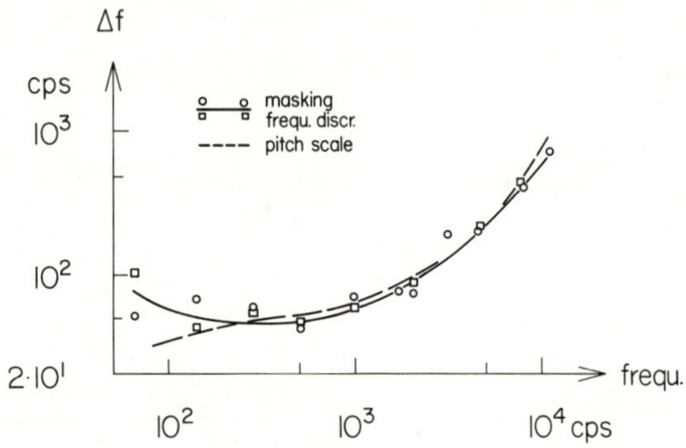

Fig. 6.17. Critical band Δf, the bandwidth of noise contributing to masking of a tone at the center of the band, as a function of frequency. The data pertain to (*a*) direct masking experiments (circles); (*b*) frequency discrimination (squares), 20 jnd's are graphed against frequency; (*c*) pitch scale, the dashed curve shows the width in frequency of intervals that are 50 mels wide in pitch (cf. Fig. 6.20). The similarity of the curves may suggest a common basis in the auditory mechanism. From Licklider [6.31].

bandwidth against frequency. The data, which compares poorly with the mechanical tuning characteristics of the displacement of the basilar membrane, as explained in the previous section, can be related quite closely, however, to the auditory system's ability to discriminate frequency changes [6.3, 6.31].

The capacity of a listener to discriminate between tones of slightly different frequencies or intensity is often determined in order to yield the inverse measure: the *differential threshold*, or the just noticeable difference (jnd) which is also a statistical measure. Although it is not easy to determine the jnd's, since abrupt changes cause the spectrum of the stimulus to spread, both the difference thresholds for frequency and intensity have been obtained. From Fig. 6.18 we see that the jnd in frequency is approximately 3 Hz below 1000 Hz, above which the relative jnd$(= \Delta f/f)$ is approximately 0.002 (i.e., a constant, so long as the intensity is moderate). Figure 6.19 shows the difference limen for intensity, with logarithmic scaling. The relative jnd is a constant (0.5–1 dB within limited bands in intensity and frequency; i.e., 30–70 dB *re* 0.0002 dyne/cm^2, and 60–10,000 Hz).

Pitch, the subjective attribute of a tonal sensation pertaining to the stimulus frequency, can be quantitated (e.g., by the method of magnitude estimation) and is expressed in *mels*. Arbitrarily, 1000 mels is assigned to the pitch

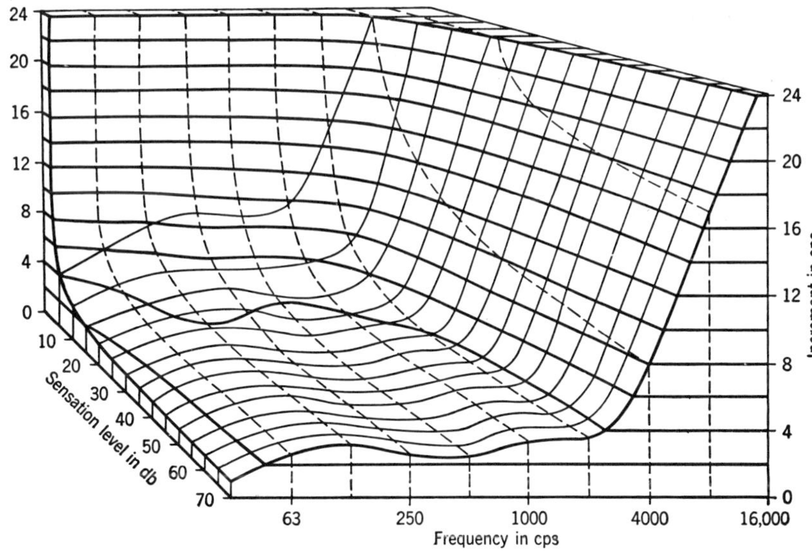

Fig. 6.18. Surface showing the differential frequency-discrimination threshold as a function of frequency and intensity of the standard tone. Near absolute threshold (rear part of figure) and at high frequencies (right-hand part of figure), frequency discrimination is rather poor. However, changes of the order of 3 Hz may be detected in stimuli below 2000 Hz and above 30 dB *re* absolute threshold (i.e., 0.0002 dyne/cm^2). From Licklider [6.31].

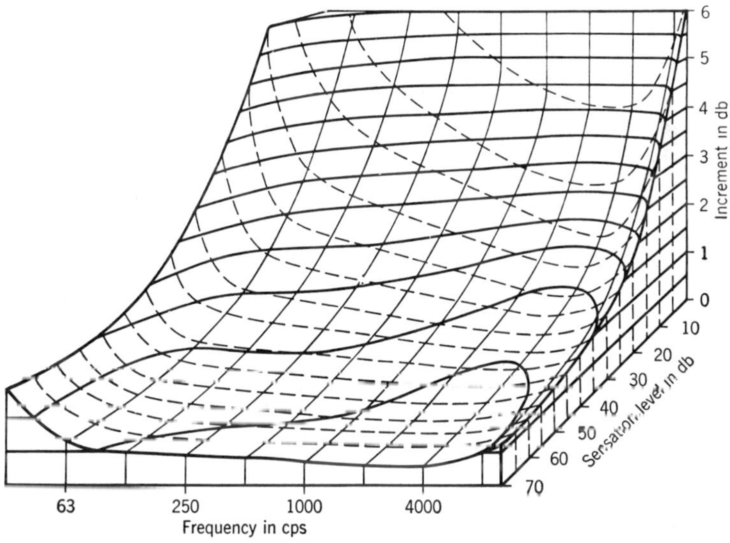

Fig. 6.19. Surface showing the differential intensity discrimination threshold as a function of frequency and intensity of a standard stimulus. The just noticeable difference (jnd) is given in decibels. For example, at 1 kHz and 30 dB above absolute threshold, stimulus intensity must be raised by 1 dB before the average listener can detect a change. From Licklider [6.31]

sensation of a tone of 1000 Hz, 40 dB above its audibility threshold. The tone judged to have half the pitch is assigned 500 mels; it corresponds to approximately 400 Hz. The subjective pitch level increases at a progressively faster rate when stimulus frequency increases logarithmically. Pitch in mels increases by approximately 60%/octave between 100 and 4000 Hz. The pitch function (Fig. 6.20) is virtually independent of intensity. It also can be directly related [6.3, 6.31] to the integral of the relative difference limen of frequency detection.

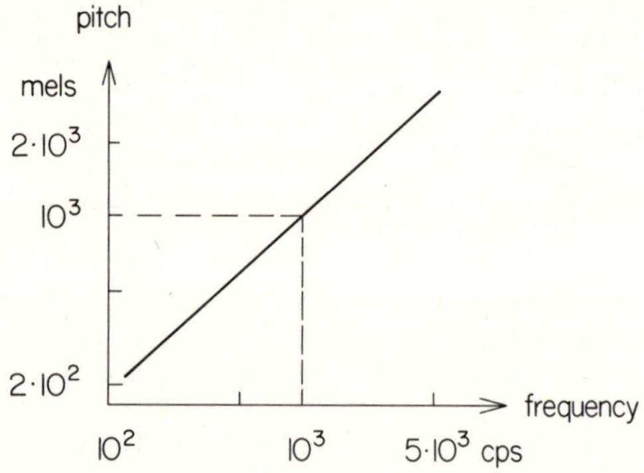

Fig. 6.20. Pitch, the subjective experience of the "frequency" of a tonal stimulus as a function of frequency. The pitch scale is standardized to 1000 mels at 1000 Hz (40 dB *re* threshold). It is a power function of f (i.e., a straight line in the log–log diagram, at least over the frequency band shown). Deviations from the line due to variations in intensity are small. After Licklider [6.31].

The subjective experience of *loudness* is measured in *sones*; 1 sone is assigned to 1000 Hz at 40 dB above threshold; 2 sones is judged to be twice as loud, 0.5 sone half as loud as 1 sone, and so on. Loudness depends not only on frequency, as does pitch, but also on the intensity of the stimulus. Figure 6.21 plots the loudness surface of an average listener. Thus a tone of 100 Hz at 75 dB (*re* 0.0002 dyne/cm^2) is judged to have loudness equal to a tone of 500 Hz at 65 dB. Hellman and Zwislocki investigated the effect of masking by noise on the sensation of loudness [6.32]. It is interesting to note, with Stevens [6.33], that the loudness–intensity curves approximate straight lines in log–log plots if intensity is measured with reference to the psychological zero, the threshold of audibility. This is S. S. Stevens's "power law," found to hold in many other sensory modalities, according to which sensation is proportional to the stimulus (above threshold) raised to the nth power. The n takes on different values for different sensors; for monaural loudness it is 0.55, for binaural loudness, 0.60. There exists only one intensity level at which binaural doubles monaural sensation. Note also that the pitch

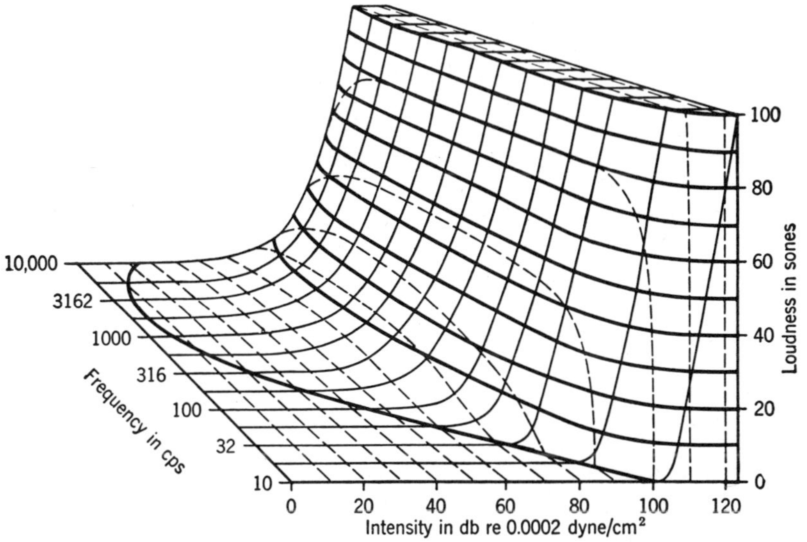

Fig. 6.21. Surface showing loudness (i.e., the subjective experience of intensity) as a function of frequency and intensity of the sinusoidal stimulus. Note that the loudness–intensity relation, at a fixed frequency, becomes a straight line when plotted on log–log paper; that is, it is described by a power function (see Stevens [6.33]). From Licklider [6.31].

scale becomes a straight line in a log–log graph. Stevens himself pointed out, however, that *deviations from a power law* might indeed point the way to the discovery of new factors in sensory processes. See also p. 169.

It should not be surprising that tonal experience has attributes other than pitch and loudness. The auditory system should not be compared with a simple frequency analyzer. *Volume* is found to be an independent attribute, although a volume scale has not been worked out. Tones of different frequencies and intensity may appear to be of equal "bigness" or volume. Licklider [6.31] shows an intensity–frequency graph containing equal-volume curves. *Density and brightness* of complex sounds let us differentiate a clarinet from a recorder. It is difficult, however, to establish scales for these attributes describing experiences of many different listeners.

Masking of one tone by another, a phenomenon known to exist peripherally (see above), is also found on the psychophysical level. Tones at lower frequencies mask those at higher frequencies more effectively than vice versa, and this is especially true if the tones are of comparable frequency. Very little masking occurs if one tone is applied to one ear, the other to the contralateral one. The rate at which masking increases with increasing intensity depends on the frequencies involved. If masking is not complete, different regions (in an intensity–frequency plot) exist for different sensations such as mixture of tones, beats, and difference tone with or without primary and secondary tones. Temporal effects become quite important if the stimuli are short wave trains [6.31].

2. Sensation of Complex Stimuli; Temporal Effects

Many more interesting phenomena are found when a listener is exposed to bursts of sine waves, noise, or maintained complex sounds. The *loudness* scale for complex sounds differs from that of tones. As might be expected from the observation on masking, the sensation of several simultaneous tones is not the sum of the individual loudnesses. But the loudness scale of noise lies above that of tones, as in Fig. 6.21. *Pitch* takes on a different meaning in the case of complex tones. Here it is found that the missing fundamental is experienced for tones that contain a largest common factor. Even if the component frequencies are incommensurable, the semiperiodically recurring waveforms of their envelope determine the pitch sensation.

Similarly, when a sound consists of three components of frequencies F, $(F + f)$, and $(F - f)$ (where $f < F$) the sensation of that tone varies as the phase of the "carrier" at F changes. When adhering to the hypothesis that the auditory system functions like a spectral analyzer, we would not admit this "phase sensitivity." On the other hand, the amplitude function of the total signal assumes different shapes when the phase of F is varied. It is known from physiological data on neuron activity in the cortex that information must be available about rapid changes of stimuli (see previous section). It is thus not difficult to hypothesize that certain nerve action patterns are responsible for sensing envelopes of signals. Indeed, periodic noise bursts have a pitch at their repetition rate. We do not need to look for nonlinearities in the ear's structures creating subharmonics.

Temporal integration becomes important in the sensation of short stimuli. Absolute, difference, and masked thresholds increase when the duration of a signal decreases. For short bursts of sine waves, a "click" is detected at the onset (i.e., loudness increases rapidly and then decays again). The tone burst assumes a pitch if 2 to 3 full cycles or more are presented (at frequencies below 1 kHz) or if the burst lasts more than 5 to 10 msec. The end of the burst again is perceived as a click. (It is interesting to note that a continuous frequency–time–intensity spectrum for a tone burst, such as is seen in the visible speech apparatus, is sensed in the visual system in a way comparable to the sensing of the acoustic stimulus by the auditory system.) To keep the loudness constant, von Békésy (quoted in [6.31]) varied the duration (T) and intensity (I) of moderately long tone bursts as follows:

$$\log I = -k \log T + \log I_0$$

or

$$\text{sensation } S \propto I_0 T_0 = \frac{IT}{T^{1-k}}$$

The latter expression shows that the energy of the stimulus must be kept almost, but not quite, constant in order to cause constant sensation.

The localization of a sound source depends not only on auditory stimuli but also on information from other sensory systems, at least outside the laboratory. Slight movements of the head are important for estimating the elevation of a sound source and for avoiding confusion between sources in the front and in the rear. Head movements are probably also important in the simpler task of deciding about the direction from which a sound comes in the horizontal plane. When a listener wears head phones that are driven individually with sounds of different frequencies or intensities, the source appears to be close to or within his head. This does not occur when locating the position of a loudspeaker in an echoless room. This "externalization" probably is achieved on the basis of information gained by moving one's head.

Besides these complicating factors, localization of pure tone sources is achieved on the basis of phase and intensity differences between the stimuli at the two ears. Below 1500 cycles, the threshold of detection of phase differences is of the order of 3°; 3° at 100 Hz correspond to 80 μsec. Differences in arrival of clicklike stimuli may be smaller, approximately 30 μsec. (i.e., remarkably small). Intensity differences of the order of 1 dB (1.12%) are the threshold of perception used in localization of higher frequencies. Licklider [6.31] quotes Stevens's result on localization of a tone source in the horizontal plane. Errors are below 12° (in space) for $f < 500$ Hz, they increase to approximately 20° with frequency increasing to 3000 Hz, and then fall off again to 12° at 10,000 Hz. This achievement can be explained only by mechanisms that allow for subtle trading of phase and intensity information.

E. REFERENCES

6.1. Von Békésy, G. *Experiments in Hearing*. McGraw-Hill Book Co., New York, 1960.

6.2. Wiener, F. M. and D. A. Ross. The pressure distribution in the auditory canal in a progressive sound field. *J Acoust Soc Amer* **18**: 401–408, 1946.

6.3. Zwislocki, J. Analysis of some auditory characteristics, pp. 1 97 in: *Handbook of Mathematical Psychology*, Vol. 3. R. D. Luce, R. R. Bush and E. Galanter (Eds.). John Wiley and Sons, New York, 1965.

6.4. Von Békésy, G. and W. A. Rosenblith. The mechanical properties of the ear, pp. 1075–1115 in *Handbook of Experimental Psychology*. S. S. Stevens (Ed.). John Wiley and Sons, New York, 1951.

6.5. Zwislocki, J. Analysis of the middle-ear function. Part I: Input impedance. *J Acoust Soc Amer* **34**: 1514–1523, 1962.

6.6. Onchi, Y. Mechanism of the middle ear. *J Acoust Soc Amer* **33**: 794–805, 1961.

6.7. Von Békésy, G. Ueber die Messung der Schwingungsamplitude der Gehoerknoechelchen mittels einer kapazitiven Sonde. *Akust Z* **6**: 1–16, 1941.

6.8. Stuhlman, Jr. O. *An Introduction to Biophysics*. John Wiley and Sons, New York, 1943.

6.9. Dallos, P. J. Study of the acoustic reflex feedback loop. *IEEE Trans Bio-Med Engin* **BME-11**: 2–7, 1964.

6.10. Von Békésy, G. On the elasticity of the cochlear partition. *J Acoust Soc Amer* **20**: 227–241, 1948.

6.11. Wever, E. G. *Theory of Hearing*. John Wiley and Sons, New York, 1949.

6.12. Huggins, W. H. A Theory of Hearing. Electronics Research Directorate, Air Force Cambridge Research Center, Tech. Rpt. 53–14, Cambridge, Mass., 1953.

6.13. Zwislocki, J. Theory of the acoustical action of the cochlea. *J Acoust Soc Amer* **22**: 778–784, 1950.

6.14. Zwislocki, J. Theorie der Schneckenmechanik. *Acta Otolaryng (Stockholm)*, Suppl. **72**: 7–76, 1948.

6.15. Peterson, L. C. and B. P. Bogert. A dynamical theory of the cochlea. *J Acoust Soc Amer* **22**: 369–381, 1950.

6.16. Kiang, N. Y.-S., T. Watanabe, E. C. Thomas, and L. F. Clark. Stimulus coding in the cat's auditory nerve. *Ann Otol (St. Louis)* **71**: 1009–1026, 1962.

6.17. Huggins, W. H. and J. C. R. Licklider. Place mechanisms of auditory frequency analysis. *J Acoust Soc Amer* **23**: 290–299, 1951.

6.18. Huggins, W. H. A phase principle for complex-frequency analysis and its implications in auditory theory. *J Acoust Soc Amer* **24**: 582–589, 1952.

6.19. Tasaki, I. Nerve impulses in individual auditory nerve fibers of guinea pig. *J Neurophysiol* **17**: 97–122, 1954.

6.20. Weiss, T. F. A model of the peripheral auditory system. *Kybernetik* **3**: 153–175, 1966.

6.21. Galambos, R. and H. Davis. Inhibition of activity in single auditory nerve fibers by acoustic stimulation. *J Neurophysiol* **7**: 287–303, 1944.

6.22. Greenwood, D. D. and N. Maruyama. Excitatory and inhibitory response areas of auditory neurons in the cochlear nucleus. *J Neurophysiol* **28**: 863–892, 1965.

6.23. Goldberg, J. M., H. O. Adrian, and F. D. Smith. Response of neurons of the superior olivary complex of the cat to acoustic stimuli of long duration. *J Neurophysiol* **27**: 706–749, 1964.

6.24. Hall, II, J. L. Binaural interaction in the accessory superior-olivary nucleus of the cat. *J Acoust Soc Amer* **37**: 814–823, 1965.

6.25. Dewson, J. H. Efferent olivocochlear bundle: some relationships to noise masking and to stimulus attenuation. *J Neurophysiol* **30**: 817–832, 1967.

6.26. Galambos, R. Processing of auditory information, pp. 171–204 in *Brain and Behavior*, Vol. I. M. A. B. Brazier (Ed). American Institute of Biological Sciences, Washington, D.C., 1961.

6.27. Erulkar, S. D., J. E. Rose, and P. W. Davies. Single unit activity in the auditory cortex of the cat. *Bull Hopkins Hosp* **99**: 55–86, 1956.

6.28. Goldstein, M. H. Single unit studies of cortical coding of simple acoustic stimuli, pp. 131–151 in *Physiological and Biochemical Aspects of Nervous Integration*. F. D. Carlson (Ed.). Prentice-Hall, Englewood Cliffs, N.J., 1968.

6.29. Neff, W. D. Neural mechanisms of auditory discrimination, pp. 259–278 in *Sensory Communication*. W. A. Rosenblith (Ed.). John Wiley and Sons, New York, 1961.

6.30. Schwartzkopff, J. Hearing. *Ann Rev Physiol* **29**: 485–512, 1967.

6.31. Licklider, J. C. R. Basic correlates of the auditory stimulus, pp. 985–1039 in *Handbook of Experimental Psychology*. S. S. Stevens (Ed.). John Wiley and Sons, New York, 1951.

6.32. Hellman, R. P. and J. Zwislocki. Loudness function of a 1000-cps tone in presence of a masking noise. *J Acoust Soc Amer* **36**: 1618–1627, 1964.

6.33. Stevens, S. S. The surprising simplicity of sensory metrics. *Amer Psychol* **17**: 29–39, 1962.

6.34. Fletcher, H. A space–time pattern theory of hearing. *J Acoust Soc Amer* **1**: 311–343, 1930.

Part III

SENSORY–MOTOR SYSTEMS

CHAPTER SEVEN

Autonomic, Involuntary Sensory–Motor Systems

A. INTRODUCTION: GENERAL COMMENTS ON BIOCONTROL

In this chapter we treat examples of involuntary sensory–motor reflex systems, moving into the area of biocontrol. General principles of control systems in the animate world can be approached by considering the progressive complexity of biological linkages.

In *open* systems every process or physiological function involves continual or intermittent flows—inflow of material or information, bearing energy and specific organization, respectively, and the outflow of these, accumulated or transformed by the processes and systems involved. Glucose flow through the cell wall, blood flow through the heart, inflow of information to the central nervous system (there transformed into problem-solving activity, the outflow), are but a few examples. Many open systems are important in homeostasis. The mechanisms are very diverse.

Regulating mechanisms of diffusion processes may be fitted by linearly or proportionally linked models if concentration (rather than electrochemical potential) gradients drive the components and if no chemical reactions are involved (see Chapter 2). So-called active processes are involved if flows exist against electrochemical potential gradients. Extrinsic energy is then required for long-term stability. Thus pumping rates of the sodium pump are determined not only by concentrations at source and sink, but by energy fed to the pump. The steady states of these open systems show stationarity of flows rather than equilibria, as pointed out originally by von Bertalanffy [1.11].

Nonlinearities play significant roles in open regulatory systems. Stability against external disturbances often is achieved without feedback or other closed-loop systems. It may result from highly curvilinear functional relationships, or when one process operates on the enhancement or suppression of another. Such couplings are very common in the animate world. Examples are found in autoregulation of blood flow in vascular beds or in heterometric autoregulation of cardiac output (Chapter 15). Again, stability in these cases means dynamic equilibrium. Couplings between flows frequently are nonlinear and modulated. In nerve membrane (Chapter 2) the rate constants of fluxes depend on voltage, resulting—when triggered—in regenerative action on conductance (positive "feedback" between voltage and conductance). On the other hand, internal periodicity in pacemaker cells (e.g., of the heart) exists because of back-coupling. Other types of linkages in open systems involve time delays due to transport, potential drops, or gain of entropy. These phenomena cause a smudging of temporal and spatial patterns by multiple paths. Linkage mechanisms also may exhibit limited storage aspects, permitting congestion or overflow (i.e., an upper discontinuity of capacity). Control action in such open systems depends on unpredictable external information.

In the *open* biological *regulatory* system, at a particular level of organization, either the system itself is maintained at some reference level (e.g., homeostasis of body temperature, discussed in Chapter 19; electrolyte concentration; erythrocyte concentration), or it follows in time a prescribed control curve (eye tracking, Chapter 12, for example). In both cases, however, systems theory teaches us to include in the analysis the disturbances and influences of the outside, the environment, or the external load, as well as the stress applied to the system. Control curves can be time or position *functions* of a reference variable, and they move in an arbitrary way. Classical control theory rarely suffices for the mathematical treatment. In homeostatic as well as in "follow-up" systems, rate-dependent sensors commonly introduce essential dynamic nonlinearity; feed-forward, gain modulation, or parametric driving are ubiquitous. Control may be achieved through adaptation. Parameters thus may depend on the load, and setpoints or reference levels often are perturbed to obtain optimal task fulfilling (e.g., blood pressure control under stress, see Chapter 16). Optimal task fulfilling, however, may not involve energy criteria at all. In contrast to load adaptation, systems often reorganize to become task-adaptive. Predictive operators may be "switched" into the system if learning is possible. Control of hand movements (Chapter 11) can be highly precise, within a certain bandwidth of frequencies, if rhythmic predictable movements are made.

Intricate coupling, obviously also existing between closed regulatory systems, results in the need for a hierarchy of control. For example, cardiac output is regulated and partitioned to supply oxygen according to demand, to eliminate catabolites, and to serve thermocontrol. Not only must there be several controllers of flow, but also an *order* of activation, whereby subsidiary mechanisms serving one purpose (e.g., increasing cooling) are called on when cardiac output suffices for the other demands on it. In disease, deeper hierarchic levels are seen. In cardiac shock, cerebral blood flow gets preference over muscle perfusion. Perhaps the most central problem, on which biomedical engineers and physiologists have not focused quantitatively, is the hierarchic ordering of multiple controls on a single quantity (like flow) or function (like eye movement). An example is presented in Chapter 11.

Further challenges to the biomedical engineer will evolve. First, definitions of controlled and controlling variables must be sought from the viewpoint of the system's task. Second, if only overall responses (e.g., closed-loop performance) are known, it is impossible to establish a block diagram revealing the individual contributions of subsystems *and* of their linkages to overall behavior. Fortunately, many clues are available. In some cases open-loop performance and the degree of stability can be obtained (see Chapter 12). Many sensors and effector subsystems can be separated and their performances evaluated quantitatively. The baroreceptors, the Pacinian corpuscle, and the neuromuscular spindles are examples of the former; for the latter we can cite whole-muscle dynamics, the transformation of vagus nerve activity into heart rate, or the response in release of cortisol by the adrenal cortex to changes in adrenocorticotropic hormone (ACTH) (Chapters 6, 10, 16, 20). Thus block diagrams and signal-flow diagrams are employed to show relations between those variables, as well as those couplings between subsystems which are significant and/or essential for task fulfilling. This is important when it becomes necessary to differentiate models. Furthermore, the systems physiologist can devise, at many levels of organization, biologically significant measures for degree of stability and for its changes with age and disease, for the dynamics of reorganization in multiple controls, or for the ordering in hierarchical regulation and its dysfunction in pathophysiological states. Clearly, these not only contribute to understanding of systems behavior but also to diagnosis and treatment.

Relatively few regulatory systems have been investigated and analyzed quantitatively. Even fewer studies concern modes of coupling between higher and lower levels of control. The "classical" theory on feedback servomechanisms has been applied extensively (see [0.8, 0.9]. Both analog and digital computers have been employed. However, full advantage has not been taken of recent work in optimal control theory which may be particularly suited to study of many biological regulatory systems.

B. THE PUPIL–LIGHT REFLEX; UNDIRECTIONAL RATE SENSITIVITY

The overall response of the pupil–light reflex loop can be studied almost ideally in intact, nonanesthetized animals and men. The differential reflectivity of iris and pupil for infrared light allows us to measure pupil area at these wavelengths without interfering with the system. Furthermore, the feedback loop can easily be opened by projecting onto the retina a light beam narrower than the pupil opening [7.1]. The pupil reflex, contributing to the adaptiveness of the visual system, is fast. Within 0.5 to 1.0 sec the light intensity entering the eye can be changed, maximally, by a factor of 20 to 25. The retinal adaptation of the threshold of vision is much slower (10 min from light to dark, 10 sec from dark to light), but it has a dynamic range of up to 10^7.

Afferent nerve fibers of the pupil reflex loop, carrying information on light intensity, belong to the optic nerve. They branch off the main optical pathways before these enter the lateral geniculate body and pass to the pretectal nucleus, where connections to the contralateral hemisphere are made. Then they enter the oculomotor nucleus. Two sets of *efferent* pathways are known. One passes by way of the inferior ciliospinal centers to the dilator muscles of the iris, where increased activity of these sympathetic fibers causes the pupil to open. The other, belonging to the parasympathetic system, innervates the iris sphincter. Direct stimulation of the Edinger-Westphal nucleus causes the sphincter to constrict. Modulation of the activity of these latter (parasympathetic) fibers allows for both constriction and dilation; this may mean man does not possess a true agonist–antagonist system. There is evidence that the steady-state dilation is activated mainly by the sympathetic pathways, whereas steady-state closure and the transients are controlled by the parasympathetic ones [7.2, 7.3]. Note that in near vision the pupil is also coupled to the accommodative reflex, although the converse is not true. Each has its own feedback loop. Since there is but one pupillomotor center, however, the pupil reflex is "consensual." See [7.4] for more detailed description of the neural pathways.

Thus we can draw a schematic block diagram as in Fig. 7.1. The input of the system is the light intensity (L). The task of the system, however, is not to follow L. Because of the multiplier unit (retinal intensity $I = L \times$ area of pupil), the intensity L is modulated so that variations in retinal illumination are minimized. This is done by sensing the "output," the (net) neural activity A of the optic nerve, and by adjusting D, the pupil diameter. Information content of the optical image, carried by other variables than average flux, is disregarded.

Clearly the "black boxes" can be arranged in different order. Stark [7.1], for example, let the light intensity reaching the retina be the controlled variable. He further regarded L not as the input but rather as a disturbance

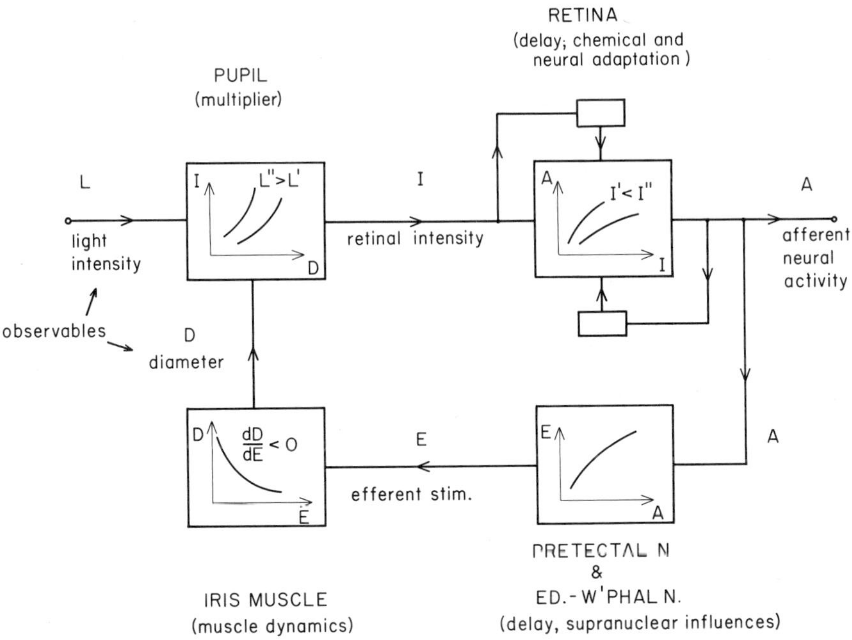

Fig. 7.1. Information flow diagram showing the main subsystems of the pupil–light reflex loop. The graphs given in the four "black boxes" represent their *steady-state* input–output relation in a qualitative manner. The pupil transfer function is equivalent to a multiplier: $I = L \times \pi D^2/4$. The two unlabeled pathways controlling retinal function (adaptation, spatial summation, etc.) indicate that both light intensity and neural activity determine the "transfer function" of the retina. The negative slope of the D–E relation of the iris muscle causes the "sign inversion" needed in the negative feedback loop. As explained in the text, the loop is opened conveniently between the "black boxes" of the iris and the pupil multiplier. Open-loop performance is obtained by relating responses in D to stimuli in L.

introduced into the forward path of the feedback system. With respect to Fig. 7.1, we should emphasize two points. First, the characteristics as shown for the individual components give their input–output relations in a qualitative manner only. Second, not all dynamic aspects of the *sub*systems are known in detail. We can only state with certainty that the muscle dynamics of the sphincter, the directional rate sensitivity and delay of the receptors, and the (synaptic) delays in the central nervous system must play important roles in this regard.

However, we can study again in a qualitative way the possible effects, both chemical and neural, of the slow adaptation at the retinal level. This is shown in the lower half of Fig. 7.2. By simple graphical construction, developed by Jones [7.5], the operating points of the subsystems are obtained. Start from

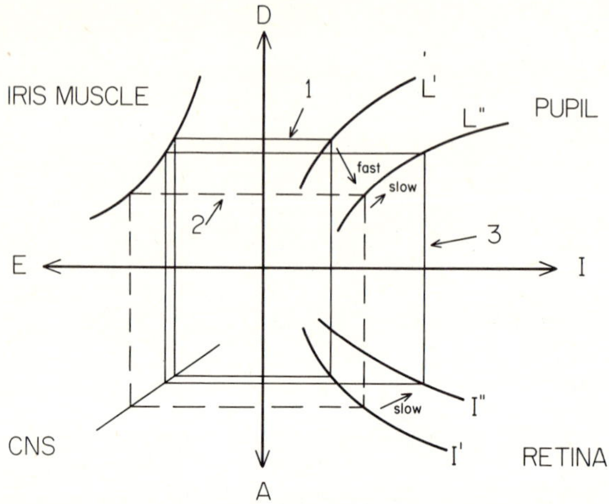

Fig. 7.2. Operating conditions of the pupil–light reflex, obtained by graphical construction (rectangles *1*, *2*, *3*) determining simultaneous equilibrium or steady-state points of the four subsystems involved. Fast changes from rectangles *1* to *2* move the pupil onto curve L''. Slow retinal adaptation changes the "gain" of the retina from curve I' to I'', which causes all other systems to attain the state defined by rectangle *3*. Clearly the sequence of cases *1*, *2*, and *3* occurs if light intensity is suddenly increased. See Fig. 7.1 for definitions of symbols. After Jones [7.5].

the initial solution, at the given light intensity, defined by rectangle *1* in Fig. 7.2. A fast increase in light intensity shifts the operating points to rectangle *2*, because of the jump from L' to L'' in the graph depicting the iris functions. After approximately 10 sec, retinal adaptation has taken place which results in a decrease of the *loop gain.* Then characteristic I'' in the retinal diagram applies, and a new solution, rectangle *3*, results. This brings the iris diameter D back to a higher level that is determined by the static gain of the system. Clearly the negative sign of the feedback occurs because increased efferent muscle stimulation E, due to increased light on the retina I, causes a sphincter to *contract*, thus decreasing retinal illuminance.

As pointed out earlier, discussion of contributions of individual components to the system cannot be carried further because of present lack of quantitative knowledge. On the other hand, *overall* behavior of the closed or open feedback loop is amenable to analysis. At the outset, existing nonlinearities should be noted. Their effects are obvious when responses to stepwise increases and decreases in light intensity, or to light pulses, are observed. Similarly, responses to sinusoidal intensity variations are distorted. In the latter case, however, we might be tempted to term the distortions small and to continue with a linear analysis, unless drastic changes in such a "describing function" are established by varying *mean* light intensity over large ranges.

Figure 7.3 presents experimental results reported by Stark [7.6] and by Clynes [7.3]. Stepwise decreases in light stimulus elicit very small and gradual changes in pupil area. After a delay of approximately 0.25 sec, responses to stepwise increases in light show a large overshoot before they settle at the level given by the D.C. response. The pulse responses correspond; light flashes cause vigorous temporary pupil contractions. Responses to "dark flashes" *also* are contractions, somewhat smaller, which are in fact responses

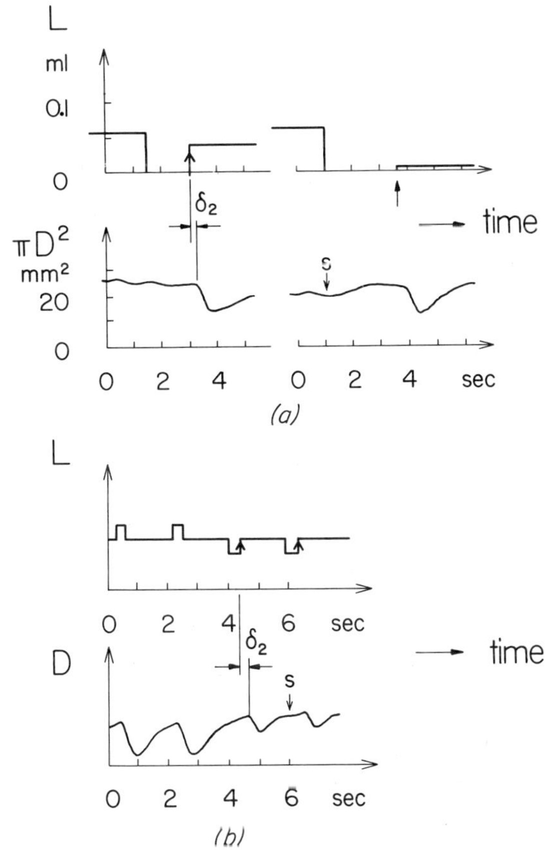

Fig. 7.3. Pupillary responses to positive and negative steps in light intensity (L) and to impulses demonstrating dynamic nonlinearities of the system. Note that vigorous responses are elicited, with delay δ_2, only by positive-going changes in L. This is called undirectional rate sensitivity. Negative rates of change cause only small slow changes, marked *s*, due to a proportional controller. In (*b*), the response to positive-going flanks of the "darkness" flashes is smaller than that to light flashes, since the static diameter–light curve is nonlinear, and since the action of the proportional controller is encountered, too. After Clynes [7.3] and (*b*) Stark [7.6].

to increase in light at the lagging edge of a darkness pulse (as marked by arrows in Fig. 7.3). We must therefore conclude, with Clynes [7.2, 7.3], that the fast-acting parts of the system exhibit *undirectional rate sensitivity* (URS). Only positive light steps cause immediate, rapid pupil contractions. Clynes [7.2] separated these from the symmetric responses of the system sensing changes in steady, average light intensity. The following two differential equations result [7.2, 7.3], and they predict all responses with satisfactory accuracy:

Open loop, reacting to changes in steady light:

$$\frac{dD}{d(t+\delta_1)} + \frac{1}{\tau_1} D = -\frac{b}{\tau_1} L(t) \tag{7.1}$$

Open loop, reacting rapidly to positive light changes:

$$\frac{d^2D}{d(t+\delta_2)^2} + \frac{2}{\tau_2}\frac{dD}{d(t+\delta_2)} + \frac{1}{\tau_2^2} D = -\frac{a}{\tau_2^2}\left[\frac{dL}{dt}\right]^+ \tag{7.2}$$

where D = diameter of pupil, a function of time = $D(t)$
$L(t)$ = light intensity
δ_1 = delay of slowly reacting, symmetric system; ca. 0.35 sec
τ_1 = time constant, ca. 0.4 sec
a, b = sensitivity "constants," depending on retinal adaptation
δ_2 = delay of rapidly reacting system exhibiting unidirectional sensitivity; ca. 0.25 sec
τ_2 = time constant of fast system; note that $1/\tau_2$ is a double pole of the system

In eq. 7.2 the right side is nonzero for positive dL/dt only. Hence we also can give the Laplace transform of the system's performance if we introduce [7.3] a special rate-sensitive operator. If we let

$$\Omega = \begin{cases} 1 & \text{if } \dfrac{dL}{dt} \geqslant 0 \\[2ex] 0 & \text{if } \dfrac{dL}{dt} < 0 \end{cases}$$

then

$$\frac{D(s)}{L(s)} = -b\,\frac{e^{-\delta_1 s}}{1+\tau_1 s} - \Omega a\,\frac{s e^{-\delta_2 s}}{(1+\tau_2 s)^2} \tag{7.3}$$

Equation 7.3 does not incorporate static, or steady-state, nonlinearities. It shows that the principle of superposition applies in each of the two domains defined by the states of Ω. For negative dL/dt, solution of eq. 7.3 depends only on the system's parameters, that is, only transients are seen, since the forcing function is identically zero. This is important when "oscillations" are elicited by placing a small, stationary light beam on the edge of the pupil. That portion of the beam which enters the eye causes the pupil to constrict. This in turn reduces retinal illumination to zero (assuming large enough initial intensity entering the eye), whereupon the iris dilates with a small delay. In other words, the feedback loop is opened periodically. The function $D(t)$ then is a composite of transients of eqs. 7.1 and 7.2. This is unlike a true feedback oscillation which is sustained just because the feedback loop is closed. A stability study such as Llaurado's [7.7] is therefore inapplicable to the pupil servomechanism, although his methods are useful for other systems showing transport lags.

If there existed static nonlinearities only, such as indicated in Fig. 7.1, we probably could treat them as nonessential deviations from linear laws, here as well as in many other biological systems. We then could indeed carry out a small signal analysis (i.e., assume linearity in a first-order approximation and improve results by the method of perturbations). The higher harmonics created by a sinusoidal stimulus, applied to the system containing *static* nonlinearities, would be found to depend strongly on the amplitude of the stimulus. On the other hand, *dynamic* nonlinearities such as those introduced by unidirectional rate sensitivity are essential nonlinearities, and these cannot be linearized. (When insisting on sinusoidal analysis, we realize that the distortions depend also on the frequency of the stimulus; see below.)

A further consequence of URS is that responses to *impulselike* stimuli of opposite polarity do not cancel each other. They rather reinforce each other, so that the response to a doublet pulse (the time derivative of a Dirac impulse) is twice the response to a unit impulse, independent of the polarity of the doublet. As a corollary, once an impulse response is elicited, it cannot be abolished by an equal and opposite impulse applied to the input.

Frequency-dependent rectification is an additional phenomenon introduced by dynamic nonlinearities. Periodic stimulation at one frequency (e.g., 2 Hz) causes the pupil to contract and dilate with a smaller amplitude, centered at a lower average opening, than when exposed to a wave train of lower frequency (0.25 Hz) and the same amplitude. This is clearly demonstrated in Fig. 7.4. This phenomenon occurs within the low-pass frequency band of the left side of eq. 7.2. Figure 7.4 also shows how an amplitude-dependent rectification effect occurs under dynamic conditions. Furthermore, during the first few cycles of a finite pulse train at the input, a "Treppe" or staircase phenomenon is seen (i.e., a gradual buildup of a D.C. shift, observed in several other biological systems). Indeed in certain frequency ranges the D.C. shift, or mean level of the pupil opening, depends on the product of

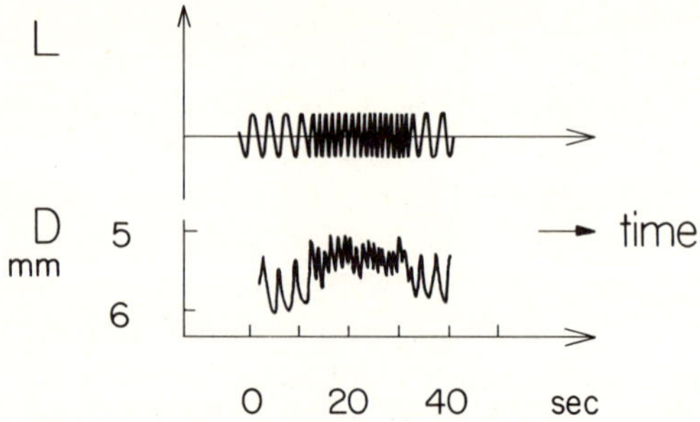

Fig. 7.4. Response of pupil to sinusoidal light variations of different frequencies. Note that the response is not sinusoidal due to the dynamic nonlinearity inherent in unidirectional rate sensitivity. This nonlinearity causes a rectification effect seen in the change in mean diameter during the period of rapid stimuli. This represents a widening of the effective bandwidth of the system. After Clynes [7.3].

frequency and amplitude. This, however, is a measure of maximum velocity of the input, at least for sinusoidal stimuli. Thus, Clynes pointed out, the system senses frequency (the pupil opens and closes at the same frequency as the stimulus waxes and wanes) as well as peak velocity (average diameter of iris). He further showed that the latter information still is conveyed when the input frequency is well above the "cutoff frequency" of the system ($1/\tau_2$). Hence the effective frequency response is of wider bandwidth than might be determined from a Bode diagram (i.e., amplitude and phase versus frequency, applicable to systems without essential nonlinearities).

Some of the most striking observations concerning URS are summarized in this section because this type of nonlinearity is found in many biological systems. As the next section reveals, this nonlinearity plays a role in the reflex arc coupling respiration and heart rate. The positive rate sensitivity of the baroreceptors in circulation is discussed in Chapter 16. As Chapter 4 demonstrated, conductance changes in the membrane of the primary receptors of the *Limulus* eye, when stimulated by a light step, follow the time course of a rate-sensitive element. The responses of the Pacinian corpuscle, as well as of the neuromuscular spindle, are strongly rate sensitive. This phenomenon is partly due to the (mechanical) dynamics of the transducer structure [7.8]. In other cases the mechanisms are not understood. Clynes [7.3] discussed systems in which chemical reactions are important links in the chain of events at the transducer. There, URS results, since negative chemical concentrations cannot exist. (See also Chapter 16: effect of vagal activity on heart rate.)

Regardless of whether the underlying mechanisms are known, the resulting dynamic asymmetries can be analyzed. They are of physiological importance at the next higher level of organization, as is shown with this example of the pupil–light reflex.

C. THE PNEUMOCARDIAC REFLEX

Analysis of the pneumocardiac reflex, manifested by the waxing and waning of heart rate with respiration, was worked out by Clynes [7.9, 7.10]. He showed how unidirectional rate sensitivity requires two channels of sensory information transmission if stimuli changes of both positive and negative sign are to be processed. His approach is generally useful to the model builder. A set of differential equations representing an initial "working hypothesis," derived to explain available data, serves in designing experiments that yield new data and in predicting their outcome. In the case under discussion, Clynes explains with his first hypothesis the temporal pattern of heart rates caused by positive and negative "steps" in chest diameter (e.g., "hold breath—inhale—hold breath"). This initial set of differential equations cannot predict, however, the heart rate responses to pulselike changes in respiration ("hold breath—exhale and inhale rapidly—hold breath"). Using the data on these impulse responses, the model can be improved to allow prediction of observations on heart rate over a wide variety of respiration patterns. Further steps in the development incorporate tests of the hypothesis using data on many different individuals. In the course of sharpening a hypothesis, it becomes easier to predict general aspects of the organization of the system as well as component properties; that is, further questions relating to the system can be asked more succinctly.

Heart rate is controlled by many regulatory loops. Afferent information about average and pulsatile blood pressure from baroreceptors in the aortic arch and the carotid sinus reaches the vasomotor center in the medulla. Efferently, parasympathetic and sympathetic nerves exert control over the heart rate. These are the dominant, direct-acting cardiovascular reflex loops. (These and additional factors are discussed in Chapter 16). In addition, continuous waxing and waning of heart rate, roughly synchronous with inspiration and expiration, are conspicuous in men and animals. Lung perfusion is enhanced when ventilation is maximal. Fluctuations in heart rate usually elicit similar changes in blood pressure. Clearly this must be caused by a fast-acting linkage between the respiratory and cardiac motor systems, which therefore can be studied separately from such long-range effects as those determined by the P_{O_2} and P_{CO_2} regulatory loops.

A few basic observations on the coupling of respiration with heart rate allow us to describe more carefully, although still qualitatively, the probable chain of events. Artificial respiration during apnea causes heart rate fluctuations. This means that proprioceptors in the lungs and chest wall influence the vasomotor center directly. The possibility that signals from the pneumotaxic and apneustic centers may flow to the vasomotor center, in a "feed-forward" manner, is not rigorously precluded by this observation. However, the information from the peripheral proprioceptors is dominant. Furthermore, abrupt breathing pulses cause equally large or larger transients in heart rate than does the normal respiratory cycle. Shifts in blood distribution, however, must be larger during the latter. The hemodynamic states of the left atrium and the pulmonary veins therefore have little direct influence on heart rate.

Hence it is very promising to carry out a computer simulation postulating the following two mechanisms: (*a*) proprioceptive receptors in chest wall and lungs are the sensors whose afferent signals strongly influence the vasomotor centers in the medulla, and (*b*) the efferent signals affect the heart rate, by way of the parasympathetic fibers of the vagus nerve. The entire system causes transient cardiac acceleration during inspiration and a transient slowing of the heart rate with expiration (see Fig. 7.5). The time constants involved are of

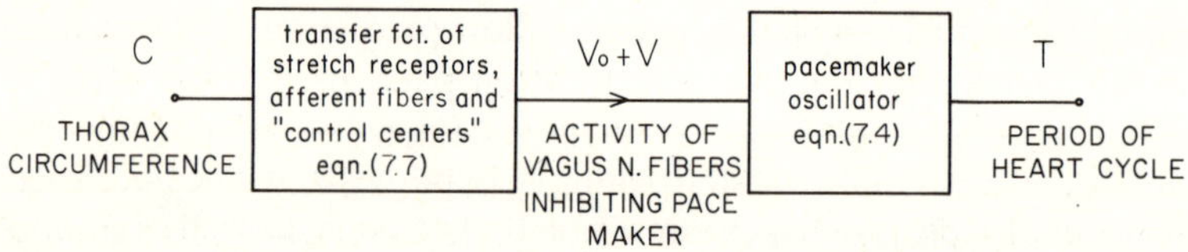

Fig. 7.5. Functional subsystems of the pneumocardiac reflex causing fast heart rates during inspiration and slow ones during expiration. The vasomotor center action in the medulla of the brain is included in the subunit on the left. Its dependence on information inflow from baroreceptors and other centers is assumed not to play any role in resting man during the study of this reflex. The two sets of equations (7.4 and 7.7), which include unidirectional rate sensitivity, suffice to predict heart rate changes due to a great variety of breathing patterns. After Clynes [7.9].

the order of a few seconds. We therefore can consider this system independently of the regulatory loops that control blood P_{O_2}, P_{CO_2} and pH. These loops also act on heart rate; their effective time constants are longer, however, due to the large storage capacities in lung, blood, and tissues (see Chapter 18).

Detailed description of the transfer function relating heart rate with parasympathetic (and sympathetic) activity is deferred until Chapter 16, where the manifold cardiovascular reflexes are discussed. Here we follow Clynes's approach [7.9] in giving a phenomenological equation fitting the observed data on vagal activity and heart rate, rather than explaining these in terms of mechanisms. This enables us to concentrate on the particular performance of the whole system.

The law describing the heart rate as a function of efferent vagal inhibition may be given by a differential equation of an oscillator. Its frequency (ν) is that of the pacemaker at the sino-atrial (SA) node and can be made dependent on changes in net firing rate (V) of the parasympathetic fibers innervating the right atrium as follows:

$$\frac{d^2y}{dt^2} + [(2\pi\nu_0)^2 - (V_0 + V)]y = 0 \tag{7.4}$$

where
y = oscillatory output of pacemaker whose frequency ν equals the heart rate
t = time
ν_0 = heart rate when vagal inhibition is absent
V_0 = a signal proportional to average vagal "tone"
V = change in vagal tone due to respiration; a function of time

Equation 7.4, describing parametric forcing of the pacemaker frequency, enables us to give the *steady-state* relation between output rate ν and $V(t)$:

$$\nu = \frac{1}{2\pi}[(2\pi\nu_0)^2 - (V_0 + V)]^{1/2} \qquad \text{(real only)} \tag{7.5}$$

Clearly ν_0 is the maximum frequency of the "free-running" sinoatrial node (ca. 120/min). At the other extreme, when $(2\pi\nu_0)^2 = V_0 + V$, the frequency of the oscillator becomes zero, corresponding to the (temporary) arrest of the heart when the vagal nerve is maximally stimulated at approximately 12 pulses/sec or at a higher rate. Figure 7.6 shows the heart period ($T = 1/\nu$) for the intermediate range of vagal tone. Comparison with Fig. 16.4 indicates that the basic characteristics of this steady-state relation are fairly well represented by eq. 7.4, especially at large vagal firing rates where Warner's model fails to predict cardiac arrest. Heart periods during *transients* in $V(t)$ also lie on the curve in Fig. 7.6. Details of the temporal pattern of the firing of the SA node, however, depend on the timing of the onset of the transient within the heart cycle. Equation 7.4 is of convenient form for solution by analog computer. Other forms of voltage-controlled oscillators, showing the

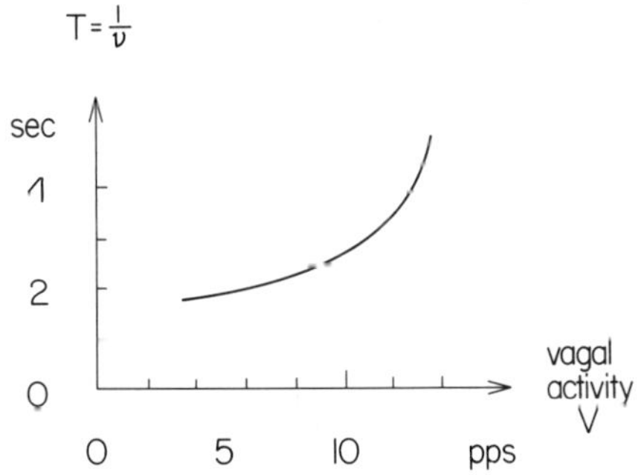

Fig. 7.6. Steady-state solution of eq. 7.4 relating the equilibrium heart period T with efferent vagal activity. Clynes's model of vagal control fits experimental data well in the range of heart periods shown. Very short periods (high heart rates) are predicted better by Warner's model, see Chapter 16.

basic characteristics of Fig. 7.6, certainly are admissible. It also should be noted that the asymmetric, rate-sensitive response of the heart rate to changes in vagal tone (Fig. 16.5) is not included in this model. Clynes [7.9] demonstrated that the dynamic nonlinearities in the proprioceptors and the neural centers dominate in shaping the entire system's transfer function.

Clynes arrived at a transfer function relating thorax dimension to change in vagal inhibition (i.e., the first "block" in Fig. 7.5) by systematically improving his hypotheses, using more diverse experimental data. We refer the reader to the original papers [7.9, 7.10] for developmental details. However, three basic features of the natural response of heart rate to abrupt respiratory changes enable us to predict with satisfactory accuracy the functions that must be included in the "subsystem" under discussion. They are:

1. The heart rate changes exhibit a biphasic time course for monophasic changes in chest diameter (Fig. 7.7*a*). Cardiac acceleration is followed by deceleration. Since the dynamic solutions of eq. 7.4, describing the second block in Fig. 7.5, do not allow for any such "distortions" (i.e., double differentiation) of its input, the first block must include such an operator. Furthermore, the temporal spread of these biphasic variations in T indicates two lag networks. We therefore can postulate that the ratio of vagal activity

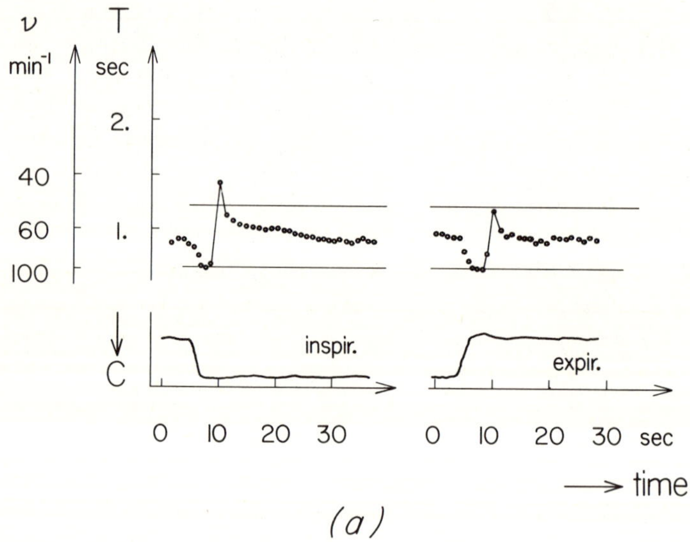

Fig. 7.7*a*. Overall response of the pneumocardiac reflex arc observed in man. Stepwise changes in thorax circumference (C, lower traces) cause biphasic variations in heart period (T, upper traces). Note also the delay. A scale for v, the heart rate in beats per minute, is given for comparison. For discussion see text. From Clynes [7.9].

(V) and thorax circumference (C), in the Laplace domain, must be of the following form:

$$\frac{V(s)}{C(s)} = \frac{\pm ks^2}{(1 + s\tau_1)(1 + s\tau_2)} \tag{7.6}$$

2. In addition, Fig. 7.7*a* shows that the transients in T, and therefore also those in V, have the same polarity for both inspiratory and expiratory steps in C.

3. As Fig. 7.7*b* reveals, the responses to pulselike stimuli (rapid "in-out" or "out-in" breathing) are essentially the same as those for stepwise changes in C. Two slight differences are noted: The transients following rapid decreases in C are smaller and they are spread out more in time than those due to rapid increases. These aspects appear to be modeled with satisfactory accuracy by the following two equations, relating changes in vagal inhibition (V) and chest circumference (C):

$$\frac{V(s)}{C(s)} = \begin{cases} \dfrac{-ks^2}{(1 + \tau_1 s)(1 + \tau_2 s)} & \text{for } \dfrac{dC}{dt} > 0 \quad \text{inspiration} \quad (7.7\text{a}) \\[2ex] \dfrac{k's^2}{(1 + \tau_1 s)(1 + \tau_2 s)(1 + \tau_3 s)} & \text{for } \dfrac{dC}{dt} < 0 \quad \text{expiration} \quad (7.7\text{b}) \end{cases}$$

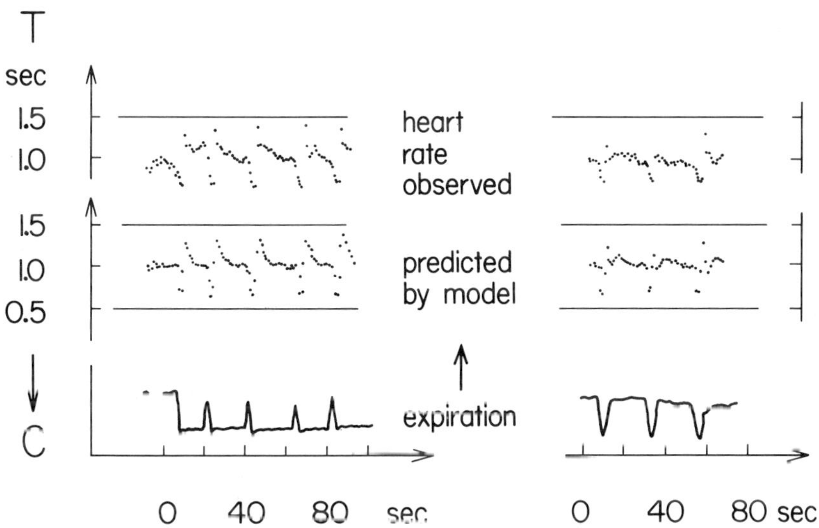

(*b*)

Fig. 7.7*b*. Responses of heart period (T) to impulselike changes in thorax circumference (C). Uppermost curves are observed in man, middle traces are predicted by the model of eq. 7.7. Note the similarity of responses on the left to those on the right. From Clynes [7.3, 7.9].

The change in sign takes care of the experimental finding described under item 2. The additional lag network $(1 + \tau_3 s)^{-1}$ and the factor $k'(< k)$ in eq. 7.7b introduce the differences in expiratory transients quoted under 3. These equations can be given in the Laplace domain (as were those for the pupil light reflex; eq. 7.3), because the behavior of the system is linear in each of the two domains specified: $dC/dt > 0$ and $dC/dt < 0$.

The equations given so far (7.4 and 7.7) are designed to simulate "test runs" involving abrupt changes in respiratory state as shown in Fig. 7.7. Whether these relations are adequate in explaining a wider class of respiratory signals must be tested by extensive experimentation. Fortunately, this can be done easily in resting man. The question of uniqueness of the model certainly cannot be resolved. However, it is necessary to delineate the applicability and to test the predictive power of the model with respect to ranges of inputs, time span, and so on. Clynes [7.9, 7.10] demonstrated that the model does predict sinus arrythmia due to normal breathing at many different rates, due to regular and irregular breathing patterns, and due to deep-held inspiration and held expiration, as well as due to "impulse breathing." For one of the subjects thus tested the parameters were adjusted to the following values: $\tau_1 = 1.2$ sec, $\tau_2 = 4.6$ sec, $\tau_3 = 3.2$ sec, $\nu_0 = 1.05$ Hz.

The final form of the model suggests that *two* basic reflexes are involved in the genesis of the sinus arrythmia. One must be triggered by proprioceptors sensing increases, the other by those sensing decreases in thorax diameter and the resulting stresses. Thus both channels involve unidirectional rate sensitivity. The waxing and waning of heart rate during normal breathing therefore is not an underdamped oscillation but rather a sequence of transients. A similar conclusion was reached in Section 7.A for the pupillary "oscillations," when step and impulse responses of the iris diameter were carefully analyzed. In the system under discussion here, we must make the further distinction between the large inspiratory transients and the smaller ones following expiration. In fact, during normal breathing the inspiratory responses are dominant. Expiratory "waves" are seen only when expiration is performed more vigorously than during normal breathing at rest. The exact nature of the time course of heart rate depends on the rate of respiration, which in turn determines to what extent the transient can develop.

The double differentiator (s^2 in eq. 7.7), introduced solely to match the observed waveforms, should not be interpreted in terms of a sensor of inertial forces. The acceleration of the chest wall is negligible compared with visco-elastic forces. For interpreting our results it is sufficient to note that both proprioceptive systems involved in this reflex must exhibit URS, thus accounting partly for the transfer function of eq. 7.7. As an example, the inspiratory channel then is given, in parts, by

$$\begin{aligned} &\frac{s}{1+\tau_1 s} \quad \text{for} \quad \frac{dC}{dt} > 0 \\ &0 \quad \text{for} \quad \frac{dC}{dt} < 0 \end{aligned} \tag{7.8}$$

The remaining portion, $s(1 + \tau_2 s)^{-1}$, representing a linear system, must be assigned to central nervous action. It is not known, however, whether τ_2 is the variable depending strongly on the state of the vasomotor centers of the CNS. Such hypotheses, which are consequences of successful model development, thus suggest experimental work at the neurophysiological level.

Parts of the assumptions have been verified by Hamlin et al. [7.11]. These workers established, in lightly anesthetized dogs, that the sinus arrhythmia is of purely vagal origin. The sinus arrhythmia was found to disappear temporarily during apnea and completely after administration of atropine and when the vagus nerve was severed. Furthermore, the variations in heart period were observed when the enddiastolic ventricular pressures were at a constant level. This is evidence which strongly supports Clynes's hypothesis on the origin of the heart rate variations synchronous with respiration.

D. SUMMARY

This chapter presents two regulatory mechanisms, mediated by the autonomic nervous systems, with smooth and cardiac muscle in the role of the effector. Many more reflexes of this nature, with open or closed loops, exist in the body, but only a few have been analyzed quantitatively and from a systems point of view. However, unidirectional rate sensitivity was introduced and mathematical tools for its description were presented. URS, a basic mechanism that is highly effective for optimizing task performance, is encountered at many levels of the metabolic, cardiovascular, respiratory, sensory, and central nervous systems.

The chapter also presents principles of simulation, or modeling, of biological systems. These methods were applied to simple systems. (We refer the reader to the work of Denison [7.12] for a more detailed analysis of the push-pull iris control in the cat, by sympathetic and parasympathetic fibers.) Complex regulatory systems are discussed in Chapters 16, 18, 19, and 20.

E. REFERENCES

7.1. Stark, L. Stability, oscillations, and noise in the human pupil servomechanism. *Proc IRE* **47**: 1925–1939, 1959.

7.2. Clynes, M. Computer dynamic analysis of the pupil light reflex, pp. 356–358 in *Proc 3rd Intl Conf on Medical Electronics*, Institution of Electrical Engineers, London, 1960.

7.3. Clynes, M. Unidirectional rate sensitivity: A biocybernetic law of reflex and humoral systems as physiologic channels of control and communication. *Ann NY Acad Sci* **92**: 946–969, 1961.

7.4. Bartley, S. H. The psychophysiology of vision, pp. 921–984 in *Handbook of Experimental Psychology*. S. S. Stevens (Ed.). John Wiley and Sons, New York, 1951.

7.5. Jones, R. W. Some properties of physiological regulators, pp. 674–678 in *Automatic and Remote Control*, Vol. II. J. F. Coales (Ed.). Butterworths, London, 1961.

7.6. Stark, L. *Neurological Control Systems*. Plenum Press, New York, 1968. Section 2, p. 194.

7.7. Llaurado, J. G. and H. H. Sun. Modified methods for studying stability in biological feedback control systems with transportation delays. *Med Electron Biol Engin* **2**: 179–184, 1964.

7.8. Loewenstein, W. R. and R. Skalak. Mechanical transmission in a Pacinian corpuscle. An analysis and a theory. *J Physiol (London)* **182**: 346–378, 1966.

7.9. Clynes, M. Respiratory control of heart rate: Laws derived from analog computer simulation. *IRE Trans Med Electron* **ME-7**: 2–14, 1960.

7.10. Clynes, M. Respiratory sinus arrhythmia: Laws derived from computer simulation. *J Appl Physiol* **15**: 863–874, 1960.

7.11. Hamlin, R. L., C. R. Smith, and D. L. Smetzer. Sinus arrhythmia in the dog. *Amer J Physiol* **210**: 321–328, 1966.

7.12. (a) Denison, B. L. A mathematical model for the isolated cat iris sphincter muscle. *Proc Ann Conf on Engineering in Medicine and Biology* **8**: 194 (no. 28.1), 1966. (b) A mathematical model for the isolated cat iris dilator. *Proc Ann Conf on Engineering in Medicine and Biology* **9**, (no. 24.5), 1967.

CHAPTER EIGHT

The Vestibulo-Ocular Reflex Arc: An Involuntary Sensory–Motor System

It is not within the scope of this book to discuss all sensory–motor systems involving skeletal muscles. Selecting a few such reflex and control systems, we plan to elucidate some of the generalizations presented in the Introduction to Chapter 7 and to discuss specific cases of goal-seeking mechanisms, adaptiveness, and reorganization of systems, and other properties encountered in biocontrol. Methods and concepts used in analyses of these systems originated largely in the engineering sciences. They must often be brought to bear when functional relationships between several complex systems are quantitatively studied to gain insight into the performance of the entire assembly of systems. Thus we study transfer functions of elements of the whole system, couplings between these elements, system performance, and stability. To cite an example, we do not inquire into the mechanisms of transduction of mechanical into electrical energy at the level of the individual hair cell in the labyrinth organ. Rather, we ask whether the net afferent information from the semicircular canal organs depends on acceleration or velocity of the skull, and how any general body movement is resolved into its spatial components. In other words, the entire system is studied at the level of organization where all the signals and (sub)system properties are essential for the performance of the system as a whole in fulfilling its function (see Chapter 1).

During many tasks which the visual system fulfills, the optical image of a "target" must be fixed on the retina, at least for a short span of time. This is quite readily achieved when both viewer and target are stationary with respect to each other. Visual targets that are in motion while the observer is stationary in space are nonetheless tracked with precision owing to mechanisms inherent in the visual system; of these, eye tracking and optokinetic nystagmus are discussed in Chapter 12. The performance of these mechanisms begins to deteriorate when the relative velocity between target and eye is larger than approximately 30 deg/sec. Tracking mechanisms likewise are effective when the body is in slow motion and the target is fixed. When the body is in rapid motion, vision would be blurred if it were not for the vestibular organ sensing such motion and causing compensatory eye movements to allow for image fixation. This vestibulo-ocular reflex plays important roles in the animal world where, for example, image fixation is indispensable during high-speed attacks or escapes. Man's vision also depends on the vestibulo-ocular reflex, at least in some special cases—for example, when running, driving an automobile over a bumpy road, or piloting an airplane through turbulent air (Melvill Jones and Milsum [8.1, 8.2]). The vestibular organ operates as an inertial accelerometer. It also plays an important role in the control of righting and balancing movements. We refer the reader to physiology books [2.2] for general discussions and to Partridge [8.3] for a quantitative investigation in this area.

The function of the entire vestibulo-ocular reflex arc is demonstrated easily. A person who attempts to read a text while shaking his head rapidly is more successful with reading than he is when the text is moved, at the same frequency, in front of his eyes. This is because in the former case the vestibular organ senses the head motions and causes, in turn, eye movements which, being opposite, therefore compensate for the "disturbance." The overall behavior of the system is readily understood and has been studied extensively (see, e.g., [8.4]). Definition and analysis of some of the subsystems involved are more difficult, as is noted subsequently. We have mentioned situations in which eye movement control by the vestibular organ is useful. At other times, the reflex could be detrimental and therefore must be blocked. For example, when a person turns his head quickly to look at a target on the side, his eyes move, from the onset of motion, in the direction of the target. The vestibulo-ocular reflex would, if not blocked, keep acting on the eyes to drive them away from the new target. It is not known how or at which level in the central nervous system the reflex arc is opened.

The flow of information during normal operation of the reflex is elucidated by Fig. 8.1. The task of the system is to minimize the tracking error $(\phi_t - \phi_e)$; that is, the difference between the angles of target (ϕ_t) and direction of gaze (ϕ_e), due to external disturbances. The ϕ_e depends on skull orientation ϕ_s, which in turn is linked to ϕ_b, defining the attitude of the body in space. Therefore, any disturbance imposed on the body (or the skull only) by the environment affects the direction of the eye's axis. In an open-loop fashion (see heavy lines in Fig. 8.1), the vestibulo-ocular reflex arc corrects for these disturbances. Fig. 8.1 does not show the other functions of the vestibular organ in righting, for example, nor does it show the usual pathways for eye movement control involving the visual systems. The main components of the vestibulo-ocular reflex arc and its overall performance are analyzed now.

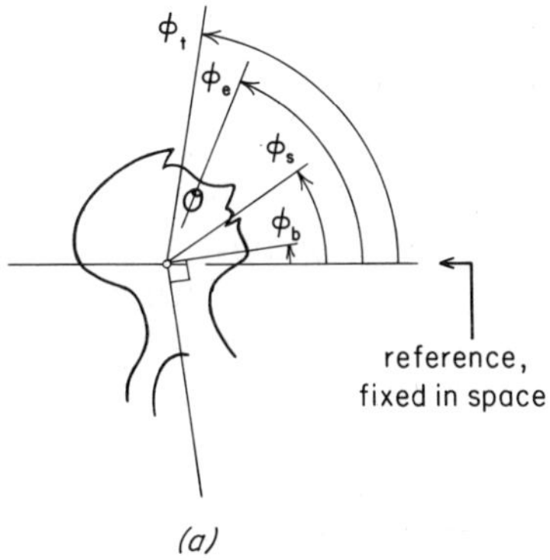

Fig. 8.1*a*. Definition of angles used in the overall analysis of the vestibulo-ocular reflex. All angles are measured with reference to a fixed axis in space (the horizontal). The position of the body (i.e., the thorax) is expressed in terms of ϕ_b, the angle between the reference and an axis perpendicular to the orientation of the body; ϕ_s determines orientation of the skull, ϕ_e is the direction of gaze, and ϕ_t measures the elevation of the target. Only angles of elevation are shown. Rotational angles are omitted, for simplicity, although they are needed for complete definition of relative orientation between skull, eye, and target. After Melvill Jones and Milsum [8.2].

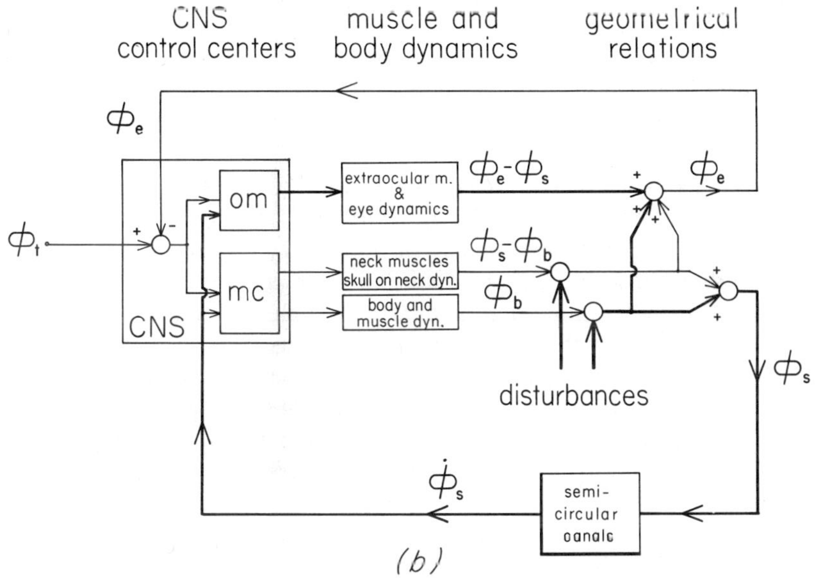

Fig. 8.1*b*. Information flow diagram showing the relations between the relevant variables needed in the overall analysis of the performance of the vestibulo-ocular reflex. The neural centers controlling eye motion (*om*) and those controlling body and neck muscles (*mc*) act to minimize the disparity between the directions of the target (ϕ_t) and of gaze (ϕ_e). The disparity is due to disturbances. The geometrical relations are determined by the information flow paths in the right half of the figure. The heavy lines (disturbances = input, direction of gaze = output) represent the reflex arc, an open-loop system. The reflex arc is blocked during volitional head (and body) movements; see text. After Melvill Jones and Milsum [8.2].

A. THE VESTIBULAR ORGAN

Figure 8.2 is a schematic view of the vestibular apparatus and its orientation within the skull. We refer the reader to physiology books [2.2] for anatomical aspects and discussion of various parts, such as the utricle containing the otolith organs, and connections to cranial cavity and cochlea. Here we restrict attention to the semicircular canals as they, together with the sensors in the ampullae, form the transducer for angular acceleration (Fig. 8.3). The orientations of the semicircular canals are nearly orthogonal, which allows for detection of rotation around any axis in space. A detailed, although more schematized, view of one canal is given in Fig. 8.3*a*. The semicircular canal and the utricle are filled with endolymph. The cupula, within the ampulla, can be compared with a flap valve or a hinged vane providing a tight seal across the ampulla and being constrained by elastic supports. (Dye particles injected into the ampulla on one side do not traverse to the other side of the cupula.) Hair cells, located in the crista underneath the cupula, are sensitive to the position of the vane. Wersäll et al. [8.5] discuss further details of the transducing mechanism in hair cells. Rectification may take place owing to directional organization of the stereo- and kinocilia of the hair cells. In man, the sensors in posterior and superior canals detect rotation of endolymph within the canal away from the ampulla; the lateral canal organ senses ampullopetal displacements. Together with the contralateral organ, however, the transducer system can be considered symmetrical (see vectors in Fig. 8.2). We therefore assume that the afferent neural information is a linear measure of relative endolymph displacement, or the angle $(\phi_s - \alpha_f)$ of Fig. 8.3*b*.

To gain insight into the performance of the mechanical part of the semicircular canals, we assume that a canal is circular and that the utricle does not affect significantly the hydraulics of the system. Because the inside radius of the canal is small compared with its radius of curvature, the analysis can be simplified. We also idealize the mechanics of the cupula, following the work by Van Egmond et al. [8.6], in Fig. 8.3*b*. The equation of motion of the cupula and the endolymph can be given for purely angular motion of the skull as:

$$mR\frac{d^2\alpha_f}{dt^2} + rR\frac{d}{dt}(\alpha_f - \phi_s) + kR(\alpha_f - \phi_s) = 0 \tag{8.1}$$

where $R(\alpha_f - \phi_s)$ = relative displacement of fluid within canal
R = radius of semicircular canal
m = mass of fluid plus cupula
r = resistance to flow within canal
k = effective restoring force per displacement of cupula

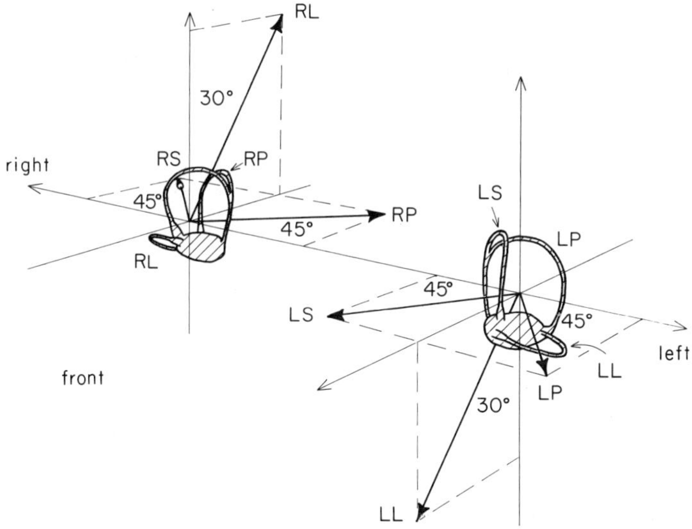

Fig. 8.2. Schematic view of the two vestibular apparatus indicating the orientation of the semicircular canals (heavy arrows) with respect to the skull. Semicircular canals are identified, together with the axes of rotation to which they are most sensitive, as follows: RL = right lateral, RP = right posterior, RS = right superior, LL = left lateral, LP = left posterior, LS = left superior semicircular canal.

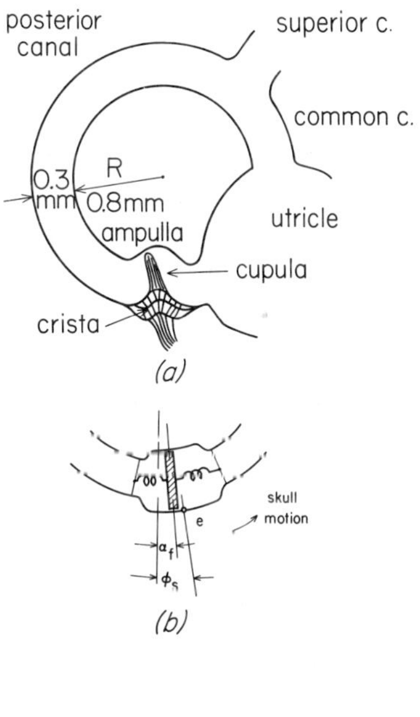

Fig. 8.3. (*a*) Schematic view of the essential elements of one of the posterior semicircular canals. The important element of the transducer, within the so-called ampulla, is the cupula, which is an elastically hinged "flap" consisting of a gelatinous mass. The endolymph within the canal cannot flow past the cupula. The cilia of the haircells (forming the crista) project into the cupula. Thus if the cupula is tilted by fluid pressures within the canal, the hair cells are stimulated. (*b*) Mechanical analog of the cupula system used in the analysis of the transduction characteristics of the canal: ϕ_s is the orientation of the skull, α_f is the relative angle between the cupula and the skull, $(\alpha_f - \phi_s)$ is proportional to the displacement of the fluid within the canal. Little circle marked e indicates the equilibrium position of the cupula.

This assumes rotation about the axis perpendicular to the plane of the canal. Clearly, the fluid and the cupula system are treated as a simple mass–spring–dashpot system. The assumption that we can compute frictional forces linearly proportional to velocity is justified if flow in the channel is laminar. This indeed is the case, as estimations of typical Reynolds numbers indicate. The similarity number applicable to oscillatory flow (developed in Chapter 14), which also gives the ratio of inertial to frictional forces within the liquid, is found to be very small in the semicircular canals, too.

It is useful to rewrite eq. 8.1 by introducing the actual displacement of the cupula δ. Let

$$\delta = \lambda(\alpha_f - \phi_s) \tag{8.2}$$

where $\lambda = a/A$, the ratio of cross-sectional areas of the canal (a) and of the ampulla (A)

Then eq. 8.1 transforms into

$$\left[m\frac{d^2}{dt^2} + r\frac{d}{dt} + k\right]\delta(t) = -\lambda m\frac{d^2\phi_s}{dt^2} \tag{8.3}$$

From this follows the input–output relation for the canal. In the Laplace domain

$$\frac{\delta(s)}{\phi_s(s)} = \frac{-\lambda s^2}{s^2 + (r/m)s + (k/m)} \tag{8.4}$$

As mentioned previously, frictional forces are overwhelming, (i.e., the system is heavily damped). Experimental findings [8.1, 8.7] indicate, in fact, that the two poles of eq. 8.4 are both real, one being very large and the other very small. It therefore becomes permissible to approximate the transfer function of eq. 8.4 by

$$\frac{\delta(s)}{\phi_s(s)} = \frac{-\lambda s^2}{(s + k/r)(s + r/m)} \tag{8.5}$$

where

$$k/r \simeq 0.1 \text{ sec}^{-1}$$
$$r/m \simeq 300 \text{ sec}^{-1}$$

The characteristics of the system in the time domain are demonstrated in Fig. 8.4*a*, which presents a response to a ramp in head position. Formally, if

$$\phi_s(s) = \Lambda\frac{1}{s^2} \qquad \text{(a ramp in time)}$$

then

$$\delta(t) = -\lambda\Lambda\frac{m}{r}(e^{-tk/r} - e^{-tr/m}) \simeq -\lambda\Lambda\frac{m}{r}e^{-tk/r} \tag{8.6}$$

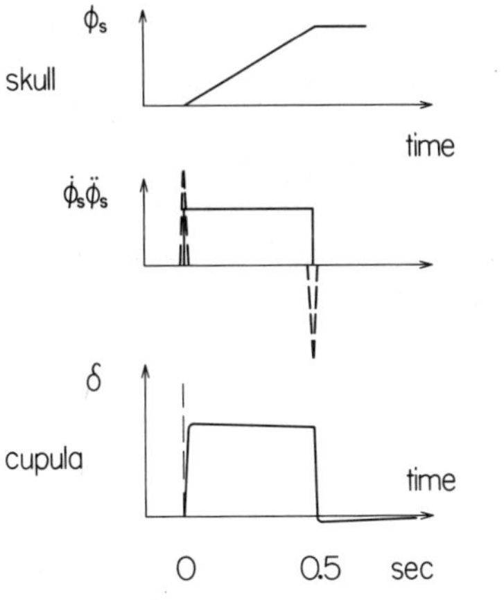

(a)

Fig. 8.4*a*. Response of cupula displacement (δ) to skull rotation (ϕ_s). If the skull is rotated in a rampwise fashion during a certain time (top graph), the cupula displacement is approximately proportional to skull velocity. The afferent neural signal is proportional to δ. The sensor system in the semicircular canal therefore acts as an integrating accelerometer. This occurs because the mass–spring–friction system is heavily overdamped. This performance is seen only within a limited frequency band [see (*b*)].

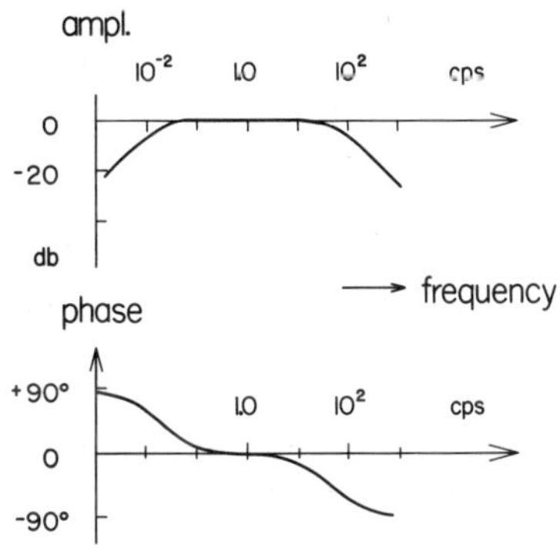

(b)

Fig. 8.4*b*. Transfer function of semicircular canal organs, giving the ratio of cupula displacement and skull *velocity* as a function of frequency. The transducer senses skull velocity only between 0.05 and 20 Hz. Above 20 Hz the afferent neural signal is less than needed for eye motion compensation. Vision then becomes blurred. After Melvill Jones and Milsum [8.2].

Because the system is heavily overdamped, time constant $r/k \gg m/r$, and the output of the semicircular canal organ essentially is a step function. But this is a signal proportional to velocity of the stimulus. In fact, the transducer behaves like an integrating accelerometer over a wide frequency band. Melvill Jones [8.1, 8.7] therefore gives the transfer function as the ratio of cupula displacement and skull velocity:

$$\frac{\delta(s)}{s\phi_s(s)} = \frac{-\lambda s}{(s + k/r)(s + r/m)} \tag{8.7}$$

This function is graphed in Fig. 8.4*b* whence it is seen that all inputs at frequencies between 0.05 and 20.0 Hz are integrated. On exposure to vibrations at higher frequencies, vision becomes blurred. On the other hand, the long time constant, of the order of 10 sec, is responsible for dizziness due to continued, postrotational nystagmus after the body has been in motion.

Direct observations of activities of single neurons in the vestibular nuclei validate the foregoing analysis [8.1]. The firing frequency of these neurons varies sinusoidally, in response to a sinusoidal stimulus. Neural information, however, exhibits a phase lead of 90° in the passband of the transducer. Hence the afferent neural signal indeed carries velocity information.

So far we have assumed that the semicircular canals are exposed to rotation about their axes, as in Fig. 8.2. If the axis of skull rotation is not the same as that of a canal but is parallel to it, the *same* analysis applies. If the axis of skull rotation is not parallel to any of the three canal axes, all three canals are stimulated. The amplitudes of these stimuli are proportional to the angular velocity of the skull and vary with the cosines of the angles between the axis of the canals and that of rotation. In this case we can inquire whether any interference between canals should be expected. All canals communicate through the utricle; the superior and the posterior canals, in addition, share the common canal. Fluid motion in one canal therefore may, at least in principle, induce some motion of endolymph in the others. Although there has been no theoretical investigation into this question, observations indicate the probable absence of "cross talk" between individual canals.

The preceding analysis of transducer action is based on the additional assumption that the semicircular canals lie in one plane. This probably is not very restrictive. Even if the assumption did not hold, it would be possible to define an effective or functional canal plane as that plane in which those axes of rotation lie, around which the canal can be accelerated without exhibiting any endolymph flow. Conversely, rotations around axes normal to the effective plane cause maximal excitation.

There is further interest [8.8] in the extent to which the semicircular canals can serve the body as *linear* accelerometers. In theory, the system may detect linear accelerations if there is a density difference between cupula tissue and endolymph. For linear accelerations a differential equation of the same form as eq. 8.3 applies; the solution then is of the form given in eq. 8.6, where the mass m is replaced by Δm, which is the difference between the mass of the cupula and an equal volume of endolymph. Thus a linear acceleration $\ddot{x}$ can be estimated which causes the same cupula displacement as a rotational $R\ddot{\alpha}_s$; that is, $\ddot{x} = (m/\Delta m)R\ddot{\alpha}_s$. The ratio $m/\Delta m$ is not known. But we can assume a possible case and estimate the magnitude of linear acceleration needed to cause endolymph motion comparable to that occurring during normal physiological head movements. Let the density of cupula tissue be 10% larger than that of endolymph. Then, if the volume of endolymph in the canal is 50 times that of the cupula, $m/\Delta m$ is about 500. This represents quite a large "amplification factor." If the skull is rotated to reach a velocity of 90°/sec within 0.1 sec, the term $R\ddot{\alpha}_x = 0.63$ cm/sec^2. Hence $\ddot{x} \simeq 350$ cm/sec^2, which corresponds to the deceleration experienced when being stopped, at a speed of 6 mph, within 0.8 sec. Such a "jolt" may occur in natural behavior and would stimulate the semicircular canals. Whether it elicits eye movements has not been investigated.

B. THE MECHANICS OF EYE MOVEMENT

The effector side of the reflex arc (i.e., the system of extraocular muscles and the eyeball) is discussed in this section mainly from the static viewpoint. In particular, we inquire into neural connections necessary to achieve co-ordinate transformation and into the geometry of motion of the globe with three degrees of freedom. The additional problem of how the extraocular motor system achieves its high degree of dynamic performance is postponed until Chapter 12. In other words, we concentrate first on the organizational aspects of eye movement control as required by the vestibulo-ocular reflex to compensate for external disturbances.

1. Nature of Neural Commands to Extraocular Muscles

The viscoelastic properties of orbital tissues, whose anatomy is found in specialized books [8.9], dictate the performance of the extraocular muscle system. In order to maintain a particular direction of gaze, the eye muscles need to exert a steady torque. To change eye position rapidly, large viscous forces must be overcome. The oculomotor centers achieve this by special structuring of the neural commands to the agonists and probably by strong inhibition of antagonists (see Chapter 12). This of course is true for volitional as well as for reflexly controlled eye movements.

First we show how these observations affect the study of the vestibulo-ocular reflex system. In particular, the requirement for a steady force to maintain an eye position allows us to infer system's characteristics. The sensors, the semicircular canals, furnish information about the angular velocity necessary for visual fixation. The effectors, on the other hand, require a signal proportional to angle. It follows that there must exist neural structures that transform the information on velocity into information about position. In other words, they must perform an integration with respect to time. The identification of the neural structures performing the integration is one of the open problems in vestibular physiology [8.10]. Evidence for their existence and some indications of their dynamic behavior are discussed in Section 8.C, together with the overall performance of the reflex arc. These requirements for particular neural commands to the extraocular muscles pertain, of course, to all three muscle pairs. Requirements for coordination among the actions of the three pairs is discussed next.

2. Geometry of Eye Movements [8.10]

The important features of the anatomy of the muscle system, relevant to our study, are schematized in Fig. 8.5, which gives the spatial relation of the planes in which the eye muscles act. Note that these planes can be placed with certainty only when the eyes assume the primary direction of gaze; that is, when the optical axes point straight forward, parallel to the x-axis. The perpendiculars to the planes (Fig. 8.6) are the axes of rotation. These axes are

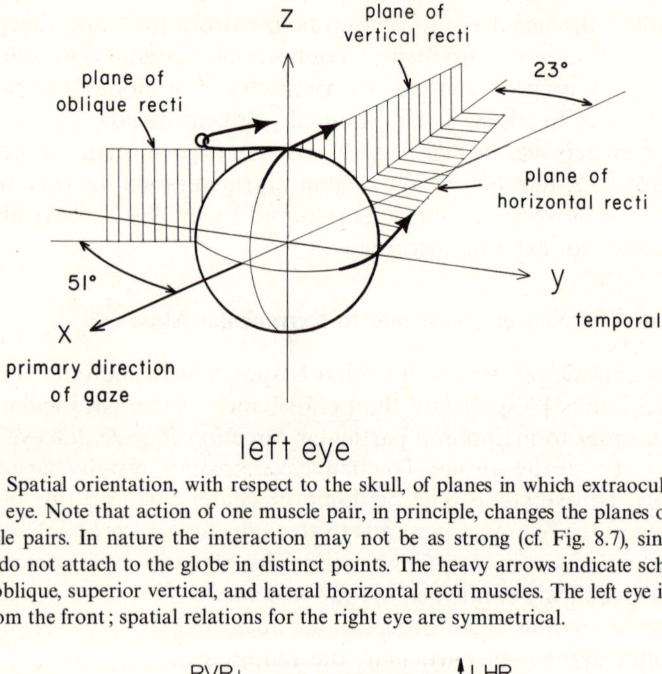

Fig. 8.5. Spatial orientation, with respect to the skull, of planes in which extraocular muscles act on the eye. Note that action of one muscle pair, in principle, changes the planes of the other two muscle pairs. In nature the interaction may not be as strong (cf. Fig. 8.7), since muscles probably do not attach to the globe in distinct points. The heavy arrows indicate schematically superior oblique, superior vertical, and lateral horizontal recti muscles. The left eye is shown as viewed from the front; spatial relations for the right eye are symmetrical.

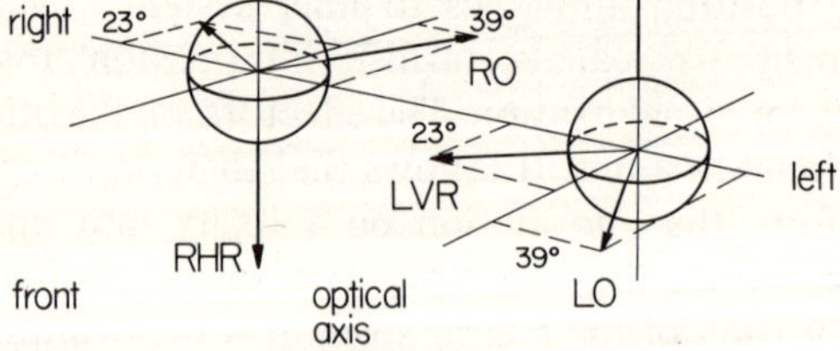

Fig. 8.6. Axis of eye globe rotation elicited by the three pairs of extraocular muscles. R = right, L = left, HR = horizontal (i.e., medial and lateral) recti muscles; VR = vertical (i.e., superior and inferior) recti; O = superior and inferior oblique muscles. Note that the diagram applies strictly only during primary gaze direction (i.e., parallel to x-axis, see Fig. 8.5). Particularly, rotation about LO (or RO) tilts the other two axes. See text for discussion of this interaction between the muscle pairs. The arrows are drawn according to the right-hand screw rule for contractions of superior oblique, superior, and lateral recti, respectively.

given according to the rule of right-hand screws for shortening of superior recti, superior obliques, and lateral recti muscles.

Since the effective location of attachment of the extraocular muscles is not known, let us consider likely cases. If they were inserted into the globe at one point, we would predict that rotation about one axis by one muscle pair would change the axes of rotation of the other two. Thus if the gaze is elevated by a certain angle, by action of the vertical (i.e., superior and inferior) recti, the points of attachment of the horizontal (i.e., lateral and medial) recti would be lifted out of the x–y plane. This in turn would cause a tilt of the axis for the horizontal recti in the posterior direction, because the points of insertion of the horizontal recti are anterior to the center of the globe. The tilt, however, is rather small, because the muscles are relatively long ($\simeq$4 cm). Assume an elevation of gaze by 30°. The points of insertion of the horizontal recti lie approximately 0.6 cm in front of the center; they are lifted out of the x–y plane by $0.6 \times \sin 30°$ cm $= 0.3$ cm. The plane of the horizontal recti, and with it their axis of rotation, is therefore tilted in theory by arcsin $(0.3/4.0) = 4.3°$. If the horizontal recti rotate the globe in this new situation, the optical axis is not traveling along a straight horizontal line any more. When observed from the front (i.e., projected onto the y–z plane), the center of the pupil moves along an ellipse, as in Fig. 8.7 (curve VH). The deviation from the straight horizontal line, however, is small.

Rotation by horizontal or vertical recti out of the primary direction also affects the axis of the (superior and inferior) oblique muscles. Theory predicts small interference in this case, too. In practice, the effects may be even smaller, as an experimental study by Tokumasu [8.11] seems to indicate. The effective location of insertion of the extraocular muscles to the globe probably is not concentrated at one point, but spread along a line.

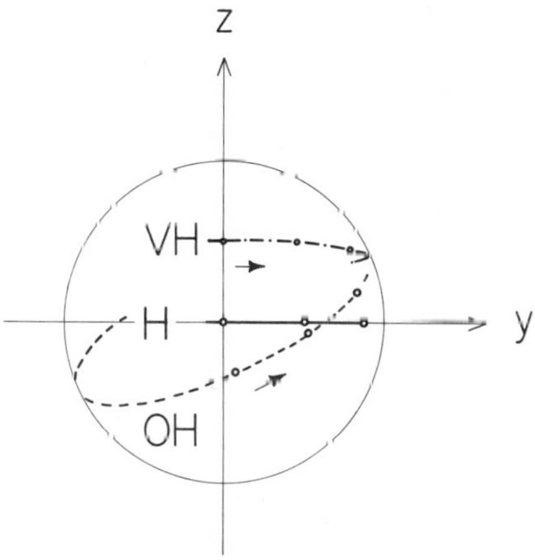

Fig. 8.7. Frontal view of the theoretical motion of the center of the pupil due to action of horizontal recti muscles (contraction of lateral rectus) as influenced by the tonus of the other muscle pairs. If vertical recti first elevate the gaze by 30°, the horizontal pair does not act purely "horizontally" any more. The ellipse VH shows the resulting trajectory. Similarly, if the oblique pair first rotates the globe by 30°, the horizontal pair then moves the direction of gaze along path OH. Horizontal movement after intorsion therefore requires more complex muscle action. Generally, all eye movements require complex, concomitant muscle action due to the interaction between the three pairs. See text.

However, when the oblique muscle pair acts, it not only rotates the globe about the optical axis but it rotates the lines of insertion of the two other muscle pairs by the same angle. The extent of this interference may be serious even though there is no distinct point of attachment, but rather an effective line, which may move about on the globe when it is moved. A theoretical curve, along which the center of the pupil would be moved by the horizontal recti, *after* so-called intorsion by 30° (due to superior oblique muscle contracting), appears in Fig. 8.7 (curve *OH*). If the planes of the horizontal and vertical recti were rotated, in torsion, by as much as indicated by curve *OH*, the neural commands to these muscles (needed for a certain change of eye position) would become functions of the initial position. This would require nonlinear neural weighting operators in the oculomotor centers [8.10].

A further complication would arise because rotations of a sphere about axes that are not fixed in space but are moving with the sphere, are not commutative. For example, if horizontal recti rotate the globe first by angle α and if subsequently the oblique muscles cause torsion by angle β, a particular final position is achieved. If torsion by β is effected first, and followed by rotation α due to the horizontal pair, a *different* final position results. It is not known how the living system copes with these difficulties. It is likely that the nonlinear effects and those mentioned last are small and are not corrected when the eyes are moved under vestibular control only. Therefore, visual fixation may be incomplete. Additional oculomotor controls, depending on information reaching the visual centers, may be brought to bear in order to improve fixation. The latter assistance, however, can only be effective at relatively low frequencies (Chapter 12).

3. Transformation of Coordinates

It is possible to predict the composition of neural commands to extraocular muscles assuming that the movements required for visual fixation under vestibular control are small. In particular, signals to the oblique muscles must be assumed small, which follows from the foregoing discussion and in view of lack of detailed knowledge on the system's capability to correct for nonlinearities.

In any case, the axes of the semicircular canals are known not to be parallel with those of eye rotation (Figs 8.2, 8.6). For example, if skull rotation about an axis parallel to that of the lateral canals is ω_s, the needed compensatory eye rotation about the vertical axis is only $\omega_s \cos 30°$. In this manner, the overall transformation, in terms of weighted sums, can be established between the net signals from the semicircular canals and the commands to the extraocular muscles. Let the net input signals to the reflex arc be the difference between the neural information from the left labyrinth organ and that on the right side. Then, if complete visual fixation is to be achieved, the commands to the extraocular muscles must be related to the inputs by the following equations:

$$\begin{vmatrix} HR \\ O \\ VR \end{vmatrix} \propto \begin{vmatrix} \cos 30^\circ & 0 & 0 \\ -\sin 30^\circ \cos 39^\circ & -\cos 6^\circ & -\cos 84^\circ \\ -\sin 30^\circ \cos 67^\circ & -\cos 112^\circ & -\cos 22^\circ \end{vmatrix} \cdot \begin{vmatrix} LL - RL \\ LP - RS \\ LS - RP \end{vmatrix} \tag{8.8}$$

where HR = command to horizontal recti muscles (lateral rectus shortens; medial rectus lengthens)

O = command to oblique muscles (superior oblique muscle shortens; inferior oblique lengthens)

VR = command to vertical recti muscles (superior rectus shortens; inferior rectus lengthens)

$LL - RL$ = difference of signals from left and right lateral semicircular canals

$LP - RS$ = difference of signals from left posterior semicircular canal and from right superior semicircular canal

$LS - RP$ = difference of signals from left superior and right posterior semicircular canals

These obviously are equations showing only weighting factors due to the relative orientations of sensors and effectors in space. Transfer functions expressing the dynamic behavior of the reflex arc are discussed in the next section.

C. PERFORMANCE OF THE WHOLE REFLEX ARC

The nature of the compensatory eye movements elicited by the vestibulo-ocular reflex is illustrated in Fig. 8.8. The stimulus is, in this case, a steady skull velocity, beginning at time zero. Thus the output of the sensors in the labyrinth is a slowly decaying exponential (Fig. 8.4). This results, after the necessary stage of integration, in a command to the eye muscles which is a "ramp," at least during the first few seconds. However, the extent to which the eyes can be rotated is limited. The system therefore interrupts the slow compensation movement by saccade-like "jumps," bringing the eyes back into their original positions (see response curve in Fig. 8.8). Visual fixation is possible between these saccades, at least during the beginning of the experiment, as we see when all the slow phases of the nystagmus are added ("sum

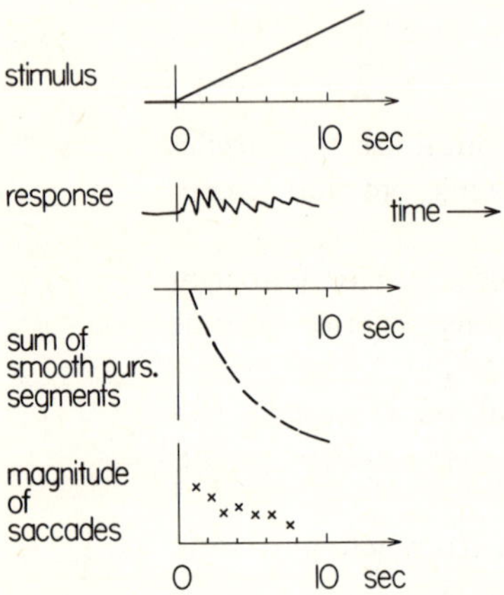

Fig. 8.8. Response of the vestibulo-ocular reflex arc to constant head rotation, observed in a slightly anesthetized cat. The uppermost curve shows the stimulus in terms of orientation (angle) of the skull. In this example, the cat's entire body was rotated at constant velocity. The second curve from the top, the response, shows the typical time function of eye orientation. The compensatory eye movement elicited by the vestibulo-ocular reflex is interrupted by saccades in the opposite direction. Due to the main time constant of the semicircular canal, the compensatory eye movements are equal and opposite only at the beginning of the maneuver. This is seen in the third curve from the top, where all smooth pursuit (compensation) movements are added by graphical construction. The figure at the bottom shows temporal changes in magnitude of saccades in the response shown (scale approximately doubled). Note that the angle scale in the top graph differs from that used in the lower graphs. After Sugie and Melvill Jones [8.12].

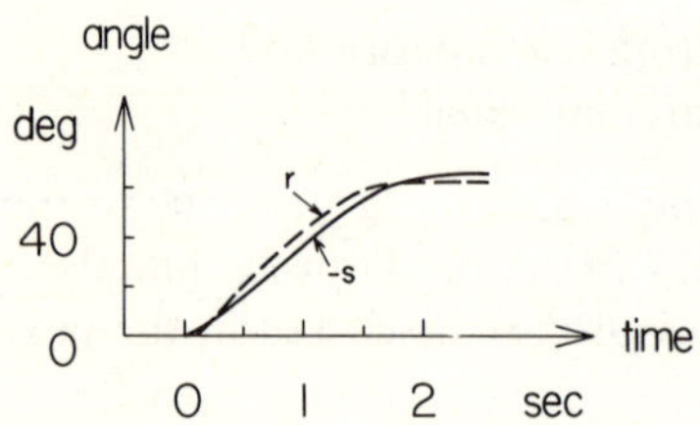

Fig. 8.9. Response of the vestibulo-ocular reflex system to a stimulus (constant velocity) considerably shorter than the integrator's time constant. The accuracy of compensation can be judged directly from this graph in which, as in Fig. 8.8, the smooth pursuit segments of the response (r) are added (graphically) and plotted together with the inversed stimulus ($-s$). Note that the stimulus is 40°/sec, close to the upper limit of tracking velocity; r is obtained with closed eyes. From Wendt [8.4].

of smooth pursuit segments" curve in Fig. 8.8). The sum of all slow-phase movements corresponds, as expected, to the integral of the decaying exponential signal from the sensors. Accuracy in compensating for skull rotation, during the initial 2 sec, can be judged from Fig. 8.9. Wendt [8.4] obtained these data in man by again summing all slow-phase movements occurring during a 2-sec ramp stimulus. Response under pure vestibulo-ocular reflex action (i.e., eyelids closed) exhibits a position error of the order of 30%. When the eyes are open, the tracking ability of the visual system may reduce this greatly, provided rotational velocities are below 30°/sec.

Another example of vestibulo-ocular reflex action appears in Fig. 8.10, where the stimulus is sinusoidal and the response again is nystagmoid (i.e., a composite of slow and fast movements [8.12]). The frequency in this case is

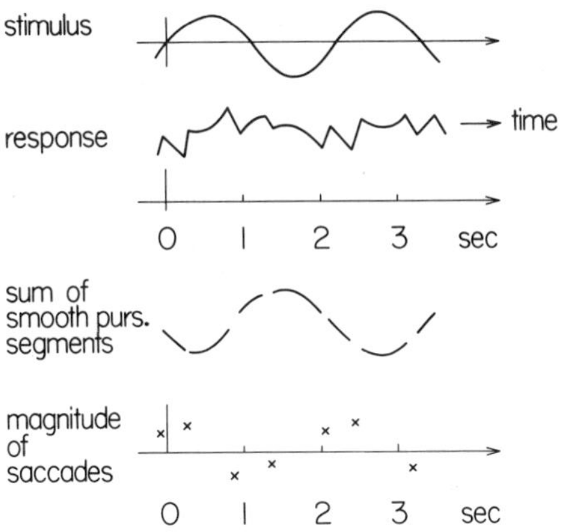

Fig. 8.10. Response of the vestibulo-ocular reflex system to a sinusoidal stimulus (top). The response (second from top) again comprises fast (saccadic) and slow (smooth pursuit, elicited by reflex) phases. The sum of all compensatory movements is equal and opposite to the stimulus. Vision is unblurred during the slow phases. The bottom graph indicates that the magnitude of saccades may be proportional to angular velocity. After Sugie and Melvill Jones [8.12].

within the passband of the semicircular canals. It is obvious, also from Fig. 8.8., that the time periods between the saccades are not constant but rather, in a first approximation, inversely proportional to velocity of the skull. The shortest slow-phase period seems to be of the order of 0.2 sec; the longest period is approximately 1 sec. From this it follows that the neural integrator, which must lie between the sensor and the oculomotor centers, need not exhibit a long time constant. In fact, as Robinson [8.10] pointed out, it suffices if it integrates over one second. This is indicated in the functional block diagram of the reflex arc in Fig. 8.11. The property of the "integrator" is given as $k/(s + s_3)$; s_3 may be, according to the foregoing discussions between 0.3 and 1.0 sec^{-1}. If the integrator is reset each time the fast phase of nystagmus is elicited, it fulfills its task satisfactorily. This presupposes that the constant k, which contributes to the gain in the reflex arc, assumes an optimal value. The criteria of optimization, however, are not known. The system may tend to minimize the errors in velocity, as was suggested by Outerbridge and Wettstein [8.13].

Sugie and Melvill Jones [8.12] suggested a model differing from Robinson's in that their integrator is not reset but receives a second input in addition to the velocity signal. The second signal is a train of pulses whose amplitudes are proportional to the velocity signal but opposite in sign to it. The proportionality factor plays a critical role. All these models need further study and validation by experimentation. Figure 8.11 therefore summarizes in general fashion the functional components of the entire system. The sensor's action has been studied and is well understood, as discussed in Section 8.A. Both the necessity for the additional integrator and postulations on its characteristics were inferred from eye muscle mechanics and overall response. Requirements for the coordinate transform can be given from theoretical considerations

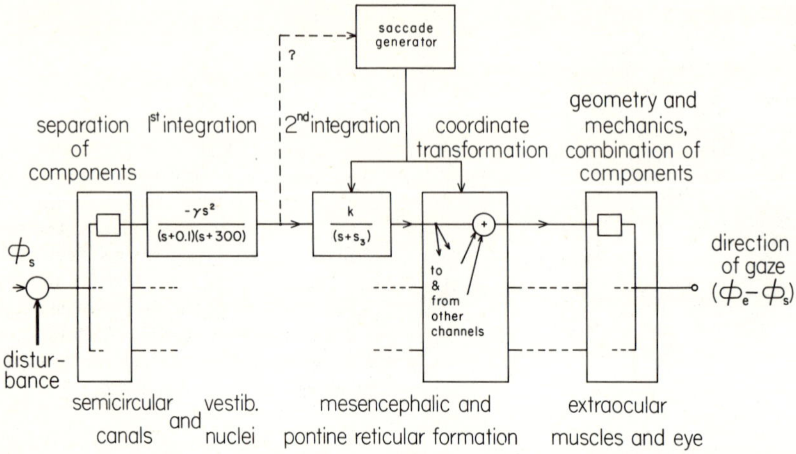

Fig. 8.11. Subsystems of the vestibulo-ocular reflex arc. The separation into angular components of the disturbance (far left) occurs according to the geometry of semicircular canals (see Fig. 8.2). Each canal, whose responses correspond to those of an integrating accelerometer, produces a (neural) signal proportional to the velocity of the disturbance. This needs to be integrated by a second integrator (i.e., a neural structure), since the output of the system must compensate for an angular *position.* The value of s_3 can be between 0.3 and 1.0 sec^{-1} (see text). The coordinate transformation is required, since the particular orientation of the planes of the extraocular muscles differs from that of the semicircular canals (see eq. 8.8). The "black box" on the far right is given by the extraocular muscle mechanics; its output is the orientation of the eye. As is seen in Figs. 8.8 to 8.10, a saccade generator is coupled into the system. It may receive information on the velocity of the disturbance. Nothing is known about adjustment of the gain of this open-loop controller. Neither is it understood at which stage the gain is set to zero, as it must be during volitional head movements combined with volitional changes in gaze direction.

(Section 8.B). The nature of the generator of the fast phases of nystagmus seems to be closely linked with skull velocity, and this information is provided by the sensors. The exact strategy used for timing these reset movements is not known. Furthermore, there is no information available about the time course of the neural command for an individual fast-phase movement. As Chapter 12 reveals, the highly viscous load is moved rapidly because the extraocular muscles initially are stimulated very intensely until a certain position is reached, after which neural activity is but a fraction of the original.

Clearly, the many open questions mentioned here challenge the neurophysiologist who would identify neural structures and their functions. Together with the systems engineer, he also finds intriguing problems pertaining to gain adjustment and optimization criteria. It is worth repeating that it often is easier to define some of these problems because we are seeking models for all subsystems of the reflex arc.

D. REFERENCES

8.1. Melvill Jones, G. Vestibular inaptitude in the environments of flight and space. *J Laryng* **80**: 207–221, 1966.

8.2. Melvill Jones, G. and J. H. Milsum. Spatial and dynamic aspects of visual fixation. *IEEE Trans Bio-Med Engin* **BME-12**: 54–62, 1965.

8.3. Partridge, L. D. and J. H. Kim. Dynamic characteristics of response in a vestibulomotor reflex. *J Neurophysiol* **32**: 485–495, 1969.

8.4. Wendt, G. R. Vestibular functions, pp. 1191–1223 in *Handbook of Experimental Psychology*. S. S. Stevens (Ed.). John Wiley and Sons, New York, 1951.

8.5. Wersäll, J., A. Flock and P.–G. Lundquist. Structural basis for directional sensitivity in cochlear and vestibular sensory receptors, in *Sensory Receptors. Symp Quant Biol* **30**: 115–132, 1965.

8.6. Van Egmond, A. A. J., J. J. Groen, and L. B. W. Jongkees. The mechanics of the semicircular canal. *J Physiol (London)* **110**: 1–17, 1949.

8.7. Melvill Jones, G. and K. E. Spells. A theoretical and comparative study of the functional dependence of the semicircular canal upon its physical dimensions. *Proc Roy Soc* [*Biol*] **157**: 403–419, 1963.

8.8. ter Braak, J. W. G. Kann der Bogengangapparat durch geradlinige Beschleunigung gereizt werden? *Pflügers Arch Ges Physiol* **238**: 327–332, 1937.

8.9. Davson, H. *The Eye*, Vol. 3, 2nd ed. Academic Press, New York, 1969.

8.10. Robinson, D. A. The oculomotor control system: a review. *Proc IEEE* **56**: 1032–1049, 1968.

8.11. Tokumasu, K., K. Goto, and B. Cohen. Eye movements produced by the superior oblique muscle. *Arch Ophthal (Chicago)* **73**: 851–862, 1965.

8.12. Sugie, N. and Melvill Jones, G. A mathematical model of the vestibulo-ocular system. *Digest, 6th Intl Conf on Medical Electronics and Biological Engineering*, pp. 422–423 (no. 23.6), 1965, Tokyo.

8.13. Outerbridge, J. S. and U. Wettstein. Optimization in the vestibulo-ocular system. *Digest, 7th Intl Conf on Medical and Biological Engineering*, p. 291 (no. 19.9), 1967, Stockholm.

CHAPTER NINE

Skeletal Muscle; Microscopic Level

A theory of muscle contraction must take into account and correlate a great number of discoveries made by electron microscopists, biochemists, and physiologists investigating thermal and mechanical events during and after stimulation and contraction of muscle. The two muscle models presented in this chapter allow for an integrated, quantitative discussion of mechanisms underlying contraction of *vertebrate skeletal muscle* at the level of the contractile element of the fibril. There are, of course, many other problems in various branches of muscle anatomy and physiology which can be treated separately and which also require physical and mathematical analysis.

The sliding filament theory of contraction is today probably the most widely accepted theory because there is a great body of supporting observations. Yet many problems remain unsolved, and others are being uncovered. Some of these are of geometric–physical nature. The sliding filament hypothesis says that the hexagonal array of actin filaments and the triangular array of myosin filaments, seen within each sarcomere, slide into each other because of forces between the two sets of filaments. Interaction between the actin and the myosin rodlets probably is brought about by formation of links, or crossbridges, between spatially and directionally organized active sites on the two sets of filaments. This is shown schematically in Fig. 9.1. Since excitation spreads from the Z-lines, or Z-discs, to the two adjacent sets of actin rodlets, the contractile unit of muscle consists of two half-sarcomeres extending from the middle of one H-zone to the middle of the next. A. F. Huxley [9.1] discussed early findings leading to this theory. Many of these observations were clarified later, for example, by direct correlation of sarcomere length and isometric force [9.2]. In another review, Wilkie [9.3] discusses more recent work in this area.

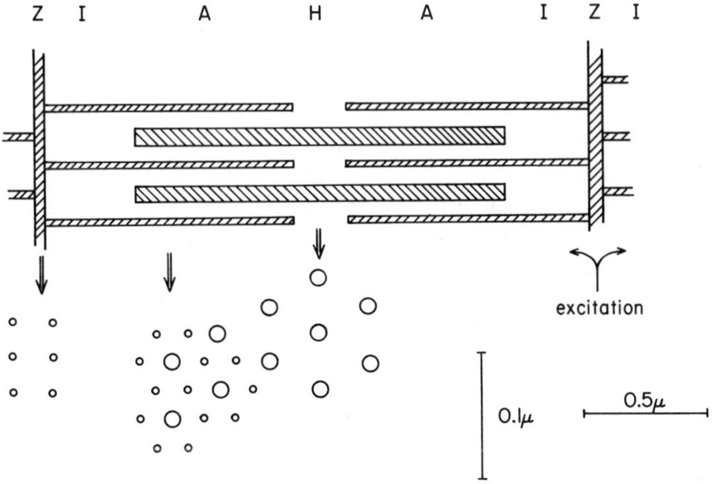

Fig. 9.1. Longitudinal section (top) and cross sectional views (bottom) of a sarcomere, schematized after electron micrographs, without detailed structures of filaments and crossbridges. The actin or thin filaments, connecting to the *Z*-bands, probably are thin double helices. In electron micrographs, the heavy myosin filaments show projections near their ends (left and right) and a region free of such projections in the middle. These projections presumably are the active sites. The cross-sectional views of the lower half, which correspond closely to transverse electron micrographs, reveal directly how the two filaments interdigitate. Sections at the level of the *Z*-bands show a near-square array of "*Z*-filaments." It has been postulated that the thin filaments change from hexagonal into rectangular arrays near the *Z*-discs. Each filament may then split into four threads, each of which joins three others from other actin filaments to form a pyramid at whose apex, on the other side of the *Z*-disc, an actin filament originates pointing into the adjacent sarcomere. Dimensions are given according to the two scales at the lower right. Thickness of the filaments is 5 times exaggerated with respect to their lengths.

If the theory of sliding filaments or interdigitation is accepted, tentatively or finally, the effects of redistribution of sarcoplasma within the sarcomere during contraction become of interest. Similarly, it is not clear how effectively "mechanical" cross-links, passing laterally from one filament to another, can exert forces in axial or longitudinal direction, as noted by Zierler [9.4]. The role of the *Z*-filaments, forming the *Z*-disc, during excitation and contraction has not been investigated in detail either. Moreover, it will be interesting to see whether structural differences between slow and fast muscle fibers explain differences in their dynamic performances.

The amount of information available on *chemical events*, chemical properties of actin and myosin, and the role of ATP (adenosine-5′-triphosphate) as energy source is vast. We refer the reader to Zierler's [9.4] lucid and concise summary of the important processes and hypotheses. Obviously, a muscle model is incomplete from the systems viewpoint if its emphasis is on the chain of chemical events, as was the one worked out by Davies [9.5]. Thermal and mechanical events must be temporally correlated and integrated with chemical processes. The kinetics and control mechanisms of the latter, therefore, are important. The example given by Chance [9.6] of a study inquiring into these aspects can be of interest to the bioengineer.

However, from the holistic viewpoint, we must treat the muscle as an energy transducer. It converts directly chemical energy into mechanical and thermal energy. Caplan [9.7] showed how very general energy transducers can be formulated by combining an energy store at the input and excitation-triggered events at the output of strongly *coupled systems*. The spontaneously occurring chemical processes and thermal and mechanical events thus are brought to interaction. Mathematical treatment requires the tools of irreversible thermodynamics of open systems, presented in Chapter 2. Caplan demonstrated that certain characteristics of muscle performance indeed are found in a particular class of such generalized energy converters.

The approach chosen in this chapter differs from Caplan's. *Mechanisms* of actin–myosin interaction are postulated to explain the particular nature of energy outputs found in muscle, given certain kinds of chemical processes. Thus we are not addressing ourselves to description of whole muscle and its function, which is the topic of Chapter 10. The models presented here concern, in fact, that part of muscle which is the actual source of energy.

The dynamic performance of this *actively contractile part* of the muscle fiber is studied most easily during (tetanic) isotonic contractions. In this condition all sarcomeres of a fiber are stressed to the same, constant degree. Furthermore, we usually assume that the shortening velocitv of the whole fiber reflects directly that of individual contractile elements. The "static" performance (i.e., the force produced at a fixed length of these elements at zero velocity) can be obtained from isometric studies. As is discussed in Chapter 10, more information than is obtained from these two types of experiments is needed to explain and predict performance of whole muscle in any condition. But it obviously is of great importance to know quantitatively the performance of the contractile elements. This information enables us to proceed to the next higher level—the whole muscle.

A. BASIC OBSERVATIONS ON PERFORMANCE, HILL'S HYPERBOLA, AND TOTAL POWER

The sliding filament hypothesis rests strongly on electronmicrographic evidence [9.8]. The *static performance* of contracting muscle fibers also appears to support it. Gordon et al. [9.2] determined the fiber force maintained (in isometric tetanus) as a function of sarcomere length. They found that this function could be correlated directly with the degree of interdigitation of actin and myosin filaments. For long sarcomere lengths, the force maintained is linearly proportional to overlap (see Fig. 9.2). This is the range between no overlap (zero force) and complete interdigitation (maximum force), that is, when the sarcomere length changes from approximately 3.5 to 2.1 μm. In this region the force apparently is proportional to the number of sites on the filaments available for bridge formation. In a second and narrower region, near 2-μm sarcomere length, tension is greatest and does not depend strongly

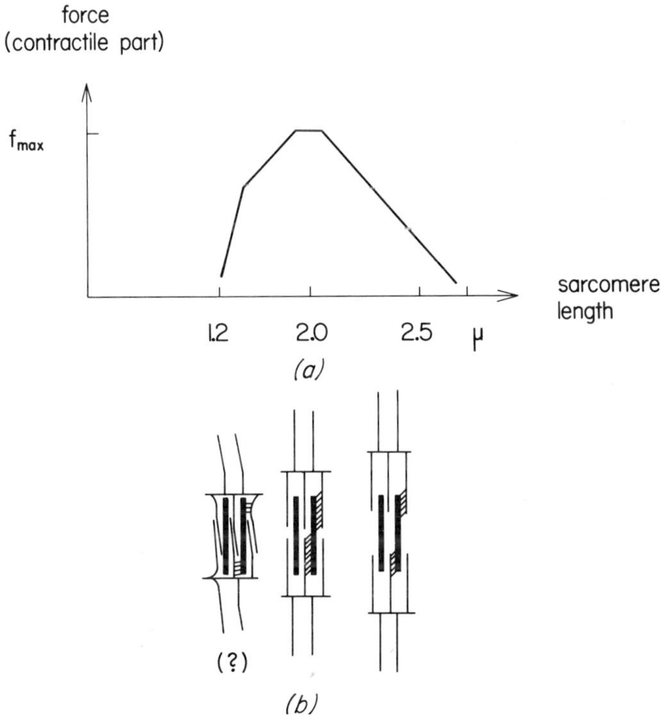

Fig. 9.2. (*a*) Relation between isometric tension and length of one sarcomere. (*b*) Hypothetical geometries of the sarcomeres at three different lengths, explaining the length–tension relation. This "sliding filament" model fits the data by Gordon et al. [9.2] obtained in the frog's semitendinosus muscle. Only part of the crossbridges connecting the thick filaments on the right are shown. For large sarcomere extensions, the tension is proportional to the number of crossbridges (i.e., the degree of overlap of thick and thin filaments). Experimentally, the forces exerted by the passive structures surrounding the sarcomeres add onto the actively maintained force of the contractile elements. With the sarcomeres 2 μ long, the force developed is maximum. This is defined as the "resting length." Shorter sarcomeres are unable to maintain the large force, probably because of deformation of the sarcomeres near the Z-line or interference of actin filaments with each other, as indicated in the lower left drawing. After Gordon et al. [9.2].

on length. Here all sites ready for cross-link formation may be engaged. However, when actin filaments from the two halves of a sarcomere begin to meet and to overlap, force decreases if length decreases. This probably is indicative of the directionality of the forces between the two filaments due to asymmetry of the myosin molecule. Below lengths of approximately 1.5 μm the force drops precipitously. The exact reason for this is not understood. Perhaps the sarcomere deforms and the (actin) filament arrays are so distorted that bridges cannot form if both actin filaments overlap and myosin rodlets butt against the Z-discs. Figure 9.2 indicates this, as well as the possibility of interference among overlapping actin rodlets.

It should be noted that it is not easy to obtain experimental data that lie on the straight lines, as in Fig. 9.2. Some of the sarcomeres within a fiber studied may not contract isometrically. Moreover, at large sarcomere lengths, mechanical properties of the sarcolemma and the tissues connecting the fibrils are superimposed on those of the contractile parts. *Dynamic measurements* (e.g., during *isometric* twitches) yield results that seem to be partly explained by the sliding filament theory, too. The peak force and total extra heat depend similarly on fiber length, as does tetanic tension. ("Total extra" heat is the component released in addition to resting and recovery heats.) Both exhibit a maximum at the so-called resting length, and both decay with increasing or decreasing fiber length. The largest stress (i.e., the force per cross-sectional area of fiber or the force times length divided by fiber volume) is of the order of 1.0 to 1.2×10^6 dynes/cm^2 at resting length and approximately 75% of this at 0.8 and 1.2 times resting length. Heat released during the twitch, divided by the muscle mass, has its peak at 1.0 to 1.2×10^5 dynes/cm^2 (1 dyne/cm^2 $= 2.4 \times 10^{-8}$ g-cal/cm^3). This heat depends on fiber length in the same way as does stress, such that the following ratio, the so-called isometric heat coefficient, turns out to be a constant [9.9, 9.10]:

$$\frac{\text{peak twitch stress}}{\text{heat released per mass, per twitch}} \simeq 10$$

for fiber lengths between 0.8 and 1.2 times resting length.

Dynamic measurements on fibers loaded *isotonically* and stimulated tetanically indicate that muscle contracts more slowly as load becomes larger, which is the result that would be expected intuitively. A graph of initial (or maximum) shortening velocity versus load indicates a hyperbolic relationship between the two observables (Fig. 9.3). A. V. Hill [9.11] proposed the following mathematical expression fitting this characteristic dependence of peak velocity (v) on the constant force (f) produced by the muscle:

$$(f + a)(v + b) = \text{constant} \tag{9.1}$$

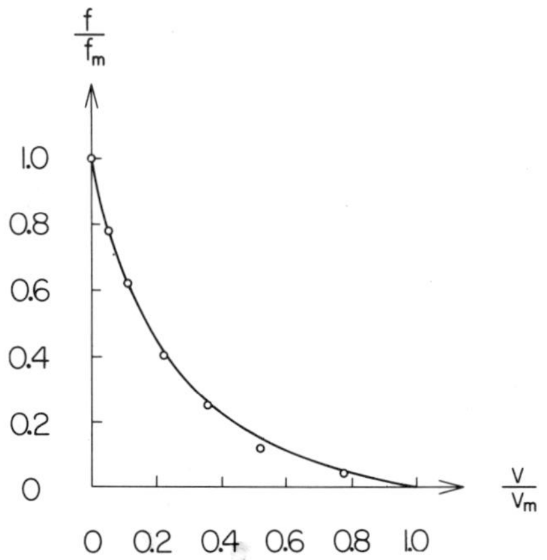

Fig. 9.3. Hill's hyperbola: the force–velocity relationship as obtained from isotonic tetanically stimulated muscle. The circles give Hill's original data [9.11], which were observed in the sartorius muscle of the frog, at its resting length at 0°C. They are fitted by $(v/v_m + 0.25)(f/f_m + 0.25) = 0.3125$. When v_m and f_m are used to normalize the data, the hyperbola also predicts force–velocity relations at lengths other than the resting length. (See Fig. 9.2 and text for definition of resting length.) For this particular muscle, v_m and f_m are 5.2 cm/sec and 65 g*, respectively.

where

$$f = \text{force}$$
$$v = \text{velocity}$$
$$a, b = \text{constants}$$

Hill originally arrived at this equation by analyzing liberated power, mechanical and thermal, during an isotonic contraction. From his early investigations it appeared that the shortening heat (the "extra" heat released in isotonic run, in addition to the heat in isometric studies) was proportional to shortening distance only. This also may be expressed by letting the rate of extra heat production be proportional to velocity. On the other hand (total extra) power was found to decrease linearly with increasing load. Formally this gives

$$\text{mechanical power} + \text{(extra) rate of heat release} = fv + a'v = b(f_m - f) \tag{9.2}$$

where f_m is the load under which muscle does not shorten any more.

Equation 9.2 is transformed into one resembling eq. 9.1 by adding $a'b$ and bf to both sides:

$$(f + a')(v + b) = b(a' + f_m) \tag{9.3}$$

Hill [9.12] found later, however, that a' in eq. 9.2 is not a constant but dependent on load as follows:

$$a'(f) = 0.16f_m + 0.18f \tag{9.4}$$

It is therefore impossible to obtain the constants in eq. 9.1 from thermal observations only. However, eq. 9.1 proves to be very useful in that it fits, as an empiric expression, the mechanical performance of several kinds of skeletal muscle fibers. Wilkie [9.13] found that the velocity of intact human muscle depends in the same manner on the force produced. Moreover, individual hyperbolas of the muscle tested are geometrically similar. Normalization of eq. 9.1 by maximum force (f_m) and maximum velocity (v_m) leads to the widely applicable form of Hill's equation (see Fig. 9.3.)

$$\left(\frac{f}{f_m} + \frac{1}{4}\right)\left(\frac{v}{v_m} + \frac{1}{4}\right) = \frac{5}{16} \tag{9.5}$$

because, for many different types of muscle,

$$\frac{a}{f_m} = \frac{b}{v_m} \simeq \frac{1}{4} \tag{9.6}$$

The physical meaning of v_m, of course, is that maximum velocity at which the muscle contracts without any load at all. It is given by the internal mechanisms of contraction. The f_m is the maximum (i.e., the tetanic isometric) force developed against a load so large that $v = 0$. As mentioned earlier, f_m is a function of fiber length. Abbott and Wilkie [9.14] showed, however, that eq. 9.5 approximates performance not only at the so-called resting length, as Hill had demonstrated, but also at other initial lengths. In other words, both f_m and v_m vary as a function of muscle extension in such a way that the force–velocity relation of eq. 9.5 holds over a considerable range of initial lengths. Performance of the contractile element of muscle therefore is completely defined, for tetanic stimulation, if the isometric length–tension diagram (Fig. 9.2) is given together with eq. 9.5. The force produced then can be given exactly for all lengths and all velocities. We use this model throughout the chapter. In Chapter 10 it is seen that different expressions for force as a function of length and velocity also are useful if performance of the whole muscle, as part of an entire muscle system, is considered.

It should be noted that detailed studies [9.15] reveal small but systematic deviations from these basic observations. If a muscle is made to contract isotonically from large initial lengths, the developed velocities are smaller than those obtained from smaller starting extensions. Velocities, of course, must be compared at certain muscle lengths; differences can be of the order of 5 to 15%. This probably means that velocity also is slightly dependent on duration of stimulation [9.4].

In order to obtain total power production during isotonic contraction, or at least during its beginning, we must remember that more heat is released in this case than in an isometric run. Although there exist contradictory data and views on types of heat released [9.4], we formulate the rate of total heat release using eq. 9.4:

$$\dot{Q} = q_0 + (0.16f_m + 0.18f)v \tag{9.7}$$

where $\dot{Q}$ = thermal power
q_0 = rate of heat production in isometric condition ($v = 0$)

This form is intended as a phenomenological description rather than one detailing different mechanisms of heat production. Total power ($\dot{W}$) is obtained by adding the mechanical term (fv) to eq. 9.7. If we simplify the sum by letting $a' = a$, that is,

$$\dot{W} = \dot{q}_0 + 0.25 f_m v + fv \tag{9.8}$$

we do not change the general character of $\dot{W}$'s dependence on load and velocity.

As we see later, the more accurate equation (9.7) could be used in the following development. Some parameters in the models discussed should be adjusted to correspond to more recent findings. For our purpose it appears important, however, to demonstrate principles underlying the models rather than to arrive at parameter values (whose numerical values always are subject to improvement). We therefore also introduce an early finding by Hill concerning the term $\dot{q}_0$. Hill approximated

$$\dot{q}_0 \simeq \frac{f_m v_m}{16} \tag{9.9}$$

which enables us to give $\dot{W}$ in a different form, namely:

$$\dot{W} = \frac{f_m v_m}{16}\left[1 + 4\frac{v}{v_m} + 16\frac{f}{f_m}\frac{v}{v_m}\right] \tag{9.10}$$

Equation 9.10 can be combined with eq. 9.5 to yield $\dot{W}$ dependent on either v/v_m or f/f_m.

$$\dot{W} = \frac{f_m v_m}{16}\,\frac{1 + 24\, v/v_m}{1 + 4\, v/v_m} \tag{9.11}$$

This function (Fig. 9.4) plays an important role, together with Hill's hyperbola and the expressions for rate of heat release, in the following discussion of muscle models.

B. THERMODYNAMICS, PODOLSKY'S MODEL, AND MUSCLE EFFICIENCY

In addition to phenomenological description of the dynamics of muscle performance, summarized in the preceding section, we can obtain laws governing some aspects of the mechanochemical processes from thermodynamics. From the first law, the principle of conservation of energy within

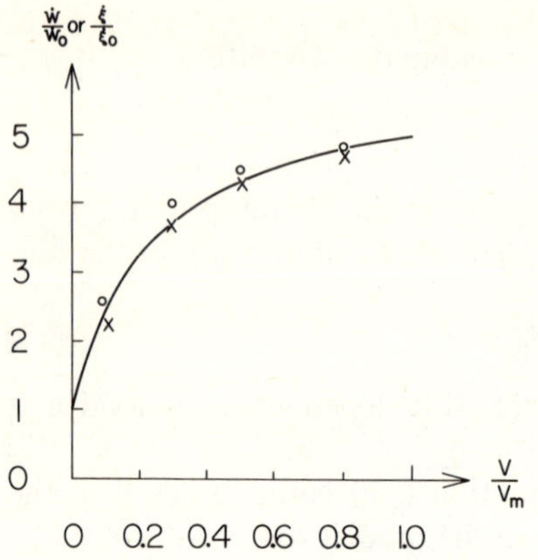

Fig. 9.4. Relative power liberation. The rate of total energy liberation (i.e., the thermal and mechanical power during shortening) is normalized by energy liberation during an isometric case ($\dot{W}_0$ or $\dot{\xi}_0$) and plotted against normalized velocity. The solid line follows from Hill's observations (eq. 9.11). Crosses are from Podolsky's model, circles from Huxley's. The two models predict $\dot{\xi}/\dot{\xi}_0$ equally satisfactorily; Huxley's hypothesis is more universal, however, since it also predicts well mechanical power and the force–velocity relation.

a closed system, it follows that changes of internal energy (u) must be given by exchange of energy with the outside. Formally, we have

$$\Delta u = -Q - W_m \tag{9.12}$$

where
Δu = internal energy change
Q = heat flowing *out* of system
W_m = work done by system

The work done by the system includes both external work and that due to (internal) pressure and volume changes. Thus, if we assume pressure constant, we have

$$\Delta u = -Q - w_m - P\Delta V \tag{9.13}$$

where
w_m = external work
P = pressure
V = volume

From this it follows that the enthalpy change ΔH is given by

$$\Delta H \equiv \Delta u + P\Delta V = -Q - w_m \qquad \text{(isobaric)} \tag{9.14}$$

But for muscle we know that chemical reactions supply the energy for the mechanical work. The rate at which the reaction proceeds determines the rate of energy output; that is,

$$\dot{\xi}\widetilde{\Delta H} = -\dot{Q} - \dot{w}_m \tag{9.15}$$

where
ξ = extent to which reaction proceeds
$\dot{\xi}$ = rate of reaction
$\widetilde{\Delta H}$ = ΔH per unit rate of reaction

Podolsky [9.16] suggested that the rate $\dot{\xi}$ may be given by that of the slowest reaction of a whole chain of coupled reactions, if all its members except first and last are assumed to be cyclic. In this situation the amount of chemical energy made available is proportional to the extent of the first reaction. Since all reactions are coupled, however, the rate of any of them can limit the whole system. Thermodynamic properties of intermediate reactions in the chain do not affect the net energy flux if the reactions are cyclic in the sense that they are back in their initial states before the energy outputs of the whole system have been measured.

Thus the parameter ξ, the extent by which the first reaction proceeds during excitation, is the only one needed to describe and quantitate all chemical changes. During recovery, coupling between an additional reaction and the first one of the chain takes place. This proceeds to the extent ξ and lets the first reaction system assume its initial state.

Podolsky [9.16] showed that $\dot{\xi}$ exhibits a dependency on contraction velocity (of the sarcomere), which can be matched satisfactorily by the function given in eq. 9.11, assuming simple laws of interaction between interdigitating filaments. First we note that the rate of interaction (i.e., of cross-bridge formation) must be proportional to the rate $\dot{\xi}$ of the chemical reaction. Second, we assume that there exists an interaction probability. In the vicinity of each active site A (see Fig. 9.5) there exists a spatial function

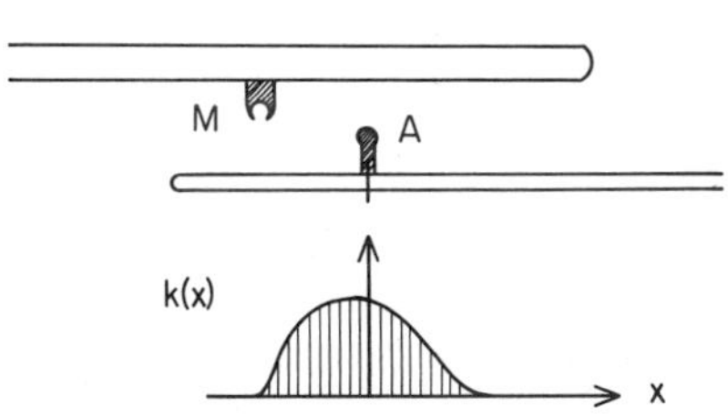

Fig. 9.5. Podolsky's model of sarcomere contraction. It is assumed that sites exist on the myosin and actin filaments (M and A, respectively) which interact with a certain probability $k(x)$, a function of the distance between the two sites. Since the model covers only energetics of the force development, no further assumptions need be made. As explained in the text, the exact form of $k(x)$ is unimportant. After Podolsky [9.16].

$k(x)$ which gives the probability of a crossbridge forming per unit time if a similarly active site (M) on the other rodlet is at a distance x from A. We also assume that if interaction occurs, it occurs only once (during a passage of M by a site A). If n is the probability that an interaction has occurred during the past ∂t, we can find the probability ∂n that an interaction will occur during the next ∂t. For a given distance x between M and A sites, we have

$$\partial n = [1 - n(x)]k(x)\,\partial t \tag{9.16}$$

But time can be eliminated by introducing the relative velocity between the filaments, v_{rel}. With

$$\frac{\partial n}{\partial t} = v_{rel} \frac{\partial n}{\partial x} \tag{9.17}$$

eq. 9.16 becomes an ordinary differential equation:

$$\frac{dn}{1-n} = \frac{1}{v_{rel}} k(x)\, dx \tag{9.18}$$

which can be integrated directly. If we do this over the entire range of x where $k(x)$ is nonzero, we obtain the probability of interaction having occurred during the last entire passage:

$$n = 1 - e^{-\lambda/v_{rel}}, \tag{9.19}$$

where $\lambda = \int_R k(x)\, dx$

R = range of x for which $k(x)$ is nonzero

Hence interaction occurs certainly ($n = 1$) if $v_{rel} = 0$ and, if $v_{rel} \simeq \infty$, there is no time for crossbridge formation (i.e., $n = 0$). Furthermore, n is not dependent on the particular shape of $k(x)$. The latter may well be asymmetric, favoring directional action.

If the distance between A sites is a, each M site will be presented with v_{rel}/a of A sites per unit time. Since the muscle shortens at a velocity $v \propto 2v_{rel}$, the rate of interaction at an M site becomes $n\, v/(2a)$. If we let the density of M sites be m/cm^3, then the rate of bridge formation ν per volume of muscle mass is

$$\nu = \frac{mv}{2a} n = \frac{mv}{2a}(1 - e^{-2\lambda/v}) \tag{9.20}$$

This rate, however, is the same as the extra liberation of chemical energy, as mentioned before.

$$\dot{\xi} - \dot{\xi}_0 = \frac{mv}{2a}(1 - e^{-2\lambda/v}) \tag{9.21}$$

where $\dot{\xi}_0$ = reaction rate in isometric condition (per unit muscle mass)

Hence

$$\frac{\dot{\xi}}{\dot{\xi}_0} = 1 + \frac{mv_m}{2a\dot{\xi}_0} \frac{v}{v_m} [1 - e^{-(2\lambda/v_m)(v_m/v)}] \tag{9.22}$$

This expression can be made to fit satisfactorily that of eq. 9.11 by suitable choice of parameters, as in Fig. 9.4. In particular, $m\ v_m/(2a\dot{\xi}_0) = 13$ and $v_m/(2\lambda) = 2.8$. Verification of the model obviously hinges on determination of various quantities represented in these parameters.

The rate of energy conversion as given in eq. 9.15 does not give us any insight into *efficiency* of the conversion process. If we dealt with a mechanical system we could define η_{mech}, the "mechanical efficiency," from eq. 9.12

$$\eta_{\text{mech}} = \frac{W_m}{W_m + Q} \tag{9.23a}$$

Or, if we separate work done to change volume, we get from eq. 9.14

$$\eta_{\text{mech}} = \frac{w_m}{-\Delta H} \quad \text{or} \quad \frac{\dot{w}_m}{-\dot{\xi}\widetilde{\Delta H}} \tag{9.23b}$$

But in the mechanochemical system, maximum possible external work is given by free energy, which includes entropy change. From the second law of thermodynamics it follows that, in isothermal condition, the *maximum* energy that can be obtained from a chemical reaction is

$$W_{\text{max}} = -\Delta(u - TS) \tag{9.24}$$

The left-hand side again contains external work and that due to volume changes (in isobaric system), that is,

$$w_{m,\text{max}} = -\Delta(u - TS) - P\Delta V \tag{9.25}$$

But the Gibbs potential G is defined by $u - TS + PV$. Hence, in the isothermal and isobaric case, we have

$$w_{m,\text{max}} = -\Delta G \tag{9.26}$$

The thermodynamic efficiency therefore becomes

$$\eta = \frac{w_m}{-\Delta G} \quad \text{or} \quad \frac{\dot{w}_m}{-\dot{\xi}\widetilde{\Delta G}} \tag{9.27}$$

The relation between mechanical efficiency and η is thus

$$\eta = \eta_{\text{mech}} \frac{\widetilde{\Delta H}}{\widetilde{\Delta G}} \tag{9.28}$$

The entropy change could be estimated only if maximum work were obtained from the system. This implies that its heat gain also would be maximal (i.e., equal to $T\Delta S$). Since the change in free energy in muscle is unknown, however, it is impossible to compute η.

C. HUXLEY'S MODEL OF CONTRACTION

Podolsky's model, presented in the previous section, emphasizes predicting the kinetics of coupled chemical and mechanical processes. Details of the force-producing mechanisms remain to be specified. Huxley [9.1] proposed a hypothesis on the nature of forces between actin and myosin filaments. He also considered separately the rates of bridge formation and breaking. His model thus becomes more universal. It predicts the rate of heat production as well as muscle tension as a function of positive *and* negative velocity. (Negative velocities occur when active muscle is stretched.)

Crossbridge formation, as in Podolsky's model, is assumed to take place between specialized sites on actin filaments (A sites) and on myosin rodlets (M sites). The latter sites are supposed to be mobile, within a certain range ($\pm h$) of their equilibrium positions. This is shown schematically in Fig. 9.6.

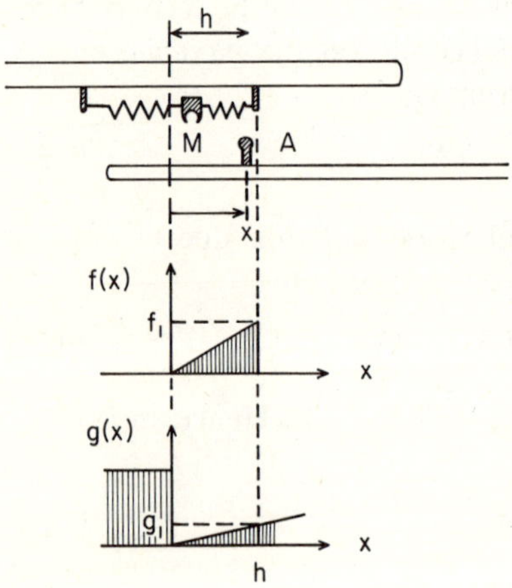

Fig. 9.6. Huxley's model of the force-producing mechanism. If a crossbridge between the active sites M and A has been formed, it transmits a force between the filaments which is proportional to distance x, as symbolized by the "spring suspension" of M. The probability that interaction occurs is defined by $f(x)$; the probability that a crossbridge is broken is given by $g(x)$; M and A can latch up only within the range $\pm h$. See text for the significance of f_1 and g_1. Note that x in this model defines the distance of site A from M's resting position. After Huxley [9.1].

An elastic force, proportional to displacement of the M site from its resting position (x), will act to shorten the sarcomere once the A and M sites have interacted. During excitation, this cross-link formation is thought to occur spontaneously at a rate f, and temporarily, since chemical compounds are available which are sufficiently energy rich to break the link at a rate g. A bridge therefore can be formed as soon as the A site is within the "reach" of the M site. Directionality is obtained by letting f, the rate of bridge formation, be larger than g for positive $x \leqslant h$. For $x < 0$, f vanishes and g must take on rather large values to ensure rapid destruction of the link once the M site has pulled the A site past its mid-position. (See Fig. 9.6 for f and g as functions of x.) Beyond that, the "spring" force counteracts shortening forces of other bridges. However, it takes some time before a link can be opened. At high shortening velocities the fraction of bridges still in existence for A sites at negative x's, with respect to their M sites, is relatively large. Nor is bridge formation instantaneous. This diminishes the number of links formed per time. Hence smaller tension develops at higher velocities.

Very many links are thought to close and open so that the overall force is developed smoothly. Furthermore, spacing of individual reacting sites is so assumed that an A position can be within the reach of only one M site at any one time. However, M sites may well be spaced differently from the A's.

On the basis of these assumptions, the rate of crossbridge formation and hence the tension developed can be computed. This is done using the same considerations as in Podolsky's model (eq. 9.16). Taking into account rates of both link opening and closing, we have

$$\frac{\partial n}{\partial t} = [1 - n]f(x) - ng(x) \tag{9.29}$$

where $f(x) -$ rate of link formation, per sec
$g(x) =$ rate of link breaking, per sec (see Fig. 9.6)

Equation 9.29 gives the rate of change of the fraction n of all M sites that have combined with A sites at position x. Introducing the shortening velocity $(= -2v_{\text{rel}})$ and eq. 9.17, we find

$$-\frac{1}{2}v\frac{dn}{dx} = f(x) - [f(x) + g(x)]n \tag{9.30}$$

This equation must be integrated separately over the three regions of x where $f(x)$ or $g(x)$ have different forms. Using the boundary conditions

$$\begin{aligned} n(h) &= 0 \qquad \text{and} \qquad n(+0) = n(-0) \\ n(x) &= 0 \qquad \text{for} \quad x > h \end{aligned} \tag{9.31a}$$

we obtain

$$n(x) = \frac{f_1}{f_1 + g_1}[1 - e^{-(\Psi/v)(1 - x^2/h^2)}] \qquad 0 \leqslant x \leqslant h \tag{9.31b}$$

$$n(x) = \frac{f_1}{f_1 + g_1}[1 - e^{-\Psi/v}]e^{2g_2x/v} \qquad x < 0 \tag{9.31c}$$

where

$$\begin{aligned} \Psi &= h(f_1 + g_1) \\ f_1 &= f(h) \\ g_1 &= g(h) \\ g_2 &= g(x),\ x < 0 \end{aligned}$$

From terms in the brackets in eq. 9.31b and 9.31c we realize that n decreases (over the entire range $0 < x < h$) if v increases. At a fixed velocity, n is maximal near $x = 0$. It decreases slowly for small positive x, more rapidly near $x = h$, and at this point it vanishes. For negative x, n decreases exponentially.

The *average* force developed at one M site can now be obtained from the ratio of work done (when "springs" between the reactive sites shorten) and the distance a between A sites. If k is the effective "spring constant" we have

$$f' = \frac{k}{a}\int_{-\infty}^{+\infty} n(x)\,dx$$

$$= \frac{kh^2}{2}\frac{f_1 h}{a\Psi}\left\{1 - \frac{v}{\Psi}[1 - e^{-\Psi/v}]\left[1 + \frac{v}{2\Psi}\left(\frac{f_1 + g_1}{g_2}\right)^2\right]\right\} \tag{9.32}$$

The factor $kh^2/2$ is maximum mechanical work, per M site. Using again the density of these sites, m, mechanical power is obtained as

$$\dot{w}_m = mf'v = w_{m,\max}\frac{hf_1}{a}\frac{v}{\Psi}\left\{1 - \frac{v}{\Psi}[1 - e^{-\Psi/v}]\left[1 + \frac{v}{2\Psi}\left(\frac{f_1 + g_1}{g_2}\right)^2\right]\right\} \tag{9.33}$$

The total rate of (chemical) energy release can also be deduced. The average frequency at which M and A sites react must be the rate at which the cyclic reactions proceed. The M sites enter the reaction only after their bridges previously formed have been broken. The fraction $(1 - n)$ of all M's therefore is available. If energy released per cross-link formation and opening is called e, total energy liberation per volume is

$$\dot{W} = em\frac{v}{a}\int f(x)[1 - n(x)]\,dt \qquad \text{or}$$

$$= \frac{em}{a}\int_{-\infty}^{+\infty} f(x)[1 - n(x)]\,dx$$

$$= \frac{e}{a}\frac{mh^2}{2}\frac{f_1 g_1}{\Psi}\left[1 + \frac{f_1}{g_1}\frac{v}{\Psi}(1 - e^{-\Psi/v})\right] \tag{9.34}$$

This expression is of exactly the same form as eq. 9.22 of Podolsky's model. The predicted relative power release $(\dot{\xi}/\dot{\xi}_0)$ therefore can be made to match equally well the one observed by Hill and the one obtained by Podolsky. We can fix, for example, the value of parameter Ψ so that the maximum contraction velocity $v_m = 4\Psi$; that is, $\Psi = b$ of eq. 9.6. Then v/Ψ of the preceding equations equals $4v/v_m$, and f_1/g_1 must approximate 4.5. In order to fix the other parameters we require, using eq. 9.32, that $v = v_m$ if the muscle is unloaded (i.e., $f' = 0$). From this we find $g_2/(g_1 + f_1) = 3.92$. Conversely, when $v = 0$, the force in eq. 9.32 must be maximal; that is,

$$f_0 = \frac{m}{2a} \frac{f_1}{f_1 + g_1} w_{m,\max} \qquad (9.35)$$

When this is multiplied by v_m it must equal $16\dot{q}_0$ (by eq. 9.9) or $16\dot{\xi}_0$ (by eqs. 9.22 and 9.34). However, $\dot{\xi}_0$ is $ew_{m,\max} f_1 g_1/\Psi$ in this model. Hence we arrive at the ratio $w_{m,\max}/e = 0.72$. An independent estimation of $w_{m,\max}$, the maximum energy in a crossbridge, therefore would allow for partial verification of the model.

When these numerical values of model parameters are introduced in equations for total power and for mechanical power released, observations by Hill are fitted remarkably well (i.e., within the experimental errors). As a consequence, thermal power is also predicted well. Further evaluation of the model may yield estimations of a and h, the distances between reactive sites. This is not without difficulties, however, since we have assumed a two-dimensional arrangement, and since g_1 cannot be obtained theoretically.

These considerations are important when the model is tested. Huxley [9.1] discussed these points thoroughly, together with questions concerning the tension necessary to lengthen activated muscle and the heat liberated in this condition. The former is predicted correctly by the model (ca. 20% of tension produced by muscle at equal but opposite velocity); the latter cannot be accounted for by the model. Furthermore, Huxley pointed out that factors other than the ones included in the model may, in nature, determine the resistance to shortening. In particular, this may be noted when the speed of contraction of unloaded muscle is compared with the prediction by the model. Equation 9.32 lets the force decay, with increasing velocity, almost exactly as Hill found it in muscle (Hill's hyperbola). But this was achieved by fixing the rate parameters (f_1/g_1, g_2/g_1) in order to ensure the match. Attempts were not made to determine their values from other independent data.

Huxley's model of contraction thus exhibits many typical aspects of a particular type of model in physiology. Most phenomena observed in muscle are elucidated on the basis of hypothesized, plausible mechanisms and some structural assumptions. Variables significant for the function of muscle (tension, velocity, heat liberation) and their interrelations are predicted by processes, and their couplings are described on a lower organizational level. Furthermore (and not less important), several questions concerning verification of the hypothesis are raised. Additional experiments and calculations suggest themselves when the model is evaluated. For example, three-dimensional filament arrays and spirally arranged sites, or frictional forces between filaments resisting shortening, may explain some discrepancies between muscle and model.

D. REFERENCES

9.1. Huxley, A. F. Muscle structure and theories of contraction. *Progr Biophys* **7**: 255–318, 1957.

9.2. Gordon, A. M., A. F. Huxley, and F. J. Julian. The length–tension diagram of single vertebrate striated muscle fibres. *J Physiol (London)* **171**: 28P–30P, 1964.

9.3. Wilkie, D. R. Muscle. *Ann Rev Physiol* **28**: 17–38, 1966.

9.4. Zierler, K. L. Mechanism of muscle contraction and its energetics, pp. 1128–1171 in *Medical Physiology*, Vol. II, 12th ed. V. B. Mountcastle (Ed.). C. V. Mosby Co., St. Louis, 1968.

9.5. Davies, R. E. A molecular theory of muscle contraction: Calcium-dependent contractions with hydrogen bond formation plus ATP-dependent extensions of part of the myosin–actin cross bridges. *Nature (London)* **199**: 1068–1074, 1963.

9.6. Chance, B. Transients in metabolism, pp. 334–343 in *Theoretical and Mathematical Biology*. T. H. Waterman and H. J. Morowitz (Eds.). Blaisdell Publishing Co., New York, 1965.

9.7. Caplan, S. R. A characteristic of self-regulated linear energy converters. The Hill force–velocity relation for muscle. *J Theor Biol* **11**: 63–86, 1966.

9.8. Huxley, H. E. The mechanism of muscular contraction. *Sci Amer* **213**: (no. 6), 18–27, 1965.

9.9. Hill, A. V. The relation between force developed and energy liberated in an isometric twitch. *Proc Roy Soc* [*Biol*] **149**: 58–62, 1958.

9.10. Fales, J. T., W. J. Crawford, and K. L. Zierler. Gradient-layer calorimetry of muscle: relation between length, twitch tension, and heat. *Amer J Physiol* **213**: 1427–1432, 1967.

9.11. Hill, A. V. The heat of shortening and the dynamic constants of muscle. *Proc Roy Soc* [*Biol*] **126**: 136–195, 1938.

9.12. Hill, A. V. The effect of load on the heat of shortening of muscle. *Proc Roy Soc* [*Biol*] **159**: 297–318, 1964.

9.13. Wilkie, D. R. The relation between force and velocity in human muscle. *J Physiol (London)* **110**: 249–280, 1950.

9.14. Abbott, B. C. and D. R. Wilkie. The relation between velocity of shortening and the tension–length curve of skeletal muscle. *J Physiol (London)* **120**: 214–223, 1953.

9.15. Bahler, A. Modeling of mammalian skeletal muscle. *IEEE Trans Bio-Med Engin* **BME-15**: 249–257, 1968.

9.16. Podolsky, R. J. Thermodynamics of muscle, pp. 359–385 in *The Structure and Function of Muscle*, Vol. 2. G. H. Bourne (Ed.). Academic Press, New York, 1960.

CHAPTER TEN

Muscle–Spindle Feedback Loop and Muscle Systems

A. INTRODUCTION; TOOLS OF ANALYSIS

The significant variables in the description of dynamic behavior of whole skeletal muscle are (1) the average level of excitation, or mean firing rate of the motoneuron pool, (2) the net force of the actively contracting part of the muscle, (3) the force exerted on the muscle by its load (equaled by the force exerted *by* the muscle), and (4) the changes in overall length of the muscle system, or its shortening velocity. The details of heat production or of the summation of forces of recruited fibers, important for investigations at the next lower level, are quite irrelevant and may be neglected in studies of the performance of whole muscle in fulfilling its task. In this chapter, we concern ourselves with the functional relationships between the four variables given, as influenced by the passive viscoelastic property of muscle tissue. The first four sections treat the kinematics for small changes in the muscle variables. It is therefore possible to apply the tools of *signal-flow diagram* analysis. This is particularly useful here, since the flow diagrams facilitate inspection of desired input–output relations, and since the elimination of uninteresting variables is achieved with a minimum of algebra. Relations of little significance are implicit but do not encumber the calculations. In other words, signal-flow diagrams yield an efficient method of analysis.

By the basic rules of signal-flow analysis, variables are represented by *nodes*, which are connected by directed *branches*, as in the simple examples of Fig. 10.1. The branches are marked to express the functional relations between the variables. In other words, when a variable is transmitted or "flows" along a branch, its value is multiplied or operated on by the "transmittance" assigned to the particular branch. Thus we can connect the motor

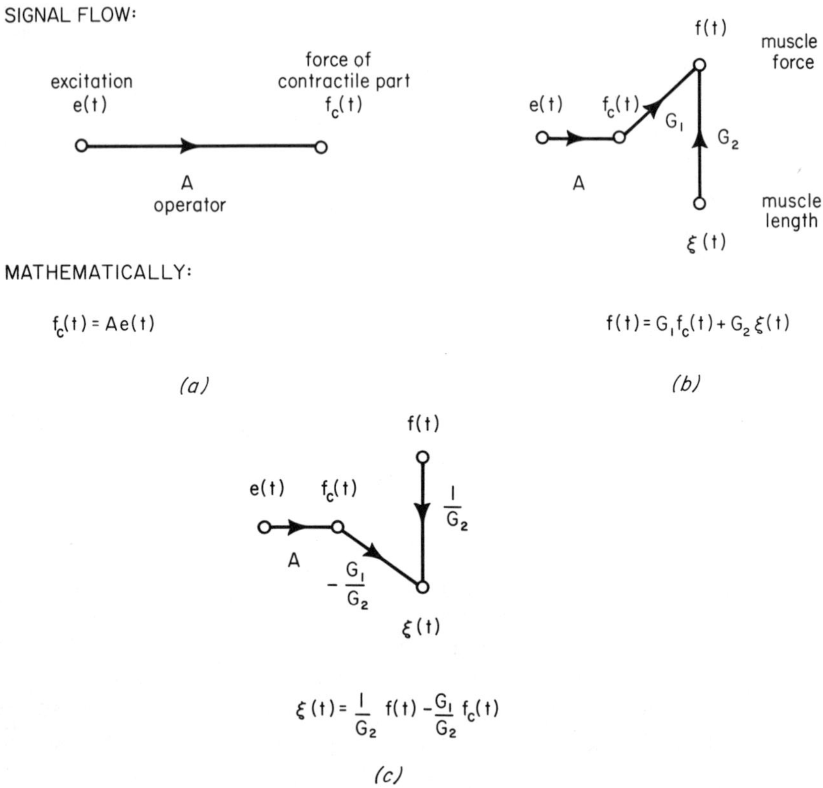

Fig. 10.1. Signal-flow diagrams, together with their mathematical equivalents (lower row), of a muscle. (*a*) Relation between net activity $e(t)$ of the motoneuron pool controlling the "force generator" of muscle (i.e., its contractile elements) and $f_c(t)$. (*b*) Complete muscle: relation between $e(t)$ and force $f(t)$ of a muscle as it depends on $f_c(t)$ and $\xi(t)$, the length of the muscle. The operators G_1 and G_2 are developed in Section B; cf. Fig. 10.4. (*c*) Complete muscle: signal-flow diagram needed when length of muscle $\xi(t)$ is the dependent variable.

nerve activity $e(t)$ and the force of the effective contractile element $f_c(t)$, as in Fig. 10.1*a*. The diagram stands for the equation $f_c(t) = A\,e(t)$; the operator A is defined later. The transmittance of the branch A can be linear or nonlinear. Generally, transmittances are component values (e.g., impedances in electric network or a spring constant in a mechanical system), characteristics of filters, or subsystems including transducers.

The *value of a variable* is equal to the sum of all signals entering its node. Figure 10.1*b* is the signal-flow diagram for the entire muscle for the case when the net muscle force $f(t)$ is observed as dependent on $e(t)$ and the muscle length $\xi(t)$. This diagram obviously implies the algebraic expressions

$$f_c(t) = Ae(t) \tag{10.1}$$

$$f(t) = G_1 f_c(t) + G_2\, \xi(t) \tag{10.2}$$

where G_1 and G_2 are operators, to be determined.

In this system we have not made $f_c(t)$ a function of $\xi(t)$, since we are going to allow only small signals (i.e., small changes from resting length of the muscle).

Clearly, in drawing the signal-flow diagram we are forced to decide whether an observable is a dependent or an independent variable. This often is helpful in systems physiology. Although it would be mathematically correct to write $e(t) = f_c(t)/A$—requiring an arrow from $f_c(t)$ to $e(t)$ in the flow chart—the functional relation between the two variables would be clouded in this alternate formulation. Similarly, if $\xi(t)$ is chosen to be the dependent variable, Fig. 10.1*c* must be used. There $\xi(t)$ receives branches. Again the transmittances G_1 and G_2 may be nonlinear; for example, G_1 may be dependent on $f_c(t)$ [but not on $\xi(t)$]. In Section 10.B G_1 and G_2 are determined. We can then study the effect of feedback through the muscle spindle system.

The *rules for reducing* a many-branched signal-flow diagram follow, in a straightforward manner, from the ones just given. The most basic are listed in Table 10.1. Note that loops, of the sort shown in case 4 in the table, do not necessarily mean feedback in the sense of control theory. Case 6 demonstrates that *paths* (continuous sequences of directed branches) that are destroyed in the process of reduction must be replaced by new branches (or paths). Note that generally there is no one-to-one correspondence between the signal-flow chart and the set of mathematical equations describing the system.

Another interesting example of the usefulness of signal-flow diagrams is that they indicate directly possible analog computer realization of a given transfer function $H(s)$. Let

$$H(s) = \frac{O(s)}{I(s)} = \frac{a_0 + a_1 s + a_2 s^2 + \ldots + a_n s^n}{b_0 + b_1 s + b_2 s^2 + \ldots + b_m s^m} \tag{10.3}$$

where

O = output

I = input

s = Laplace operator

This equation can be rewritten as follows, after dividing both numerator and denominator by s^m if $m > n$ (or by s^n if $n > m$):

$$O(s) = \left(\frac{a_n}{s^{m-n}} + \frac{a_{n-1}}{s^{m-n+1}} + \ldots + \frac{a_0}{s^m}\right) T(s) \tag{10.4}$$

where

$$T(s) = \frac{I(s)}{b_m} \frac{1}{1 + b_{m-1}/b_m s + \ldots + b_0/s^m} \tag{10.5}$$

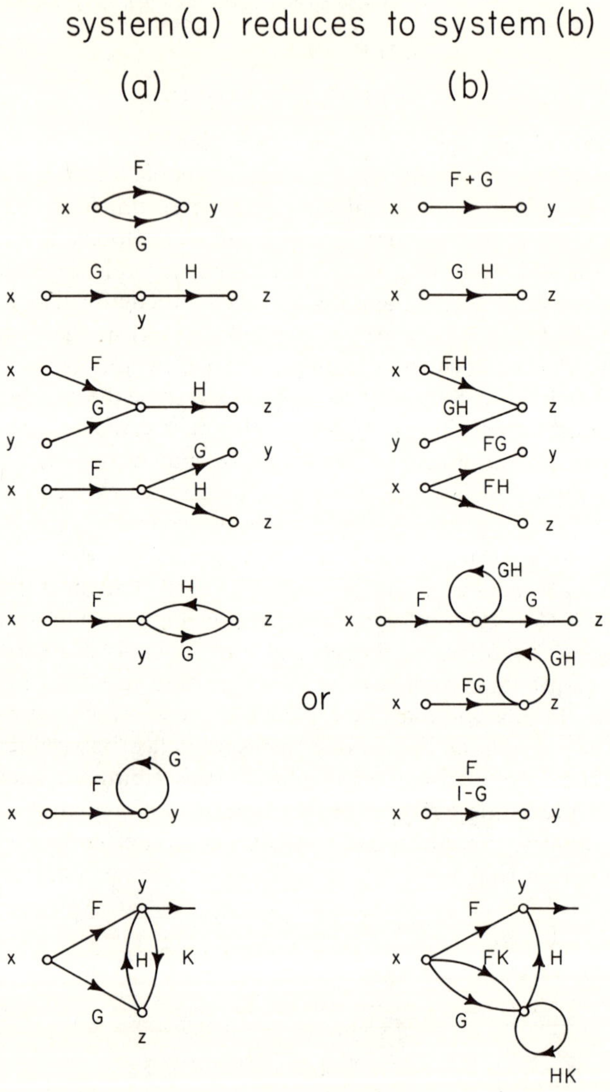

TABLE 10.1 RULES FOR REDUCING A MANY–BRANCHED SIGNAL–FLOW DIAGRAM

Equation 10.4 is in a form that can be set up on an analog computer, since $O(s)$ is a linear combination of multiple integrals of $T(s)$. The latter, however, can be obtained elegantly. This is seen when eq. 10.5 is given as

$$T(s) = \frac{I(s)}{b_m} - \left(\frac{b_{m-1}}{b_m}\frac{1}{s} + \frac{b_{m-2}}{b_m}\frac{1}{s^2} + \ldots + \frac{b_0}{s^m}\right)T(s) \qquad (10.6)$$

Figure 10.2 presents the signal-flow diagram of a computer programmed to solve eqs. 10.4 and 10.6 simultaneously. Readers interested in additional techniques and theorems are referred to Mason et al. [1.12], Robichaud et al. [10.1], or Truxal [10.2].

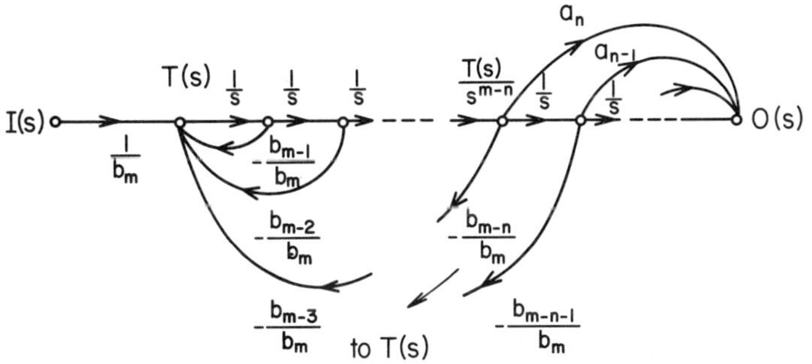

Fig. 10.2. Signal-flow diagram of a general, linear transfer function (eq. 10.3) which makes it possible to program an analog computer by inspection.

B. RELATION BETWEEN FORCE AND LENGTH OF MUSCLE: THE VOIGT SERIES SPRING MODEL

The short-term behavior and fast responses of muscle can conveniently be modeled by the simple mechanical system in Fig. 10.3. This type of analogy was discussed first by Levin and Wyman [10.3] and has since been used extensively. The element labeled f_c is the "force source," the contractile part of the muscle exerting force f_c. The spring and dashpot (Voigt or Kelvin body, in parallel with f_c) have the constants K_1 and R; K_2 is the constant of the spring connected in series with the three elements that are parallel to each other. The parallel spring is often thought to represent the elastic property of the sarcolemma and connective tissue and the series spring that of the tendons, although there does not exist clear evidence for this. The parallel dashpot limits the shortening velocity of the muscle.

It is well known, thanks to Hill's pioneering work, how the velocity of shortening of muscle fibers is indeed limited. This has been hypothesized to be inherent in the force-generating mechanism (see Chapter 9). Wilkie [10.4]

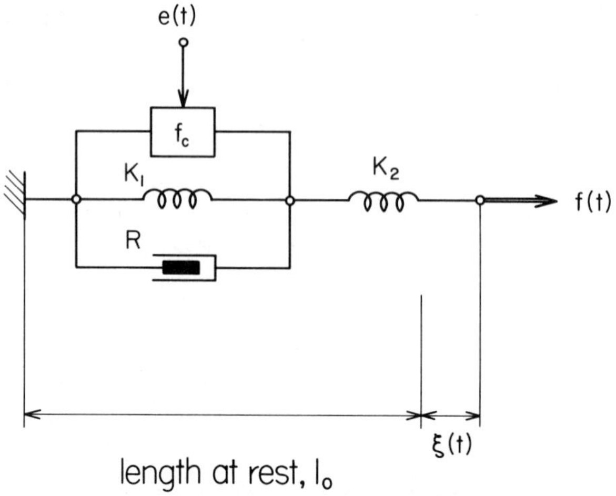

Fig. 10.3. Mechanical system that models many aspects of muscle performance: K_1 and K_2 are springs, R is a dashpot. The "black box" f_c is an ideal force generator of the actively contractile part of muscle. Its output is proportional to the neural input $e(t)$, the activity of the muscle's motoneurons. The element labeled f_c is freely extensible if the muscle is passive $[e(t) = 0]$. For small ξ's, f_c can be assumed to be independent of ξ. Generally, the springs are nonlinear. For small length changes, however, they can be assumed to be linear. Similarly, if the dashpot (R) is assumed to be linear, the velocity-dependent features of muscle performance are only approximated. This model is used not only for skeletal muscle but also for the cardiac muscle (Chapter 15).

demonstrated that Hill's equation predicts equally well the force–velocity relation of intact muscle systems. The observation that small loads can be moved more rapidly than large loads—assuming constant excitation of the muscle—thus is conveniently modeled by the ideal force generator f_c in parallel with the dashpot. Note that Hill's hyperbola, or rather the force–velocity hyperbola for the entire muscle, is approximated by a straight line if R is assumed to be constant.

Here we do not assume the task of defining which parts of the model are assigned to which anatomical part of the muscle. We rather take the synthetic viewpoint, since we shall continue to incorporate the model into more complex systems. For this we can be satisfied if the elements in Fig. 10.3 predict the important phenomena observed at this level, as they influence the performance of the entire muscle system.

In a small signal analysis the nonlinearities in the springs, which are described later, can be neglected and the frictional forces can be assumed to be proportional to velocity. Therefore, $\xi(t)$, defined in the following as the length change (see Fig. 10.3), depends on f_c and on f (the force exerted on the muscle by the load) as follows, when given in the Laplace domain:

$$\xi(s) = \frac{-f_c(s)}{K_1(1 + sT)} + \frac{f(s)}{K_3}\frac{1 + sKT}{1 + sT} \tag{10.7}$$

where

$$T = R/K_1$$
$$K = K_1/(K_1 + K_2)$$
$$K_3 = K_1K_2/(K_1 + K_2)$$

The corresponding signal-flow diagrams appear in Fig. 10.4, where we show (*a*) the flow analysis that is useful when isotonic responses are studied (eq. 10.7) and (*b*) the diagram to be used when isometric behavior is of interest or when the force is observed as a function of length and f_c. Now we are ready to compare Figs. 10.4 and 10.1 to obtain G_1 and G_2 for this linearized representation of the muscle.

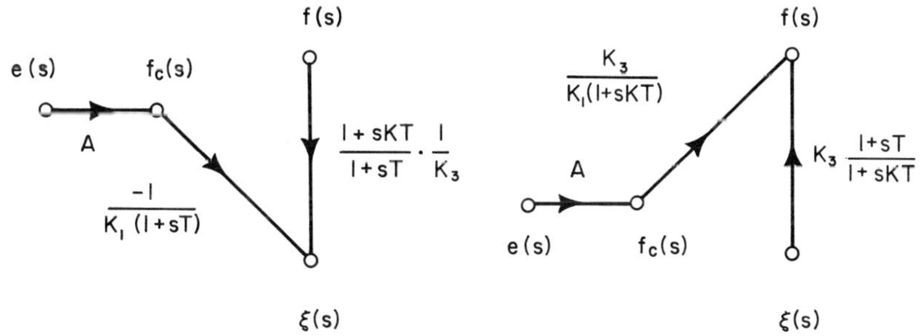

Fig. 10.4. Signal-flow diagram describing (small signal) performance of the muscle model in Fig. 10.3. Depending on whether muscle length or muscle force is the dependent variable, the diagram on the left or the one on the right must be used. e = excitation of muscle, f_c = output of ideal force source, f = force exerted by muscle, ξ = length changes (from resting length), s = Laplace operator. See text (eq. 10.7) for definitions of K, K_3, and T.

The performance of the model appears schematically in Fig. 10.5 for the two cases used most often in muscle physiology. Isotonic and isometric "twitches" are graphed which compare satisfactorily with the twitches of muscle, considering that the "stimulus" applied to the model was assumed to be an ideally sharp pulse. The active state of muscle, during which f_c exists, has finite times of rise and decay as indicated in Fig. 10.5. Hence the velocity and the time derivative of tension do not change abruptly, assuming that they are governed by a first-order differential equation. (The precise time course of f_c is discussed below.) Conversely, the model would also show this with a "sloppy" stimulus, even though the mass of muscle is neglected. Responses of the model to instantaneous stretch applied to the passive muscle ($f_c = 0$) also compare well with those observed in nature, although the long-term behavior deviates. Creep and stress relaxation cannot be seen in the model.

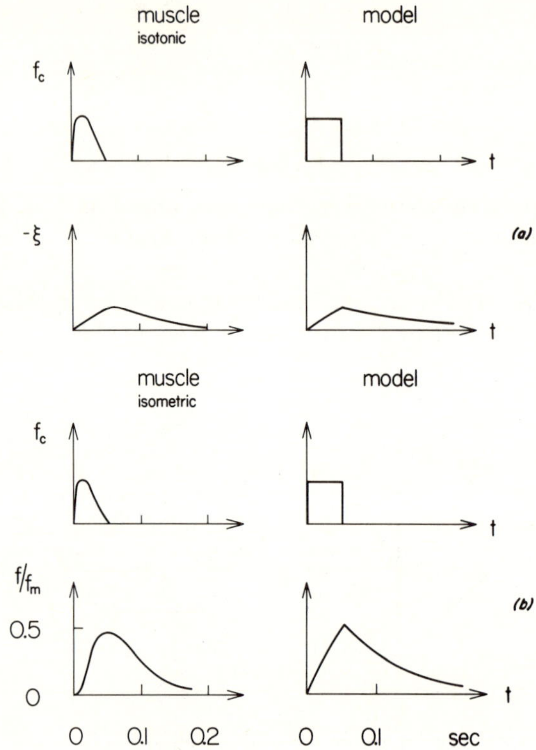

Fig. 10.5. (*a*) Isotonic and (*b*) isometric twitches in the muscle model of Fig. 10.3 and in real muscle. The time course of the force generator (i.e. of the contractile element) appears in the upper row in (*a*) and (*b*). The muscle is said to be in its active state as long as f_c exists. (See Section 10.E for a detailed discussion of the time course of f_c.) The resulting total muscle force is calculated readily for the isometric case from the flow graphs of Fig. 10.4 if f_c is assumed to be an impulse, as here. In spite of the simplified mechanical properties and in spite of the simple $f_c(t)$, $f(t)$ compares rather well with forces measured in real muscle, an example of which is shown in the lower row of (*b*). The same good comparison is seen in the isotonic case (*a*) for which, however, the time constants determining the rise and fall of $\xi(t)$ are much longer than those involved in the isometric case. The isotonic twitch in muscle shown is a schematic curve. Note that K in eq. 10.7 is much smaller than unity. See text for further discussion.

These are slow phenomena and would have to be modeled by more complex mechanical systems, as Buchthal and Rosenfalck [10.5] have observed.

No attempt has been made so far to quantitate the model in detail. From Fig. 10.5 it is concluded that by suitable choice of components the performance of the model can be matched satisfactorily to the behavior of muscle. The choice, however, depends on the muscle to be represented (e.g., whether fast or slow myofibers dominate the performance). The emphasis here is on describing the method of attack and elucidating the main functional relations between net activity of the α-motoneurons (e), force (f), and length of the muscle (ξ + resting length).

C. THE NEUROMUSCULAR SPINDLE

It is natural to choose a mechanical system equivalent to the one in Fig. 10.3 for the model with which the action of all neuromuscular spindles is explained. This is given in Fig. 10.6*a*. The series spring is thought to represent the

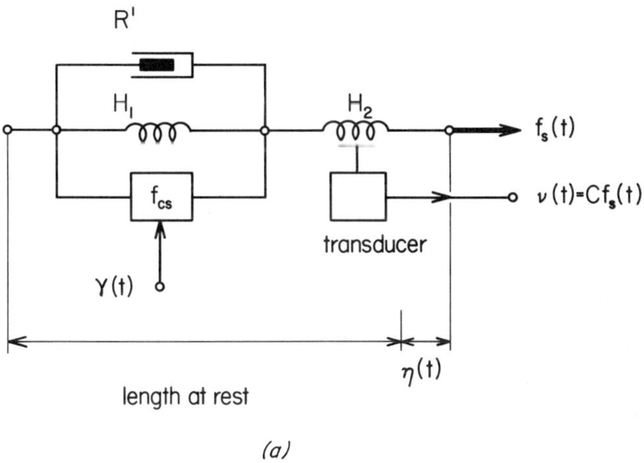

Fig. 10.6*a*. Model of neuromuscular spindle performing, essentially, as a muscle length sensor. Series-elastic element H_2 is a lumped representation of the nucleated bags of the spindles containing annulospiral nerve endings. Dashpot R', spring H_1, and force generator f_{cs} model the contractile parts of the spindle (the so-called intrafusal fibers), which are innervated by the γ-motoneuron system. The output, $\nu(t)$, from the annulospiral or primary endings (transducer) is proportional to the stress f_s within the nucleated bag, which in turn depends on muscle length *and* on f_{cs} (the force produced by intrafusal fibers). The distension $\eta(t)$ is beyond resting length of the spindle.

properties of the noncontractile regions containing the nucleated bags and free nerve endings within the spindle. The intrafusal fibers of the spindles are lumped into the parallel arrangement of spring H_1, dashpot R', and the tractor element f_{cs} innervated by the γ-motoneurons. We introduce here only the transducing action of the (primary) annulospiral endings. The flower-spray type endings behave differently because their threshold is higher and

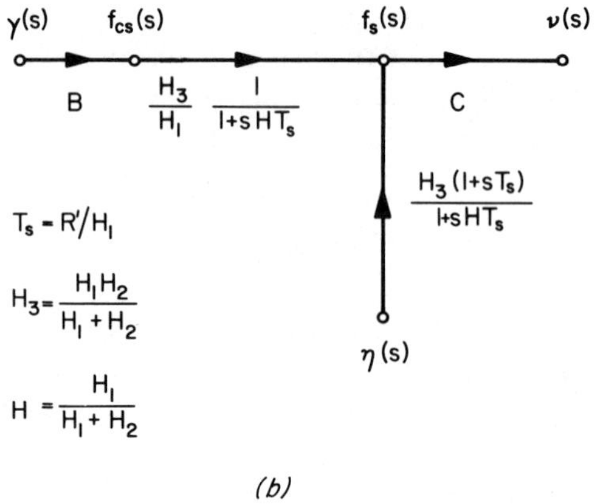

Fig. 10.6*b*. Signal-flow diagram defining (small signal) performance of the muscle spindle model in (*a*). C is the proportionality factor relating stress on the spindle and the afferent neural signal $\nu(s)$; s is the Laplace operator.

because they are innervated by fibers of slower propagation velocity (group II). Thus the transducer in Fig. 10.6*a*, representing the annulospiral nerve terminals, converts the stress in the series elastic spring (H_2) linearly into the (afferent) discharge rate (ν). Hence ν is also proportional to the strain of H_2. A linear relation ($\upsilon = Cf_s$) between firing rate and force on the spindle certainly suffices for the small signal case. At large stresses we should expect a saturation phenomena, although it is not known over what range the net output of the *many* existing spindles is proportional to stress within a muscle. The signal-flow diagram of the spindle model (Fig. 10.6*b*) is similar to that of Fig. 10.4 except that here the output of significance is the (net) firing rate ν.

It is interesting to compute the output frequency when the spindle is stretched instantaneously (with γ-motor innervation kept constant). The typical "phasic" and "static" responses are obtained. Let $\gamma = \Gamma_0$ and $\eta(s) = Y_0/s$, then in the Laplace domain:

$$\nu(s) = C\left\{\left(\Gamma_0 B\frac{H_3}{H^2\tau_s} + \frac{H_3}{H}Y_0\right)\frac{1}{1/H\tau_s + s} + \frac{H_3 Y_0}{H\tau_s}\frac{1}{s}\frac{1}{1/H\tau_s + s}\right\} \tag{10.8}$$

where $\tau_s = R'/H_1$.

The corresponding time function is

$$\nu(t) = C\left\{H_3 Y_0 + \left[B\Gamma_0\frac{H_3}{H^2\tau_s} + H_3 Y_0\left(\frac{1}{H} - 1\right)\right]e^{-t/H\tau_s}\right\} \tag{10.9}$$

As Fig. 10.7*a* reveals, the firing rate given by eq. 10.9 assumes a high value initially, then decays exponentially (phasic response) to reach a steady level at about 25% of the initial maximum. This type of response was discussed in detail by Houk [10.6]. Furthermore, it can be shown that the response to a displacement ramp (i.e., a velocity pulse, such as seen in Fig. 10.7*b*) compares well with those observed by Shepherd and Ottoson [10.7] and by Bessou and Laporte [10.8] in isolated spindles. (See also Matthews [10.9].) Obviously, the analysis given here applies only for small signals and for positive v.

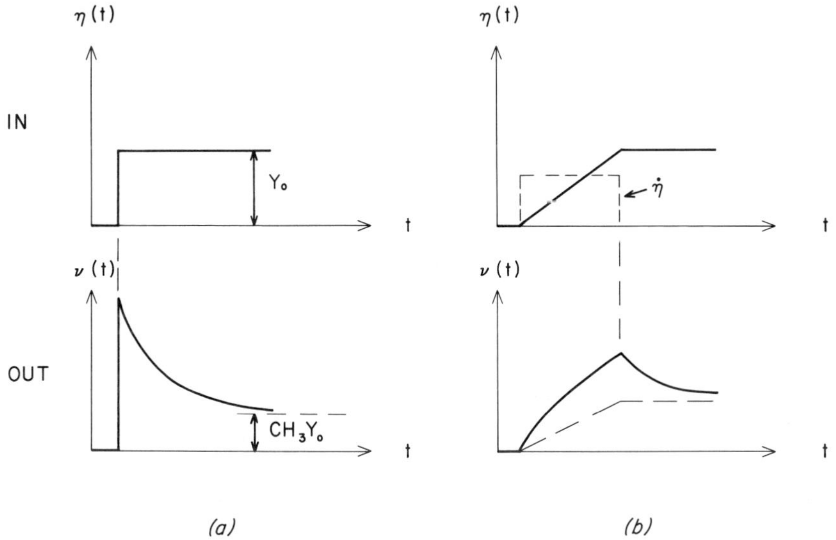

Fig. 10.7. Performance of muscle spindle. (*a*) Response (schematic) of an afferent neural signal $v(t)$ from primary nerve endings to an abrupt change in length $\eta(t)$ of Y_0. The combination of viscous and elastic forces causes a large phasic (velocity sensitive) response in addition to the proportional one ($CH_3 Y_0$; for definition of terms see Fig. 10.6). According to eq. 10.9, only the amplitude of the phasic response depends on $\gamma(t)$, the neural input to intrafusal (spindle) fibers. Direct observations on spindles indicate that the proportional response partly depends on $\gamma(t)$, too. (*b*) Response (schematic) of $v(t)$ to a ramplike stimulus in $\eta(t)$, a stimulus often used in physiology to demonstrate sensitivity of a spindle to velocity. Note that the activity $\gamma(t)$ on the gamma system innervating the intrafusal fibers acts as a gain factor in this case, as well.

It would be an injustice to the many workers in the field to claim that we have included in our model all that is known about neuromuscular spindles [10.8, 10.11, 10.14]. We introduced only those parameters which are of first importance. Even so, the model points to several difficulties that arise and to lack of data available for attempts to quantitate its elements. These questions, however, pertain to functionally important relations (e.g., transfer functions) and may therefore lead to experimental studies contributing directly to functional analysis. Thus the model building can be an important phase of a physiological investigation.

D. MUSCLE–SPINDLE FEEDBACK LOOP

Physiologists have long been aware that the spindles essentially are (mechanically) in parallel with the muscle. For example, the discharge rate of the afferent fibers does not change significantly while the muscle is made to contract isometrically, except for an initial phase during which the rate decreases before it settles near the original level. This initial silent phase is due to the inability of the spindles to sense the entire length of the muscle (a

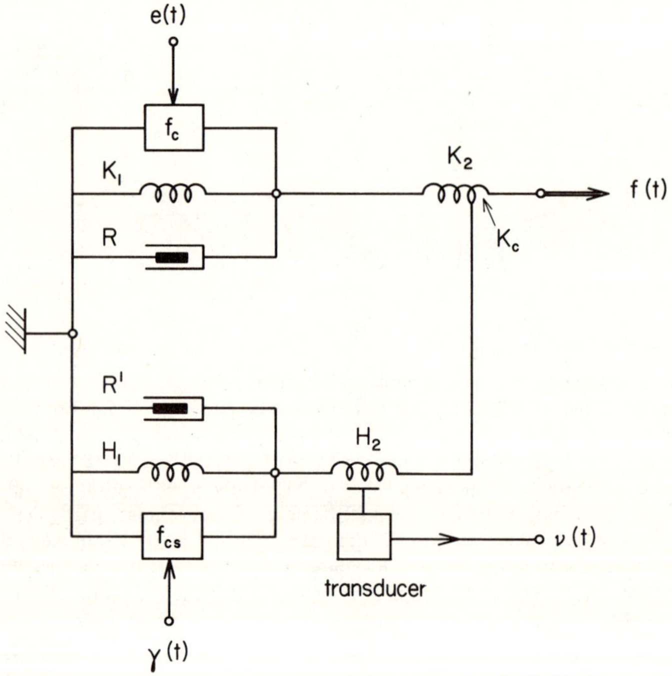

Fig. 10.8. Muscle model (upper half) combined with a model representing all muscle spindles (lower half). Since all spindles sense only parts of the entire muscle (fiber) length, the spindle system is connected, in this model, to the muscle at a point within the series elastic element. Here K_c is the part of K_2 through which the model spindle is connected to the tendon; K_c explains the silent period (see text). The system shown is useful for small signal analysis only. For definition of symbols see Figs. 10.3 and 10.6.

fact clear from anatomical evidence). In the model this is quite easily taken into account when the system of the spindle (representing all spindles) is connected to that of the muscle, as in Fig. 10.8. Part of the series elastic element of the muscle is also in series with the spindle.

The output of the (stress) transducer within the spindle is carried by way of group Ia afferent fibers to the dorsal root of the spinal cord; then it excites directly the motoneurons belonging to the pool innervating the same muscle, thus closing the feedback loop. This feedback is negative in that increased activity at the annulospiral endings, due to augmented stretch, causes the muscle to shorten. This in turn releases the stress on the spindle, assuming that the γ efferent excitation stays at a constant level. The afferent signal from the spindles of a muscle have the opposite effect on the antagonistic muscle. This apparently is achieved by way of internuncial cells connecting to a motoneuron of the antagonist's side. The time delay is small, since only one additional synaptic transmission is engaged.

Thus we can draw the signal-flow diagram for the feedback loop involving one muscle and its spindles. Figure 10.9 illustrates how the coupling between muscle and spindle through K_c is achieved and how the output frequency v of the spindles contributes to e, the gross excitatory rate of the motoneuron pool. Delays in the feedback chain are neglected. The entire system has two

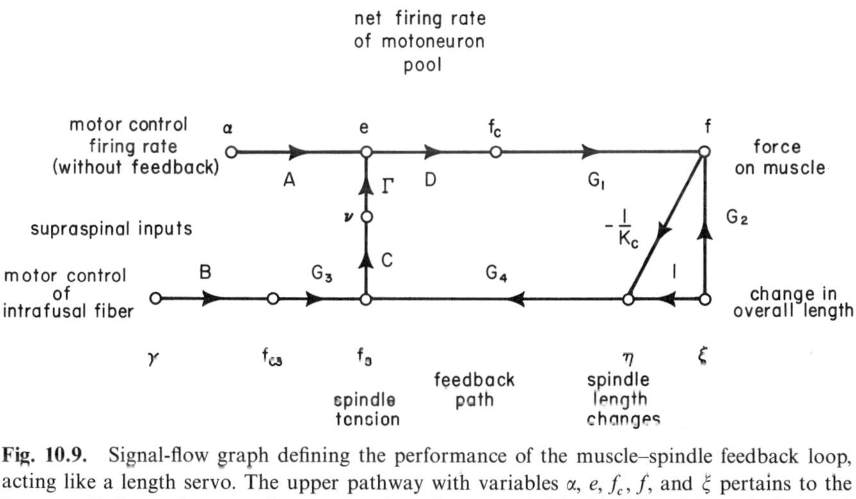

Fig. 10.9. Signal-flow graph defining the performance of the muscle–spindle feedback loop, acting like a length servo. The upper pathway with variables α, e, f_c, f, and ξ pertains to the (uncontrolled) muscle system. It corresponds to the system in Fig. 10.4. Variables η, f_s, f_{cs}, γ, and v in the lower pathway pertain to the spindle system (see Fig. 10.6b, from which the definitions of G_3 and G_4 are also obvious). The operator Γ determines the weight with which the afferent neural information (v, from the spindle system) influences the activity (e) of the motoneuron pool; K_c is defined in Fig. 10.8. See Fig. 10.10 for coupling of this system with an external load.

inputs, and its output is either the change in muscle length or the force, as was covered previously for the uncontrolled muscle. In voluntary contraction both α and γ inputs may so increase that upon shortening of the muscle, the spindle also is actively shortened, thus increasing (or maintaining) the frequency v in order not to counteract the purposeful movement. Sudden closure of the feedback loop may occur when coarse movements are being stopped. Furthermore, the γ input biases the spindle's output to such an extent that the effective stiffness of the muscle is optimal for the muscle's task, such as maintaining length under variable load. Note that for constant α firing rate, a change in γ input alone also elicits contraction.

If the muscle is loaded, the difference between the external force (f_0) and that of the muscle (f) is equaled by possible frictional, elastic, and inertial forces due to the motion of the load. This becomes

$$f_0 - f = Z(\omega)\dot{\xi} \qquad \text{(sinusoidal } f\text{'s and } \xi) \tag{10.10}$$

where $Z(\omega)$ is the mechanical impedance of load, or in the Laplace domain:

$$f_0 - f = sZ(s)\xi$$

The complete signal-flow diagram appears in Fig. 10.10*a*, its reduced forms in Figs. 10.10*b* and 10.10*c*. In all these diagrams ξ, the change in length of the muscle, is chosen to be the dependent output variable.

The effective (static) stiffness is an immediate result from this analysis. For slow and small changes in f_0 the following change in length results:

$$d\xi = \frac{(G_2 + sZ)^{-1}}{1 + CD\Gamma G_1 G_4 (G_2 + sZ)^{-1}} \, df_0 \equiv \frac{df_0}{\text{stiffness}} \tag{10.11}$$

If the spindle feedback loop is opened (e.g., by letting $\Gamma = 0$), then

$$d\xi = (G_2 + sZ)^{-1} \, df_0 \tag{10.12}$$

Hence the term ($CD\Gamma G_4$) is the feedback factor which increases the stiffness, or tonus, of the muscle. In other words, the sensitivity of ξ to f_0, the external force, is decreased. Note that eqs. 10.11 and 10.12 do not include the element K_c, which essentially increases the "gain" in the forward path. This does not matter much, in theory, so long as the feedback factor is relatively large.

The feedback from the spindle which results in increased impedance to changes in length is of course the basis of the so-called stretch reflex. It has the properties of a length servo, as Hammond et al. [10.10] pointed out. These authors used the term "follow-up servo," for the spindle obviously does not sense the length of the muscle but only differences in length between muscle and spindle, depending on or biased by the γ activity. Changes in muscle length due to changes in load are minimized automatically, which constitutes

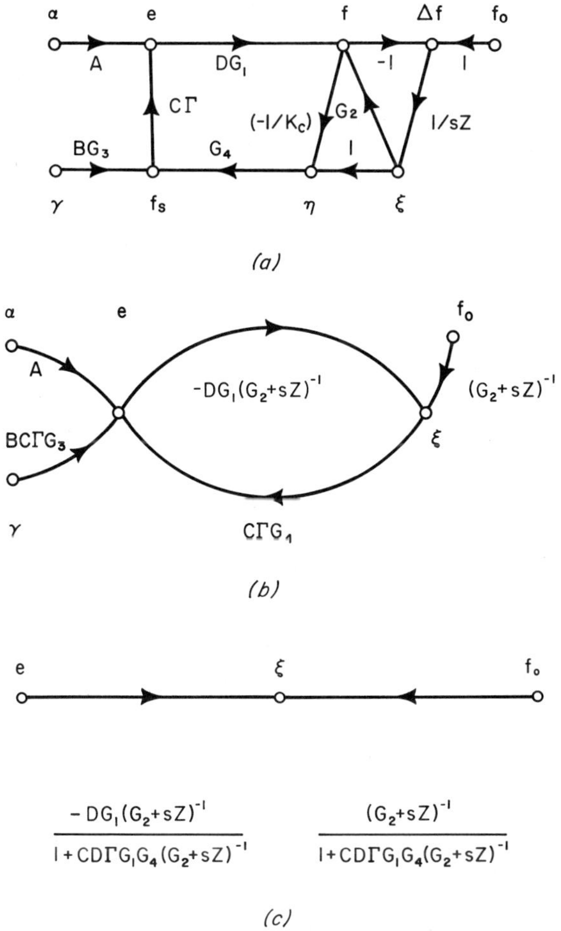

Fig. 10.10. The closed muscle–spindle feedback loop as affected by an external force f_0 and an external load (impedance Z). (*a*) The signal-flow diagram of Fig. 10.9 is reproduced, in slightly reduced form, and combined with an external load according to eq. 10.10. For definitions of G_3 and G_4 see legends of Fig. 10.9 and Fig. 10.6*b*. (*b*) First stage of signal-flow graph reduction, showing forward and feedback paths separately. All variables are eliminated which are not of interest in the study of the overall performance of the muscle–spindle length (or follow-up) servo system. See text for derivation of stiffness, or tonus. (*c*) Complete reduction of the signal-flow graph defining a muscle–spindle feedback loop to show the relation between *e*, the activity of the motoneuron pool, ξ, the length changes of the muscle, and f_0, the external force applied to the system. Although the effect of the feedback is easily seen in this diagram (cf. eqs. 10.11, 10.12), studies of postural control, and so on, must involve the entire agonist–antagonist system, of which the system here is only a part. Note that *e* depends on activities of both α- and γ-type motoneurons, and that the analysis applies to small changes in *e*, ξ, and f_0 only.

the functional advantage. Thus, as an example, posture can be maintained without maintenance of a steady level of α-motoneuron activity, which would have to be corrected (at a supraspinal level) according to some afferent information when the load changes. In disease (spasticity), loop gain can be exceedingly high and not easily changed, so that voluntary movements may be nearly impossible.

Evidence (quoted in [10.10]) for the muscles being under this type of servo-regulation in postural control came from investigations of the silent phase. The silent phase may be seen when the load suddenly is decreased or when a twitch is caused in a muscle that is already partly contracted under volitional control. Matthews [10.11] correctly investigated the question of how the loop gain under discussion can be determined. He showed evidence for apparently very small loop gains. This was obtained from single-unit responses. It seems likely, however, that tests of the overall performance will yield better insight into this problem.

Using the results of the signal-flow analysis (Fig. 10.10), it is possible to study the small signal performance of an *agonist–antagonist muscle system.* To do this we introduce the inhibitory action of the spindle afferent of one muscle onto the motoneuron pool of the other:

$$-I = \frac{\text{change in } e \text{ of synergist}}{\text{change in } f_s \text{ of antagonist spindle}}$$

Furthermore, it is reasonable to introduce the angular position ϕ of the limb (and load) acted upon and the lever ratio R by which the length changes of the two muscles are related with ϕ, as well as the ratio of the forces (Fig. 10.11*b*). The signal-flow diagram of Fig. 10.11*a* is easily reduced (Fig. 10.11*c*) so that the performance of the entire system becomes analyzable. It can be shown that the tonus of the complete circuit is considerably larger, as would be expected. For dynamic responses, however, it is important to investigate the time course of f_c, the forces of the active tractor elements of the two muscles. If the system has to move a load abruptly from one position into another, it may happen that the synergist not only has to work against the load and the passive spring–dashpot network, but also against f_c of the antagonist. This would lengthen the response time unless f_c of the antagonist dropped very rapidly from its previous level to the new level. It is possible that the muscular control centers are "preprogrammed" to cope with this situation (e.g., by an initial strong burst of excitation to the synergist followed by the maintained lower level of activity). These aspects, however, which follow from the inspection of the model have not been investigated experimentally.

As explained previously, the variables $e_{1,2}$ (Fig. 10.11) are a measure of the net neural activity of the efferent, α-motoneuron fibers. Therefore, e depends on the "lateral" interaction among individual α-motoneurons of the pool as

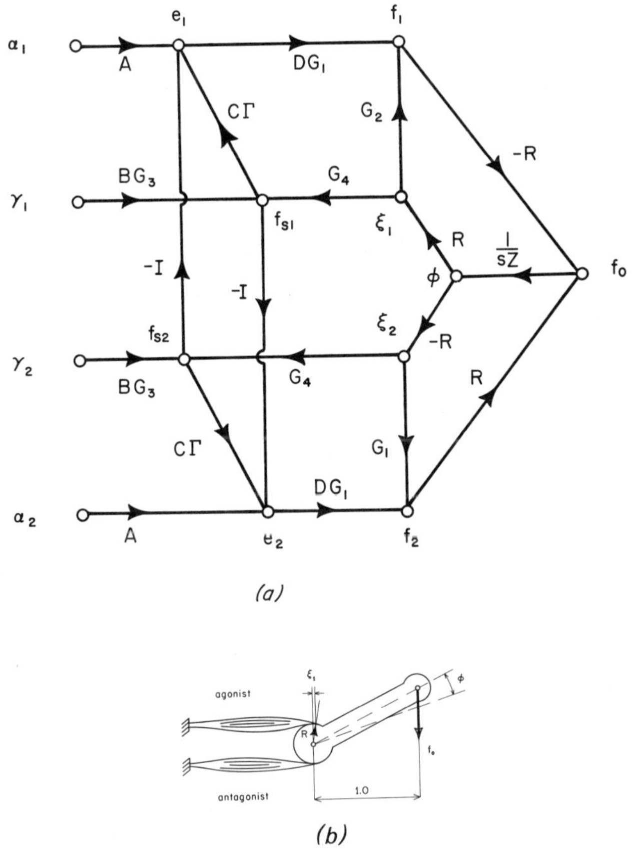

Fig. 10.11. Signal-flow diagrams for a muscle pair including the spindle feedback path and mutual inhibitory interaction by the path labeled *I* (see text). The signals (lower-case letters) are subscripted as follows: *1*, agonist; *2*, antagonist. Definition of symbols as in Figs. 10.4 to 10.10. (*a*) Combination of two systems of the type in Fig. 10.10. Change in position of the load is given in terms of angle ϕ (of the limb moved). The forces should be replaced by torques. For simplicity, the effective lever arm of the external force f_0 is normalized to be unity. Hence the "Δf" (cf. Fig. 10.10 and eq 10.10) is computed by multiplying the muscle forces by the associated lever arm length R (a fraction of "unity"). Generally, the agonist has an R which differs from that of the antagonist. See (*b*) for definitions. For simplicity, we assume here that $R_1 = R_2 = R$. (*b*) Simplest geometry of muscle movement showing the influence of lever arm lengths. For small angular displacements (ϕ) all lever arms can be assumed to be constants. Whether agonist and antagonist have equal R's depends on the anatomy of the particular skeletomuscular system studied. Note that, in real cases, tendons are not attached to bones in distinct points but over certain areas. It is not known whether the *effective* point of attachment shifts during a movement.

See next page for Fig. 10.11*c*.

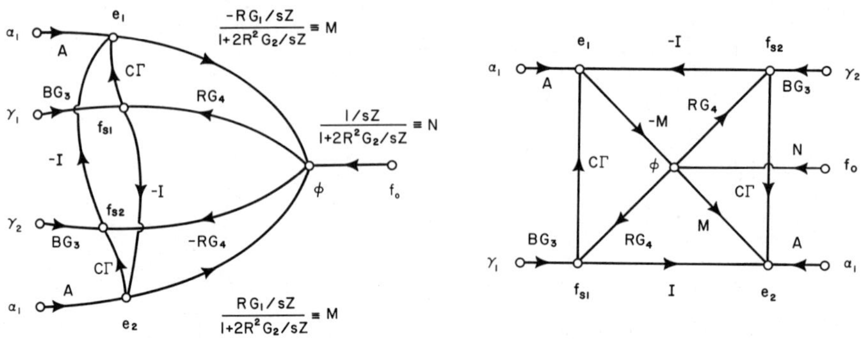

(c)

Fig. 10.11c. Final stages of reduction of the signal-flow diagram of an agonist–antagonist muscle pair.

well as on the action of the local Renshaw interneuron loop limiting the discharge rate of each motoneuron. We are not certain to what extent the myotatic or tendon jerk reflex, often called a monosynaptic reflex, depends on these spinal mechanisms. For sufficiently fast tendon "jerks" the α-motoneurons may initially be unaffected by their Renshaw cells, since their action can come in only after a delay of 2 to 5 msec [10.10]. In other words, the gain at the level of the motoneuron in reality may not be constant thus violating the assumptions in the model of Fig. 10.10. Similarly, we have excluded possible nonlinear integration of supraspinal input to the α-motoneurons and the afferent information from the spindles. Although it suffices to write $e = A\alpha + \Gamma v$ when small signals only are studied, we should remember the finding of Rall [10.12] that subthreshold postsynaptic potentials (PSPs), due to synapses near one another, are not exactly additive. The result, in terms of the soma potential, of two simultaneous PSPs is less than the sum of the individual PSPs. Furthermore, Jenik [10.13] determined the ranges of PSP strengths and frequencies in which discharge frequency of a neuron may be the product of the two input frequencies (α and v). Multiplication is likely to occur, as well as summation. Granit [10.14] indeed cites experimental evidence for both summation and multiplication.

A model similarly simplified was worked out by Adolph [10.15]. This author also considered small signals only. He introduced, however, the inhibitory action of the Golgi tendon apparatus on the motoneurons while studying isometric behavior and the influence of random components in the transfer function Γ (Γ = change in α-motoneuron activity *per* afferent firing rate v of the spindle), and he carried out in detail the stability analysis.

E. LARGE SIGNAL BEHAVIOR OF MUSCLE

In this section we investigate the performance of muscle under large changes in length and force. Nonlinear phenomena are encountered, and many can be analyzed graphically.* First we note that the series and parallel elastic elements, called K_1 and K_2 in Fig. 10.3, are "hard" springs that stiffen when lengthened. This is observed when the static length–tension relation of a passive, "resting" muscle is measured. Second, the active tractor element produces a (maximum, isometric) force f_c which is a function of its extension. This is seen, of course, in the active length–tension characteristic of *isometric*, tetanically stimulated muscle as discussed in the physiology books. These findings are summarized in Fig. 10.12 where, in a qualitative manner, the hard spring characteristics (K_1 and K_2) as well as $f_c(\xi)$ are plotted versus total muscle length. We do not attempt to relate *in detail* the latter function in

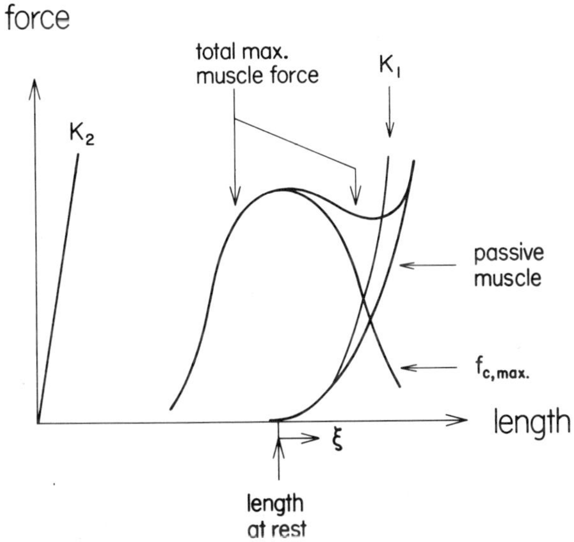

Fig. 10.12 Force–length characteristics of passive muscle and of maximally contracting active muscle ($f_{c,\max}$). At small lengths, passive muscle can be very easily distended. The two springs (K_1 = parallel element; K_2 = series spring), acting essentially in series, develop large elastic forces only at lengths greater than resting length (curve labeled "passive muscle"). Active muscle, in isometric condition, develops a force $f_{c,\max}$ which depends on length in a fashion similar to individual fibers; see Fig. 9.2. Total maximal muscle force (tetanic isometric condition) is given by the sum of passive and active performances.

* The authors are indebted to Dr. D. A. Robinson, who developed many of the concepts presented here.

terms of forces observed in a muscle fiber as a function of sarcomere length [10.16]. There is no particular need for this, since it is possible to study entire muscle systems, including spindle feedback, using descriptions of the observed behavior in the form of curves such as those of Fig. 10.12.

In addition to the isometric characteristics, the force–velocity relations are needed. In view of the foregoing statements, we would expect these to depend on muscle length also. Carlson [10.17] described an elegant approach to measuring the kinematics of muscle. His results from isotonic frog sartorius muscle in fused tetanus are replotted in Fig. 10.13, where the force developed appears as a function of both extension and shortening velocity. (Three-dimensional representations of the "state space," introduced by Fry [10.18], came into use early in respiratory physiology.) The surface in Fig. 10.13 defines f, ξ, and velocity v for the whole muscle during tetanic stimulation or peak activation. The force–velocity relations of the whole muscle at different

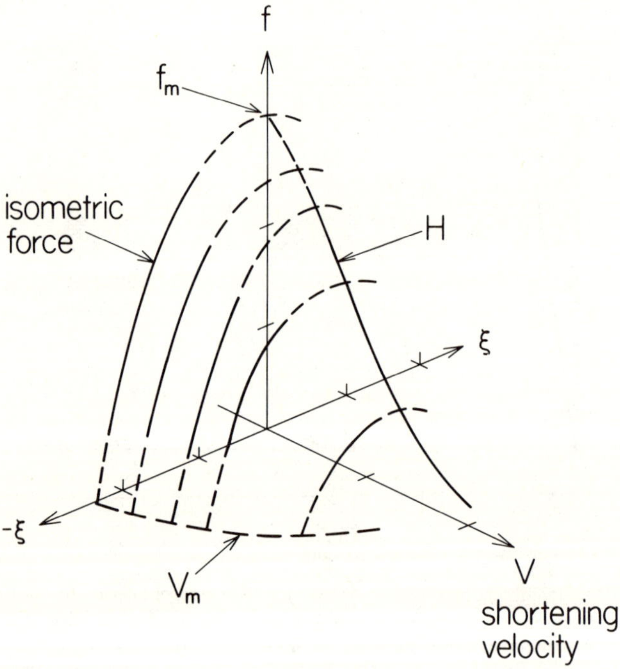

Fig. 10.13. Force–length–velocity characteristics of maximally stimulated muscle, as reported by Carlson [10.17]. The curve in the f–ξ plane corresponds to the one in Fig. 10.12. The f–v relations, at constant ξ's, are given by Hill's hyperbolas (e.g., curve H at $\xi = 0$). Note that muscle is not completely characterized unless data are also obtained for positive values of $d\xi/dt\ (= -v)$. Similar surfaces are obtained for muscles stimulated at less than maximal level, as is indicated in Fig. 10.15.

lengths are given by hyperbolalike curves. Furthermore, we see that the maximum unloaded shortening velocity is a function of muscle length. Carlson [10.17] did not report data on the slope of the surface for positive $\dot{\xi}$ (or $-v$), information, which also should be obtained.

The force–length–velocity relations of mammalian muscle (gracilis of the rat) were studied by Bahler [10.19], who also described an improved instrument for this type of study [10.20]. It is not easy to cause muscle to contract very slowly (e.g., under heavy load), or to observe the maximum velocities occurring only when the muscle is completely unloaded, and these problems constitute the main experimental difficulties in force–length–velocity studies. The data for the tetanic surface, therefore, cannot be made to connect as tightly as one would wish with the isometric force–length curve in the f–ξ plane. Similarly, the curve labeled v_m in Fig. 10.13 is obtained by extrapolation only. It is, in principle, possible to subtract from the surface the forces exerted by the elastic elements and the viscous forces seen in dynamic tests of *passive* muscle. Then the surface defining the properties of the contractile compartment, including its velocity-limiting elements, would be obtained.

Carlson [10.17] gave an analytic expression approximating the surface (Fig. 10.13) he obtained over a limited range of lengths and velocities. He showed that the dependencies of force on length and on velocity v can be separated as follows:

$$f(\xi, v) = f_1(\xi) - f_2(v) = f_m - A\xi^2 - \bar{R}\,\frac{v}{B + v} \tag{10.13}$$

where $v = -\dot{\xi}$, shortening velocity
f_m = maximum isometric force at resting length
$A, B, \bar{R}$ = constants

This particular form suggests itself since the isopleths (seen, e.g., as the projections of force–length curves for different constant velocities on the f–ξ plane) are similar. They represent $f_1(\xi)$ to which, apparently, only constants (f_2 at different, constants v's) are added. Of course, $f_1(\xi)$ is the maximum isometric force–length relation. When eq. 10.13 is normalized by $f_1(\xi)$ and the maximum shortening velocity (v_m) of the unloaded muscle, it becomes

$$\frac{f}{f_1(\xi)} = 1 - Q\,\frac{v/v_m}{Q - 1 + v/v_m} \tag{10.14a}$$

where $Q - \bar{R}/f_1(\xi)$
$v_m = B/(Q - 1)$

It is interesting to note that eq. 10.14a reduces to Hill's hyperbola (see Chapter 9), since we find directly

$$\frac{f}{f_1}(Q - 1) + \frac{v}{v_m}(Q - 1) + \frac{f}{f_1}\frac{v}{v_m} = Q - 1 \tag{10.14b}$$

or

$$\left[(Q-1)+\frac{f}{f_1}\right]\left[(Q-1)+\frac{v}{v_m}\right]=(Q-1)Q \qquad (10.15)$$

Here Q and $(Q-1)$ are functions of ξ only, expressing the fact that Hill's hyperbolas, observed at lengths smaller than resting length, are less convex (towards the abscissa). In other words, v_m also decreases with increasing negative ξ, as was seen already in Fig. 10.13.

The simple muscle model of Fig. 10.3 now can be modified by introducing the nonlinearities of the elastic elements, the dependence of f_c on distension, and the nonlinearity of the dashpot. We then should inquire about the time function of f_c in situations that do not involve continuous stimulation. For example, it is logical to investigate $f_c(t)$ during a twitch, since many aspects of muscle performance have been looked at in this condition. The basic four-element model makes it easy to design an experiment to determine the tractor force as a function of time. During an isometric twitch, the observed force equals the difference between f_c and the force on the viscous element due to "internal" velocity. The length of the contractile part obviously is the total muscle length minus the length of the series spring. The latter, however, can

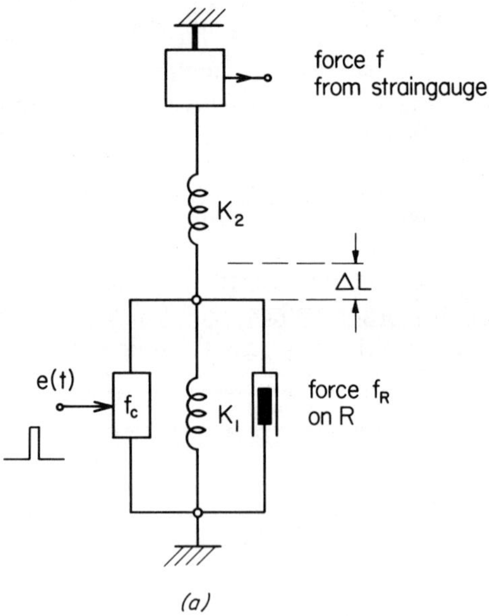

Fig. 10.14a. Experiment to determine the time course of the active state during an isometric twitch, caused by a pulselike stimulus $e(t)$, during which force f is monitored.

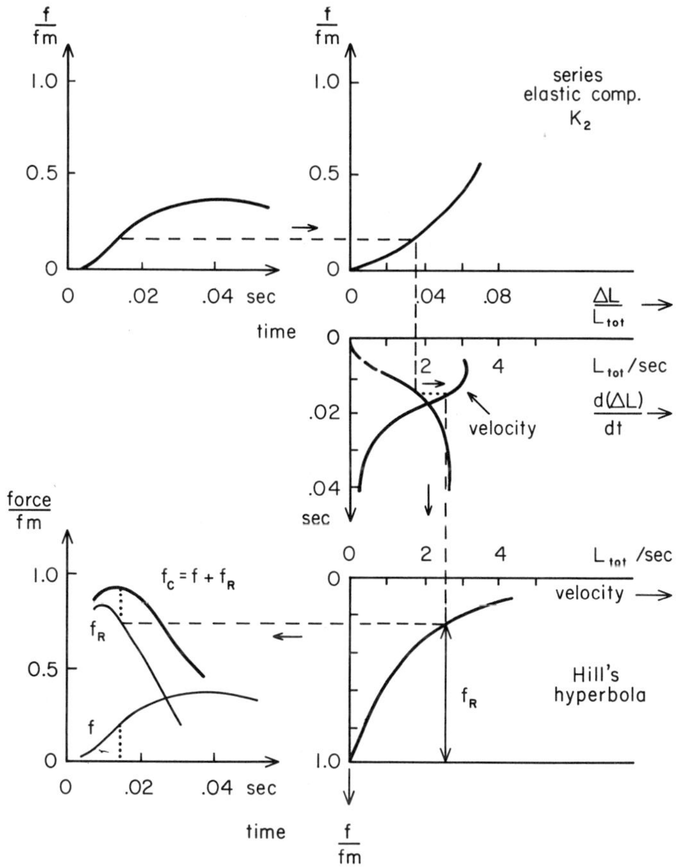

(b)

Fig. 10.14*b*. Graphical construction of $f_c(t)$, the "active state," from $f(t)$ during twitch (upper left). Since K_2, the characteristics of the series element, can be determined beforehand (upper right), ΔL, the change in length of the contractile part of the muscle, can be graphed as a function of time (curve with interrupted beginning; middle right). By graphical differentiation, the velocity of the contractile part can be determined in terms of muscle lengths (L_{tot})/sec. This then allows, with the aid of Hill's hyperbola (determined independently at the length at which the twitch is observed; lower right), construction of the force on the dashpot $f_R(t)$ (lower left). By adding $f(t)$ graphically to $f_R(t)$ (dotted line), the force of the tractor element is obtained. This $f_c(t)$ we define here as the active state. Note that the force of the parallel spring element does not appear in this figure (i.e., the procedure is useful at relatively short muscle length). The principle can be extended to long lengths. After Bahler et al. [10.21].

be determined independently by the quick-release method. Therefore, the force exerted by the "dashpot" at all its lengths and velocities remains to be established. This, however, can be directly obtained in the force–velocity curve as the *difference* between the maximum force ($v = 0$) and the one at a given v. In this we neglect the changes in length of the viscous elements, which are small in the isometric twitch due to the large stiffness of the series spring. Similarly, the force of the soft parallel spring may be neglected in a first-order approximation. Bahler, Fales, and Zierler [10.21] carried out this study and their method for obtaining $f_c(t)$ is reproduced in Fig. 10.14. The subtraction of ΔL, the length of K_2, from the overall length and the addition of viscous forces (f_2 in eq. 10.13) to the total force f may be done graphically. The graph in the lower left-hand corner of Fig. 10.14*b* shows the desired result (i.e., the time course of the active state), which here is defined as the time function of the force of the contractile element during a twitch. The contractile element essentially produces f_m, the maximum force, during peak activation.

Most of the foregoing discussion describes $f(\xi, v)$ for fused tetanus or for maximum maintained neural activity (e). In order to establish a more general relation between the "input" e and the outputs, we must determine how the surface in the (ξ, f, v) space varies with e below its extreme level. This has been partly achieved by Robinson [10.22], who measured isometric tension in an extraocular muscle (lateral rectus) of cats as a function of both length and stimulus. By stimulating the isometric muscle at fractions of the maximal frequency (causing fused tetanus), Robinson determined the length–force characteristic of less than maximally excited muscle. The results are presented in Fig. 10.15, where both the total force and the force of the contractile elements are graphed. The latter is easily obtained from the former after subtracting the tension of the passive muscle (i.e., springs K_1 and K_2 in series). The parameter kept constant along each curve is the level of excitation e (impulses/sec).

The *static* (isometric) relation between excitation, load, and length can further be explored by replotting the data in a graph of e versus ξ, using the force as parameter. Such a plot for the muscle data obtained by Robinson is given in Fig. 10.16. Since these curves are derived from isometric muscle, they could be used to derive the e-dependence of the contractile elements of the muscle. It is possible to determine the length–tension curve for the series-elastic elements, as mentioned before, by the quick-release method whereby the almost instantaneous retraction of initially isometrically contracting muscle is observed as a function of the loads. Hence the length at which a certain excitation would cause a certain force f_e could be obtained by subtracting the length of the series-elastic component at that force which, generally, is small (ca. 0.1 rest length). This then would result in a graph similar to Fig. 10.16 for the contractile portion of the muscle.

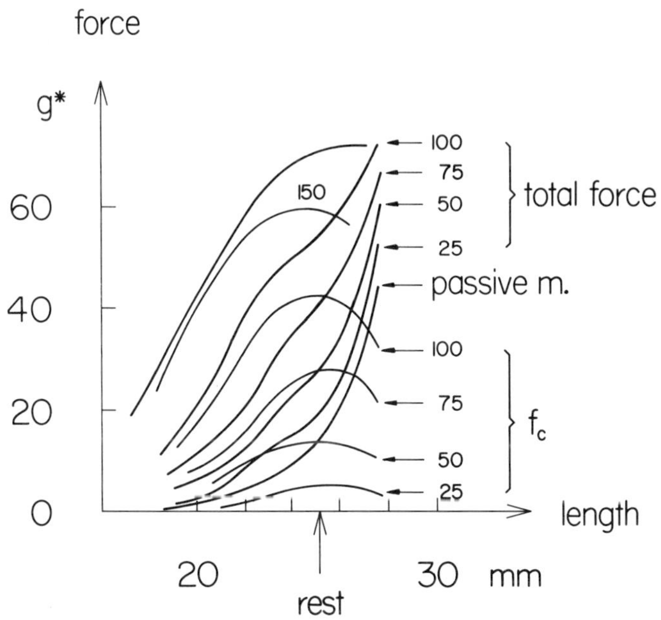

Fig. 10.15. Isometric force–length curves of a cat's lateral rectus muscle for different levels of stimulation (25, . . . , 150 pulses/sec). The total forces (heavy lines) are the observed ones. By subtraction of passive muscle properties, the extra forces (fine lines, f_c) developed by the actively contracting element are obtained; g* = gram force. From Robinson [10.22].

We can, however, discuss the action of the spindle feedback loop on the whole muscle in the excitation–length diagram (Fig. 10.16). Assume that we know the relation between average intrafusal fiber forces and spindle length, at different levels of γ innervation. This then can be converted into average afferent spindle output versus length with γ activity as a parameter. Data of this type, for individual spindles, were obtained, for example, by Crowe and Matthews [10.23]. For the discussion of whole muscle, information on the gross, effective afferent activity of all spindles is obviously needed. It probably is similar to the activity of the individual annulospiral ending and can be superimposed onto the e–ξ diagram of the uncontrolled muscle. This is illustrated schematically in Fig. 10.17, where we assume that there is a steady "background" activity of the motoneuron pool onto which a signal proportional to the afferent feedback firing rate is added. Thus e becomes the net level of stimulation (i.e., "background" activity plus a fraction of v). The example in Fig. 10.17 indicates how a particular level e_R would cause the muscle, under load f_2, to shorten to ζ_R, *given* the level γ_2 of intrafusal fiber

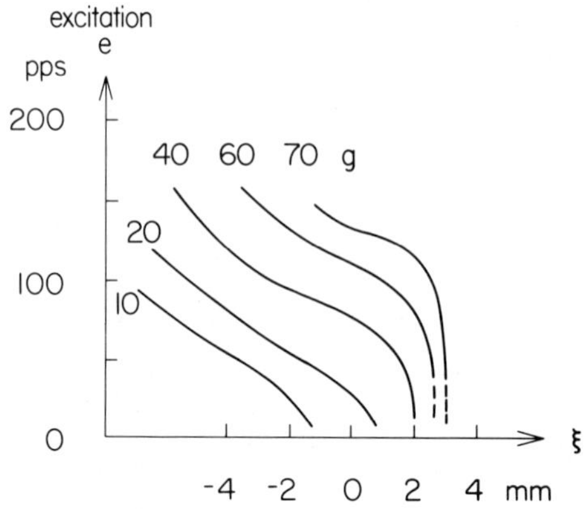

Fig. 10.16. Excitation–length diagram useful when muscle length is studied as the dependent variable. The same data as given in Fig. 10.15 are replotted so that changes in excitation, given here in terms of pulses per second (pps) needed in artificial stimuli to create certain length changes, can be evaluated for a desired change in length. Such a diagram is useful in delineating the range of linearity ($-\xi \propto e$) assumed to hold in small signal analyses (Sections 10.A–10.D); a further application is discussed in Fig. 10.17. Here g* = gram force.

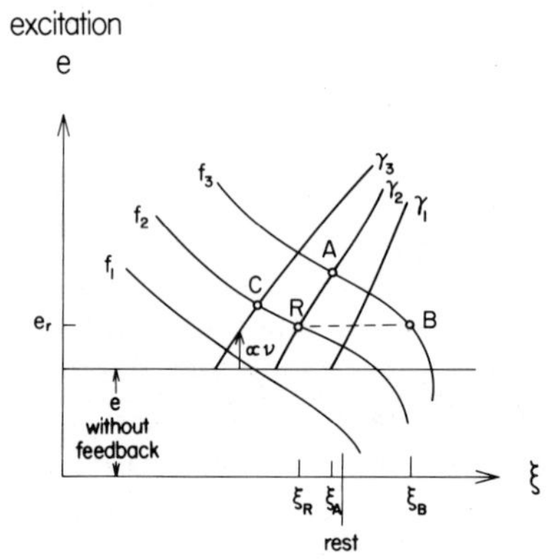

Fig. 10.17. Static performance of muscle–spindle feedback loop. On the graph of Fig. 10.16 ($f_1 = 20$, $f_2 = 40$, $f_3 = 60$ g*) are superimposed the transducer characteristics of the spindle system. The latter is given by the net afferent neural activity (proportional to v), as a function of ξ, with γ (the neural activity of the γ system innervating intrafusal fibers) as a parameter. The spindle curves are hypothetical. They are added, at the α motoneurons, to a neural activity level labeled "without feedback." Given load f_2 and γ_2, the muscle assumes length ξ_R. An increase of the load by 20 g* would shift the muscle length to ξ_A. [Static tonus is given by $f_3 - f_2/(\xi_A - \xi_R)$.] Without spindle feedback, ξ_B would occur at the same e_r. Point C would be attained if only γ changed. Note that dynamic action of the spindle feedback loop cannot be obtained from this graph. After D. A. Robinson (personal communication).

excitation. If the load now is increased to f_3, the muscle comes to rest at point A in the diagram, assuming that the γ level stays constant.

On the other hand, without the spindle feedback, the same load change would lengthen the muscle to point B. Clearly, $(\xi_A - \xi_R)$ is considerably less than $(\xi_B - \xi_R)$, which is a manifestation of the length servo action or of the increased muscle "stiffness" due to the feedback loop. It should be noted that this consideration concerns the behavior of the feedback loop in the static case or the one for slow load changes. Owing to the strong phasic response of the spindle (Fig. 10.7), the position servo can be more effective for rapid load changes than for slow ones. Little is known quantitatively, however, about the kinetics of the entire system. Figure 10.17 also indicates that an increase in γ alone (e.g., from γ_2 to γ_3) would cause the muscle to shorten to point C. This, of course, can be achieved by an increased activity of the motoneurons, too. Clearly, the relative increase in tonus, or changes in α or γ activity needed for a certain amount of shortening, depend very much on the location of the operating point (R) within these families of curves. In certain regions the spindle feedback is more effective in stiffening the muscle system than in others. Changes in e necessary to cause the muscle to contract by a certain distance are equally dependent on the initial operating point. It is not known in which parts of the plane a muscle operates during such a task as postural control, or in which regions of the e–ξ plane optimal performance is achieved. The models presented in this chapter should, however, indicate the nature of questions that must be asked when defining, investigating, and quantitating muscle performance.

F. REFERENCES

10.1. Robichaud, L. P. A., M. Boisvert, and J. Robert. *Signal Flow Graphs and Applications*. Prentice-Hall, Englewood Cliffs, N.J., 1962.

10.2. Truxal, J. G. *Automatic Feedback Control System Synthesis*. McGraw-Hill Book Co., New York, 1955.

10.3. Levin, A. and J. Wyman. The viscous elastic properties of muscle. *Proc Roy Soc [Biol]* **101**: 218–243, 1927.

10.4. Wilkie, D. R. The relation between force and velocity in human muscle. *J Physiol (London)* **110**: 249–280, 1950.

10.5. Buchthal, F. and P. Rosenfalck. Elastic properties of striated muscle, pp. 73–93 in *Tissue Elasticity*. J. W. Remington (Ed.). American Physiological Society, Washington, D.C., 1957.

10.6. Houk, J. C. A model of adaptation in amphibian spindle receptors. *J Theor Biol* **12**: 196–215, 1966.

10.7. Shepherd, G. M. and D. Ottoson. Response of the isolated muscle spindle to different rates of stretching, in *Sensory Receptors. Symp Quant Biol* **30**: 95–103, 1966.

10.8. Bessou, P. and Y. Laporte. Responses from primary and secondary endings of the same neuromuscular spindle in the tenuissimus muscle of the cat, pp. 105–119 in *Symposium on Muscle Receptors*. D Barker (Ed.). University of Hong Kong Press, Hong Kong, 1961.

10.9. Matthews, P. B. C. Muscle spindles and their motor control. *Physiol Rev* **44**: 219–288, 1964.

10.10. Hammond, P. H., P. A. Merton, and G. G. Sutton. Nervous gradation of muscular contraction. *Brit Med Bull* **12**: 214–218, 1956.

10.11. Matthews, P. B. C. Vibration and the stretch reflex, pp. 40–55 in *Myotactic, Kinesthetic, and Vestibular Mechanisms*, CIBA Symposium. A. V. S. DeReuck and J. Knight (Eds.). Churchill, London, 1967.

10.12. Rall, W. Theoretical significance of dendritic trees for neuronal input–output relations, pp. 73–97 in *Neural Theory and Modeling*. R. F. Reiss (Ed.). Stanford University Press, Stanford, Calif., 1964.

10.13. Jenik, F. Pulse processing by neuron models, pp. 190–212 in *Neural Theory and Modeling*. R. F. Reiss (Ed.). Stanford University Press, Stanford, Calif., 1964.

10.14. Granit, R. Effects of stretch and contraction on the membrane of motoneurones, pp. 37–50 in *Muscular Afferents and Motor Control*. R. Granit (Ed.). John Wiley and Sons, New York, 1966.

10.15. Adolph, A. R. Feedback in physiological systems: An application of feedback analysis and stochastic models to neurophysiology. *Bull Math Biophys* **21**: 195–216, 1959.

10.16. Gordon, A. M., A. F. Huxley, and F. J. Julian. The length–tension diagram of single vertebrate striated muscle fibres. *J Physiol (London)* **171**: 28P–30P, 1964.

10.17. Carlson, F. D. Kinematic studies on mechanical properties of muscle, pp. 55–72 in *Tissue Elasticity*. J. W. Remington (Ed.). American Physiological Society, Washington, D.C., 1957.

10.18. Fry, D. L. and R. E. Hyatt. Pulmonary mechanics, a unified analysis of the relationship between pressure, volume and gasflow in the lungs of normal and diseased human subjects. *Amer J Med* **29**: 672–689, 1960.

10.19. Bahler, A. S. A quantitative investigation into the dynamics of an *in vitro* mammalian muscle. Ph.D. thesis, The Johns Hopkins University, 1966.

10.20. Bahler, A. S. and J. T. Fales. A flexible lever system for quantitative measurements of mammalian muscle dynamics. *J Appl Physiol* **21**: 1421–1426, 1966.

10.21. Bahler, A. S., J. T. Fales, and K. L. Zierler. The active state of mammalian skeletal muscle. *J Gen Physiol* **50**: 2239–2253, 1967.

10.22. Robinson, D. A. The mechanics of human saccadic eye movement. *J Physiol (London)* **174**: 245–264, 1964.

10.23. Crowe, A. C. and P. B. C. Matthews. The effects of stimulation of static and dynamic fusimotor fibres on the response to stretching of the primary endings of muscle spindles. *J Physiol (London)* **174**: 109–131, 1964.

CHAPTER ELEVEN

Motor Control: Manual Tracking

A. INTRODUCTION; GENERAL PROPERTIES

This chapter covers certain aspects of the performance of those entire neuromuscular systems controlling the movements of body and extremities. We gain insight into the intricacy of motor control and into the approaches that can be taken in quantitative investigations in this field. Essential tasks of the system are emphasized: efficient performance of mechanical work and of precise, purposeful motion. Such a task as maintaining muscle length under variable loads, partly fulfilled by the muscle–spindle feedback loop discussed in Chapter 10, is obviously simple and plays a minor role. Many more complex functions can be performed—the body is balanced while walking or running; we can move heavy loads very carefully; we may be able to hit a tennis ball precisely while running; a pianist plays rapid fortissimo passages as correctly and in as finely controlled a fashion as slower and softer ones. In all cases the central nervous system plays an essential role in integrating afferent information from somesthetic and exterosensory systems and knowledge (acquired previously and during action) about the tasks and their solutions, and in controlling efficiently the muscle systems. As the simple examples discussed in this chapter reveal, concepts such as adaptiveness, prediction, sampled data processing, or systems reorganization must be introduced when characterizing overall performance. These concepts, useful when analyzing physiological control systems, were discussed in general fashion in the introduction to Chapter 7. Here we attempt to quantitate properties like adaptiveness as a function of load or control of movements as a function of frequency of a repetitive motion.

A few simple observations show us that the motor system must comprise many different levels of control and that it actually changes its organization depending on the task and on movements carried out. We know that some movements, or entire sequences of movements, can be executed precisely without conscious control. Motions of hand and fingers when we sign our own name belong in this class. Sequences of neural commands controlling the muscles appear to be of the nature of a preprogrammed routine that has been acquired and learned by practice. On the other hand, if we had to copy somebody else's signature, different processes would be involved. First, the feedback path by way of the visual system is necessary so that movements of the pen are monitored and the patterns of the given signature and those of the copy can be transmitted to the central nervous system (CNS). Second, continuous pattern comparators and error detectors, and perhaps reduced error aftereffects, must be brought to bear in the CNS if the task is to be fulfilled. (See Chapter 4 for a general discussion of these central mechanisms.) In other activities—dancing is a good example—the control centers again reorganize to assume different functions. Once a particular rhythm has been recognized as one attribute of the movements to be executed, internal " clock " mechanisms play a role. These clocks enable us to stay perfectly in phase with a periodic command by adding predictive power to the system. Another example of reorganization, occurring partly peripherally, is Stark's [11.1] observation that the γ-fiber input to the muscle system may be used intermittently when rapid and rather unprecise movements have to be slowed, so that slow, precise motions are possible near and at the target. Obviously, there are many other activities in which both force and precision must be continuously and precisely controlled. Some of these are mastered only after a great deal of training.

Thus investigations of the motor control system should comprise different phases. First, we might want to describe the performance of certain subsystems whose mode of organization can be assumed to be constant during a particular set of movements. Second, questions can be asked about the capability of higher systems to override or influence lower ones in order to achieve reorganization. These inquiries should also concern the criteria of optimization which are fulfilled during such changes.

In the following, we describe experiments permitting quantitation of essential parameters of highly coordinated movements. Aspects of both sets of questions mentioned enter the discussion.

B. HAND TRACKING TASKS AND NORMAL TYPES OF PERFORMANCE

In tracking experiments a *command* signal is displayed which indicates the required task to the person tested. He then attempts to match the command as fast and as accurately as possible by an appropriate *response* (Fig. 11.1*a*). The command signal may be given as the position of an electron beam on the face of an oscilloscope, on which the test subject must superimpose a second beam. Stark [11.1] used the position of a pointer driven by a motor, which the subject had to follow with another pointer attached to a handle controlled by the person tested. Smith [11.2] and Johns [11.3] used, in a similar fashion, a row of lamps arranged along a circle for the display of the command. The test subject's task was to place a pointer in coincidence with the lamp that was lit at any instant. In these manual tracking tasks, the muscle effector system is relatively well defined. Pronator and supinator muscles of the forearm rotate the wrist. This involves sufficiently complex control mechanisms

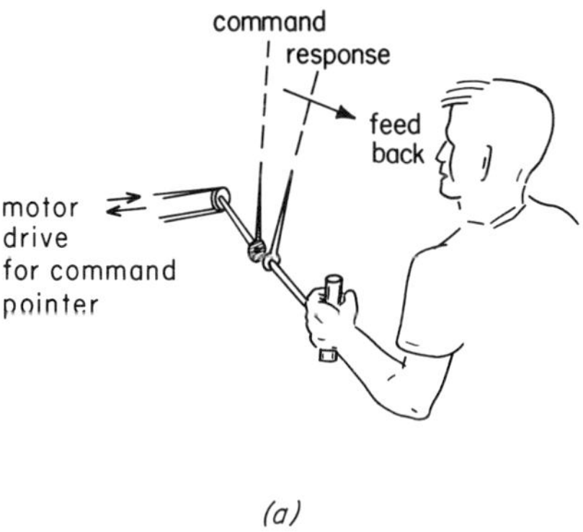

Fig. 11.1*a*. Experimental setup to test hand tracking performance. The "command pointer" (shaded) is driven by a motor in stepwise fashion. If steps are at random intervals and of random amplitude, the actual motor system performance (step response) is observed independently of learning and prediction operators. The test subject's task is to respond to new commands as fast as possible and to superimpose the "response pointer" onto the other. The feedback loop is closed by the visual system. Propriosensors in the forearm, however, may furnish important information for feedback, too.

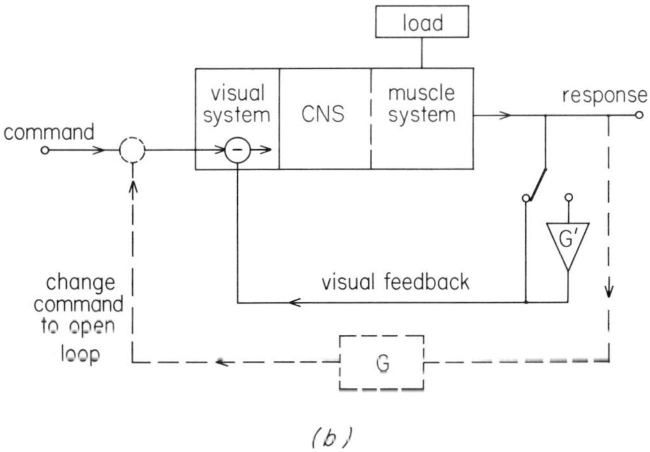

Fig. 11.1*b*. Block diagram of a feedback system tested in hand tracking experiments. Solid lines, with the switch in the position shown, represent the normal situation. The system may be tested by adding, by throw of a switch, additional gain (G') into the feedback loop. Or, as shown by dashed lines, the natural loop can be opened or thrown into instability by placing an artificial feedback path around the entire (natural) system.

to elucidate some of the features of the system. However, previously acquired skills or visual acuity play a minor role. As we explain below, the experimental method allows several tests of the system. Predictable as well as unpredictable responses can be required from the subjects, or the muscle system can be loaded mechanically by adding mass or friction to the handle driven by the subject. Then, fine regulation of both length changes *and* forces is required.

It is revealing to study the entire system, the interrelation between its parts, and the flow of information in a block diagram, as in Fig. 11.1*b*. We deal with a closed feedback loop whose input is the command. If the test person controls the position of a pointer, it is this output which is compared, by the visual system, with the input. The deviation between the two is minimized or brought to naught by the hand movement control system. If the subject controls the position of an electron beam on an oscilloscope, his response is measured by the setting of the potentiometer. If the latter affects beam position directly and in constant fashion, the system is equivalent to one in which a pointer is coupled directly to the handle driven by the subject.

Any addition of amplifier or filter between potentiometer and oscilloscope allows for elegant manipulation of the system. This path can be considered to be part of the feedback (Fig. 11.1). It is possible, for example, to open the loop momentarily by adding a stage of moderate gain in the (visual) feedback path by the throw of a switch. The subject, unaware of the increased gain, will overshoot the target. When he tries to correct his error, he will undershoot. For a short time an oscillation is observed which is quenched rapidly, since the entire system adapts to the new task. The new task consists, of course, in tracking the command beam by way of the amplifier that was switched into the loop. It can be easily fulfilled because the subject is able to adapt internally the feedback factor to the new task (*input* adaptation). A similar phenomena is seen [11.1] when the feedback loop is opened by an additional positive feedback path, as is given by the interrupted lines in Fig. 11.1. It becomes clear below, however, that the visual feedback path is not the only one. Tactile sensors in the tissues of hand and arm provide important information to render the system *load* adaptive.

For determination of the dynamics of volitional motor control, when independent of prediction factors in the CNS, the test subject may be asked to track a sequence of commands in the form of *steps* [11.2] repeated at regular intervals. They may be spaced by 1 sec, which is sufficient time to complete one individual tracking task. The execution of one step is not affected by the previous one. Amplitudes of the steps as well as their directions can be made *random* numbers. Thus the subject never employs his learning and predicting ability. With each step he must decide whether to turn the handle right or left and how far. The nature of the task, however, stays constant, and therefore the experimenter gains insight into the step response of the pronation–supination system. Response times, errors, and relations between speed of motion and amplitude of command can be observed and quantitated. A typical tracing of the response of one individual test subject to an unpredictable step is shown in Fig. 11.2. The mechanical system used in this case was not constrained by any extraneous load.

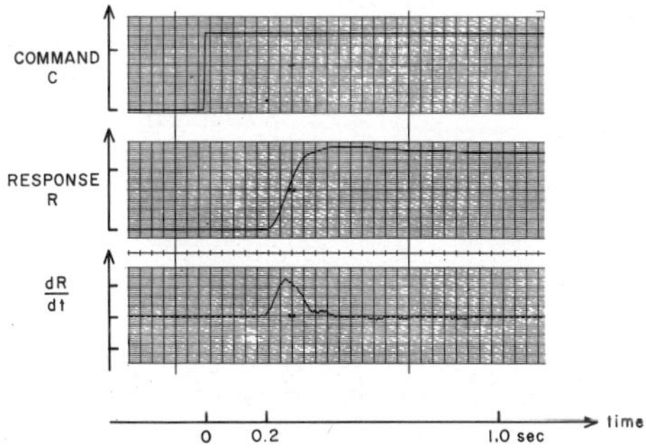

Fig. 11.2. A typical response (middle trace) to a partly unpredictable command (top trace). The bottom trace is the velocity of the response. The scales on the ordinates: one division = 1 rad for C and R and 10 rad/sec for dR/dt. The relatively high peak velocity probably occurred because only time of occurrence was unpredictable. Amplitude was predictable.

Two types of *errors* in tracking performance have to be differentiated. First, a subject responds to a command after a certain delay which, in turn, depends on two factors: the occurrence of a command and the position of the target. Psychologists call this delay the disjunctive reaction time. Johns [11.3] found this reaction time to be 0.21 ± 0.028 sec. He defined lag error as the product of reaction time and the amplitude of the step command. This is area L in Fig. 11.3, where a tracking response is schematically shown. Second, there is the kinetic error which depends on the speed and accuracy with which the target is tracked. It is conveniently defined as area K in Fig. 11.3. The *total* error ($L + K$) in normal, young subjects was found by Johns [11.3] to be approximately 13 rad sec. The command steps covered a range of ± 0.9 rad in this study. The error generally decreases by approximately 6%, when the testing sequence comprising 50 step responses is repeated 3 to 6 times. This reflects familiarization with the task, but it does not mean that the persons tested can learn the random sequence. (Attempts to do so cause larger errors.)

Differences between individuals are about four times larger than variations in performances of one man on different days. The variance in day-to-day performances is approximately 2.3%. The total error in normal subjects increases slightly with age (ca. 6%/20 years). These observations are important if the method is used as a clinical test of motor control. They are

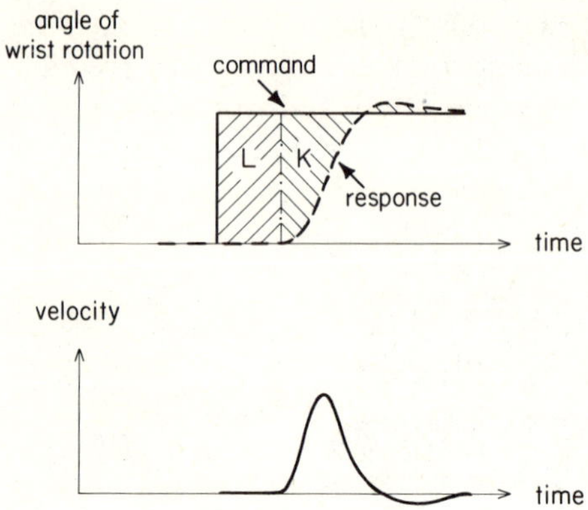

Fig. 11.3. Definition of lag error (L) and kinetic error (K). The L is due to so-called disjunctive reaction time. Responses do not always show overshoot. The steady-state error (not shown) is, in most cases, zero. After Johns and Draper [11.3].

equally interesting to engineers investigating man–machine systems [11.4]. In both cases a highly reproducible method of quantitating coordinated movements is desired.

Analysis of results obtained in normal subjects indicates many interesting facts. Johns [11.3] computed correlations between kinetic errors and amplitudes and between maximum velocities and amplitudes from observations with 12 subjects carrying out 52 step responses of randomly distributed amplitudes. With increasing step amplitudes, peak velocities also increase but less than proportionately. The kinetic error, as defined here, also increases when amplitude grows, as is expected. In fact, the error increases more than twofold if step size is doubled. The linear regression equations between amplitude A (in radians) and velocity v_{max} (in rad/sec) are:

$v_{max} = 1.2 + 4.1A\ (\pm 0.012)$
correlation coefficient $r = 0.81$
standard deviation of the estimate of $v_{max} = 0.85$ rad/sec

and between amplitude A and kinetic error E_k we have

$E_k = 0.011 + 0.15A\ (\pm 0.044)$
correlation coefficient $r = 0.82$

This is quite unlike the performance of a linear, man-made feedback system. Control engineers who concern themselves with optimization problems know, however, that such nonlinear effects as force saturation or limits in the rate of reversal of force must be considered also in man-made controllers. In particular, the brachistochrone problem has long been known to be solved by a "bang-bang" system. In this, minimal response time is achieved by (a)

first applying full force in one direction, (*b*) computing the time at which the force must be reversed, and (*c*) applying full force in the opposite direction. In the design of machines, it is not always feasible to realize this. It may be impractical to provide enough amplification to saturate fully some of the elements involved. Similarly, the decision of step (*b*) is generally difficult. Extensive theoretical considerations concerning optimization therefore have been developed.

Smith [11.2] quotes nonlinear functions of the error, useful in special cases, upon which a computer must base its decisions in step (*b*). However, *energy* criteria can be given only if the particular nature of the load is known. Smith observed that muscle systems in extreme manual tracking situations apparently fulfill these criteria. When moving very large masses, the system works in the " bang-bang " mode just mentioned; when working against large frictional loads, a different mode is assumed so that large negative undershoots in error (as would occur in the " bang-bang " system working against friction) are avoided.

Little is known about the criteria of biological optimization of the neuromuscular system in tracking tasks under discussion in this chapter (Johns [11.3]). Some pertinent features of the system are observed when small perturbations are imposed. Added inertia, for example, has surprisingly little effect on performance, as mentioned before. Acceleration and peak velocity essentially are unaltered up to loads of seven times the inertia of the forearm. The subjects, however, require a few trials to adapt to the increased load—or to the original task after the load is removed. Further insight might be gained by analyzing Stark's [11.1] findings concerning responses to unpredictable command signals other than step functions.

Stark used the sum of three sinusoidal signals of different and incommensurate frequencies to create a command signal that is essentially unpredictable. Responses can be so filtered that the response to one particular sinusoid is obtained which, in turn, can be compared in phase and amplitude with the corresponding component of the input command. This approach presupposes that the system under observation is linear. Stark tested this hypothesis in parts by varying the amplitude of the command. It appears that amplitude of the response increases linearly with amplitude of the command, at least over the range tested (10–50° of wrist rotation). There is, of course, scatter in the data which is not negligible. There are not sufficient data available to warrant statistical analysis that would yield better insight. For example, we could test whether errors in estimates of slopes are sufficiently small to allow linearization in one particular task.

Stark [11.1] did not elaborate on this question. He simply used linear systems analysis and worked out Nyquist and Bode plots to represent command–response relations. These partly enable us to see to what extent

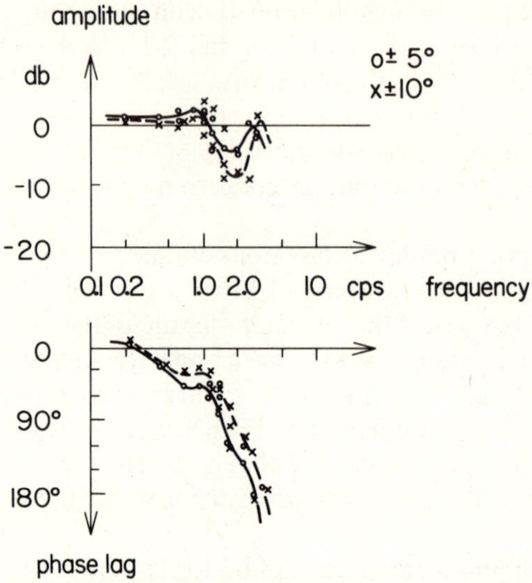

Fig. 11.4. Bode plots of responses to sinusoidal commands made unpredictable by addition of two signals at incommensurate frequencies. Solid lines and circles represent response to command amplitude of $\pm 5°$; dashed lines and crosses pertain to command of $\pm 10°$. The general nature of frequency response remains unchanged, but increases in command amplitude cause shift in the response. This is indicative of nonlinear behavior (see text). Redrawn from Stark [11.1].

the motor control system performs as a linear system. In Fig. 11.4 we reproduce amplitude and phase functions of two sets of data. In the first case, the frequency component of the unpredictable signal had an amplitude of $\pm 5°$ of wrist rotation. In the second case, the amplitude was $\pm 10°$. Scatter is not excessive, and thus we can conclude that there exists a difference between the two responses. This means that nonlinearities are present. It is clear, however, that the responses are essentially similar. Amplitudes begin to decrease at 1 Hz. The phase lag increases continually at frequencies above 0.2 Hz and is 90° at approximately 1.5 Hz.

In passing we mention Stark's [11.1] findings on tracking responses to *predictable* command signals. Pure sinusoidal commands are tracked essentially without error up to 2 Hz. It is particularly noteworthy that the test subject tracks without phase lag up to this frequency. This clearly indicates that different neural mechanisms are brought to bear in this case and that we should think in terms of reorganization of the control centers. At the onset of a predictable sinusoidal tracking task, a person is able, within 0.5 to 1.0 sec, to adapt and to achieve a perfect in-phase response, provided the frequency is below 2 to 3 Hz. This is close to the effective upper band limit of

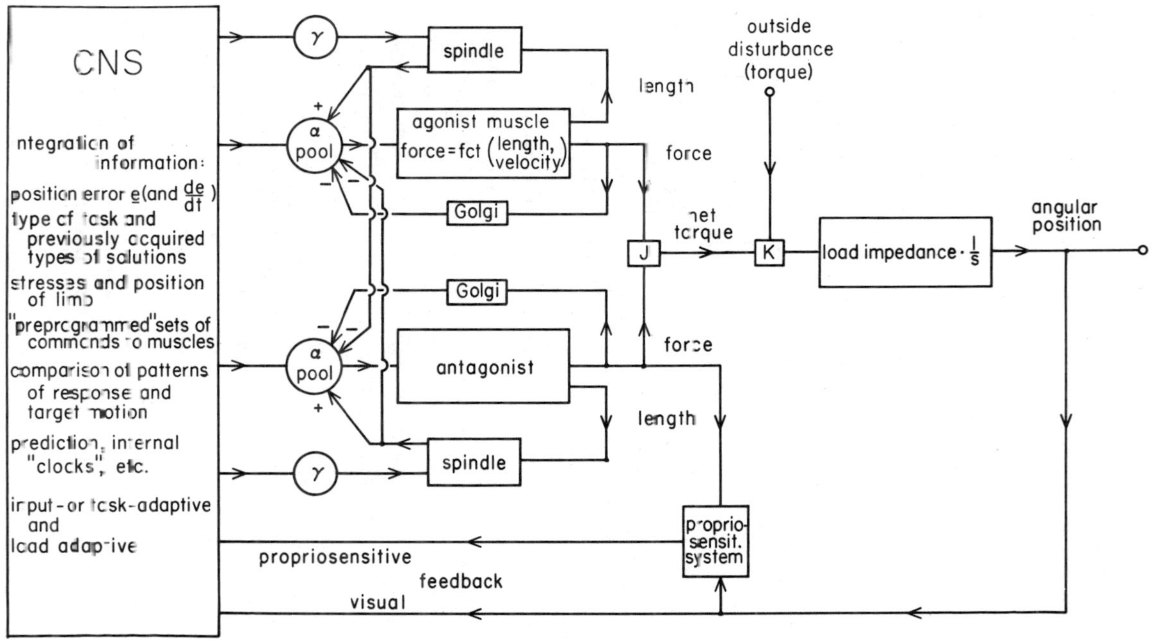

Fig. 11.5. Information flow diagram of neuromuscular control system. Different modes of operation of control centers in the central nervous system (CNS) are summarized in the large block at the left. Efferent paths to both agonist and antagonists are shown, as well as the systems organization at the periphery. See Chapter 10 for tools suited to analyze muscle–spindle feedback loops. Boxes J and K represent the geometric relationships determining torque applied to load, given the muscle forces; s is the Laplace operator. Two important kinds of feedback paths are shown: (1) the position of the limb is fed back by way of the visual system, (2) limb position as well as stresses within tissue are signaled back to the CNS by way of mechanoreceptors. The importance of the various information flow paths varies and depends on the task to be fulfilled. The system reorganizes to perform optimally. Criteria of optimization and the dynamics of reorganization are understood in only a very few cases, as discussed in the text.

the neuromuscular system. If a person is asked to turn his wrist as fast as possible (without following any command at all), he can do this at 3 to 5 Hz maximum if the amplitude is kept quite large ($\pm 20°$). This upper frequency depends, as Stark showed, on the motivation of the test subject. If he is asked to produce as rapid an oscillation as possible, he apparently does it in a "bang-bang" mode which allows him to increase the frequency to 5 to 6 Hz [11.1]. This, however, is probably not the mode assumed in fulfilling the tracking task. On the other hand, if asked to imagine a command signal at the highest frequency he could follow, he produces a (nonsinusoidal) oscillation at approximately 2 Hz. Thus Stark [11.1] demonstrated in an elegant way the extreme dynamic characteristics of various organizational forms of the system.

The information flow diagram of Fig. 11.5 indicates how complex the connectivity of the entire system can be. We have cited evidence for both task adaptivity and load adaptivity. It is therefore very difficult to use results from tests elucidating the properties of parts of the system to predict the overall performance. As Stark [11.1] points out, a subsystem such as the spindle-length servo system is brought into action only in some tasks and probably only temporarily. Analogously, responses of the entire system to some commands seldom allow us to predict responses in other cases. It is found, for example, that the steady-state position error in step function tracking is zero. This does not mean, however, that we deal *strictly* with a velocity servo. Stark found the system performing rather poorly in maintaining constant velocities, at least when a subject was asked to follow unpredictable ramps. If the test person is familiar with the ramp-tracking task, however, he can maintain constant velocities fairly well. It has indeed become clear [11.1] that the system may include a true integrator. The open loop impulse response has the general appearance of a step.

These examples show which aspects of the performance of the entire system should be investigated. The dynamics of the plasticity of the system after the loop has been opened, *or* changes in the error during the phase of familiarization to a particular task, are characteristics of the entire system which are of interest and which can be quantitated. Criteria of optimization, already mentioned, represent additional unknowns. Intermittency of the motor system, as observed by Stark [11.1] in ramp-tracking tasks, should be studied from this viewpoint. Stark has already pointed out that this type of action cannot be compared with a true sampled-data system. The intermittency is irregular and asynchronous and may be determined by repetitive changes in organization of control loops. These, however, may base on particular error reducing schemes. Of further interest is the minimum gain of an additional feedback loop around the entire system, which throws a subject into temporary oscillation.

C. ABNORMAL TRACKING PERFORMANCE

Both Stark [11.1] and Johns [11.5] used manual tracking tasks to quantitate abnormalities of motor control in patients with Parkinsonism. It is of clinical importance to have a quantitative method of defining the severity of the disease as well as improvements following therapy. Analysis of disordered movements may yield insight into the pathophysiology and perhaps even into basic questions pertaining to the organization of the motor control system.

The Parkinson syndrome may include all or some of the following features: large resistance to passive, imposed movements (" rigidity ") [11.6], slowness of active movements, loss of coordination of movements, resting tremors at 4 to 6 Hz and more rapid action tremors, changes in posture, and changes or disorders in sequences of motions in writing and speaking.

Using the same definitions as in the previous section (Fig. 11.3), Johns [11.5] found that the total error in step function tracking is, in Parkinson patients, four to five times that of normal subjects. Continued testing causes this error to increase by up to 35 % in severe cases of the illness. This probably is due to fatigue. The lag error amounts, in the abnormal case, to 23 % of total error. Reaction time is approximately 0.4 sec (i.e., twice the normal time). Kinetic error, however, is considerably larger in patients with Parkinsonism. It is observed, first of all, that maximum velocities are limited. For large step amplitudes this is more pronounced than for small ones. The regression line is found to be

$$v_{\max} = 0.6 + 0.51A$$

with ± 1.1 rad/sec standard deviation in the estimate of $v_{\max}$. In the average, abnormal peak velocities are approximately 25 % of normal. Careful analysis of the time course of the time derivative of the error in abnormal subjects shows the following. Slowness of response is due to intermittency of motion. We do not observe a superposition of " purposeful " motion and its opposing movements, which would exhibit positive and negative velocities. However, purposeful movements are interrupted by pauses at a frequency of 6 to 9 Hz. The rapidly oscillating velocities stay positive in most cases and are motion dependent. The motion remains unidirectional, and peak velocities are seen in midcourse as in the normal case. This has been partly verified by electro myographers, who found intermittent, synchronous bursts of action potentials in both agonist and antagonist. This may be what causes action tremors. Thus slowness of motion essentially is given by this intermittency. Peak accelerations in Parkinsonism, as seen in this tracking task, do not differ much from normal ones. (The *average* force which a patient can produce, however, is of course below that of a normal subject [11.1].)

Another particular feature of motor control in Parkinsonism is that a patient often is unable to change a pattern of movements before one task (i.e., one step response) is completed. A subject continues to approach the previous target position even though a new command has been presented. This "psychological refractory period," which is longer than the reaction time and which causes a person to refuse to accept the next task before the current one is fulfilled, is longer in Parkinsonism. Whether this can be explained by slowness due to intermittency of action is not known.

Stark concluded, from his observations on the limited bandwidth of the abnormal system and on increased rigidity opposing passive movements, that the Parkinsonism patient cannot open up the muscle–spindle feedback loop as is needed in rapid volitional movements. We have seen here that, if this hypothesized mechanism does exist, it must occur in intermittent fashion.

D. REFERENCES

11.1. Stark, L. *Neurological Control Systems.* Plenum Press, New York, 1968.

11.2. Smith, O. J. M. Nonlinear computations in the human controller. *IRE Trans Bio-Med Electron* **BME-9**: 125–128, 1962.

11.3. Johns, R. J. and I. T. Draper. The control of movement in normal subjects. *Bull Hopkins Hosp* **115**: 447–464, 1964.

11.4. (a) Elkind, J. I. and C. D. Forgie. Characteristics of the human operator in simple manual control systems. *IRE Trans Automatic Control* **4**: 44–55, 1959. (b) For a survey, see McCormick, E. J. *Human Factors Engineering*, 3rd ed. McGraw-Hill Book Co., New York, 1970.

11.5. Draper, I. T. and R. J. Johns. The disordered movement in Parkinsonism and the effect of drug treatment. *Bull Hopkins Hosp* **115**: 465–480, 1964.

11.6. Webster, D. D. A method of measuring the dynamic characteristics of muscle rigidity, strength, and tremor in the upper extremity. *IRE Trans Med Electron* **ME-6**: 159–164, 1959.

CHAPTER TWELVE

Motor Control: Eye Tracking

Understanding the neuromuscular system controlling eye movements is based on some of the same concepts needed in the discussion of hand tracking performance in Chapter 11. The task of centering an optical image on the fovea, and keeping it centered if there exists relative motion between head and target, is fulfilled optimally owing to a multilevel system comprising neuromuscular reflex loops, volitional control centers, and predictors. The system is able to reorganize for adaptation to particular tasks, as summarized in the following. The eyes can be rotated very quickly from one position to another, sometimes by a rapid succession of steps (*saccades*), so that a target image is brought at least into the vicinity of the fovea. Volitional eye movements as they are executed in reading or in "scanning" a landscape consist of such saccades only. In general searching-and-*tracking* tasks, however, the eyes can also move steadily. Continuous eye rotation, or *smooth pursuit* movement, is controlled by different brain structures (related to the visual centers). It occurs only when a moving target is tracked. The visual system is highly developed for the latter task. In some cases, smooth movements may be interrupted by saccades in the opposite direction so that tracking can begin again (optokinetic nystagmus). In other cases, they are interrupted by saccades in the same direction so that tracking accuracy is improved.

Moreover, the vestibulo-ocular reflex loop is "switched" into the oculomotor control system if the body (or the head only) is accelerated, as was discussed in Chapter 8. Furthermore, physiological nystagmus, another mode of eye movements, contributes significantly to acuity of the visual system. Minute eye movements (by ca. ± 5 min of arc, horizontal and vertical) ensure that the optical image, when centered on the fovea, is continually shifted to different receptor cells. Thus bleaching is avoided. Furthermore, very small targets cause temporal changes in the stimulus at a rate of 30 to 80 Hz, and these changes are significant attributes of coding in the visual system (see Chapter 5).

When the eyes move together as they generally do, their movements are said to be conjugate. When they are tracking an object that approaches or recedes from the viewer, however, quite different convergence movements are required. In fact, it is clear today that there are five distinct types of eye movements: the saccades, smooth pursuit motions, convergence movements, movements elicited by the vestibulo-ocular reflex, and physiological nystagmus. In addition, propriosensors in the neck muscles can drive the eyes, through centers at low levels in the brain, in the direction in which the head is moved. Different neural structures, linked hierarchically, control these various types of motion. Commands from higher centers may combine with or override those from lower ones [12.1]. The study of eye movements therefore is also of interest to neurologists who can determine, for example, the extent of brain damages by testing responses of the different reflex loops.

In this chapter we concern ourselves with the overall performance of the oculomotor system in tracking tasks and with the behavior of some of its parts. The performance of the feedback loops involved here is quite well understood and has been described quantitatively. Robinson [12.1] gave an excellent review of the subject. Precise location of the different nervous structures forming parts of the system, and exact definition of their roles and interactions, remain goals of the workers in the field.

In the following we analyze first the kinetics of the load of the oculomotor system, that is, the eyeball and associated structures. Second, we discuss the neural commands required to elicit eye movements. Third, we present data on the behavior of the entire system in tracking situations.

A. DYNAMICS OF EYE MOVEMENTS

The classes of tasks of the oculomotor system are not quite as varied as those encountered by general motor control because the mechanical impedance of the load of the system (i.e., of eyeball and orbital tissues) is a constant. For this reason it has been possible to characterize the oculomotor systems satisfactorily. It is unnecessary to consider in detail the weighting of neural commands, nonlinear in eye position, which is required because of the complex anatomy of the extraocular muscle system and the geometry of eye rotation (see Chapter 8). Here, in order to understand eye movement control, we concentrate on factors significant at this level: the torque–rotation transfer function of the load, the character of neural commands necessary for certain movements, and in a later section, feedback properties and stability of the systems.

The impedance of the eye is largely determined by orbital tissue. The forces are probably the largest that are encountered when the globe is rotated and the optic nerve is moved through orbital fat. These forces are mainly viscous. Elastic components also exist which cause the eye to resume its center position after an externally applied torque is released (assuming passive muscles). Conversely, owing to the elastic forces, the muscles have to apply a constant force to maintain a certain direction of gaze. From the functional viewpoint, it is important to remember that the agonistic extraocular muscles not only "see" the tissue impedance but that they must, in addition, overcome the forces exerted by their antagonists. These forces are, at best, given by the *passive* muscle properties.

The total impedance is computed from responses in eye position to stepwise changes in externally applied torques (Robinson [12.2]) or to sinusoidally varying torques (Thomas [12.3]). Once impedance is determined, a satisfactory estimate of the relation between *active state* muscle force and eye position can be obtained, using the model for active muscle performance discussed in Chapter 10.

The elastic restoring force on the globe is approximately 1.3 $g^*/^\circ$ (g^* = gram force) [12.1]. In dynamic situations, however, the viscous forces over-

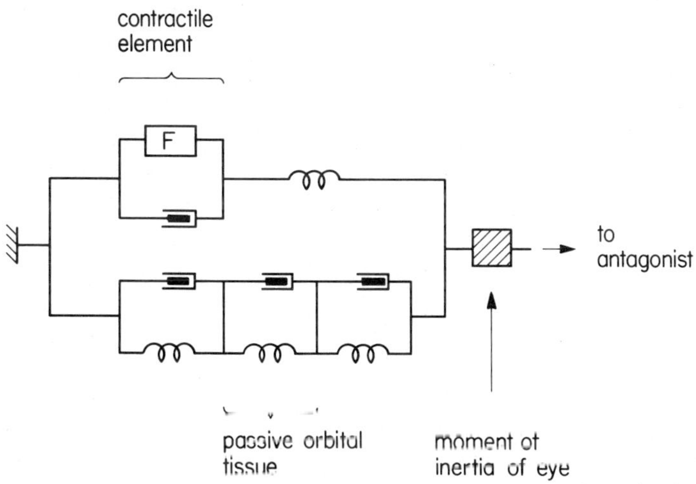

Fig. 12.1. Model of eye muscles, orbital tissues, and mass of globe which determine the torque–angle relationship of the eye. The active part of the eye muscle is represented by a force source (F) in parallel with a dashpot. The single spring in the upper half of the model represents the series-elastic muscle element. The three spring–dashpot combinations represent orbital tissues and parallel elastic element of muscle. The impedance of the viscoelastic networks dominates the overall behavior. The impedance, against which the active muscle acts, includes also the viscoelastic elements of the antagonist (not shown). For numerical values of elements, see text. After Robinson [12.1].

whelm the elastic ones. The step response (i.e., changes in eye position following an instantaneous change in torque) therefore is slow. The rheological behavior of orbital tissue is such that a composite of 2 to 3 exponential functions can be detected, with time constants of approximately 0.5, 0.3, and 0.02 sec. When the globe is loaded by a mass, increasing the moment of inertia by a factor of 100, the step response is oscillatory and heavily damped, showing maximum overshoot of only 18 %. This indeed demonstrates that normally the mass of the eyeball does not contribute significantly to the impedance [12.2]. Using the simple four-element muscle model of Chapter 10 and information from the step response, Robinson [12.2] was able to approximate the eye mechanics satisfactorily by the system presented in Fig. 12.1. Here the "parallel elastic" element of muscle is lumped with other tissues constraining the globe. These are represented by three spring–dashpot combinations. In the Laplace domain, the passive eye impedance (i.e., the torque–angular velocity relation) is

$$Z^{-1}(s) = \frac{0.67s(1 + 0.22s)(1 + 0.02s)}{(1 + 0.30s)(1 + 0.06s)(1 + 0.0042s + 1.03 \cdot 10^{-5} s^2)} \qquad (12.1)$$

The units of $Z^{-1}(s)$ are °/(g* cm sec). The impedance determined by Thomas [12.3] is essentially the same as that of eq. 12.1 in the frequency range above 16 Hz, as discussed by Robinson [12.1].

Some important observations yielding insight into the nature of nervous commands to eye muscles, and thus into the modes of action of the control system, follow from the foregoing statement. During *saccades* (Fig. 12.2) the eye is rotated much more rapidly than it would be if the applied active-state force had been a step function in time. Depending on the angle traversed, one would estimate, from the saccade waveform, a bandwidth of the system's transfer function of 5 to 20 Hz. Equation 12.1 indicates, however, that the −3-dB level below the D.C. response is reached at approximately 1 Hz. Thus the neural command, and in turn also the active muscle force, must be very large initially in order to accelerate the eye within a short time. When the globe axis is close to or in its final position, the active state tension of the muscle drops precipitously to the value necessary to keep the globe in that position, as in Fig. 12.2. Small overshoots in position are seen, but only in small saccades. The steady-state force also contains a small component to compensate for stress relaxation in the orbital tissue. However, it is the initial excess force, called preemphasis by Robinson, which effectively widens the bandwidth of the system.

With increasing saccade magnitude, from 5 to 40°, the duration of preemphasis increases from 30 to 100 msec, as does the duration of the saccade. The active state excess force is constant at 25 to 30 g*cm. This is because the muscles are driven maximally, in a "bang-bang" mode, to overcome the

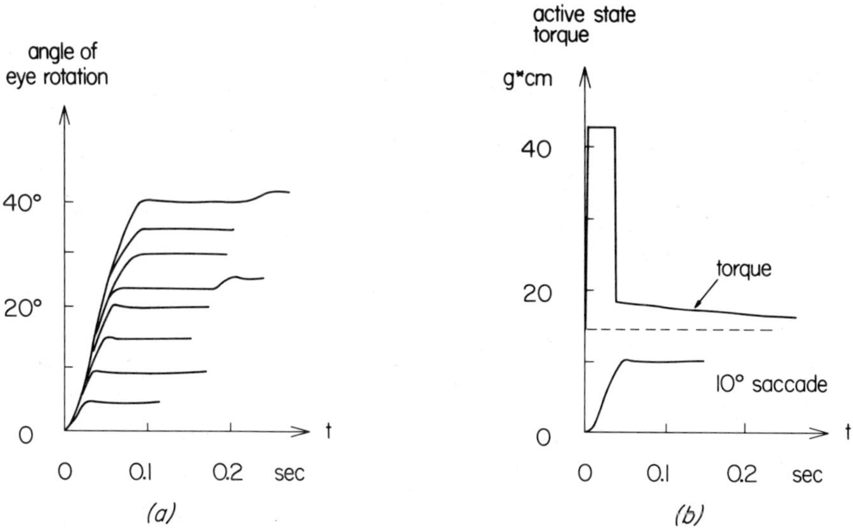

Fig. 12.2. (*a*) Eye position versus time for saccadic movements of different magnitudes. (*b*) Active state torque for saccade computed using position function (shown in lower half) and impedance of Fig. 12.1. The high viscoelastic impedance is overcome by large preemphasis of torque during which the eye globe is rotated rapidly. The steady-state torque balances elastic forces and compensates for stress relaxation.

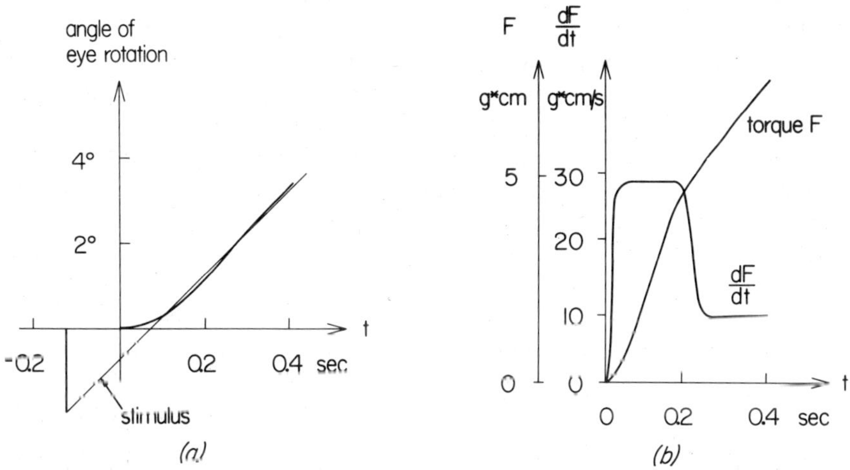

Fig. 12.3. (*a*) Eye position, as a function of time, for pure smooth pursuit movement elicited by Rashbass's "step-ramp" stimulus. (*b*) Torque and rate of torque computed from position function and impedance of orbital tissues. There also appears some preemphasis in this case; the control, however, differs from that of saccadic movements. After Robinson [12.2, 12.4].

viscous load. In fact, the peak force indicates the saturation level, which also determines peak angular velocity. Electromyograms of the extraocular muscles verify the preemphasis. Strong bursts of activity are observed during saccades. Similar but quantitatively different excess forces are found at the onset of *smooth* pursuit movements (Fig. 12.3). Within approximately 130 msec [12.4] the eye is accelerated to the steady-state velocity. The amount of preemphasis increases when tracking velocity increases, in contrast to saccadic preemphasis.

Thus the strategy in overcoming the large viscous components of eye impedance is the same in both saccadic and smooth eye movements. It is, of course, the strategy assumed by the central "control centers." The fine structures of the preemphases therefore differ significantly. Robinson [12.4] collected many additional data giving strong evidence for the existence of two different motor-control centers, one for saccades and another for smooth pursuit and following movements. Some of these differences between the two systems are discussed in the next section.

B. EYE TRACKING OF UNPREDICTABLE TARGETS, WITH SACCADES AND SMOOTH PURSUIT MOVEMENTS

Early observations, reported first by Westheimer [12.5], seemed to indicate that the *saccadic motor control* acts like a sampled data system. If a test person is asked to track a target that is displaced in a pulselike fashion during 0.1 sec, his eyes respond after a delay (i.e., central latency) of 0.2 sec. This occurs although the target has returned, by that time, to its starting position. The system requires an additional 0.2 sec to resume the original position. This shows that once the system has sensed an error, it decides to minimize it, regardless of possible changes in error during latency. Wheeless et al. [12.6] later found that the situation is, in fact, more complex.

Assume a stimulus consisting of a target displacement of 6° to the left, lasting $T = 0.2$ sec, followed by a displacement by 6° to the right ("pulse-step" stimulus). Under such conditions, responses include the pulse in only 70% of all cases; 30% consist of the step to the right only, without the pulse. If $T = 0.1$ or 0.05 sec, the fractions of responses that would be predicted by a classical sampled-data model (i.e., pulse-step responses) are only 23 and 8%, respectively. This indicates that saccadic motor control is not a simple position–error-actuated system. Position *and* velocity of the target in between sampling times do influence the response. The earlier the error changes after a particular "sampling instant," the more strongly is the next response affected. Nevertheless, the concept of a sampled–data system is useful.

In particular, Young and Stark [12.7] (see also [11.1]) simulated and explained several findings on the basis of such a model. They included a periodic sampler measuring the position error at a rate of 5 Hz, a dead zone element suppressing any response for errors below $\pm 0.5^\circ$, a delay unit, and the muscle–eye system. As Robinson [12.1] pointed out, their success lay in the prediction of the system performance with external feedback (see below).

The performance of the saccadic system in tracking normal step functions can easily be derived from Fig. 12.2. Let us analyze the kinetic error E_k, as defined in the previous chapter: E_k is given by the area between the command (a step function) and the response. The average duration (ΔT) of a saccade depends on the magnitude ($\Delta\phi$, rad) of the saccade as follows [12.2]:

$$\Delta T \simeq 0.03 + 0.115\Delta\phi \text{ (sec)} \qquad \text{for} \quad \Delta\phi \geqslant 0.10 \text{ rad} \tag{12.2}$$

The kinetic error therefore depends on $\Delta\phi$ in quadratic fashion, and we have approximately

$$E_k \simeq (0.015 + 0.0575\Delta\phi)\Delta\phi \text{ (rad sec)} \tag{12.3}$$

However, if we define an average angular velocity of the eye by $\omega = \Delta\phi/\Delta T$, we find the area under the *velocity* error curve to be linearly dependent on the size of the saccade:

$$E_{k,\text{vel}} = \text{velocity error during saccade} \simeq \Delta\phi \text{ (rad)} \tag{12.4}$$

Note that the foregoing expressions pertain to saccades in the temporal direction which are faster, by approximately 10%, than those in the nasal direction. It is also remarkable that velocities reached by the extraocular muscles are the highest found in any muscle. The angular velocity of the eye, during saccades, seems to reach its maximum at approximately 400 °/sec.

The method of testing stability and open loop performance is the same here as in the hand tracking task discussed in Chapter 11 (see Fig. 12.4). The

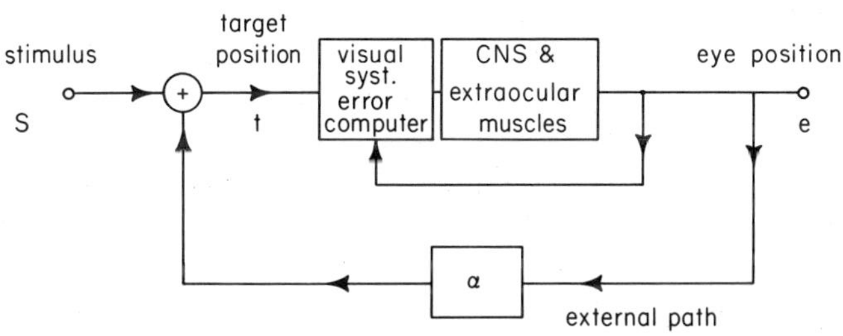

Fig. 12.4. Method of testing characteristics of the oculomotor control system. An external feedback path is placed around the natural system. By adding fraction α of response to input, the control loop can be opened ($\alpha = 1.0$) or thrown into oscillation by making the overall feedback factor large and negative.

natural system has a feedback factor of -1.0. If, however, a fraction α of the response is fed back and added, with positive or negative sign, to the stimulus, the total feedback factor is $\beta = 1.0 - \alpha$. The value of β can be varied from zero to such negative numbers that the system becomes unstable. In the open-loop condition ($\alpha = 1.0$, $\beta = 0$) any small movement of the target causes a saccade, which in turn drives the target away from the fixation point. After this, in a series of fairly regular steps (Fig. 12.5), the eye seeks in vain to catch up with the target, which keeps running away. This is typical of a sampled data system.

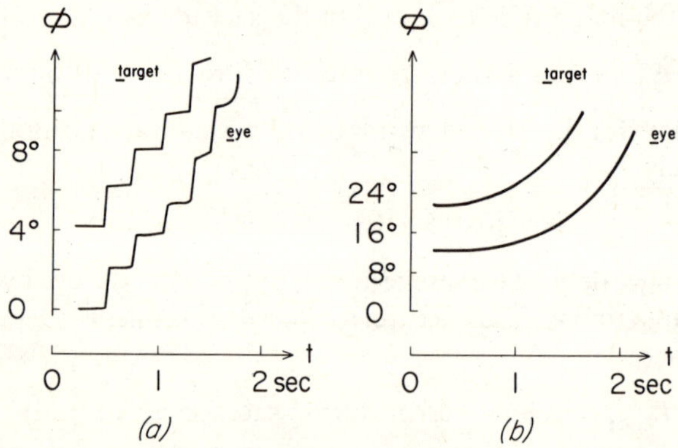

Fig. 12.5. Open-loop performance ($\alpha = 1.0$) of saccadic (*a*) and smooth pursuit (*b*) systems. Target (*t*) position essentially is equal to eye response (*e*), since no special stimulus is needed to throw the system into the runaway condition.

On the other hand, if β is -2.0 or less, any spontaneous eye motion by $\Delta\Psi$ drives the target in the opposite direction to $-\Delta\Psi$ or farther. If the system then, after 0.2 sec, attempts to track, the target is moved back to its original position. This causes a square wave oscillation at approximately 2.5 Hz which dies out either because the eye moves slightly less than by $\Delta\Psi$ or because predictive elements are brought to bear. In the latter case, the test subject would predict the oscillation, after a few saccades, and then be able to suppress it. Subsequently, however, there is nothing to predict and the burst-like oscillation begins again. Only when β is as small as -5 [12.4] do the oscillations grow indefinitely, indicating that the system has become unstable (Fig. 12.6).

With this we have summarized the most important features of the saccadic system. Some detailed observations on saccade intervals shorter than 0.2 sec were discussed by Robinson [12.1]. Wheeless et al. [12.8] studied the influence of target luminance on the system. We learn in the following how strongly the saccadic differs from the smooth pursuit system.

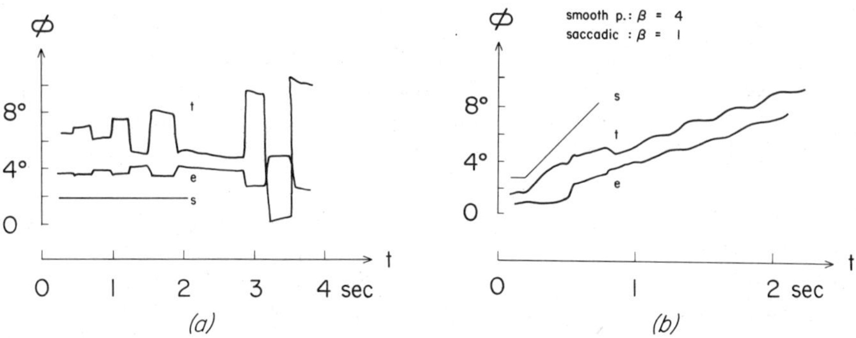

Fig. 12.6. Instabilities of saccadic (*a*) and smooth pursuit (*b*) systems. With $\alpha = -6$ the *saccadic* system is thrown into oscillations that have the form of a growing square wave at 2 to 2.5 Hz. Owing to predictive operators the system can suppress such oscillations. Here *t* is the target position, equal to $-\alpha e$ (eye position); *s* is the stimulus. Complete instability is reached if negative feedback factor $\beta = 5$. The *smooth pursuit* system is tested separately by putting a low-pass filter into the external feedback path. Under this condition, the system exhibits (damped) oscillation of 3.3 Hz for $\beta = 4$. Oscillations are insured if negative feedback factor $\beta = 8$. After Robinson [12.2, 12.4].

The study of *smooth pursuit oculomotor control* is possible rigorously only if the saccades, which occur concurrently in most of the simple tracking tasks, can be avoided. Rashbass [12.9] described the "step-ramp" stimulus (Fig. 12.3) which makes this possible. The eye executes a pure smooth pursuit movement, provided the ramp crosses the initial (zero) position at a time less than 0.2 sec. Both position and velocity of the error during the first 200 msec are such that this holds over quite a wide range of step and ramp amplitudes. This finding provides additional evidence against the pure sampled data saccadic system and allows us to study the smooth pursuit system separately.

Data on the performance in tracking such Rashbass ramps give strong support to the hypothesis of two neurologically distinct motor controls. Average latency of the smooth pursuit system has been found to be 0.125 sec—significantly less than that of the saccadic system (0.2–0.25 sec). The maximum velocity with which the system can follow a ramp smoothly is of the order of 30°/sec, which is approximately 10% of extreme rates seen during saccades. At the beginning of a Rashbass ramp response, the eye is accelerated within approximately 0.2 sec to a velocity that is very close to that of the target. This is initiated during a time when the position error is large and negative. The smooth pursuit system therefore must be velocity-error actuated.

Robinson demonstrated in his elegant study [12.4] that, in fact, velocity can be changed quite rapidly (although not as quickly as during saccades). The system does respond to two velocity changes spaced by as little as 75 msec. Thus it can be treated exactly like a continuous system (as opposed to sampled data). Its behavior would correspond to this even if it included a periodic error sampler at 13 Hz, as opposed to 5 Hz for saccades, because the bandwidth of the system does not allow for unequivocal distinction of the two possibilities. Robinson also showed that the *open-loop* performance is what would be expected from a continuous control circuit. The eye "runs away" smoothly as shown in Fig. 12.5.

Another significant difference from the saccadic system is noted when the negative feedback, around the system for smooth movements, is increased. It is found [12.4] that β must be at least -8 to cause continued oscillations. These are sinusoidal, at 3.3 Hz, as opposed to square waves of 2 to 2.5 Hz for the saccadic system (with $\beta = -5$) (see Fig. 12.6).

C. SYSTEM PERFORMANCE

Knowing the behavior of the two types of motor control, we can synthesize, consider possible interactions between the two, and study the performance of the whole. In search-and-track operations the eye's smooth movements are almost always interrupted by saccades. With this alternating mode a target can be sought very rapidly and tracked smoothly to enable fixation of the image on the fovea during most of the time. Tests have not been made at the psychophysical level to determine the extent to which visual information content is diminished by and during the saccades. It is known [12.10], however, that nerve cells exist in the visual cortex which are sensitive to optical information conveyed during saccades. These, in all likelihood, contribute to optimizing the system's performance during tracking *and* processing of visual information.

It is of great interest that the occurrence in *time* of any of the foregoing movements does not influence the position in time of the others [12.4]. Thus, for example, parts of smooth movements may be elicited before, during, or after saccades, depending on velocity errors only. Conversely, whether saccades occur depends on a combination of position and velocity errors and their temporal changes, as explained earlier. It is not known whether the criteria for their occurrence are related to minimizing the *average* velocity error.

Robinson, who established the extent to which the *amplitudes* of the two types of movements influence each other, found couplings that *improve* performance during combined action. Abrupt changes in smooth velocity are observed but only coinciding with and due to saccades. The velocity immediately after the saccade may be larger than that before, but only if saccade and smooth pursuit are in the same direction. Thus we can describe more concisely than was possible in the case of hand tracking (Chapter 11) how the motor-control system of eye muscles "reorganizes" according to its task.

Owing to the saccadic subsystem, large position errors are minimized rapidly and equally fast velocity changes can be elicited. Furthermore, quickly changing errors are diminished more efficiently than those which vary slowly. Although saccades can occur only at discrete times, they boost the dynamic performance of this muscle system to a remarkable degree. Whenever the errors are small and changing slowly, however, the smooth pursuit system dominates, as it must, in order to guarantee unblurred vision. A few typical mixed responses to simple ramps are presented in Fig. 12.7.

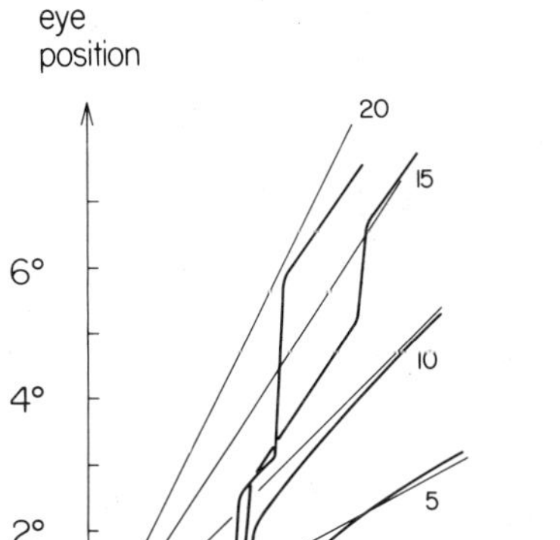

Fig. 12.7. Search-and-track performance of the oculomotor system exhibiting both saccadic and smooth movements. A combination of position and velocity errors determines the response. Temporal occurrences of the two types of movements are independent of each other. There is, however, coupling between their amplitudes. Velocities of ramp stimuli are 5, 10, . . . °/sec as marked. See text. After Robinson [12.4].

Our discussion so far has covered tracking performance with unpredictable stimuli of single steps, pulses, and ramps. By extension to *repetitive stimuli* which, to the system, become in some cases predictable, we learn more about additional forms of adaptivity. So-called predictive tracking was investigated by Young and Stark [12.7], Dallos and Jones [12.11], and others (see [12.1]). Sinusoidal inputs, although excellent for testing stationary and linear systems, do not seem to be useful in this case in the sense that knowledge of sinusoidal input–response relations does not allow extrapolation to performance with complex stimuli. On the other hand, some aspects of learning and predicting can be elucidated using square wave inputs. The eyes then move mostly in saccades.

In particular, Dallos and Jones [12.11] carefully analyzed the latencies—the time delays between alternating steps in the stimulus and those in the response. Latency changes with time in a fashion that depends in turn on the frequency

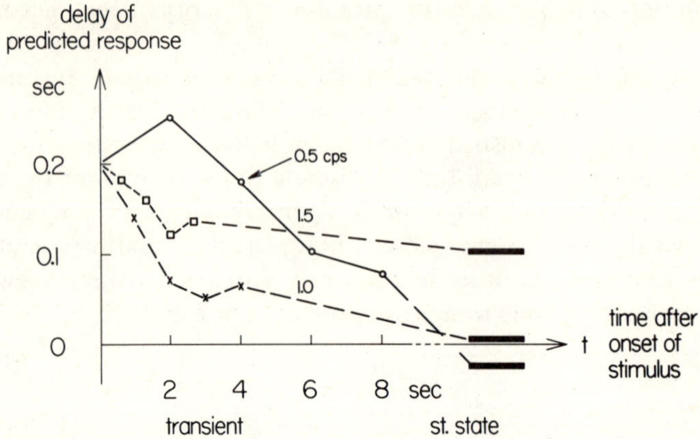

Fig. 12.8. Dynamics of reorganization of the oculomotor system when predictors are brought to bear. The latency of response to periodic square-wave stimuli is plotted against time. The steady-state latencies are marked by heavy bars. See text. After Dallos and Jones [12.11].

of the square wave, as in Fig. 12.8. Obviously, at the onset of the stimulus, the delay with which the eyes respond with the first saccade is always of the order of 0.2 sec. At very low frequencies (e.g., at 0.1 Hz) subsequent changes of sign in the stimulus occur after intervals so long that they are effectively unpredictable. Latency remains at 0.2 sec. At 0.5 Hz, however, a subject is able to reduce the latency after 4 to 5 cycles. During this time an "internal clock" is synchronized so that the next changes of sign (in stimulus) are well anticipated. In fact, as Fig. 12.8 shows, in the average the clock leads slightly in phase. Responses often occur before the stimuli. At 1 Hz, learning is very fast, and in the steady state the target is tracked almost perfectly. (The standard deviations in latency are ± 50 and ± 30 msec at 0.5 and 1.0 Hz, respectively [12.11].) At a stimulus frequency of 1.5 Hz the system learns to reduce its latency to approximately 0.1 sec but not to less. At higher frequencies prediction deteriorates and the delay between response and input approaches 0.2 sec again.

Apparently the frequency range of the internal clock is limited. From approximately 0.5 to 1.2 Hz it is easy for this predictor unit to estimate and anticipate fairly accurately the interval between periodic steps, as well as to synchronize it with the stimulus. This enables the system to learn within a certain time. Amplitude response is also improved because of the predictor. We thus are able not only to define an additional mode of eye movement control but also to describe in detail the dynamics of establishing this predictive type of motion (Fig. 12.8).

This chapter summarizes the performance of the oculomotor control center governing two of the most frequent eye movements. Many questions are left open here; for example, at the level of the subsystem we do not touch on the possible role of spindles within the extraocular muscles. Neither are the neural structures that are involved in the control circuits discussed in detail. Models of other oculomotor systems are not described (see, e.g., [12.12]). The particular types of concepts that further our understanding of multilevel optimal control systems, however, have been presented. They should be a guide in further experimental and theoretical work in the field of muscle systems analysis.

D. REFERENCES

12.1. Robinson, D. A. The oculomotor control system; a review. *Proc IEEE* **56**: 1032–1049, 1968.

12.2. Robinson, D. A. The mechanics of human saccadic eye movement. *J Physiol (London)* **174**: 245–264, 1964.

12.3. Thomas, J. G. The torque–angle transfer function of the human eye. *Kybernetik* **3**: 254–263, 1967.

12.4. Robinson, D. A. The mechanics of human smooth pursuit eye movement. *J Physiol (London)* **180**: 569–591, 1965.

12.5. Westheimer, G. Eye movement responses to a horizontally moving visual stimulus. *Arch Ophthal (Chicago)* **52**: 932–941, 1954.

12.6. Wheeless, L. L., R. M. Boynton, and G. H. Cohen. Eye-movement responses to step and pulse-step stimuli. *J Opt Soc Amer* **56**: 956–960, 1966.

12.7. Young, L. R. and L. Stark. Variable feedback experiments testing a sampled data model for eye tracking movements. *IEEE Trans Human Factors in Electronics* **HFE-4**: 38–51, 1963.

12.8. Wheeless, L. L., G. H. Cohen, and R. M. Boynton. Luminance as a parameter of the eye-movement control system. *J Opt Soc Amer* **57**: 394–400, 1967.

12.9. Rashbass, C. The relationship between saccadic and smooth tracking eye movements. *J Physiol (London)* **159**: 326–338, 1961.

12.10. Wurtz, R. H. Visual cortex neurons: response to stimuli during rapid eye movements. Science **162**: 1148–1150, 1968.

12.11. Dallos, P. J. and R. W. Jones. Learning behavior of the eye fixation system. *IEEE Trans Automatic Control* **AC-8**: 218–227, 1963.

12.12. Brodkey, J. and L. Stark. Accommodative convergence—an adaptive nonlinear control system. *IEEE Trans Systems Science and Cybernetics* **SSC-3**: 121–133, 1967.

CHAPTER THIRTEEN

Passive Mechanical Properties Of Muscle en Masse

A. INTRODUCTION

Studies in the area of biomechanics and biorheology have been useful for many different tasks in a great number of biological disciplines. Stimuli for such investigations came from needs in clinical, preventive, and aerospace *medicine*. For example, knowledge of stress distribution [13.1] and extreme loads [13.2] in bones is important for the orthopedic surgeon and for the designer of artificial limbs and joints. At low frequencies the mechanical impedance of the rib cage is of interest to cardiologists who study apex cardiography or who optimize extracorporeal cardiac massage. Transfer characteristics through the thorax of vibrations at acoustic frequencies have been studied with respect to heart sounds [13.3], a topic that is covered by the discussion below.

Other biomechanical studies finding application in surgery concern the rheology of skin, in particular its nonisotropic characteristics. Optimal wound healing is influenced by the direction of collagen fibers [13.4]. Korotkoff sound production, due to instability of the artery under the blood pressure cuff, is of interest to the physically oriented cardiologist [13.5]. Responses of entire body structures to continuous vibrations or impacts have been analyzed by engineers in automotive and aerospace industries. These studies may form the basis for optimal designs of car seats (and their padding) or help to establish levels critical to health when the body is exposed to intense vibrations. Similarly, the purely mechanical effects of ultrasound therapy can be predicted only if energy absorption by tissue is known.

In *physiology*, when the study of an organ's function is approached through its mechanical properties and behavior, the theories of the mechanical engineer must be brought to bear. Some selected examples developed in this book are: transmission of sound through external meatus and middle ear (Chapter 6), force–length–velocity relations for passive and active muscle (Chapter 10), spatial distribution of viscoelastic properties of arteries and the function of the arterial vessel system (Chapter 14), and the mechanics of the lung (Chapter 17). Many more areas may be of interest to the biomedical engineer, such as physiological changes in connective tissue [13.4] or in arterial wall properties with age [13.6]; the function of mechanoreceptors as determined by their structure [13.7] and as dependent on the propagation of mechanical stimuli on the body in the form of surface waves.

In this chapter we must concentrate on a few important aspects. We therefore use two examples to demonstrate how some tools of the biomechanical engineer can be usefully applied in the areas of physiology *and* clinical medicine. In both cases we emphasize phenomenological descriptions typical of systems analysis. When studying wave transmission in tissue (Section 13.B) we are not asking which particular structures or tissue components are responsible when a certain level of viscosity is observed. However, we are inquiring into the types of wave propagation modes and into energy absorption, given that tissue exhibits a known effective viscoelasticity. Similarly, when describing the torque–rotation characteristics of a finger joint (Section 13.C), we are approaching the behavior of the intact joint so that the dynamic load of muscle control or phenomenological changes in disease can be quantitated concisely. These two examples cover an extreme range of frequencies and signal levels, and they offer opportunity to present widely different analytical techniques.

B. WAVE PROPAGATION IN VISCOELASTIC MEDIA; SMALL SIGNAL ANALYSIS

The development in this section adheres to the work of Oestreicher, von Gierke, Franke, and others [13.8, 13.9], who correlated most elegantly theoretical with experimental findings. The mathematical treatment predicts satisfactorily the oscillatory modes and transition regions which are observed in tissue over a very wide band of frequencies (10–10^6 Hz). The analysis applies to small signal amplitudes and homogeneous, isotropic materials. The latter assumptions certainly do not hold in tissue at the microscopic level. Thus we cannot use the theory to investigate details of ultrasound scattering due to nonhomogeneities. However, tissue macroscopically exhibits an overall behavior which, for example, governs energy absorption in the frequency band discussed. Furthermore, the theory can be used to estimate transmission of vibratory energy from air (or any known coupling system) into tissue, or to predict depth of penetration of oscillations at a certain frequency. This is possible because pertinent data on viscosity and elasticity of muscular tissue are available [13.9]. Lung or bone tissues undoubtedly behave differently. The methods of attack, however, are universal. In particular, we deal with continuum mechanics applied to complex media where both compression and shear elasticity, as well as compression and shear viscosity, must be taken into account. A simpler case, omitting shear viscoelasticity, is developed in Chapter 14. We refer the interested reader to Jaeger [13.10] for basic background material on the theory of elasticity and to Wylie [0.16] for vector analysis.

We deem it useful to present some details of the derivation of the equations governing wave propagation because the concept of complex elastic moduli, not always explained in textbooks, plays an important role in biomechanics. In the Lagrangian frame (seeking a description of the motion of a particle in time as opposed to flow fields), Newton's second law is

$$\rho \frac{\partial^2 \xi_j}{\partial t^2} = \sum_{k=1}^{3} \frac{\partial \sigma_{jk}}{\partial x_k} + \rho X_j \tag{13.1}$$

where $j = 1, 2, 3$
ρ = density of medium
ξ_j = displacement in direction j
σ_{jk} = components of stress tensor
x_j = (Cartesian) coordinates of particle
X = external body force

Explaining the terms σ_{jk} we note, for example, that on a differential parallelepiped the force on the face, whose normal is parallel to coordinate 2, is the following sum: $-(\sigma_{12}\, dx_1\, dx_3 + \sigma_{22}\, dx_1\, dx_3 + \sigma_{32}\, dx_1\, dx_3)$. Furthermore, the mean pressure is $\sigma = -(\sigma_{11} + \sigma_{22} + \sigma_{33})/3$ which, however, is not comparable to fluid pressure, since it is possible that $\sigma_{11} \neq \sigma_{22}$. In addition to eq. 13.1 we need the constitutive equations for the medium, relating the stresses to the strains through the material's properties. Because of the assumption of isotropy and owing to the properties of the stress tensor, we may write

$$\sigma_{jj} = 2\left(\mu_1 + \mu_2 \frac{\partial}{\partial t}\right)\frac{\partial \xi_j}{\partial x_j} + \left(\lambda_1 + \lambda_2 \frac{\partial}{\partial t}\right)\left(\frac{\partial \xi_1}{\partial x_1} + \frac{\partial \xi_2}{\partial x_2} + \frac{\partial \xi_3}{\partial x_3}\right) \tag{13.2a}$$

$$\sigma_{jk} = \sigma_{kj} = \left(\mu_1 + \mu_2 \frac{\partial}{\partial t}\right)\left(\frac{\partial \xi_j}{\partial x_k} + \frac{\partial \xi_k}{\partial x_j}\right) \qquad j \neq k \tag{13.2b}$$

where λ_1 = coefficient of volume compressibility
λ_2 = coefficient of volume viscosity

$\mu_1 =$ coefficient of shear elasticity (i.e., ratio of shear stress and shear strain, in simple shears also called modulus of rigidity)

$\mu_2 =$ coefficient of shear viscosity (i.e., ratio of shearing stress and velocity gradient normal to displacement)

$\lambda_1 + \frac{2}{3}\mu_1 =$ bulk modulus

These constitutive equations are fairly general in that they comprise a linear relation between strain, stress, and their spatial, temporal, *and* mixed derivatives. We neglect effects due to heat conduction and elastic aftereffects by omitting higher derivatives. For small signals this appears to be acceptable.

The wave equation is obtained by introducing eqs. 13.2 into eq. 13.1:

$$\rho \frac{\partial^2 \mathbf{s}'}{\partial t^2} = \left[(\mu_1 + \lambda_1) + (\mu_2 + \lambda_2)\frac{\partial}{\partial t}\right] \text{grad div } \mathbf{s}' + \left(\mu_1 + \mu_2 \frac{\partial}{\partial t}\right)\nabla^2 \mathbf{s}' \quad (13.3)$$

where $\mathbf{s}' =$ displacement vector with components (ξ_1, ξ_2, ξ_3), a function of position and time

$X = 0$, body forces omitted

It is seen that eq. 13.3 is equivalent in principle to the wave equation of the "ordinary" theory of elasticity, if we introduce harmonic motion in time, of (angular) frequency ω

$$\mathbf{s}' = \mathbf{s}(x_1, x_2, x_3)e^{-j\omega t} \quad (13.4)$$

and complex moduli of viscoelasticity

$$\mu = \mu_1 - j\omega\mu_2$$
$$\lambda = \lambda_1 - j\omega\lambda_2 \quad (13.5)$$

Then the wave equation is

$$-\rho\omega^2 \mathbf{s} = (\mu + \lambda) \text{ grad div } \mathbf{s} + \mu\nabla^2 \mathbf{s} \quad (13.6)$$

or after division by μ

$$(\nabla^2 + h^2)\mathbf{s} = \left(1 - \frac{h^2}{k^2}\right) \text{grad div } \mathbf{s} \quad (13.7)$$

where

$$h^2 = \rho\omega^2/\mu$$
$$k^2 = \rho\omega^2/(2\mu + \lambda)$$

Oestreicher [13.8] pointed out that the classically treated modes of wave propagation are covered by eq. 13.7. For example, *plane compressional* waves are found by applying the divergence operator to eq. 13.7:

$$(\nabla^2 + k^2) \text{ div } \mathbf{s} = 0 \quad (13.8a)$$

which simplifies to

$$\frac{\partial^2 \xi}{\partial x^2} + k^2 \xi = 0 \tag{13.8b}$$

Hence the propagation constant k yields phase velocity c_{com} and attenuation (neper/distance) α_{com} as follows:

$$c_{\text{com}} = \frac{\omega}{\text{Re } k} \tag{13.9a}$$

$$\alpha_{\text{com}} = \text{Im } k \simeq \frac{2\mu_2 + \lambda_2}{2\rho c_{\text{com}}^3}\,\omega^2 \qquad \text{(low frequencies)} \tag{13.9b}$$

where real and imaginary parts are indicated by Re and Im, respectively.

Using the simplifying assumptions of the theory of elasticity $\lambda_2 = \mu_2 = 0$, we would find loss-free propagation at a speed of $[(2\mu_1 + \lambda_1)/\rho]^{1/2}$. These waves are found at high (ultrasonic) frequencies and have also been called the primary seismic waves. It is a longitudinal mode which propagates in the direction of motion of the particles.

Conversely, waves occurring in incompressible viscous fluid are *rotational.* By applying the curl operator to eq. 13.7 we get

$$(\nabla^2 + h^2) \text{ curl } \mathbf{s} = 0 \tag{13.10}$$

In particular, in the assumed medium $\mu_1 = 0$, $\lambda_1 = \infty$, $h = \omega(\rho\omega/\mu_2 j)^{1/2}$, yielding a heavily damped wave. Or, in incompressible, ideally elastic media, $\mu_2 = 0$, $\lambda_1 = \infty$, $h = \omega(\rho/\mu_1)^{1/2}$, which yields shear waves propagating in a transverse mode [velocity $= (\mu_1/\rho)^{1/2}$]. These special cases are found only in the simplest situations.

When analyzing *tissue*, however, both λ and μ must be kept complex. The geometry and boundary conditions within which the general eq. 13.7 must be solved depend on the experiment whose outcome we are trying to predict. Oestreicher, Franke, and others studied two cases: (*a*) the combination of quasi-spherical compression and shear waves excited by an oscillating sphere imbedded within an unbounded medium, and (*b*) circular surface (Rayleigh) waves, which conveniently can be excited and observed on the surface of the body. As we explain below, the experiments used in these two studies, together with additional, independent observations, allowed not only estimation of the material properties (λ, μ) by different methods but also verification, in part, of their theoretical predictions.

The pressure field around an *oscillating sphere*, of radius a, situated at the origin of the system of coordinates within an unlimited viscoelastic medium, can be calculated in closed form by choosing the following boundary conditions:

$$\xi_1 = s_0 e^{-j\omega t} \qquad \xi_2 = \xi_3 = 0 \qquad \text{if} \quad r = a$$
$$\text{outgoing wave only exists} \qquad \text{if} \quad r \to \infty$$

Oestreicher [13.8] showed that the general solution, comprising Laplace's spherical harmonics, spherical Hankel functions, and solid harmonic functions, simplifies because of symmetry and becomes

$$\mathbf{s}(x_1, r) = -A \text{ grad}\left[\frac{\mathrm{H}_1(kr)}{kr} x_1\right] + B\left[2\mathrm{H}_0(hr) \text{ grad } x_1 - \mathrm{H}_2(hr) r^3 \text{ grad} \frac{x_1}{r^3}\right] \tag{13.11}$$

where r = distance from origin
H_i = spherical Hankel function of ith order
A, B = complex constants determined by boundary conditions; dependent on h, k, and a = radius of sphere

The first term of eq. 13.11 represents the irrotational compression wave, the second term is an incompressible shear wave. The velocities of these wave components differ from those discussed previously for the plane geometry only in the vicinity of the sphere. Knowing the field **s**, it is possible to predict pressure distribution using the constitutive equation. The pressure field, however, is difficult to measure experimentally. It therefore is preferable to compute the mechanical impedance encountered by the driving system in order to compare it with measured values.

It is assumed that there exists only one (i.e., the outgoing) wave such that the measured impedance corresponds to the characteristic impedance. This is justified because damping is large, ensuring that possible reflections (from boundaries of which the theory takes no account) are attenuated heavily. Indeed Oestreicher et al. found essentially that values for the impedance measured at an oscillatory piston at the surface of the body matched the predicted impedance values. In this situation, the impedance at the oscillator is governed by the neighboring regions in the medium, and the influence of (remote) boundaries is not felt. Thus we define the characteristic impedance applicable in that case as

$$Z = \frac{\text{force on sphere in } x_1 \text{ dir.}}{\text{velocity}} = \frac{2\pi\alpha^2 \int_0^\pi \sigma_{rx_1} \sin\gamma \, d\gamma}{-j\omega s_0} \tag{13.12}$$

where $\sigma_{rx_1} = (x_1\sigma_{11} + x_2\sigma_{21} + x_3\sigma_{31})/r$, which becomes [13.8]

$$Z_0 = \frac{4\pi}{3} j\omega\rho a^3 \frac{(1 - j3/ah - 3/a^2h^2) - 2(j/ah + 1/a^2h^2)[3 - a^2k^2/(1 + jak)]}{[a^2k^2/(1 + jak)](j/ah + 1/a^2h^2) + 2 - a^2k^2/(1 + jak)} \tag{13.13}$$

The following three extreme cases should be noted:

1. At very low frequencies, that is, $\omega\mu_2 < \mu_1$,

$$Z_0 = \frac{2\pi}{3} a^3 \left[9 \left(\frac{\sqrt{\mu_1 \rho}}{a} + \frac{\mu_2}{a^2} \right) + j \left(\omega\rho - \frac{9\mu_1}{\omega a^2} \right) \right] \tag{13.14}$$

This has the form applicable to a simple dashpot–mass–spring system, where mass $= (2\pi\rho a^3)/3$ and spring constant $= 6\pi a \mu_1$. Experimentally, the damped resonance behavior is indeed found, as in Fig. 13.1. It would appear from eq. 13.14 that knowledge of resonance frequency enables the computation of μ_1, at least as long as damping is not severe.

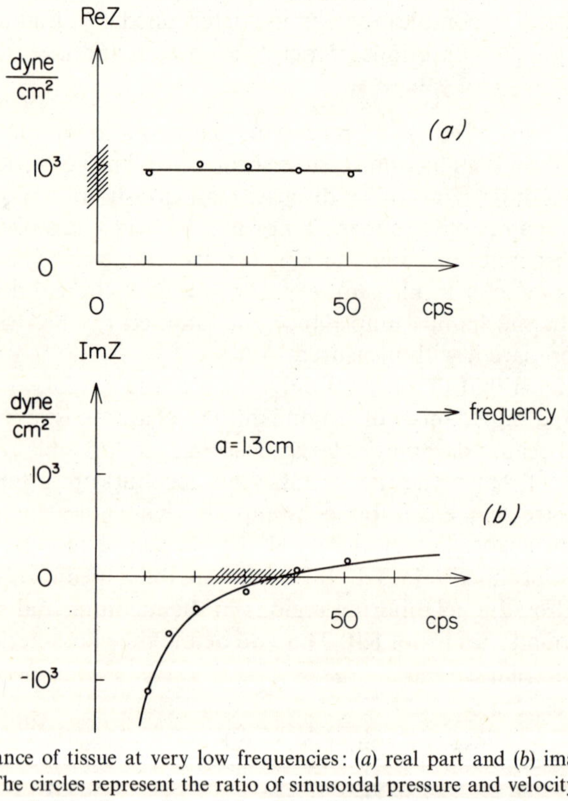

Fig. 13.1. Impedance of tissue at very low frequencies: (*a*) real part and (*b*) imaginary part versus frequency. The circles represent the ratio of sinusoidal pressure and velocity of a piston (2.6-cm diameter) applied to the surface of the thigh. The displacements imposed are small. Because of complex wave modes, the impedance depends on tissue property and piston area. The shaded area indicates the range over which the curve is displaced when piston diameter is increased to 3.8 cm. The effects of tissue properties, however, can be separated using the theory. In the frequency range, the impedance compares well with that of a mass–spring–dashpot system. Zero crossing of imaginary part is at resonance frequency. After von Gierke et al. [13.9b].

One difficulty arises because μ_1 depends on the preload (the static force upon which the oscillations are superimposed). This nonlinearity has not been investigated thoroughly. Oestreicher [13.8] therefore used other methods to obtain independent estimates for μ_1. One method employs the magnitude of indentation due to a small static load, the other the velocity of surface waves (discussed below). With these he found $\mu_1 = 3.0(\pm 1.0) \times 10^4$ dyne/cm^2 (at minimal preload). From the resistive part of Z_0 in eq. 13.14, μ_2 then can be calculated to be approximately 100 dyne sec/cm^2.

2. At high frequencies (1000–10,000 Hz), where $\omega\mu_2 > \mu_1$ and $\omega\mu_2 > 2\mu_1{}^2/a^2\rho\omega^2$, the impedance is essentially the same as in incompressible viscous fluids, that is,

$$Z_0 = \frac{2\pi}{3}\rho a^3 \left[\frac{9\sqrt{\omega\mu_2}}{a\sqrt{\rho}}\left(\frac{1}{\sqrt{2}} + \frac{\sqrt{\mu_2}}{a\sqrt{\rho\omega}}\right) + j\left(\omega + \frac{9\sqrt{\mu_1}}{a\sqrt{\rho}}\right)\right] \quad (13.15)$$

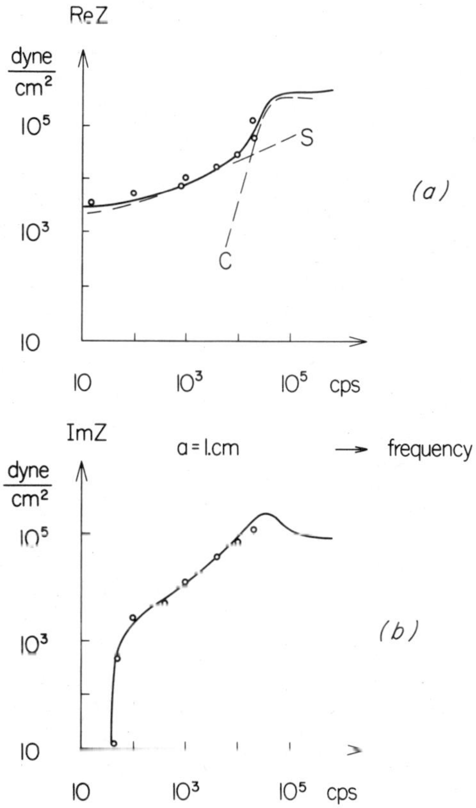

Fig. 13.2. Real part (*a*) and imaginary part (*b*) of impedance of an oscillating sphere in a viscoelastic medium of the same characteristics as tissue; sphere radius is 1 cm. The circles are measurements by von Gierke, on the surface of the body. See text for explanation of coincidence. The curves *C* (resistance of compression mode) and *S* (shear mode) as encountered in the incompressible viscous medium allow for prediction of energies absorbed in shear versus the compression mode. In tissue, generally, both modes are found concurrently. The imaginary part above the resonance frequency demonstrated in Fig. 13.1 appears in (*b*). From Oestreicher [13.8].

Clearly, μ_2 can be estimated from the frequency-dependent resistive part of Z_0. If this is done for different values of a, we find that μ_2 is 100 to 500 dyne sec/cm^2.

3. At still higher frequencies the pure (irrotational) compressional modes are encountered, as with ultrasonic waves. These travel at a velocity of approximately 1500 m/sec. Analytically, we have

$$c_{\text{us}} = \sqrt{(\lambda_1 + 2\mu_1)/\rho}$$

whence

$$\lambda_1 \simeq 2.6 \times 10^{10} \text{ dyne/cm}^2$$

Thus three of the four constants determining material properties have been established. However, λ_2 is of no influence in the frequency range considered here, although it enters the picture at frequencies above 1MHz, as do relaxational effects [13.8]. These phenomena have not been studied carefully so far. But below 1 MHz, the values of μ and λ_1 can be introduced into the general eq. 13.13, yielding the curve of Fig. 13.2. The experimentally determined values, also given, show that theory predicts remarkably well. Figure 13.2 demonstrates the frequency bands in which viscosity (causing damped shear waves) or compressibility of the medium dominate, respectively. Note that the impedance shown is dependent on the area over which forces are applied. When the areas are larger, shifts to lower values occur in resonance frequency (at very low frequencies) and in the transition region (to compressional waves).

Surface waves, which can be observed quite easily on the body's surface, have also proven useful for the determination of tissue properties. They are governed by eq. 13.7, for which the following boundary conditions hold:

$$\text{at} \quad x_3 = 0 \quad \sigma_{33} = \lambda \operatorname{div} \mathbf{s} + 2\mu \frac{\partial \xi_1}{\partial x_3} = 0$$

$$\sigma_{23} = \mu \left(\frac{\partial \xi_3}{\partial x_2} + \frac{\partial \xi_2}{\partial x_3} \right) = 0$$

$$\sigma_{13} = \mu \left(\frac{\partial \xi_3}{\partial x_1} + \frac{\partial \xi_1}{\partial x_3} \right) = 0 \tag{13.16}$$

Franke et al. [13.9a] performed a general analysis, similar to the one needed for Rayleigh waves, which applies to viscoelastic media and for circular wave fronts. As explained previously, it is normal for both compressional modes and rotational modes to exist. They appear, in the *general* solution of eq. 13.7, in the following combination:

$$\mathbf{s} = \operatorname{grad} \phi + \operatorname{grad} \operatorname{div} \psi + h^2 \psi \tag{13.17}$$

where

$$(\nabla^2 + k^2)\phi = 0 \qquad \text{compressional modes} \tag{13.18}$$

$$(\nabla^2 + h^2)\psi = 0 \qquad \text{rotational modes} \tag{13.19}$$

The solutions of the last two equations can be given, in cylindrical coordinates, as

$$\begin{aligned} \phi(r, z) &= A\mathrm{H}_0(gr)e^{-pz} \\ \psi(r, z) &= B\mathrm{H}_0(gr)e^{-qz} \end{aligned} \tag{13.20}$$

where

r = distance from origin, along surface

H_0 = usual Hankel function, zero order

$p = \sqrt{g^2 - k^2}$

$q = \sqrt{g^2 - h^2}$

The unknown quantity g is found by introducing eqs. 13.20 into the boundary conditions (eq. 13.16), whence we have the frequency equation

$$(-h^2 + 2g^2)^2 - 4pqg^2 = 0 \tag{13.21}$$

If k is small (medium nearly incompressible) then

$$g = 1.044h = 1.044\sqrt{\rho\omega^2/(\mu_1 - j\mu_2\omega)} \tag{13.22}$$

From eq. 13.20 it follows that the velocity (c_s) of the wave, when propagating in the direction of increasing r, is given by

$$c_s = \frac{\omega}{\partial/\partial r\ [\arg \mathrm{H}_0(gr)]} \tag{13.23}$$

This simplifies for large gr, since

$$r \to \infty \ldots \mathrm{H}_0(gr) \simeq \sqrt{2/\pi gr}\ e^{j(gr - \pi/4)}$$

that is,

$$r \to \infty \quad c_s = \frac{\omega}{\mathrm{Re}\ g} = \frac{0.96}{\mathrm{Re}\sqrt{\rho/\mu}} \simeq 0.96\sqrt{\mu_1/\rho} \text{ (low frequency)} \tag{13.24}$$

Equation 13.24 shows that, although both compressional and rotational waves exist near the origin of the wave, only the shear waves are observed at greater distances from the center. Franke et al. [13.9b] used this information as an additional means to determine μ_1. They found $c_s = 160$ cm/sec, hence $\mu_1 = 3.5 \times 10^4$ dyne/cm^2, which corresponds closely to values obtained by other methods.

Thus Oestreicher, von Gierke, and their co-workers were not only able to describe the different types of waves excited in most cases in viscoelastic media such as tissue, but they also evaluated tissue properties. Hence it is possible to define the frequency bands in which shear modes or compressional (acoustic) modes of wave propagation dominate (Fig. 13.2). We refer the reader to the original articles [13.8, 13.9] for mathematical details omitted here. As stated at the beginning of this section, we are dealing with a systems analysis in that we emphasize overall effects which allow us to predict, for example, energies absorbed in tissue. This may be important when the body is irradiated with ultrasound, be it for therapeutic or diagnostic purposes, or when the transmission of heart sounds through tissue is studied.

C. THE RHEOLOGY OF A JOINT; LARGE SIGNAL ANALYSIS

Tissue under large strains exhibits nonlinear behavior, which prevents us from applying mathematical tools of the kind used in the previous section. For example, the collagen fibers in connective tissue do not influence stress–strain characteristics much at low strains. At large strains, however, the fibers are aligned and individually stretched, which in turn results in high elastic moduli. This tissue again changes characteristics under excessive loads; it begins to show plastic behavior [13.4]. Stress–strain relations therefore generally are not only nonlinear but also dependent on the history of the material. Creep (i.e., lengthening of a structure under constant large stresses) is responsible for dependence of the dimensions of certain tissues, in their unstressed state, on the length of time the material was previously kept under stretch. Such effects were encountered in the discussion of skeletal muscle (Chapter 10). There they are not very significant from the functional viewpoint, so that the important mechanical properties of muscle can be modeled by simple spring–dashpot combinations.

In many other cases behavior is more complex. On the one hand, this can be described by summing contributions of individual structural components using *continuous* spectral representations of elastic, viscous, and other elements [13.11]. On the other hand, specific system's aspects may require descriptions based on combinations of *lumped* elements. These individually represent fundamental mechanical behaviors (elasticity, viscosity, Coulomb friction, etc.), independent of the structural elements involved.

The latter approach is used, in this section, for the analysis of torque–angle relations of the proximal finger joint of man (i.e., the second metacarpophalangeal joint). Figure 13.3 demonstrates how complicated the relations are [13.12, 13.13]. When the finger is rotated sinusoidally at a certain frequency, the torque needed increases progressively with increasing angle.

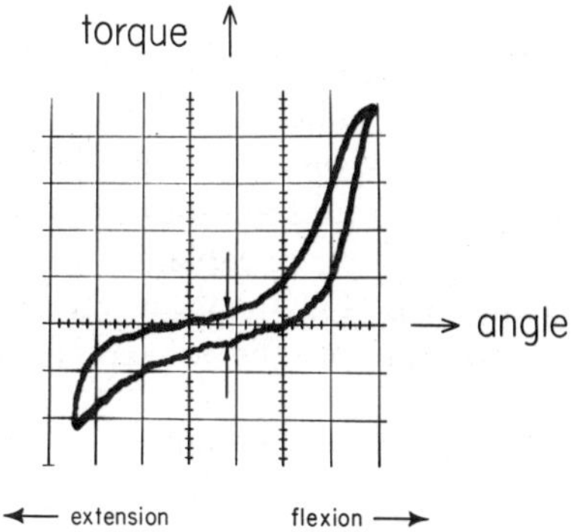

Fig. 13.3. Typical torque (ordinate) versus angle (abscissa) characteristics of the finger joint, with finger rotated sinusoidally. Note the asymmetry of the loop and strong nonlinearity of elasticity. Hysteresis is largely due to plasticity, as explained in the text. From Johns and Wright [13.13].

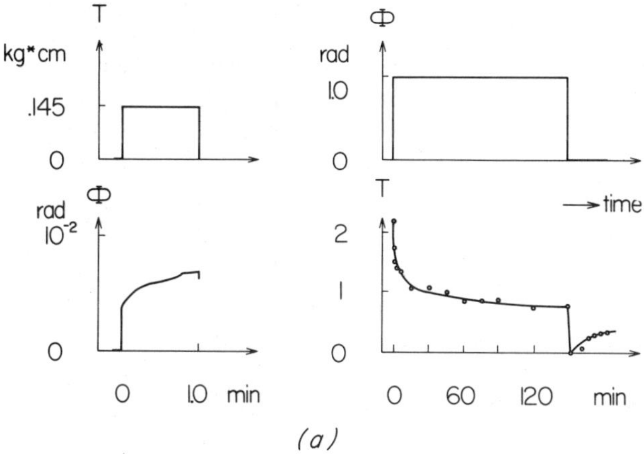

Fig. 13.4*a*. Strain retardation in finger joint position (Φ) in response to a step in torque (T), on the left, and torque relaxation after an applied step in Φ, on the right. Note strong nonlinearities. Gradual changes in time are not exponential. They are not even linear functions of the logarithm of time as seen in other tissues such as muscle. Note also the greatly different time scales. After Johns and Wright [13.13].

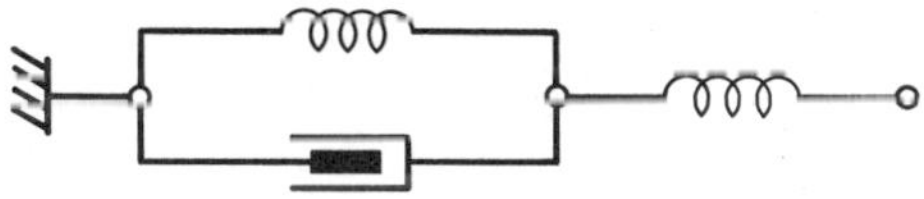

(b)

Fig. 13.4*b*. Lumped element model for the behavior observed in Fig. 13.4*a*. All elements are nonlinear (i.e., springs increase their elasticity with increasing stress); the force on the dashpot is not linearly proportional to velocity. The parallel combination of spring and dashpot is called a Voigt body.

Nonlinear ("stiff") elastic elements must be involved. Hysteresis is also observed, generally due either to (Newtonian) viscosity or to plasticity. Obviously, sinusoidal analysis alone does not yield sufficient information. More insight is gained from stress or strain step responses. For example, *strain retardation* and *stress relaxation* are observed in response to steps in stress or strain, respectively (Fig. 13.4*a*). These behaviors are explained most conveniently by the spring–dashpot combination of Fig. 13.4*b*, where all elements are nonlinear. Nonlinearity explains, for example, the nonexponential time course of strain retardation.

Creep is detected when stress is applied in a pulselike fashion (illustrated by Fig. 13.5). When stress is released after a "step response," the finger does not assume its original position. Creep occurs even at quite small step amplitudes. This is modeled by a series combination of the three-element system of Fig. 13.4*b* with a so-called Bingham body. The Bingham body exhibits Coulomb friction and an elastic force above a certain yield stress (T_0). The dashpot indicates that velocity is limited above T_0. The entire Bingham body represents the "plastic" property; its yield or creep prevents strain recovery. In normal joints, Johns and Wright [13.13] did not find one fixed level of T_0.

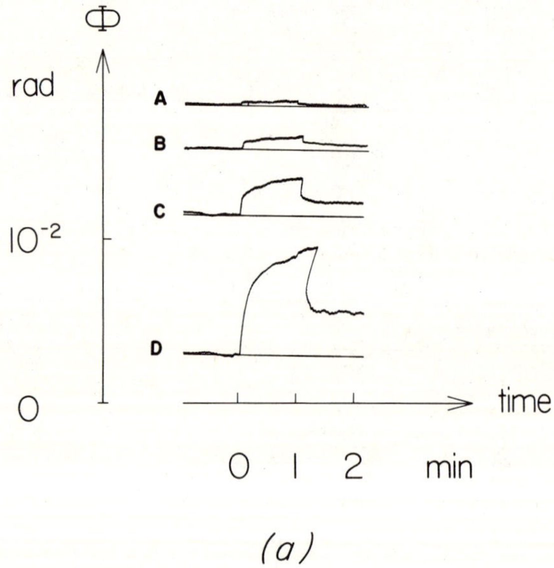

Fig. 13.5*a*. Responses in finger position (angle Φ) to pulses in torque. Amplitudes are: *A*, 14.5; *B*, 29; *C*, 72.5; *D*, 145 g*cm. The value of Φ is recorded on curvilinear coordinates versus time. Little torques are required to demonstrate creep in joint tissue. There is no strain recovery. From Johns and Wright [13.13].

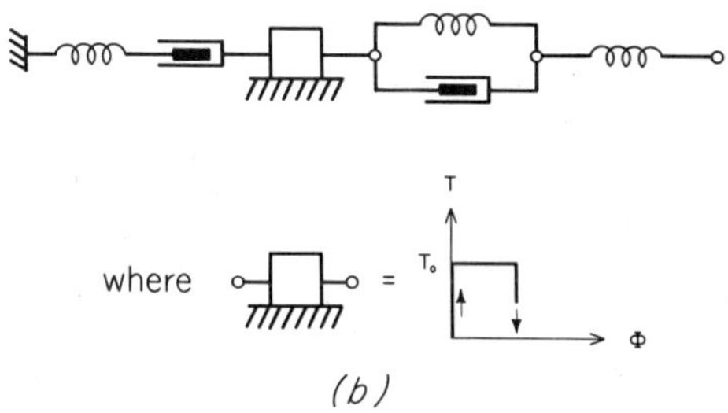

Fig. 13.5*b*. Model for the complete joint explaining the behavior in Fig. 13.5*a*. The sliding block introduces Coulomb friction above yield torque T_0, as shown in the insert. The Coulomb element–spring–dashpot system, added to the model of Fig. 13.4*b*, is called a Bingham body; it explains plastic behavior. Joints do not exhibit one discrete yield torque T_0, but a distribution of different T_0's. The model therefore should include several parallel sliding blocks and series springs. The lowest T_0 is of the order of 20 g*cm. From Johns and Wright [13.13].

Yield torques are distributed which, strictly, should be modeled by an assembly of parallel St. Venant bodies (Bingham without dashpot) of different yield torques, in series with a dashpot.

Straightforward techniques are used to separate and quantitate the different torques exerted by the various structures. When the finger is rotated sinusoidally and the torques are recorded at the rest position (i.e., when velocity is maximal), the forces encountered must be due to plastic and viscous elements only. When the maximal velocity of rotation is changed, the viscous torques also must change, by definition. Obviously, velocity should be varied by changing amplitude and frequency. Experimental results of such changes are presented in Fig. 13.6. Change in slope of the torque–amplitude characteristic is small, indicating a small (25%) contribution of Newtonian elements to the entire viscous and plastic torque: 75% are due to plastic elements, which create a torque of 40 g*cm at very small amplitudes, in the case of a normal metacarpophalangeal joint. Their torque increases slightly when amplitude is increased, demonstrating another nonlinearity. Since the plastic torques are dominant, hysteresis (Fig. 13.3) is largely due to plasticity.

Elastic torques must be measured quasi-statically. When rotating the finger a few times by a certain angle, then arresting at the point of maximum excursion and waiting for transients to disappear, the time-dependent elastic component is obtained at that (maximum) angle. It is not only a nonlinear function of strain (non-Hookean), but it shows asymmetry (Fig. 13.7). Extension requires higher torques than flexion (absolute values are given in the diagram). The total torque (solid line in Fig. 13.7) which is measured at the time of arresting includes plastic torques in addition to elastic ones. The latter are approximately three times the former at all angles. Inertial torques cannot be detected.

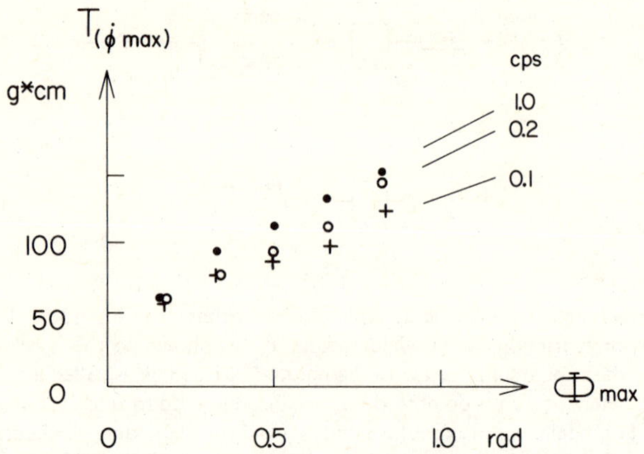

Fig. 13.6. Torque on a sinusoidally rotated finger, recorded at maximum velocity (see arrows in Fig. 13.3), as a function of amplitude at different frequencies. Torque is composed of plastic and viscous components. The viscous torque, by definition, is proportional to velocity. Since total torque changes little with frequency, plastic torque dominates the plastic component and therefore determines hysteresis (Fig. 13.3). Clearly, hysteresis is also a function of amplitude. From Johns and Wright [13.13].

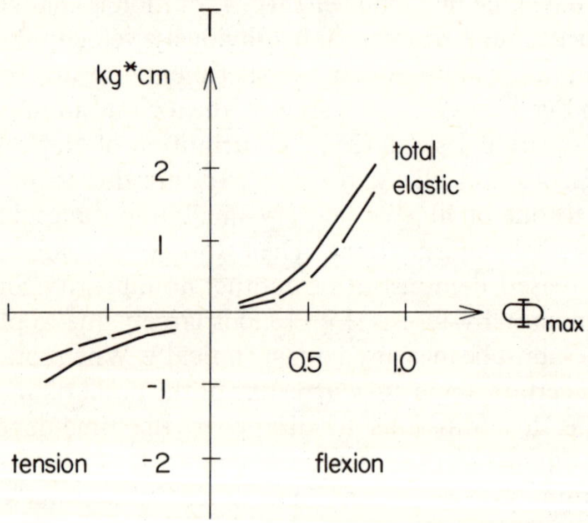

Fig. 13.7. Total torque on a finger joint (solid line) and its elastic component (interrupted line), measured at maximum excursion. Note strong nonlinearity in amplitude Φ. Details of separation of the two components are given in text. At maximum excursion, the viscous component is zero; the difference between elastic and total torque therefore is given by the plastic elements showing stress relaxation. After Johns and Wright [13.14].

Johns and Wright [13.12] demonstrated that the present technique is very suitable for quantitating effects of physiological changes and joint diseases. For example, elastic torques required to rotate a finger by ± 0.5 rad may increase by a factor of 5 to 10 between the ages of 4 and 66. Increases in temperature by 8°C cause a decrease in "stiffness" by 20%. This probably explains the subjective experience of an arthritic joint "loosening up" in warm baths [13.12]. However, elastic torques increase more than twofold in arthritic disorders (at a stage where joints are mildly deformed and limited in motion). Plastic and viscous components were not quite double the normal values. Obviously, this method of analyzing joint stiffness is powerful in quantitating the effects of treatment of diseases.

Johns and Wright [13.14] were also able to determine the contributions to stiffness made by various structural components of a joint. They investigated the wrist joint of cats, which compares well with the finger joint of a child. These contributions depend on the amplitude of motion. For small angles, the joint capsule contributes 47% of the total torque, muscles 41%, tendons 10%, and skin 2%. Tendons contribute significantly more, however, at extreme amplitudes. The separation into different force components is possible in the case of the joint. In most other tissues it cannot be achieved as accurately because the structural elements are distributed. Muscle tissue is a good example (Chapter 10). The values of its different elastic and viscous elements can be determined, but only by experiments on whole muscle. The method of "lumping" specific properties into discrete elements of a model is still powerful, as was discussed at the beginning of this chapter. The reader may also be interested in other approaches, notably the one proposed by Fung [13.15].

D. REFERENCES

13.1. Evans, F. G. Biomechanics: Stress–strain phenomena in bones, pp. 89–93 in *Medical Physics*, Vol. 3. O. Glasser (Ed.). Year Book Publishers, Chicago, 1960.

13.2. Mather, B. S. Observations on the effects of static and impact loading on the human femur. *J Biomech* **1**: 331–335, 1968.

13.3. Zalter, R., H. C. Hardy, and A. A. Luisada. Acoustic transmission characteristics of the thorax. *J Appl Physiol* **18**: 428–436, 1963.

13.4. Ridge, M. D. and V. Wright. A rheological study of skin, pp. 165–175 in *Biomechanics and Related Bio-engineering Topics*. R. M. Kenedi (Ed.). Pergamon Press, New York, 1965.

13.5. Anliker, M. and K. R. Raman. Korotkoff sounds at diastole—a phenomenon of dynamic instability of fluid-filled shells. *Int J Sol Str* **2**: 467–491, 1966.

13.6. Learoyd, B. M. and M. G. Taylor. Alterations with age in the viscoelastic properties of human arterial walls. *Circ Res* **18**: 278–292, 1966.

13.7. Loewenstein, W. R. and R. Skalak. Mechanical transmission in a Pacinian corpuscle. An analysis and a theory. *J Physiol (London)* **182**: 346–378, 1966.

13.8. Oestreicher, H. L. Field and impedance of an oscillating sphere in a viscoelastic medium with an application to biophysics. *J Acoust Soc Amer* **23**: 707–714, 1951.

13.9. (a) Franke, E., H. E. v. Gierke, H. L. Oestreicher, H. O. Parrack, and W. W. v. Wittern. On the physics of vibrations in living tissue. U.S. Air Force, Wright Air Development Division, Tech Rpt 6367, February 1951. (b) See also v. Gierke, H. E., H. L. Oestreicher, E. K. Franke, H. O. Parrack, and W. W. v. Wittern. Physics of vibrations in living tissues. *J Appl Physiol* **4**: 886–900, 1952.

13.10. Jaeger, J. C. *Elasticity, Fracture and Flow*. Methuen, London, 1956.

13.11. King, A. L. and R. W. Lawton. Elasticity of body tissues, pp. 303–316 in *Medical Physics*, Vol. 2. O. Glasser (Ed.). Year Book Publishers, Chicago, 1950.

13.12. Wright, V. and R. J. Johns. Physical factors concerned with the stiffness of normal and diseased joints. *Bull Hopkins Hosp* **106**: 215–231, 1960.

13.13. Johns, R. J. and V. Wright. An analytical description of joint stiffness. *Biorheology* **2**: 87–95, 1964.

13.14. Johns, R. J. and V. Wright. Relative importance of various tissues in joint stiffness. *J Appl Physiol* **17**: 824–828, 1962.

13.15. Fung, Y. C. B. Elasticity of soft tissues in simple elongation. *Amer J Physiol* **213**: 1532–1544, 1967.

Part IV

TRANSPORT, METABOLIC, AND CONTROL SYSTEMS

CHAPTER FOURTEEN

Mechanics of Blood Vessels: Pulsatile Arterial Flow in Vessel Systems

In this chapter we discuss physical principles governing pulsatile flow of blood and propagation of pressure and flow pulses in pulmonary and systemic arteries, as well as the physiological significance of phenomena encountered. The arteries form a vessel system of treelike architecture responsible for distributing blood throughout the body. Its smallest branches, the arterioles, are of the order of 20 μm in diameter and very numerous (10^6–10^7). They connect to the network of capillaries, which is so dense that every cell of the body is within approximately 20 μm of the nearest path of blood flow. The arterioles, as well as all larger arteries, are mechanically coupled to surrounding tissue, and their viscoelastic walls contain smooth muscles whose function is important in regulating flow of blood and pressure, as discussed in Chapter 16.

Our approach to the mathematical analysis of this system of formidable complexity is biased toward description of the factors that are important for the system in fulfilling the task just summarized. Using laws of the dynamics of fluid flow and thin cylindrical shells, the relation between oscillatory pressure gradient and flow in an individual artery is established first. Oscillatory flow velocity patterns within an artery are investigated to gain information on those factors (e.g., impedance and propagation constant) that govern wave propagation in an array of vessel segments. We then establish relations between pressure and flow at the input and output of a vessel segment as computed for sections of mismatched transmission lines. On the basis of these, we can treat entire arterial trees. Simplifying assumptions involving Newtonian viscosity of blood, completely developed velocity profiles, and linear viscoelasticity of wall material are made wherever gross description of the "subsystems" does not prevent understanding the behavior of the next more complicated system, studied theoretically and experimentally.

A. LOCAL HYDRODYNAMICS IN THE ARTERIAL CIRCULATION

The first step in discussion of pulsatile blood flow and pulse propagation in the arterial vessel tree concerns the relation of pressure and flow in one arterial segment. Most arteries are quite homogeneous between branching sites. It is therefore possible to apply the hydrodynamic model developed by Womersley [14.1] to individual segments. Womersley, in cooperation with McDonald [14.2], proposed a theory including effects of viscosity of blood, viscosity of the arterial wall material, and loading of the artery by surrounding tissue, all important parameters. Womersley's model, although neglecting several other phenomena, is useful in that fair estimates of flow from an observed pressure gradient can be obtained and in that it yields parameters needed in the description of vessel systems.

Throughout the mathematical development in this chapter it is assumed that pressures and flows are sinusoidal. Pulses produced by the ventricles of the heart often are so regular (i.e., periodic) that they can be expressed as a sum of many sinusoidal components using Fourier analysis. When the pulses are irregular, techniques described and evaluated by Taylor [14.3] involving power spectra, also resulting in sums of sinusoidal signals, should be used.

1. Fluid Motion, "Long-Wave Approximation"

The equation of motion of a small element of viscous fluid, for laminar flow, is the Navier-Stokes equation. Written in vector form, it is

$$\rho\left[\frac{\partial \mathbf{V}}{\partial t} + (\mathbf{V} \cdot \mathrm{grad})\mathbf{V}\right] = -\mathrm{grad}\, p + \mu_0 \nabla^2 \mathbf{V} \tag{14.1}$$

where ρ = density of fluid
$\mathbf{V}$ = velocity, with three orthogonal components (u, v, w)
p = pressure
μ_0 = viscosity of fluid

The first term on the left side ($\rho\, \partial \mathbf{V}/\partial t$) represents forces due to (linear) acceleration; $\rho(\mathbf{V} \cdot \mathrm{grad})\mathbf{V}$ gives accelerative forces due to convection, or changes in flow direction; grad p is the pressure gradient, and the last terms on the right side of eq. 14.1 are the frictional forces. (For a derivation, see

Prandtl and Tietjens [14.4]). The convection term is a vector with the following components:

$$(\mathbf{V} \cdot \text{grad})\mathbf{V} = \left[\left(u\frac{\partial u}{\partial x} + v\frac{\partial u}{\partial y} + w\frac{\partial u}{\partial z} \right) \quad \left(u\frac{\partial v}{\partial x} + v\frac{\partial v}{\partial y} + w\frac{\partial v}{\partial z} \right) \quad \left(u\frac{\partial w}{\partial x} + v\frac{\partial w}{\partial y} + w\frac{\partial w}{\partial z} \right) \right]$$

where x, y, and z are the three spatial (Cartesian) coordinates.

In addition to the Navier-Stokes equation, the equation of continuity must hold. Since arteries are much more distensible than blood is compressible, we can write (neglecting changes in fluid density)

$$\text{div } \mathbf{V} = 0 \tag{14.2}$$

For our purpose, eqs. 14.1 and 14.2 must be written in cylindrical coordinates. Furthermore, we assume axial symmetry. Hence $v = 0$, $\partial u/\partial y = 0$, and so on. We thus obtain for the equations of motion:

$$\rho\left(\frac{\partial u}{\partial t} + u\frac{\partial u}{\partial r} + w\frac{\partial u}{\partial z}\right) = -\frac{\partial p}{\partial r} + \mu_0\left[\frac{\partial^2 u}{\partial r^2} - \frac{u}{r^2} + \frac{1}{r}\frac{\partial u}{\partial r} + \frac{\partial^2 u}{\partial z^2}\right] \quad \text{radially} \tag{14.3a}$$

$$\rho\left(\frac{\partial w}{\partial t} + u\frac{\partial w}{\partial r} + w\frac{\partial w}{\partial z}\right) = -\frac{\partial p}{\partial z} + \mu_0\left[\frac{\partial^2 w}{\partial r^2} + \frac{1}{r}\frac{\partial w}{\partial r} + \frac{\partial^2 w}{\partial z^2}\right] \quad \text{axially} \tag{14.3b}$$

where u is the radial component of flow velocity and w is the axial component of flow velocity (see Fig. 14.1).

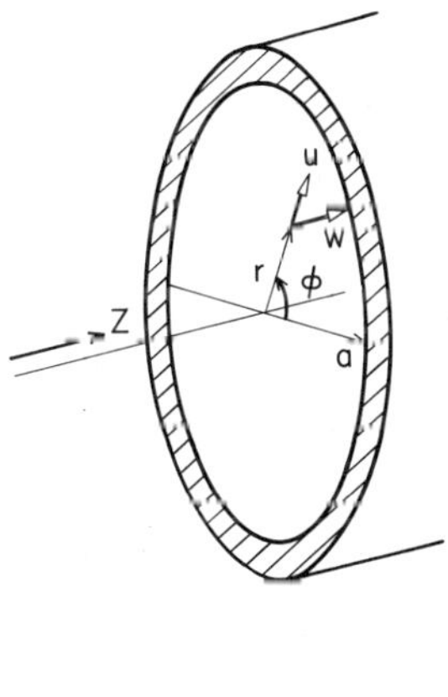

Fig. 14.1. Definitions of cylindrical coordinates r, ϕ, and z, and of flow velocities u (radial) and w (axial); a is the radius of the vessel.

For the equation of continuity we have

$$\frac{\partial u}{\partial r} + \frac{u}{r} + \frac{\partial w}{\partial z} = 0 \tag{14.4}$$

When solving these equations, we assume the signals to be waves propagating at velocity c (to be determined later) of frequency ω:

$$p = p_1(r)e^{j\omega(t-z/c)} \tag{14.5a}$$

$$u = u_1(r)e^{j\omega(t-z/c)} \qquad \text{where} \quad j = \sqrt{-1} \tag{14.5b}$$

$$w = w_1(r)e^{j\omega(t-z/c)} \tag{14.5c}$$

As Womersley [14.1] pointed out, these expressions allow us to estimate the magnitude of the terms on the left-hand side of eq. 14.3, for example,

$$\frac{\partial u}{\partial t} = j\omega u \qquad \frac{\partial u}{\partial z} = -j\frac{\omega}{c}u$$

But in (systemic) arterial circulation, velocities c range from 500 to 1000 cm/sec. This means that all terms in $\partial/\partial z$ and $\partial^2/\partial z^2$ are negligibly smaller than those in $\partial/\partial t$. Also, the wavelength (λ) of the waves under discussion is much larger than the radius of the vessel. Hence, from the equation of continuity (14.4) we learn that u must be of the order of $r\omega/c$ (or $r2\pi/\lambda$) times w, hence $u \ll w$. We therefore neglect in the following all terms stemming from $(\mathbf{V} \cdot \text{grad})\mathbf{V}$, and the equations become linear. Clearly, if the ratio of radius and λ becomes appreciable (as it might at higher harmonics of the heart beat) we must correct for some omitted terms. Techniques of taking into account (small) nonlinearities were developed by Womersley [14.1].

Equations 14.3 and 14.4 now appear simpler. After substitution of the assumed forms of solution (eq. 14.5), we obtain

$$j\omega\rho u_1 = -\frac{dp_1}{dr} + \mu_0\left(\frac{d^2u_1}{dr^2} + \frac{1}{r}\frac{du_1}{dr} - \frac{u_1}{r^2}\right) \tag{14.6a}$$

$$j\omega\rho w_1 = -\frac{dp_1}{dz} + \mu_0\left(\frac{d^2w_1}{dr^2} + \frac{1}{r}\frac{dw_1}{dr}\right) \tag{14.6b}$$

If eqs. 14.6 are multiplied by R^2 (R = inside radius of vessel at mean pressure) and divided by μ_0, and if the following abbreviations are introduced

$$\nu = \frac{\mu_0}{\rho} \qquad \text{(kinematic viscosity)}$$

$$y = \frac{r}{R} \qquad \text{(normalized radius)}$$

$$\alpha^2 = \frac{R^2\omega}{\nu}$$

we obtain

$$\frac{d^2u_1}{dy^2} + \frac{1}{y}\frac{du_1}{dy} + \left(j^3\alpha^2 - \frac{1}{y^2}\right)u_1 = \frac{1}{R}\frac{R^2}{\mu_0}\frac{dp_1}{dy} \qquad \text{radial} \qquad (14.7a)$$

$$\frac{d^2w_1}{dy^2} + \frac{1}{y}\frac{dw_1}{dy} + j^3\alpha^2 w_1 = -j\frac{\omega}{c}\frac{R^2}{\mu_0}p_1 \qquad \text{axial} \qquad (14.7b)$$

With similar substitutions, the equation of continuity becomes

$$\frac{1}{y}\frac{d}{dy}(u_1 y) = j\frac{\omega}{c}Rw_1 \qquad (14.8)$$

The left-hand sides of eqs. 14.7 are Bessel differential equations. We therefore assume that

$$p_1(y) = AJ_0(ky)$$

where J_0 is a Bessel function of first kind, order 0.

This assumption prevents us from obtaining any mode of oscillation other than the fundamental "long-wave" modes. Thus we find for radial velocity

$$u_1(y) = C_1\frac{J_1(j^{3/2}\alpha y)}{J_0(j^{3/2}\alpha)} - \frac{R}{\mu_0}\frac{Ak}{j^3\alpha^2 - k^2}J_1(ky) \qquad (14.9)$$

and for axial velocity

$$w_1(y) = C_2\frac{J_0(j^{3/2}\alpha y)}{J_0(j^{3/2}\alpha)} - j\frac{\omega}{c}\frac{R^2}{\mu_0}\frac{1}{j^3\alpha^2 - k^2}AJ_0(ky) \qquad (14.10)$$

where J_1 is a Bessel function of first kind, first order.

The $J_0(j^{3/2}\alpha)$ is introduced into the denominators of eqs. 14.9 and 14.10 to make it more convenient to match the boundary conditions.

When eqs. 14.9 and 14.10 are introduced into the continuity equation, two results are obtained:

$$k^2 = \left(j\frac{\omega R}{c}\right)^2$$

This k is a small number, particularly in comparison with α^2 (see denominators in eqs. 14.9 and 14.10). The α^2 is of the order of 10, and $\omega R/c = 2\pi R/\lambda$ is smaller than 0.1 in most blood vessels. Furthermore, $J_0(ky)$ in this case is very close to unity and $J_1(ky) \simeq (ky/2)$

We also find

$$\frac{C_1}{C_2} = \frac{k}{j^{3/2}\alpha}$$

Hence solutions of the Navier-Stokes equations giving radial distributions of flow velocities can be written as

$$u_1(y) = j\,\frac{\omega R}{2c}\left[C_1\,\frac{2J_1(j^{3/2}\alpha y)}{\alpha j^{3/2} J_0(j^{3/2}\alpha)} + y\,\frac{A}{\rho c}\right] \quad \text{radial} \tag{14.11}$$

$$w_1(y) = C_1\,\frac{J_0(j^{3/2}\alpha y)}{J_0(j^{3/2}\alpha)} + \frac{A}{\rho c} \quad \text{axial} \tag{14.12}$$

Before we can determine the unknown constants in these equations, the equation of motion of the arterial wall must be derived.

2. Equation of Motion of the Vascular Wall

It has proven satisfactory to consider the vascular wall as a thin membrane within which strains are assumed to be constants (independent of y). Womersley included forces acting on the wall that overwhelm those neglected in this very simple mechanical model. Further simplifications are possible, since the wavelength is assumed to be larger than the radius of the vessel. At equilibrium the radial force on an element of the thin wall (Fig. 14.2) must be balanced by the radial component of the tangential force:

$$\sigma_1 = \frac{h}{R}\,\sigma_2 \tag{14.13}$$

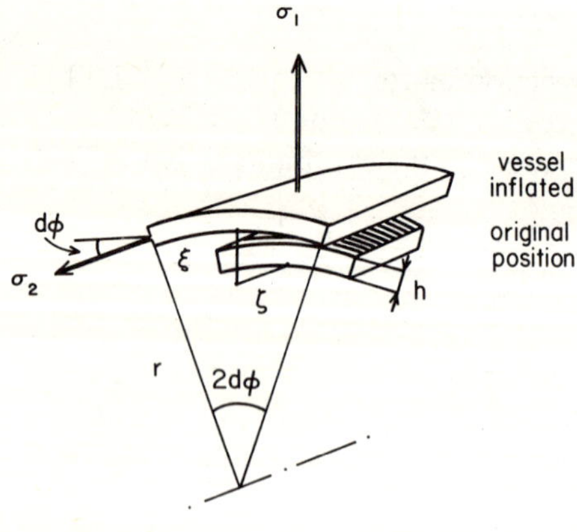

Fig. 14.2. Definitions of symbols used in analysis of motion of the vessel wall: σ_1 and σ_2 are radial and tangential stresses, respectively; the wall thickness is h; the radial and axial displacements of an element of the wall, due to stresses applied, are ξ and ζ respectively.

where σ_1 is radial stress and σ_2 is tangential stress. (Since the wall is thin, R can be used in eq. 14.13 instead of the variable r.) General relations between stresses and strains at this element are given by the three-dimensional form of Hooke's law [13.10]:

$$E\varepsilon_1 = \sigma_1 - \mu\sigma_2 - \mu\sigma_3 \tag{14.14}$$

$$E\varepsilon_2 = -\mu\sigma_1 + \sigma_2 - \mu\sigma_3 \tag{14.15}$$

$$E\varepsilon_3 = -\mu\sigma_1 - \mu\sigma_2 + \sigma_3 \tag{14.16}$$

where E = Young's modulus of elasticity
μ = Poisson's ratio

The σ_1 can be eliminated from eqs. 14.15 and 14.16 using eq. 14.13, and we then can express stresses σ_2 and σ_3 in terms of strains ε_2 and ε_3. The strains, however, are defined by

$$\varepsilon_1 = \frac{\partial \xi}{\partial r} \qquad \varepsilon_2 = \frac{\xi}{r} \qquad \varepsilon_3 = \frac{\partial \zeta}{\partial z}$$

where ξ = displacement of element in radial direction, r
ζ = displacement of element in axial direction, z

We thus obtain

$$\sigma_2 = E\frac{\varepsilon_2 + \mu\varepsilon_3}{1 - \mu^2 - (h/R)\mu(1+\mu)} \simeq \frac{E}{1-\mu^2}\left(\frac{\xi}{R} + \mu\frac{\partial \zeta}{\partial z}\right) \quad \text{tangential} \tag{14.17a}$$

therefore

$$\sigma_1 = \frac{h}{R}\frac{E}{1-\mu^2}\left(\frac{\xi}{R} + \mu\frac{\partial \zeta}{\partial z}\right) \quad \text{radial} \tag{14.17b}$$

$$\sigma_3 = \frac{E}{1-\mu^2}\left[\mu\left(1+\frac{h}{R}\right)\varepsilon_2 + \left(1+\mu\frac{2h}{R}\right)\varepsilon_3\right] \simeq \frac{E}{1-\mu^2}\left[\mu\frac{\xi}{R} + \frac{\partial \zeta}{\partial z}\right] \quad \text{axial} \tag{14.18}$$

Equations 14.17 and 14.18 relate, in a simple manner, strains with stresses within the wall. (For a much more rigorous application of Fluegge's shell theory to blood vessels, see Anliker and Raman [14.5].) Note that we have introduced E as a material's constant, although it is known that arteries are more elastic at higher than at lower pressures. However, the justification for this is that E does not vary significantly at normal blood pressures and pulse amplitudes.

Before formulating the equations of motion of the wall, relations developed so far can be used to establish a simple expression for the *static* pressure–

volume relationship for arteries. It is realistic to assume that the axial strain (ε_3) of an artery *in situ* is zero. (This assumption of axial constraint is later used in calculating the dynamics of the wall.) With $\varepsilon_3 = 0$, eq. 14.18 becomes

$$\sigma_3 = \mu \frac{E}{1 - \mu^2} \varepsilon_2 \tag{14.19}$$

Furthermore, eq. 14.16 can be expressed as

$$\sigma_3 = \sigma_2 \, \mu \left(1 + \frac{R}{h}\right) \simeq \sigma_1 \mu \frac{R}{h} \tag{14.20}$$

Combining eqs. 14.19 and 14.20, we have

$$\frac{E}{1 - \mu^2} \varepsilon_2 = \frac{R}{h} \sigma_1$$

However, σ_1 is essentially given by the transmural pressures (Δp). The relative change in cross-sectional area ($\Delta S/S$) of the vessel's lumen is

$$\frac{\Delta S}{S} = \frac{\pi(R + \xi)^2 - \pi R^2}{\pi R^2} = 2 \frac{\xi}{R} = 2\varepsilon_2 \tag{14.21}$$

Hence we can write

$$\frac{\Delta S}{S \Delta p} = \frac{1 - \mu^2}{E} \frac{2R}{h} = \frac{1}{B} \frac{2R}{h} \tag{14.22}$$

where $B = E/(1 - \mu^2)$.

This relative volume distensibility is often used when we compute c_0, the velocity of a pressure wave in a purely elastic tube filled with an ideal fluid:

$$c_0 = \sqrt{(\Delta p \cdot S)/\Delta S \; 1/\rho} \simeq \sqrt{E/(1 - \mu^2) \; h/2R\rho} \tag{14.23}$$

Equation 14.23 was derived by Bramwell and Hill [14.6], and c_0 is close to the so-called Moens-Korteweg velocity $(Eh/2R\rho)^{1/2}$ (see [14.2]). Poisson's ratio of the arterial wall μ is very close to 0.5 [14.2].

In the *equation of* (*radial*) *motion*, stress in the wall (eqs. 14.17) is added to the accelerative forces and transmural pressure (which we assume to equal, henceforth, the arterial pressure p):

$$h\rho_w \frac{\partial^2 \xi}{\partial t^2} = p - \frac{h}{R} B \left(\frac{\xi}{R} + \mu \frac{\partial \zeta}{\partial z}\right) \qquad \text{radial} \tag{14.24}$$

where ρ_w is the density of wall material.

Forces acting in the axial direction are not only the axial change in σ_3

and the inertial forces. Womersley [14.1] added the viscous drag exerted by the fluid on the wall: $(\mu\, \partial w/\partial y)_{y=1}$. (Axial changes of radial flow are neglected.) He also introduced a "tethering" force exerted by the surrounding tissue constraining the vessel in its axial motion. We let this force be proportional to ζ, the displacement in the z-direction. In summary:

$$\rho_w \frac{\partial^2 \zeta}{\partial t^2} = B \frac{\partial^2 \zeta}{\partial z^2} + \frac{\mu}{R}\frac{\partial \xi}{\partial z} + \frac{\nu \rho}{hR}\frac{\partial w}{\partial y}\bigg|_{r=R} - \rho_w \Omega^2 \zeta \tag{14.25}$$

The last term, $-\rho_w \Omega^2 \zeta$, is the "tethering" force. It can be thought to be caused by an elastic constraint ("spring constant" $= \rho_w \Omega^2$); Ω also can be interpreted as a frequency at which the wall would oscillate after an initial (axial) displacement.

From eq. 14.10 we obtain the "drag" force, which can be simplified because $k < 1$ and $(\omega R/c) \ll 1$. For convenience use the following form:

$$\frac{\partial w}{\partial y}\bigg|_{y=1} = \frac{C_1}{2}\alpha^2 j^3 F_{10}$$

where

$$F_{10} = \frac{2J_1(j^{3/2}\alpha)}{j^{3/2}\alpha J_0(j^{3/2}\alpha)} \tag{14.26}$$

In analogy to eq. 14.5 we assume sinusoidal displacements:

$$\xi = De^{j\omega(t - z/c)}$$
$$\zeta = Ee^{j\omega(t - z/c)} \tag{14.27}$$

Thus the equations of motions of the wall can be given in the following form:

$$-\omega^2 D = \frac{A}{\rho_w h} - \frac{B}{\rho_w R^2} D + \frac{\mu B}{\rho_w R} j \frac{\omega}{c} E \qquad \text{radial} \tag{14.28}$$

$$(\Omega^2 - \omega^2)E = -\frac{B}{\rho_w}\left(\frac{\omega^2}{c^2} E + \frac{\mu}{R}\frac{j\omega}{c} D\right) - \frac{\rho}{\rho_w}\frac{\nu j^3 \alpha^2}{hR}\frac{C_1}{2} F_{10} \qquad \text{axial} \tag{14.29}$$

Strain-rate dependent forces are not contained in these equations. Viscoelasticity is introduced in eq. 14.37.

3. The Frequency Equation for the Propagation Constant

We now can introduce the boundary conditions for the fluid in order to derive an expression for c, velocity of the waves, in terms of parameters of the system. At the fluid–wall interface we enforce

$$u\big|_{y=1} = \frac{\partial \xi}{\partial t} \qquad \therefore \quad j\omega D = j\frac{\omega R}{2c}\left(C_1 F_{10} + \frac{A}{\rho c}\right) \qquad \text{from eq. 14.11} \tag{14.30}$$

$$w\big|_{y=1} = \frac{\partial \zeta}{\partial t} \qquad \therefore \quad j\omega E = C_1 + \frac{A}{\rho c} \qquad \text{from eq. 14.12} \tag{14.31}$$

The four equations (14.28–14.31) can be rearranged to form the following system of linear equations:

$$\frac{1}{\rho c} A + C_1 - j\omega E = 0$$

$$\frac{j\omega R}{2\rho c^2} A + \left(\frac{j\omega R}{c}\frac{F_{10}}{2}\right) C_1 - j\omega D = 0$$

$$\frac{1}{\rho_w h} A + \left(\omega^2 - \frac{B}{\rho_w R^2}\right) D + \left(\frac{j\omega\mu B}{\rho_w R c}\right) E = 0$$

$$-\left(\frac{\rho}{\rho_w}\frac{j\omega R}{h}\frac{F_{10}}{2}\right) C_1 - \left(\frac{B}{\rho_w}\frac{j\omega\mu}{cR}\right) D + \omega^2\left(1 - \frac{\Omega^2}{\omega^2} - \frac{B}{\rho_w c^2}\right) E = 0$$

Since the right side is zero, the determinant of all coefficients (of A, C_1, D and E) must vanish. Womersley [14.1] discussed all roots of the determinant in detail. We restrict our attention to the case in which Ω^2 is larger than any other term in the determinant. This is the condition of limiting axial tethering where the vessel wall is moving radially only. If Ω^2 is very large, it suffices to compute the following subdeterminant, obtained by deleting the fourth row and fourth column of the coefficient determinant:

$$\begin{vmatrix} \dfrac{1}{\rho c} & 1 & 0 \\ \dfrac{j\omega R}{2\rho c^2} & \dfrac{j\omega R}{2c} F_{10} & -j\omega \\ \dfrac{1}{\rho_w h} & 0 & -\dfrac{B}{\rho_w R^2} \end{vmatrix} = \frac{B}{\rho_w R^2}\left[\frac{j\omega R}{2\rho c^2} - \frac{j\omega R}{2\rho c^2} F_{10}\right] - \frac{j\omega}{\rho_w h} = 0$$

This is the frequency equation determining c (previously called the wave velocity) for the condition of limiting constraint. Solving for c gives

$$c = \left(\frac{Bh}{2R\rho}\right)^{1/2}(1 - F_{10})^{1/2} = c_0(1 - F_{10})^{1/2} \tag{14.32}$$

where

$$F_{10}(\alpha) = \frac{2J_1(j^{3/2}\alpha)}{j^{3/2}\alpha J_0(j^{3/2}\alpha)}; \qquad c_0 = \text{Bramwell-Hill velocity}$$

The c is a complex number. With the abbreviations

$$c = \frac{c_0}{X - jY} \tag{14.33}$$

where

$$X(\alpha) \simeq \operatorname{Re} \frac{1}{(1 - F_{10})^{1/2}}$$

$$Y(\alpha) \simeq \operatorname{Im} \frac{1}{(1 - F_{10})^{1/2}}$$

the expression for one pressure wave, assumed in eq. 14.5, becomes

$$p(z, t) = p_1 \exp j\omega\left[t - \frac{z(X - jY)}{c_0}\right] = p_1 \exp\left(-\omega \frac{zY}{c_0}\right) \exp j\omega\left(t - \frac{zX}{c_0}\right) \tag{14.34}$$

Both $X(\alpha)$ and $Y(\alpha)$ are tabulated in [14.1] and [14.2]. Similar expression can be obtained for temporal and spatial distributions of the flow components. From eq. 14.34 it is clear that $Y(\alpha)$ introduces (exponential) decay of the amplitude of the wave (i.e., attenuation). This is due to viscosity of the blood. On the other hand, the phase velocity of the wave is $c_0 X^{-1}(\alpha)$. Deviation of the velocity from c_0 is also due to viscous effects. The complex exponent in the expression of the wave (eq. 14.34) is called the propagation constant defined by

$$\gamma \equiv j \frac{\omega}{c_0} (X - jY) \tag{14.35}$$

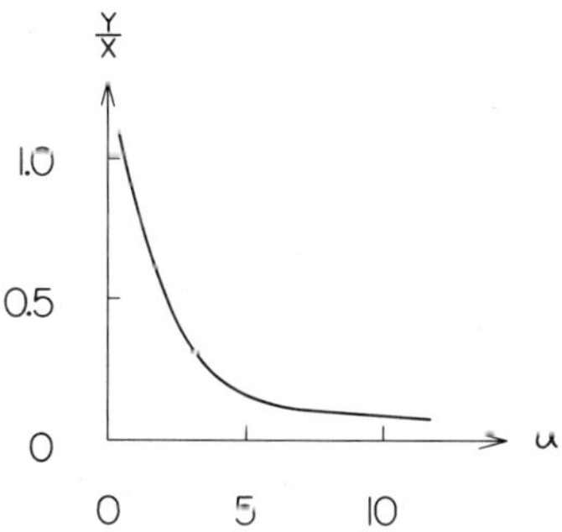

Fig. 14.3. The ratio Y/X of real and imaginary parts of the propagation constant (eq. 14.35) as a function of $\alpha(= R\sqrt{\omega/\nu})$, the velocity profile parameter. Essentially, Y determines attenuation, X wave velocity; see text. So long as Y/X is small, the wave velocity is not significantly affected by the viscosity of the wall material (eq. 14.38). Furthermore, Y/X determines the phase angle of the characteristic impedance (eq. 14.42)

Its real part is a logarithmic measure of the attenuation per distance (Neper/cm). The imaginary part of γ gives the phase shift of the signal per distance traveled (rad/cm). The ratio of real to imaginary part is plotted in Fig. 14.3.

4. Wall Viscosity

The stress–strain relations for a linear, Newtonian viscoelastic material includes a strain-rate dependent stress. In one dimension, for sinusoidal strain, this is

$$\sigma = E\varepsilon + E' \frac{d\varepsilon}{dt} = E\left(1 + j\omega \frac{E'}{E}\right)\varepsilon$$

Thus it suffices to introduce a complex Young's modulus. Poisson's ratio then also becomes complex:

$$\begin{aligned} E_d &= E + j\omega E \equiv E(1 + j\omega\Delta E) \\ \mu_d &= \mu + j\omega\mu' \equiv \mu(1 + j\omega\Delta\mu) \end{aligned} \qquad (14.36)$$

The effect of viscosity of the arterial wall material on the propagation constant is seen when these complex parameters of elasticity are introduced into eq. 14.35:

$$\gamma_d = j\omega(X - jY)\left[\frac{E(1 + j\omega\Delta E)}{[1 - \mu^2(1 + j\omega\Delta\mu)^2]}\frac{h}{2R\rho}\right]^{-1/2} \simeq j\frac{\omega}{c_0}(X - jY)(1 - j\omega W) \qquad (14.37)$$

where $W = \Delta E/2 + \Delta\mu/3 \qquad (\mu = \tfrac{1}{2})$.

Womersley [14.1] derived this result. Taylor [14.7] verified the applicability of eq. 14.37 in a hydraulic model, using rubber tubing, within the accuracy of the measurements.

The imaginary part of γ_d (the phase shift per distance) is practically not affected by viscosity within the wall:

$$\mathrm{Im}\gamma_d = \frac{\omega X}{c_0}\left(1 - \frac{Y}{X}\omega W\right) \qquad (14.38)$$

The Y/X is a small number in most arteries (see Fig. 14.3), so is W. This means that the velocity of the wave in a viscoelastic vessel is essentially the same as that in an ideally elastic arterial segment. Figure 14.4 gives c_0 as a function of $\alpha = R(\omega/\nu)^{1/2}$.

The real part of γ_d, the attenuation, is strongly influenced by W, the term engaging the viscous wall:

$$\mathrm{Re}\gamma_d = \frac{\omega}{c_0} Y\left(1 + \frac{X}{Y}\omega W\right) \qquad (14.39)$$

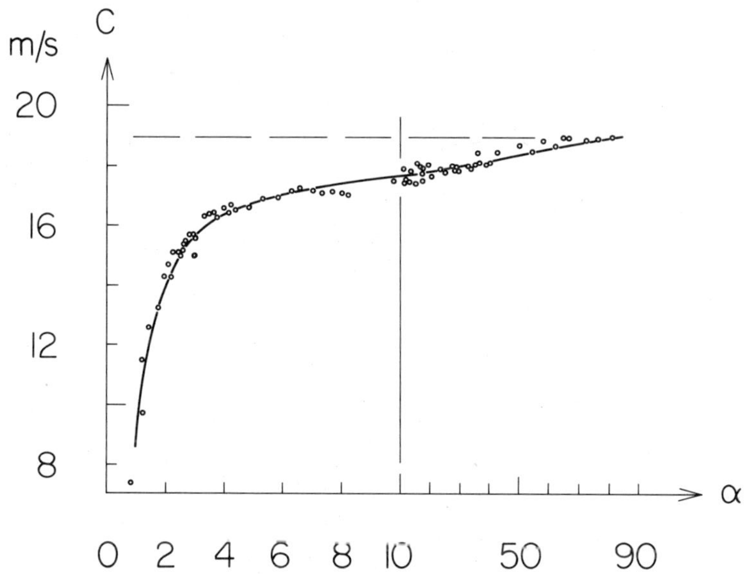

Fig. 14.4. The phase velocity of a wave in a fluid-filled viscoelastic tube as a function of α (note change of scale at $\alpha = 10$). Womersley's model is essentially verified by the data represented by dots, which Taylor [14.7] obtained in rubber tubes. The solid curve shows the theoretical prediction of $c = c_0 X^{-1}$, where $c_0 \simeq 19$ m/sec, in an ideally elastic vessel. Taylor's data, from rubber tubes, pertain of course to viscoelastic tubes. Wall viscosity, however, does not influence velocity significantly.

From Fig. 14.3 it is clear that the attenuation found in the viscoelastic artery can be several times that seen in the ideally elastic vessel ($\omega Y/c_0$), even if ωW is small.

It can be concluded from Bergel's [14.8] measurements on excised specimens that ωW is almost independent of ω and of the order of 0.1 in most of the larger vessels. McDonald and Gessner found slightly larger values (0.15–0.2) [14.9]. With these values the attenuation of a wave traveling the distance equal to one wavelength can be computed. Figure 14.5 shows this as a function of α.

5. Pressure Gradient-Flow Relation

The parameter α was used, in the foregoing discussion, as a convenient abbreviation for $R(\omega/\nu)^{1/2}$. Equation 14.12, describing the radial distribution of the axial flow velocity, reveals, however, that α is the only parameter determining the flow profiles. Equation 14.12 is

$$w_1(y) = \frac{A}{\rho c}\left[1 + \frac{C_1}{A/\rho c}\frac{J_0(j^{3/2}\alpha y)}{J_0(j^{3/2}\alpha)}\right]$$

and it can be rewritten as follows. The ratio $(C_1 \rho c)/A$ must be found from eqs. 14.28, 14.30, and 14.31 by first eliminating E:

$$\frac{c}{R\mu}D = C_1 + \left(1 + \frac{\rho R c^2}{\mu h B}\right)\frac{A}{\rho c}$$

$$j\omega D = \frac{j\omega R}{2c}F_{10}C_1 + \frac{j\omega R}{2c}\frac{A}{\rho c}$$

Hence, using the relation of eq. 14.32 we have

$$\frac{C_1}{A/\rho c} = \frac{1 - 1/2\mu + \rho R c^2/\mu h B}{1 - F_{10}/2\mu} = -\frac{1 - 1/2\mu + (1 - F_{10})/2\mu}{1 - F_{10}/2\mu} = -1$$

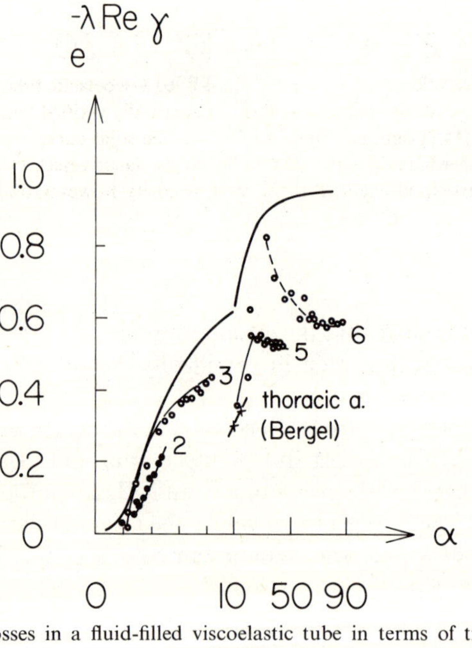

Fig. 14.5. Energy losses in a fluid-filled viscoelastic tube in terms of transmission (output amplitude/input amplitude) per wavelength travel. Since Re γ is attenuation/distance (in Nepers), exp $(-\lambda$ Re $\gamma)$ equals the transmission per wavelength. The heavy solid line gives the theoretical prediction for an ideally elastic tube (eq. 14.35). The data (dots, connected by lines) were obtained in viscoelastic vessels (eq. 14.39). The numbered sets of observations are from Taylor [14.7]. The data labeled "thoracic a. (Bergel)" are computed from Bergel's data [14.8] on viscoelastic properties of segments of the thoracic aorta. Comparable data are found in [14.9]. Attenuation is large due to wall viscosity. Very strong axial tethering of the vessel wall is assumed for the theoretical curve in both this figure and Fig. 14.4.

Therefore, in the case of limiting axial tether, the flow velocity profiles are

$$w_1(y) = \frac{A}{\rho c}\left[1 - \frac{J_0(j^{3/2}\alpha y)}{J_0(j^{3/2}\alpha)}\right] \tag{14.40}$$

(These turn out to be the same profiles as those found in a rigid tube.) Figure 14.6 presents two sets of flow profiles, at two α's, at various times during a full cycle. At least during part of the cycle, the patterns at the lower α exhibit

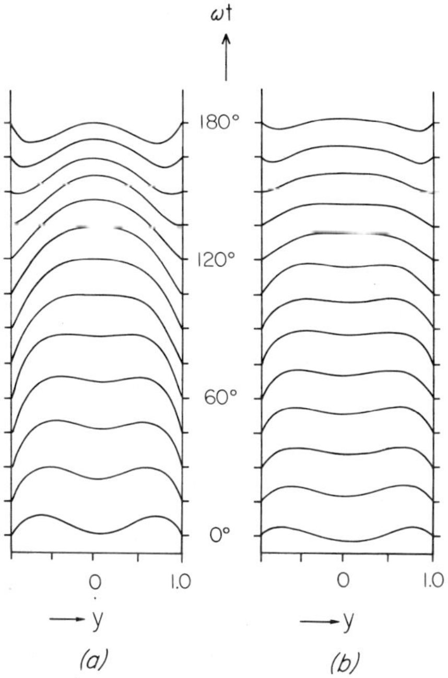

Fig. 14.6. Oscillatory flow profiles, computed for a pressure gradient varying sinusoidally in time, at different times (15°, 30°, etc.) during the cycle; y is the normalized distance from the axis of the vessel. (*a*) $\alpha = 4.7$, (*b*) $\alpha = 6.7$. In (small) arteries we would find profiles more complex than those here, since several sinusoidal components occur simultaneously. From McDonald [14.2].

great similarity to the parabolic Poiseuille flow profile ($\omega = 0$). It is interesting that this may be found at very low frequencies in large vessels and also at high frequencies in very small vessels. Here α is the "similarity number," as the Reynolds number is for steady flow. In oscillatory flow this means that the flow profile in a tube is geometrically similar to that in a tube of different size, provided their α's have the same value. In fact, α expresses the ratio of inertial forces and frictional forces in a particular flow configuration. Furthermore, α (at the heart rate) ranges from 1 to 10 in most arteries [14.2]. This, however, is the region where viscous and inertial effects are equally important (Figs. 14.4 and 14.5).

When the flow $w(y)$ is averaged over the vessel's lumen, the relation between pressure and net (volume) flow ($\dot{q}$) is obtained:

$$\dot{q} = 2\pi R^2 \int_0^1 w(y)y\, dy = \pi R^2(1 - F_{10})\frac{A}{\rho c} \tag{14.41}$$

For spatial and temporal patterns, both sides of eq. 14.41 should be multiplied by $\exp[j\omega(t - z/c)]$. This obviously holds only so long as there exists the one pressure wave introduced by eq. 14.5a. In the next section we show how eq. 14.41 is modified if a reflected wave is added to the forward propagating wave discussed here. The ratio of *forward pressure* wave and *forward flow* wave is called the characteristic impedance Z_0:

$$Z_0 \equiv \frac{A}{\dot{q}/\pi R^2} = \rho c \frac{1}{1 - F_{10}} = \rho c_0(1 - F_{10})^{-1/2} \tag{14.42a}$$

where ρc_0 is the characteristic impedance of an ideally elastic tube filled with an inviscid fluid.
Taking into account viscosity of the wall, we obtain

$$Z_{0d} = \rho c_0(1 - F_{10})^{-1/2}(1 - j\omega W)^{-1} \tag{14.42b}$$

Impedance to flow, being Z_{0d} while one wave exists, can be altered significantly by reflected waves (see Section 14.B).

Using eq. 14.40 and the relation

$$\left|\frac{\partial p}{\partial z}\right| = j\frac{\omega}{c}A$$

we can determine the relation between pressure gradient ($\partial p/\partial z$) and flow ($\dot{q}$):

$$\dot{q} = \pi R^2 \frac{1 - F_{10}}{j\omega\rho}\left(-\frac{\partial p}{\partial z}\right) \tag{14.43}$$

The ratio of ($\partial p/\partial z$) and $\dot{q}$ again has the meaning of an impedance (per length). It is the longitudinal impedance per length, Z_{long}, and is given by

$$Z_{\text{long}} \equiv -\frac{\partial p/\partial z}{\dot{q}/\pi R^2} = \frac{j\omega\rho}{1 - F_{10}} \equiv \frac{8\mu}{R^2}\frac{(j\alpha^2)/8}{1 - F_{10}} \tag{14.44}$$

The term $(1 - F_{10})$ in eq. 14.44 is due to viscosity. At very low values of α, we find $(1 - F_{10}) \simeq (j\alpha^2)/8$; that is, longitudinal impedance becomes a real number and approaches, as expected, the Poiseuille resistance ($8\mu/R^2$). In blood vessels, where α certainly is larger than unity, $1 - F_{10}$ is close to 1.0, hence Z_{long} may approach $j\omega\rho$. It is quite clear [14.1] that the relation between pressure gradient and flow is unaltered by the presence of reflected waves. Therefore, we can verify the prediction of eq. 14.44 from estimations of $\partial p/\partial z$ and direct observations of flow $\dot{q}$ (and R). This was achieved quite satisfactorily by McDonald and Womersley [14.2] in small blood vessels. In large vessels, where pressure gradient and flow are out of phase by almost 90°, errors in the phase of the estimate of $\partial p/\partial z$ can render the computation of the real part of Z_{long} (i.e., the "in-phase" component) very unreliable. This may explain, in part, deviations of experimentally obtained Z_{long}'s from their theoretical values at high α's, as reported by Fry et al. [14.10]. In particular (as we see in Section 14.B), it may be difficult to verify eq. 14.44 because $\partial p/\partial z$ is approximated by $\Delta p/\Delta z$, a ratio of differences.

The quantities $\gamma(\alpha)$ and $Z_0(\alpha)$ derived in eqs. 14.37 and 14.42 govern only wave propagation and characteristic impedance of the particular, long-wave mode of oscillation. Essentially, the continuous and spatially distributed exchange of kinetic energy (blood in motion) with potential energy (elastic forces due only to radial extension of the vessel) maintains propagation of the wave. In addition to this mode of oscillation, many others are possible, as shown by Klip [14.11].

Doubts about the applicability of Womersley's theory may be raised. It is known that the elasticity of arteries generally is pressure dependent. It is questionable also whether arteries, especially small ones, can be treated as thin-walled tubes. Assumptions that the Womersley flow profiles (Fig. 14.6) should be fully developed in short arterial segments, and that we can deal with laminar flow theory throughout, certainly ought to be scrutinized. Atabek [14.12] studied inlet length phenomena. Streeter et al. [14.13] and King [14.14] worked out techniques of computing pulsatile flow properties in nonlinear, nonuniform vessels, including frictional losses (larger than in laminar flow). In very small vessels anomalous viscosity shows its effects. This was elegantly treated by Taylor [14.15].

It appears, however, that inaccuracies inherent in observations, using current (1969) instrumentation, prevent us from separating unequivocally effects which are relatively small. Hence the simple, linear model that includes the important longitudinal constraint of the artery and the viscosities of both blood and vessel wall is satisfactory at present. This applies to sections of the vascular tree between larger branching sites (i.e., wherever arteries are reasonably uniform).

This chapter thus presents a good example of a system in which the subsystems (e.g., the details of flow profiles, stress–strain relations in tube wall) can be analytically described in such a way that the variables, significant on the higher level, are related meaningfully (e.g., in the pressure gradient–flow relation, in the ratio of pressure and flow of one wave). This process of system's description is carried further in the next section.

B. MODELING OF THE ARTERIAL SYSTEM

1. Reflections, Spatial Distribution of Pressure, Flow, and Impedance in Arterial Segments

The solutions assumed (eq. 14.5) for the differential equations governing flow in the viscoelastic, axially constrained vessel are valid in an infinitely long tube where only one wave exists, traveling in the direction of $+z$. The complete solutions include two waves:

$$p(t, z) = p_{11}e^{j\omega t - \gamma z} + p_{12}e^{j\omega t + \gamma z} \tag{14.45a}$$

$$w(t, z) = w_{11}e^{j\omega t - \gamma z} + w_{12}e^{j\omega t + \gamma z} \tag{14.45b}$$

where
p_{11} = amplitude of forward pressure wave (at $z = 0$)
p_{12} = amplitude of reflected pressure wave (at $z = 0$)
w_{11} = amplitude of forward flow wave (at $z = 0$)
w_{12} = amplitude of reflected flow wave (at $z = 0$)
γ = propagation constant, defined in eq. 14.37

All arterial segments are short. Nonuniformities exist at their peripheral termination; the branches given off by a segment are of different diameter (different α) and often of different distensibility. These facts call indeed for the complete solutions given in eq. 14.45. Relations between amplitudes of the four waves can be obtained by introducing the signals $p(t, z)$ and $w(t, z)$ into the pressure gradient–flow relation (eq. 14.43). We find

$$w_{11} = p_{11}\frac{\gamma}{Z_{\text{long}}} \qquad \text{and} \qquad w_{12} = -p_{12}\frac{\gamma}{Z_{\text{long}}}$$

But from eqs. 14.32, 14.33, and 14.35, it follows that $Z_{\text{long}}/\gamma \equiv Z_0$, the characteristic impedance. Hence the complete signals can be written in more concise form. For convenience we omit, in the rest of this chapter, the term $\exp(j\omega t)$ and concentrate on discussing the spatial distributions of amplitudes and phase. Furthermore, let us replace the distance z (measured from the origin of the arterial segment) by $-x \equiv -(L - z)$, where L is the length of the segment. Therefore x is the distance measured from the termination of the segment. Thus

$$p(x) = p'_{11}e^{\gamma x} + p'_{12}e^{-\gamma x} = p'_{11}e^{\gamma x}\left[1 + \frac{p'_{12}}{p'_{11}}e^{-2\gamma x}\right] \tag{14.46}$$

$$w(x) = \frac{1}{Z_0}[p'_{11}e^{\gamma x} - p'_{12}e^{-\gamma x}] = \frac{1}{Z_0}p'_{11}e^{\gamma x}\left[1 - \frac{p'_{12}}{p'_{11}}e^{-2\gamma x}\right] \tag{14.47}$$

where primed quantities = (original quantity) $e^{-\gamma L}$.

The term (p'_{12}/p'_{11}) represents by definition the ratio of retrograde wave and forward wave at the termination $x = 0$ $(z = L)$. This is called the terminal reflection coefficient R_0, given by

$$R_0 \equiv \frac{p'_{12}}{p'_{11}}$$

Usually R_0 is a complex number whose phase equals the phase difference between reflected wave and incident wave. The magnitude of R_0 is equal to the ratio of the amplitudes of the two waves. The ratio of retrograde wave and forward traveling wave at any point x is $R_0 \exp(-2\gamma x)$, as is seen from eqs. 14.46 or 14.47. This we call the generalized reflection coefficient, $R(x)$, determined by

$$R(x) = R_0 e^{-2\gamma x} \tag{14.48}$$

The reflection coefficients can now be introduced into expressions for the signals $p(z)$ and $w(z)$. Furthermore, impedance to flow is easily obtained in terms of $R(x)$ by dividing eq. 14.46 by eq. 14.47. Thus the following three formulas yield spatial distributions of pressure, flow, and impedance in concise form:

$$p(x) = p'_{11} e^{\gamma x}[1 + R(x)] \tag{14.49}$$

$$w(x) = \frac{1}{Z_0} p'_{11} e^{\gamma x}[1 - R(x)] \tag{14.50}$$

$$Z(x) = Z_0 \frac{1 + R(x)}{1 - R(x)} \tag{14.51}$$

where p'_{11} = forward traveling wave at input $\times\, e^{-\gamma L}$
L = length of segment
$R(x) = R_0 e^{-2\gamma x}$
R_0 = terminal reflection coefficient

The terminal reflection coefficient is determined by the load, $Z(0)$. Solving eq. 14.50 at $x = 0$ for $R_0 \equiv R(0)$ we find

$$R_0 = \frac{Z(0) - Z_0}{Z(0) + Z_0} \tag{14.52}$$

Clearly, when $Z(0) = Z_0$, R_0 vanishes. This is the (theoretical) case of a matched line on which only one wave travels. If an artery is occluded, $Z(0) = \infty$, and $R_0 = 1.0$ (i.e., the incident wave is reflected totally and in phase).

According to eq. 14.51 the *impedance* at any point in the segment $Z(x)$ can differ drastically from Z_0, the characteristic impedance. It is convenient to visualize the relation between $R(x)$ and $Z(x)/Z_0$ in a graph. Radio engineers, studying reflections on transmission lines (e.g., feeding antennas) or in wave guides, have developed this tool. It is called the Smith chart (Fig. 14.7).

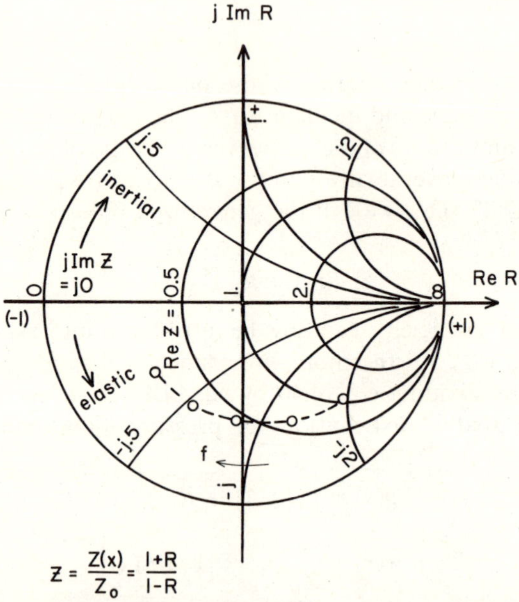

Fig. 14.7. Real and imaginary parts (circles) of the normalized impedance in the reflection coefficient plane. This figure, the so-called Smith chart, illustrates the solution of eq. 14.51. The circles whose centers are on the real axis (Re R) are the loci for impedances whose real parts have a fixed value. If an impedance has a positive imaginary part (representing an inertance) it lies in the upper half plane of the R-plane. Impedances with elastic components (negative imaginary part) are in the lower half of the figure. An acoustically closed end, causing complete in-phase reflection ($R = +1.0$), lies at the point labeled $+1$ and ∞ (i.e., $Z/Z_0 = \infty$). The five small circles in the lower half, connected by the dashed curve, give values of the input impedance and of R on a mismatched transmission line segment, as a function of frequency. The values here simulate closely the state of affairs in the femoral artery. See text for specifications of the model.

For any value of R, the magnitude of the normalized impedance $Z(x)/Z_0$ can be read from the labeled families of circles in the Smith chart. When investigating the impedance distribution in x at a fixed frequency, start by plotting the locus of $R(x)$. It is a spiral, beginning at R_0, approaching the origin of the R plane in a clockwise fashion, according to eq. 14.48. The orthogonal sets of circles of the Smith chart permit the evaluation of $(1 + R)(1 - R)^{-1}$ at each point along the spiral.

On the other hand, if the impedance $Z(x)/Z_0$ is studied at a fixed point x_1, as a function of frequency, the situation may be more complicated. The terminal reflection coefficient R_0 will be a function of ω if $Z(0)$, the load, is frequency dependent. In this case we must consider $Z(x_1, \omega)$ as a function of $R(x_1, \omega) = R_0(\omega) \exp[-2\gamma(\omega)x_1]$. If, however, the load is independent of frequency, $R(x_1, \omega) = R_0 \exp[-2\gamma(\omega)x_1]$; that is, the reflection coefficient at point x_1 moves, with increasing frequency, on almost the same spiral as the one mentioned before. The speed of rotation along the spiral is not exactly the same as when x changes, especially at low frequencies where $\gamma(\omega)$ increases more rapidly than ω.

The *example* in Fig. 14.7 represents a simple case in which the numerical values of R_0 and $\gamma(\omega)$ are chosen to approximate those in the femoral artery of a dog. The impedances determined by $[1 + R_0 \exp(-2\gamma x_1)] \times [1 - R_0 \exp(-2\gamma x_1)]^{-1}$ (i.e., by graphical readout in Fig. 14.7) are plotted against frequency in Fig. 14.8, together with those observed in the femoral artery (after O'Rourke [14.16]). The distance to an effective reflection site x_1 was assumed to be 14.5 cm; c_0 was 800 cm/sec, and attenuation per wavelength was 50%. Figure 14.8 reveals that the "transmission line" model performs satisfactorily over a limited range of frequencies. The significant discrepancies between the model and measurements in the femoral artery of a dog are (*a*) the model's impedance at low frequencies does not differ (because of reflected waves) as much from the characteristic impedance, and (*b*) the minimum impedance of the model, near 10 Hz, is much lower. The impedance pattern in the femoral artery may very well be due to a terminal R_0 that decreases with increasing frequency. When, in addition, spatial signal distributions are observed (at constant frequency) other facts emerge, notably the influence of spatial nonuniformity, that are important guides in refining the model.

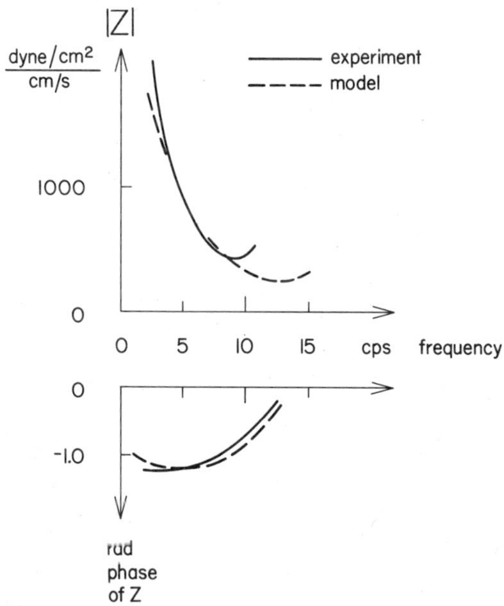

Fig. 14.8. Hydraulic input impedance of the femoral artery, as observed in a dog (solid line) and as simulated by a mismatched segment of a uniform transmission line (dashed curve; cf. dashed curve in Fig. 14.7). Impedance is given in terms of magnitude (top) and phase (bottom), both functions of frequency. The model fits the experimental data over a limited frequency range. Deviations are due to nonuniformities in geometry and properties; see text

From eq. 14.49 it is easy to deduce the effects of reflections on spatial pressure distribution $p(x)$ on a uniform segment. The forward traveling wave $p'_{11} \exp(\gamma x)$ can be considered to be a complex number. When its imaginary part is plotted against its real part, a spiral is obtained as in Fig. 14.9. Each point on the spiral (S_F) has a certain magnitude and phase, which determine amplitude and phase of the sinusoidal signal measured as a function of time at a particular site on the line. When x increases, we move counterclockwise along the expanding spiral. At larger distances from the termination the forward wave is larger in amplitude and leads the signals farther downstream. The modulus and phase of the retrograde wave—that is, $p'_{11}R_0 \exp(-\gamma x)$ (S_R in Fig. 14.9)—for which R_0 was assumed to be $(0.5 + j0.0)$, is given by a spiral approaching the origin of the signal plane clockwise. The sum of the two, labeled S_p in Fig. 14.9, represents the total signal. Obviously the amplitude waxes and wanes along the line. It always increases, on the average,

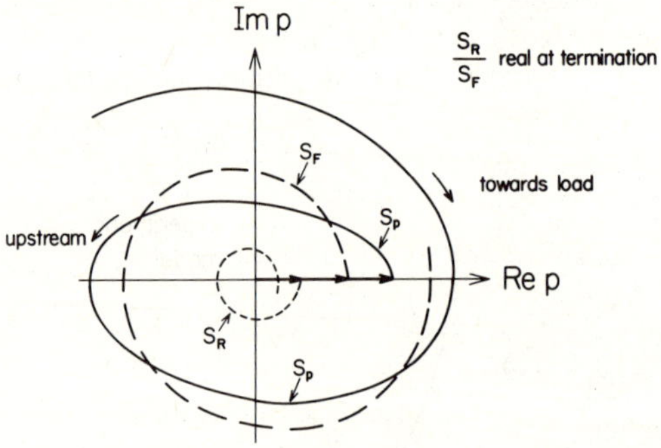

Fig. 14.9. Spatial pressure distribution, in terms of real and imaginary parts, on a mismatched, uniform transmission line at one frequency (attenuation/λ = 50%). The example assumes a 50% in-phase reflection at the termination; S_F is the incident wave, S_R the reflected one, and S_p is the sum, the observable signal. Note the maxima and minima in amplitude and the overall tendency of S_p to decrease average amplitude from source to load. (Arrows on the real axis are the signals at the termination.)

toward the input of the segment. The phase advance per Δx is large wherever the amplitude is small. Wherever the reflected wave causes the signal to be larger than the incident wave, the phase change per distance is small. In attempting to determine the velocity of the (total) wave (on the mismatched line) from the phase differences ($\Delta\phi$) between signals observed at two neighboring sites, we obviously find it varying along the line. Due to reflections, only an apparent velocity (c_{app}) can be calculated:

$$c_{\text{app}} = \frac{\omega \Delta x}{\Delta \phi}$$

The local apparent velocity ($\omega\,\partial x/\partial\phi$) can be derived from eq. 14.49. The phase of the total pressure $p(x)$ is Im ln $[p(x)]$. Hence the phase change is given by

$$\frac{\partial\phi}{\partial x} = \mathrm{Im}\,\frac{\partial p/\partial x}{p(x)} = \mathrm{Im}\gamma\,\frac{1 - R(x)}{1 + R(x)} \tag{14.53}$$

From this result and eq. 14.37, we find

$$c_{\mathrm{app}} = c_0\,\mathrm{Re}^{-1}\left[(X - jY)(1 - j\omega W)\frac{1 - R(x)}{1 + R(x)}\right] \tag{14.54a}$$

This can be approximated by noting that attenuation is relatively small (i.e, Re $\gamma <$ Im γ) and writing

$$c_{\mathrm{app}} \simeq c_0\,\mathrm{Re}^{-1}\frac{1 - R(x)}{1 + R(x)} \tag{14.54b}$$

When apparent velocity at one fixed location is computed as a function of frequency, we find patterns similar to those of the impedance. The expression on the right-hand side of eq. 14.54b can also be obtained from the Smith chart. The circle belonging to the family giving Re (Z/Z_0) = constant, and passing through $(-R)$, yields directly c_0/c_{app}.

Apparent velocities caused by reflections assumed in the example of Fig. 14.8 are very much larger than those measured in the femoral artery of a dog. The reasons are the same as those given for discrepancies between the impedance on the model below 3 Hz and above 9 Hz.

The theory presented earlier for describing distribution patterns of signals and impedances in terms of reflections is useful for basic understanding of phenomena occurring in the circulation. The transmission line theory can also be applied, for example, when the true phase velocity is to be determined from signals that encompass both orthograde and retrograde waves [14.17]. Furthermore, the local pressure gradient ($\partial p/\partial x$) can be estimated from a pressure–difference quotient ($\Delta p/\Delta x$), using the theory outlined. It can be demonstrated from eqs. 14.48 and 14.49 that

$$\frac{\partial p(x_2)}{\partial x} = \frac{p(x_2) - p(x_1)}{x_2 - x_1}\left[\frac{\gamma(x_2 - x_1)}{\sinh\gamma(x_2 - x_1)}\,\frac{1 - [p(x_1)/p(x_2)]\cosh\gamma x}{1 - p(x_1)/p(x_2)}\right] \tag{14.55}$$

On the other hand it is clear that the entire structure of the vascular tree, exhibiting large changes in cross-sectional area and considerable increases in elasticity toward the periphery, cannot be modeled by one uniform transmission line.

2. Systems of Arteries

The increase in (visco)elastic properties which accompanies the geometric change is observed when going from central to peripheral arteries. It is a significant feature that should be included in a model of the arterial system. Transformation of high peripheral resistance into low aortic impedance to flow, essential for efficient operation of the ventricle, depends strongly on these two types of spatial nonuniformities: dimensional and elastic. In the systemic circulation of the dog, the radius of the upper thoracic aorta is, on the average, 0.9 cm. The phase velocity in this vessel ranges from 5 to 6 m/sec. A more distal (and shorter) vessel, such as the femoral artery, whose radius is approximately 0.2 cm, is almost four times less distensible. The phase velocity in the femoral artery is of the order of 9 m/sec. Taylor [14.18] found the velocities in vessels of the size of the mesentery artery still higher (e.g., 15 m/sec). These vessels must exhibit elasticities at least ten times that of the central aorta. This means that separation of these small terminal arteries from the sites of the peripheral resistances at the arteriolar level is also very small in terms of wavelengths.

The pulmonary circulation, although physically smaller than the systemic system, has a very similar functional structure, since all pulmonary arteries are more distensible than the systemic ones. Phase velocity in the main pulmonary artery ranges from 1.5 to 3.0 m/sec [14.19].

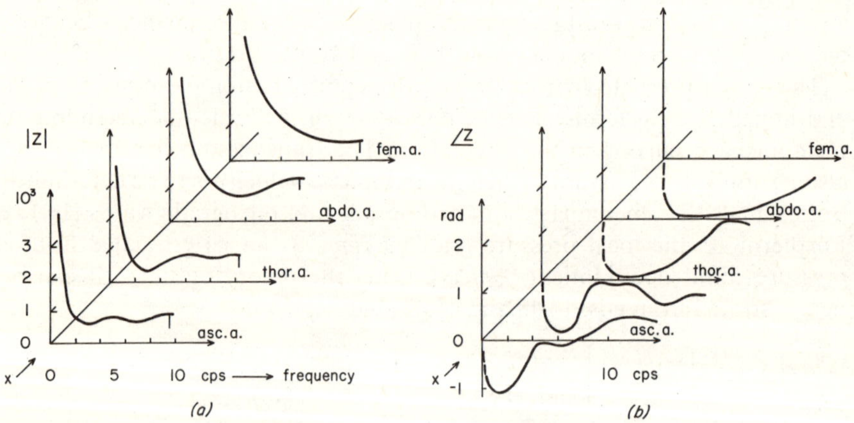

Fig. 14.10. Impedance, in terms of modulus (*a*) and phase (*b*) as a function of frequency and location along the systemic arterial circulation. The characteristic impedance, Z_0, as well as the mean level of $|Z|$ rises when going from heart to periphery. The reflection coefficients are larger peripherally than they are centrally. These data were obtained in one dog and reported by O'Rourke [14.16].

The frequency functions of the impedances, as measured by O'Rourke [14.16] at various levels in the systemic circulation of the dog, are seen in Fig. 14.10. Near the periphery the modulus of the impedance is large at low frequencies, and near 10 Hz it reaches a minimum that is approximately 50% of the characteristic impedance. The phase of the impedance is negative at most frequencies in the physiological range. Owing to the particular structure of the vascular tree, described briefly earlier, the aortic impedance is considerably lower over the entire frequency range (2–10 Hz). The modulus has its first minimum near 4 Hz and does not deviate much from the local characteristic impedance, whereas the phase approaches zero (and sometimes small positive values) also near 4 Hz. This means, as we see later, that the input impedance above 2 Hz of the whole system is to a large extent independent of the peripheral (D.C.) load. It is greatly influenced, however, by the asymmetry of the entire vessel tree. Arteries supplying viscera and legs are longer and centrally less elastic than those going to head and arms. This effect can easily be studied in a very simple model of the entire system in which two uniform lines of unequal lengths and distensibilities (and losses) are arranged asymmetrically, as in Fig. 14.11. The input impedance of this

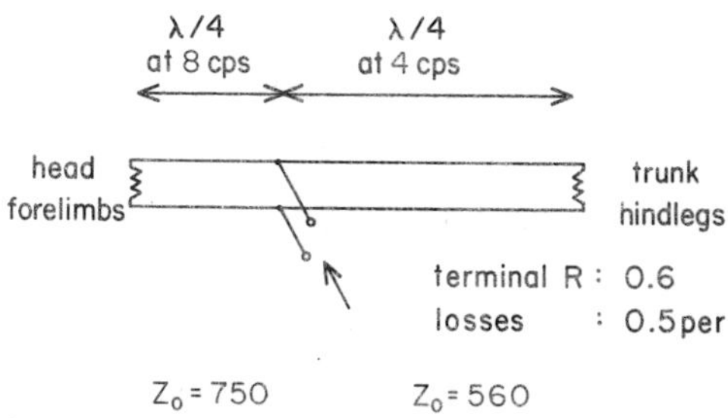

Fig. 14.11. A simple model simulating satisfactorily the input impedance of the entire systemic vascular tree. The asymmetric geometry causes significant cancellation, at the input, of reflections from the "peripheries." The first impedance minimum occurs at 4 Hz. At 8 Hz, the shorter segment exhibits its first impedance minimum. Hence the total input impedance at 8 Hz is low (even though the long arm of the asymmetric Tee assumes a high impedance level). This model demonstrates the interaction of different reflections at the input, and hence also the influence of the overall geometry on the aortic load of the left ventricle; however, it fails to explain spatial impedance distributions.

model resembles measured aortic impedances reasonably well. On the other hand the model neglects spatial nonuniformities known to exist and does not, therefore, predict the spatial impedance distribution.

Taylor [14.20 and 14.21] developed a model for whole vessel systems consisting of many segments, each uniform within itself. The segments are arranged in a "tree" similar to the one formed by the systemic arteries in

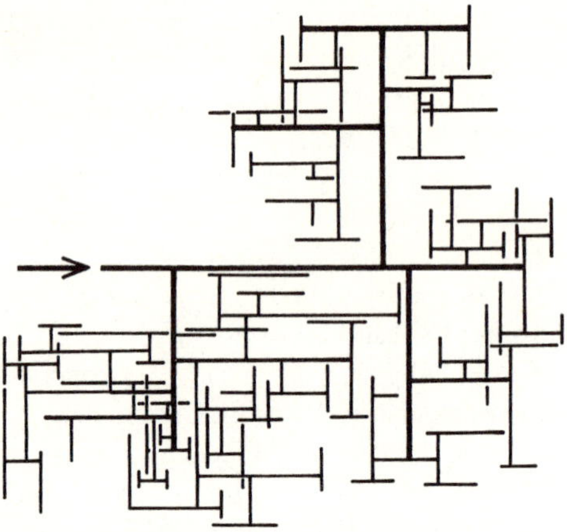

Fig. 14.12. Graph showing the geometry of one of the randomly branching vessel systems analyzed by Taylor. See text for laws governing the distribution of vessel length, diameter, and elasticity throughout the system. Asymmetry (cf. Fig. 14.11) is introduced by a smaller mean value of vessel length at the first level in the subsystem on the lower left, representing circulation to head and forelimbs. Although the model's performance was compared with data from animals, it also simulates well the basic phenomena existing in the human circulation. From Taylor [14.18].

animals, except that in the model each segment gives off only two branches while small side branches (e.g., intercostal arteries originating from the thoracic aorta) are neglected. See Fig. 14.12. Three essential characteristics of the system are introduced into the model:

1. At each level of dichotomization the lengths of the segments differ from one another. Taylor assumed that the mean length of all vessels at the nth level is the length of the first segment divided by $n + 1$. Individual lengths at any level are chosen to be distributed around their mean according to a second-order gamma probability distribution.

2. The more distal vessels are less distensible. Taylor introduced this spatial distribution of distensibilities by letting the Bramwell-Hill phase velocity (eq. 14.23) depend on n, the number of the level of branching, in the following manner:

$$\frac{c_{0n}}{c_{01}} = 3 - 2\left(\frac{2}{3}\right)^n \tag{14.56}$$

where c_{0n} = phase velocity in the nth segment
c_{01} = phase velocity in the first segment

Hence phase velocity in the most distal vessels approaches a velocity that is three times that in the first segment, as is observed in nature.

According to eq. 14.37, the propagation constant in the segments at the nth level is

$$\gamma = j \frac{\omega}{c_{0n}} (X - jY)(1 - j\omega W) \qquad (14.57a)$$

Taylor used the following similar expression to simulate the behavior of real arteries:

$$\gamma = j \frac{\omega}{c_{0n}} (X - jY)\left(\cos \frac{\phi}{2} - \sin \frac{\phi}{2}\right) \qquad (14.57b)$$

The phase angle ϕ, by which radius changes lag pressure changes (the phase lag in turn being due to wall viscosity) was made dependent on frequency as follows:

$$\phi = \phi_0(1 - e^{-2\omega}) \qquad (14.58)$$

Equation 14.58 takes into account that ϕ stays almost constant at high frequencies. In some calculations ϕ_0 was assumed to be zero to study the effect of wall viscosity.

3. Naturally, diameters of the segments become progressively smaller when n increases. Taylor expressed this by letting the sum of the cross-sectional areas ($2S_{n+1}$) at the ($n + 1$)th level be equal to d times the cross-sectional area of the vessel at the previous, nth level:

$$2S_{n+1} = dS_n$$

If $d = 1$, there is no increase in area. McDonald [14.2] reports d's of the order of 1.2. Taylor carried out calculations with different d's to investigate the effect of overall "opening up" of the vessel tree. Examples of the "architecture" of Taylor's model appear in Fig. 14.12.

The *input impedance* of this randomly branching tube system shows two regions which are influenced most strongly by elastic nonuniformities and reflections from many sites scattered in space. The effects of viscosities are small. The modulus of the input impedance is large at zero frequency (D.C.) and decreases rapidly at low frequencies. Its first minimum may amount to only 5 to 10% of the peripheral resistance. This may be compared with the behavior of one single mismatched transmission line segment at low frequency. At higher frequencies, however, input impedance stays close to the characteristic impedance (Z_{01}) of the first segment, as in Fig. 14.13. The small fluctuations about amplitude and phase of Z_{01} are due to reflections. They never become large, since reflections from different sites (bifurcations and terminations) tend to cancel each other because of their random separation from the input.

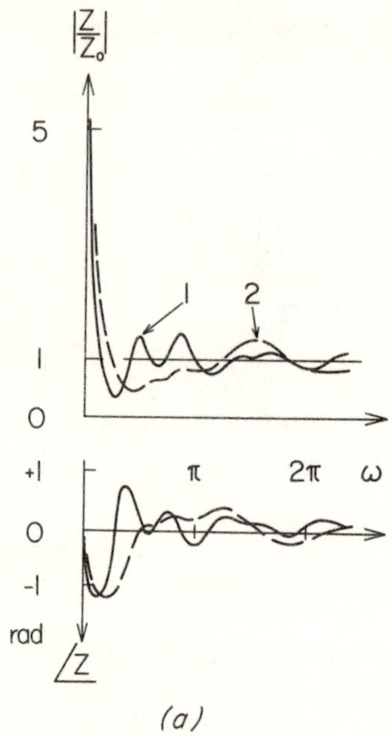

Fig. 14.13*a*. Input impedances of two randomly branching models of the type shown in Fig. 14.12 in terms of magnitude (normalized by the characteristic impedance of the first segment) (top) and phase (bottom). In these cases the fluid is assumed to be ideal; the vessel walls are assumed to exhibit a $\phi_0 = 10°$ (eq. 14.58). The terminal reflection coefficients are 0.6. Although the geometries of the two cases differ greatly, the impedance patterns are quite similar. From Taylor [14.18].

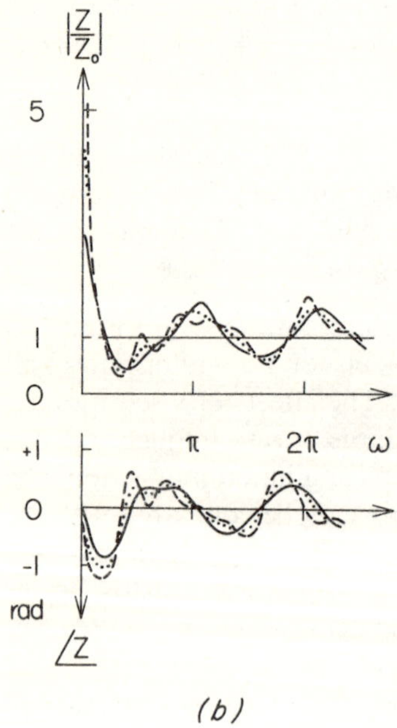

Fig. 14.13*b*. Input impedance of a randomly branching model, of the type appearing in Fig. 14.12, in terms of magnitude (normalized) (top) and phase (bottom) for different terminal reflection coefficients: solid curve, $R = 0$; dotted curve, $R = 0.2$; dashed curve, $R = 0.4$. The impedance is affected by the terminal loads only at the very low frequencies. For all higher frequencies, the loads are decoupled. From Taylor [14.18].

Figure 14.13*a* shows input impedance for two different vascular trees. The influences of the viscosities of blood and wall are also indicated. Figure 14.13*b* gives the dependence of input impedance on terminal resistors. A strong decoupling is noticed. Impedance to pulsatile flow is almost completely independent of the D.C. resistance, except that the frequency at which the first minimum occurs is slightly shifted. In other words, the low value of the characteristic impedance of the artery at the entrance essentially determines the input impedance of the entire system. This results in a low value for external work done by the ventricles. For this reason the pulsatile work amounts, in normal situations, to only an approximate 10% of the total.

The *spatial distribution of pressure* over this randomly branching system of "arterial" segments also exhibits the characteristic patterns observed in animals. Taylor [14.21] computed the amplitudes at all points along an arbitrary path through the system. These are strongly influenced by the viscous properties of blood and wall and the increase in cross-sectional area—factors less important in the input impedance. Amplitudes of pressure waves increase progressively when traveling toward the periphery. With a realistic value for the branching ratio ($d = 1.2$), the "peaking" in a lossless system can be as large as 500%. Realistic values of ϕ_0 (8°) reduce peaking. The largest value of a wave amplitude, at a point in the system effectively removed from the heart by one quarter of the wavelength, might then be 350% of the amplitude at the input.

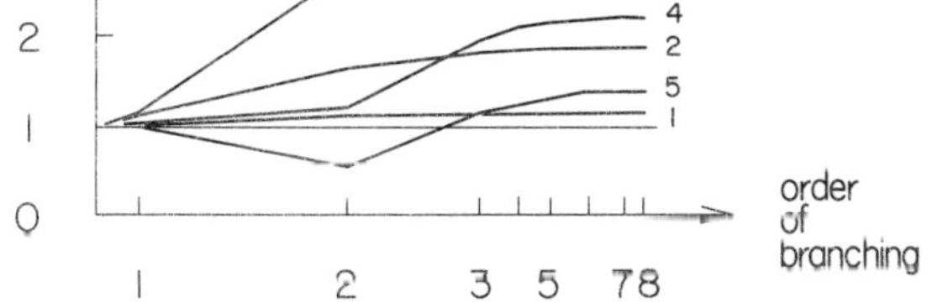

Fig. 14.14. Spatial distributions of pressure amplitudes along arbitrary paths in one of Taylor's assemblies. The parameters include fluid viscosity (α of the initial segment at fundamental frequency is 15), wall viscosity ($\phi_0 = 8°$), and an area ratio at all bifurcations of 1.2. All terminal reflections are 0.6. The first five harmonics are shown ($\omega_0 = 0.39$). Note the overall increase toward the periphery of the fourth and fifth harmonics. This occurs because of nonuniform distensibility and geometry. From Taylor [14.18].

The gradual increase for five harmonics along the path of travel is illustrated in Fig. 14.14. These results were obtained in a system for which $d = 1.2$, $\phi_0 = 8°$, $R_{term} = 0.6$, and (Womersley's) α in the first segment, at the fundamental frequency, is 15. The peaking that occurs in an elastically nonuniform line is quite different from that in one homogeneous segment. On the average, peripheral signals are larger than those near the heart. This would not occur in the uniform attenuating line, where reflections of higher harmonics do not reach the center, so that (on the average) central pressures would always be larger than peripheral ones.

In summary, Taylor's model [14.20, 14.21] explains in a general manner how widely distributed reflection sites and nonuniformities account for the phenomena observed in nature, that is, the low value of the input impedance, its independence of the peripheral loads, and peaking of the pressure pulse. If the random distribution of vessel lengths and the analytic form for elastic nonuniformity were replaced by values obtained from an animal, the general relationships still would hold. This, in part, was demonstrated by Wiener et al. [14.22], using a model of the pulmonary circulation comprising many transmission line sections.

C. REFERENCES

14.1. (a) Womersley, J. R. Oscillatory motion of a viscous liquid in a thin-walled elastic tube—I: The linear approximation of long waves. *Phil Mag* (Ser 7) **46**: 199–221, 1955. (b) Womersley, J. R. Oscillatory flow in arteries: the constrained elastic tube as a model of arterial flow and pulse transmission. *Phys Med Biol* **2**: 178–187, 1957.

14.2. McDonald, D. A. *Blood Flow in Arteries*. Williams and Wilkins Co., Baltimore, 1960.

14.3. Taylor, M. G. Use of random excitation and spectral analysis in the study of frequency-dependent parameters of the cardiovascular system. *Circ Res* **18**: 585–595, 1966.

14.4. Prandtl, L. and O. K. G. Tietjens. *Fundamentals of Hydro- and Aeromechanics*. Dover Publications, New York, 1957.

14.5. Anliker, M. and K. R. Raman. Korotkoff sounds at diastole—a phenomenon of dynamic instability of fluid-filled shells. *Int J Sol Str* **2**: 467–491, 1966.

14.6. Bramwell, J. C. and A. V. Hill. The velocity of the pulse wave in man. *Proc Roy Soc [Biol]* **93**: 298–306, 1922.

14.7. Taylor, M. G. An experimental determination of the propagation of fluid oscillations in a tube with visco-elastic wall. *Phys Med Biol* **4**: 63–82, 1959.

14.8. Bergel. D. H. The dynamic elastic properties of the arterial wall. *J Physiol (London)* **156**: 458–469, 1961.

14.9. McDonald, D. A. and U. Gessner. Wave attenuation in visco-elastic arteries, pp. 113–125 in *Hemorheology*. A. L. Copley (Ed.). Pergamon Press, New York, 1968.

14.10. Fry, D. L., D. M. Griggs, and J. C. Greenfield. In vivo studies of pulsatile blood flow: The relationship of the pressure gradient to the blood velocity, pp. 101–114 in *Pulsatile Blood Flow*. E. O. Attinger (Ed.). McGraw-Hill Book Co., New York, 1964.

14.11. Klip, W. *Velocity and Damping of "The Pulse Wave."* Martinus Nijhoff, The Hague, 1962.

14.12. Atabek, H. B. End effects, pp. 201–220 in *Pulsatile Blood Flow*. E. O. Attinger (Ed.). McGraw-Hill Book Co., New York, 1964.

14.13. Streeter, V. L., W. F. Keitzer, and D. F. Bohr. Pulsatile pressure and flow through distensible vessels. *Circ Res* **13**: 3–20, 1963.

14.14. King, A. I. A modified one-dimensional theory of flow in arteries. Winter Annual Meeting on Biomechanics and Human Factors, Publ. 66, WA/BHF-5. American Society of Mechanical Engineers, New York, 1966.

14.15. Taylor, M. G. The influence of the anomalous viscosity of blood upon its oscillatory flow. *Phys Med Biol* **3**: 273–290, 1959.

14.16. O'Rourke, M. F. Pressure and flow waves in systemic arteries and the anatomical design of the arterial system. *J Appl Physiol* **23**: 139–149, 1967

14.17. Gessner, U. and D. H. Bergel. Methods of determining the distensibility of blood vessels. *IEEE Trans Bio-Med Engin* **BME-13**: 2–10, 1966.

14.18. Taylor, M. G. Wave travel in arteries and the design of the cardiovascular system, pp. 343–372 in *Pulsatile Blood Flow*. E. O. Attinger (Ed.). McGraw-Hill Book Co., New York, 1964.

14.19. Milnor, W. R., C. R. Conti, K. B. Lewis, and M. F. O'Rourke. Pulmonary arterial pulse wave velocity and impedance in man. *Circ Res* **25**: 637–649, 1969.

14.20. Taylor, M. G. The input impedance of an assembly of randomly branching elastic tubes. *Biophys J* **6**: 29–51, 1966.

14.21. Taylor, M. G. Wave transmission through an assembly of randomly branching elastic tubes. *Biophys J* **6**: 697–716, 1966.

14.22. Wiener, F., E. Morkin, R. Skalak, and A. P. Fishman. Wave propagation in the pulmonary circulation. *Circ Res* **19**: 834–850, 1966.

CHAPTER FIFTEEN

The Dynamics of the Heart as a Muscle System

A. INTRODUCTION

The discussion of Chapter 10 on the performance of whole muscle sets the stage for developing a model for the ventricle. For myocardium, as for skeletal muscle, the system analyst must seek to define the variables and establish their interdependencies and those input–output relations which determine the performance of the organ. In Chapter 10 we showed how the structurally complex muscle, within whose cells complex chemicomechanical energy conversions occur, can be successfully treated, at the next higher level, as a force source whose output is dependent only on activity of the motoneurons, on muscle length and shortening velocity. This force source represents a compact functional "unit" that can be incorporated, at the level of entire muscle systems, into the various neural loops controlling motion. Description of the heart as a pump system is less easy, but many of the concepts introduced in Chapter 10 are useful.

In contrast to whole skeletal muscle, no recruitment of motor units can grade the forces produced by the myocardium. Once excitation begins, it spreads rapidly and triggers activation of the entire muscle mass. Yet the heart fulfills its task of supplying the organs of the body with blood according to their demands, which vary greatly. Cardiac output at rest may be 4 liters/min; in strenuous exercise, it is 15 liters/min or more. Two separate mechanisms cause increases in cardiac output. Stroke volume is made larger by more vigorous and faster contractions of the heart muscle. The rate at which contractions recur (i.e., heart rate) also goes up, and under stress it can be close to triple the resting value. These changes are only partly brought about by neural control systems.

Local regulatory mechanisms, inherent in the heart muscle, adjust pumping performance to demands astonishingly well. Stresses occurring in normal behavior are well coped with, as witness the heart-transplant patients without neurogenic control [15.1]. In a fixed inotropic state (i.e., when contractile properties of the muscle are unchanged by either neural and humoral action or frequency shifts) stroke volume and certainly stroke work increase if enddiastolic volume is enlarged. This is especially so if diastolic volume is enlarged because of increased *venous pressure* and prolonged *diastole*. This, the Frank-Starling mechanism (Guyton [15.2]), is essentially determined by the active heart muscle properties. It enables the ventricle to pump all blood offered to it by way of the venous system. It is also called heterometric autoregulation, because enddiastolic volumes vary when adjustments occur. The Frank-Starling mechanism is also influenced by factors other than active muscle characteristics, and not all these have been identified clearly. As an example, we mention the dynamics of diastolic filling, which is determined in part by the rheology of the muscle mass during diastole, when it is in the process of relaxing. The model presented in this chapter is in many aspects a crude approximation to the real system, partly due to lack of data pertinent to these problems. One of the obstacles in obtaining data lies in the experimental difficulties in observing ventricular volume as a function of time.

As indicated, the pressure and the rate of volume change during ventricular systole are functions of volume, because the force and shortening velocity of the fibers (e.g., as seen in isolated papillary muscle) depend on fiber length. In particular, maximum available pressure, as observed during isovolumetric beats (aorta occluded), and maximum attainable rate of volume change, seen in principle in an unloaded beat (ventricle emptied), are two important variables determining contractile properties of the myocardium. The definitions of these variables pose problems, however, since contractions of the myocardium are to be compared with twitches in striated muscle. Thus an analogy to isometric muscle force, which is observed in tetanized skeletal muscle, cannot possibly be obtained. We have to be satisfied by attempts to characterize maximum available pressure by peak isovolumetric pressure occurring during midsystole. The same consideration applies for maximum rate of volume change. In this approximation we assume that activation follows a particular, fixed time course actually true only in a fixed inotropic state.

In addition to intrinsic regulation of cardiac performance, *neurogenic* mechanisms exert control over contractile properties, the time course of the active state, and the heart rate. Again the efferent neural information depends on local conditions (endocardial baroreceptors and epicardial chemoreceptors are known to exist, although not all their functions are understood), as well as on the state of the peripheral circulation. The control loops affect the myocardium in two ways. The sympathetic system exerts inotropic and chronotropic influences, the parasympathetic system mostly chronotropic ones. As we know from work in papillary muscle (Sonnenblick [15.3]). positive inotropic interventions cause shifts in both peak isometric pressures and peak velocity (norepinephrine) or in peak force only. Concomitantly, the time course of the active state may change considerably. With norepinephrine, onset is faster, duration shorter. These changes in active state, undoubtedly important for cardiac function, have not been incorporated into any mathematical model. Chronotropic changes are linked, however, with inotropic ones. Koch-Weser [15.4] showed how peak ventricular pressures increase with increasing heart rate; we infer from work in papillary muscle that this result is due to a positive shift in peak velocities [15.3].

Further discussion of regulatory mechanisms is postponed until Chapter 16. We find it important first to develop a model of the isolated heart that (*a*) yields insight into the main relationships between geometry, muscle property, and pump performance, and (*b*) is suitable for combination with mathematical descriptions of hemodynamic and control phenomena. The model discussed in the following partly fulfills these requirements. It remains incomplete for reasons already given and because of other problems mentioned later. Formulation of the model, however, is of considerable assistance in defining some of these "unresolved problems." It also forces us to characterize the heart's action in terms of quantities that are important for its pump performance. In particular, we see to what extent heterometric autoregulation is explained by the properties of active myocardium.

B. THE HEAVY-WALLED VENTRICLE; STATIC RELATION BETWEEN FIBER FORCE, PRESSURE, AND VOLUME

At any one time, the cardiac muscle tissue assumes a certain distensibility. This determines, in part, the pressure–volume relationship of the chamber at that time. The pressure–volume relation depends also on the geometry of the ventricle. The formalism developed in the following allows for separate study of these two factors. It is applicable only at a fixed time (e.g., at the end of diastole) and strictly relates only to static conditions and slow changes in pressure.

In a model that suits the purpose of this study, structural details of heart muscle and the geometry of the left ventricular chamber must be simplified. Following a suggestion by Rushmer [15.5], we assume that circumferential fibers, lying in horizontal planes, contribute most of the force developed during ejection. Thus a hollow *cylinder*, with circumferentially arranged fibers, may be the simplest model. Fry et al. [15.6], who chose a spherical model, also assumed the fibers to be circumferential. Other workers, notably

Levine and Britman [15.7] and Beneken [15.8], also used spheres. Beneken, however, placed all fibers equatorially. From experimental work on determination of ventricular volume using cineangiography [15.9, 15.10] it has become clear that an ellipsoid of rotation approximates the inside shape of the chamber best. For this reason Mirsky [15.11] calculated rigorously the stress–strain relationships in a heavy-walled ellipsoid.

It has not been resolved which of these approximations fits the natural situation best in a case where the relation between functionally important variables—pressure, volume, mean wall stress—is determined (as opposed, e.g., to studies on stress distribution within the wall). Furthermore, we do not know how errors due to the various assumptions (simple geometry, isotropic wall material, uniform fiber arrangement) compare in magnitude. Although the natural fiber arrangement has been studied extensively (Streeter et al. [15.12]), it still is impossible to relate rigorously fiber forces and fiber length to wall stress and ventricular pressure.

We therefore adhere to the much simplified model in which basic mechanical relations can be studied, a model that can form the basis of experimental studies of some of the questions we have raised. We assume that the model fibers in a heavy-walled cylinder are placed circumferentially, and that the ends of the cylinder are closed and free to move. We seek the pressure–volume relationship and the stress distribution in the wall by assuming, furthermore, isotropic wall material. Stress distribution can be obtained by considering, as a starting point, the simplified equilibrium condition for an element of the wall at radius r, as in Fig. 15.1:

$$\sigma_r r - \left(\sigma_r + \frac{\partial \sigma_r}{\partial r}\,dr\right)(r + dr) + \sigma_t\,dr \simeq \left(\sigma_t - \sigma_r - r\frac{\partial \sigma_r}{\partial r}\right)dr = 0 \qquad (15.1)$$

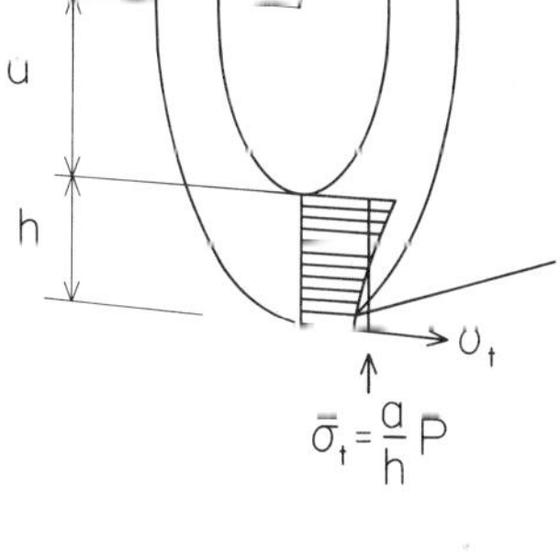

Fig. 15.1. Geometrical relationships and definitions of symbols used in the analysis of the cylindrical model of the left ventricle. The radius of the cylinder is a, its wall thickness h; r is the distance from the axis to a general element of the wall, whose thickness is dr. The stresses acting on the element are σ_r, σ_t, and σ_z (in the radial, tangential, and axial directions, respectively). Stress σ_z is not shown. At the bottom, the radial distribution of tangential stress is plotted as a function of r, as computed for a heavy-walled cylinder (isotropic material). The *average* value σ_t is proportional to p, the transmural pressure.

where σ_t = tangential stress
σ_r = radial stress

At a fixed instant in time during the heart cycle, we can assign certain elastic properties to the myocardium. From Hooke's equations (see eq. 14.14) ε_z, the axial strain, can be eliminated so that we arrive at two equations for σ_r and σ_t in terms of σ_z, the axial stress, and ε_r, ε_t, the radial and tangential strains, respectively. When these are introduced into eq. 15.1, a differential equation for ξ, the radial displacement of the element, is obtained using the following definitions:

$$\varepsilon_r = \frac{\partial \xi}{\partial r} \qquad \varepsilon_t = \frac{\xi}{r}$$

This differential equation can be solved with the boundary conditions that the radial stress σ_r at the inside of the wall is equal to the pressure within the ventricular chamber and that σ_r at the outside is zero. When $\xi(r)$ is known, the radial and the tangential strains are found:

$$\varepsilon_r = \frac{p}{E}\frac{a^2}{A^2 - a^2}\left[1 - \mu - (1 + \mu)\left(\frac{A}{r}\right)^2\right] - \frac{\mu}{E}\sigma_z \tag{15.2}$$

$$\varepsilon_t = \frac{p}{E}\frac{a^2}{A^2 - a^2}\left[1 - \mu + (1 + \mu)\left(\frac{A}{r}\right)^2\right] - \frac{\mu}{E}\sigma_z \tag{15.3}$$

where p = pressure within cylinder
E = modulus of elasticity; property assigned to myocardium at one particular instant in time
a = inside radius
A = outside radius
μ = Poisson's ratio

Since the ends of the cylinder are closed [i.e., since axial stress σ_z is balanced by $pa^2(A^2 - a^2)^{-1}$], we obtain

$$\varepsilon_r = \frac{p}{E}\frac{a^2}{A^2 - a^2}\left[1 - 2\mu - (1 + \mu)\left(\frac{A}{r}\right)^2\right] \tag{15.4}$$

$$\varepsilon_t = \frac{p}{E}\frac{a^2}{A^2 - a^2}\left[1 - 2\mu + (1 + \mu)\left(\frac{A}{r}\right)^2\right] \tag{15.5}$$

and, when introducing these two equations into Hooke's relations, we have

$$\sigma_t = p\frac{a^2}{A^2 - a^2}\left[1 + \left(\frac{A}{r}\right)^2\right] \tag{15.6}$$

Equation 15.6 gives the distribution of tangential stress within the wall. A typical case that might occur in the ventricular wall is plotted in Fig. 15.1. Notice that the *average* stress, $\bar{\sigma}_t$, equals the one assumed in the thin-walled cylinder (Laplace's relation):

$$\bar{\sigma}_t = \frac{1}{A - a} \int_a^A \sigma_t(r)\, dr = \frac{pa}{h} \tag{15.7}$$

where h is the wall thickness.

The radius at which the stress assumes the average value $\bar{\sigma}_t$ is very close to the mean radius $(A - a)/2$ if $(A - a) \leqslant 0.4a$. From eq. 15.7 it is clear that the ratio of stress and pressure $(\bar{\sigma}_t/p)$ must vary greatly during volume changes. When a increases, as during diastole, h decreases. On the other hand, the ratio of stress and net axial force, $\bar{\sigma}_t/(p\pi a^2)$, does not vary as much when intraventricular volume changes. Assuming the wall thickness to be 35% of inside radius at the end of diastole, the intraventricular volume can decrease by 40% during ejection, whereas $\bar{\sigma}_t/(p\pi a^2)$ changes only by approximately 8%. Hefner et al. [15.13] found that experimental wall stress, averaged by measuring the force across a cut in the myocardium, can be linearly related to the product $p\pi a^2$. This aspect of behavior of the cylinder model therefore approximates that of the left ventricle satisfactorily.

Equations 15.4 and 15.5 are useful because they enable us to compute the (static) volume distensibility $\Delta V/\Delta p$ of the cylinder

$$\frac{\Delta V}{\Delta p} = V[2\varepsilon_t + \varepsilon_z] \tag{15.8}$$

Using again the assumptions that the wall material is isotropic and incompressible, we get the steady-state pressure–volume relation

$$\frac{\Delta V}{\Delta p} = V \frac{3}{E} \frac{A^2}{A^2 - a^2} \tag{15.9}$$

since $\mu \simeq 0.5$ due to incompressibility and hence $\varepsilon_z = 0$.

Equation 15.9 therefore can be used, in principle, to determine E experimentally from observed ΔV's and Δp's. For this, however, estimates of intraventricular volume and wall thickness (or A and a) are needed. Generally, E would be expected to be a function of p. Thus the $\Delta V/\Delta p$ distensibility should be called the incremental volume distensibility.

In establishing the performance of our model, the relation between pressure, wall stress, and volume within the cylinder is needed. Instead of dealing with radial distributions of stress (eq. 15.6), fiber lengths, and so on, we follow the suggestions of Fry [15.6] and Bencken [15.8] and use the average values (eq. 15.7), remembering that both a and h are variables. From eq. 15.7 we get

$$p = \bar{\sigma}_t \left(\frac{A}{a} - 1 \right) \tag{15.10}$$

But if the wall is incompressible,

$$A_0^{\,2} - a_0^{\,2} = A^2 - a^2 \tag{15.11a}$$

where a_0 and A_0 are the enddiastolic inside radius and outside radius, respectively.

Hence
$$\left(\frac{A}{a} \right)^2 = 1 + \left(\frac{a_0}{a} \right)^2 \left[\left(\frac{A_0}{a_0} \right)^2 - 1 \right] \tag{15.11b}$$

Two facts should be noted with respect to eqs. 15.11. The term $(a_0/a)^2$ is equal to the ratio of enddiastolic volume V_0 to the instantaneous volume V. Second, A_0/a_0 is a constant given by the particular geometry of the heart muscle. Beneken [15.8] found that experimental values from several investigators average to

$$\frac{A_0}{a_0} \simeq 1.35 \tag{15.12}$$

Hence we can write the following pressure–tension–volume relation:

$$p = \bar{\sigma}_t F_1 \left(\frac{V}{V_0} \right) \tag{15.13}$$

where V/V_0 is the ratio of instantaneous and enddiastolic volumes and

$$F_1 \left(\frac{V}{V_0} \right) = \left\{ 1 + \frac{V_0}{V} \left[\left(\frac{A_0}{a_0} \right)^2 - 1 \right] \right\}^{1/2} - 1 \tag{15.14}$$

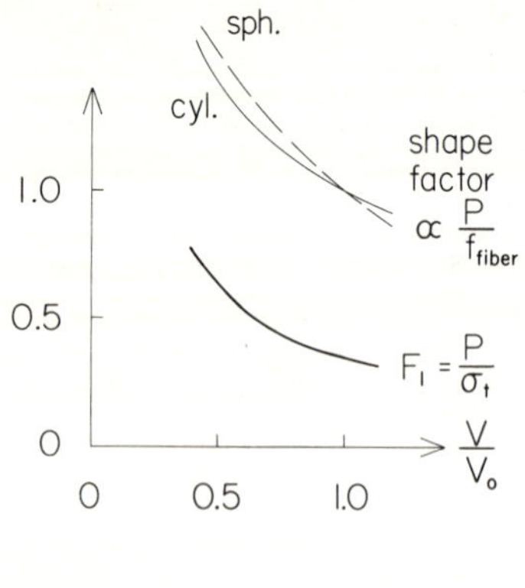

Fig. 15.2. The relation between intraventricular pressure p and tangential stress in the myocardial wall (F_1, heavy curve) as a function of ventricular volume V, normalized by the enddiastolic volume V_0. This relation is needed when fiber force, and therefore stress due to fiber force, is used to compute pressure p. Pressure is proportional to fiber force, the coefficient of proportionality dependent on the geometry of the ventricle at diastole and on instantaneous volume (eq. 15.17). The fine curve (*cyl*) gives the "shape factor" (i.e., the dependence of the relation between p and the fiber force on volume). The dashed curve (*sph*), from Beneken [15.8], gives the same factor for a spherical ventricle volume. Note the similarity with curve *cyl*.

The volume function $F_1(V/V_0)$ is plotted in Fig. 15.2.

We next develop relationships in terms of fiber force instead of wall stress. Assuming that the model fiber supporting the mean stress does not change its volume and that it is at the midpoint within the wall, we let

$$\bar{S}(A + a) = \bar{S}_u(A_u + a_u) \tag{15.15}$$

where $\bar{S}$ is the cross-sectional area of model fiber at $r = (A + a)/2$, and subscript u pertains to the unstressed state.

The wall stress thus becomes, in terms of the fiber force f

$$\bar{\sigma}_t = \frac{f}{\bar{S}_u}\frac{\bar{r}_0}{\bar{r}_u}\left(\frac{A_0}{a_0} + 1\right)^{-1}\frac{a}{a_0}\left(\frac{A}{a} + 1\right) \tag{15.16}$$

where $\bar{r}_0$ and $\bar{r}_u$ are the enddiastolic and unstressed mean radii, respectively, and subscript 0 pertains to the enddiastolic state.

We now can establish the relation between fiber force and pressure by introducing eq. 15.16 into eq. 15.10 and using eq. 15.11a:

$$p = f\frac{\bar{r}_0}{\bar{S}_u\bar{r}_u}\left[\frac{A_0}{a_0} - 1\right]\left(\frac{V_0}{V}\right)^{1/2} \tag{15.17}$$

In this equation we can consider the term $(V_0/V)^{1/2}$ as a "shape factor" influencing the p–f relationship in the same manner as F_1 affects the p–$\bar{\sigma}_t$ relation. It is plotted in Fig. 15.2, together with the "shape factor" obtained by Beneken [15.8] for a heavy-walled sphere. The factor in brackets in eq. 15.17 is a constant. Beneken found that $(\bar{r}_0/\bar{r}_u)$ can be approximated with a constant of 1.25. A direct evaluation of the entire term in the brackets is obviously needed to deal with eq. 15.17.

Equation 15.17 can be interpreted in a different manner. When it is written in the equivalent form

$$p = f\frac{\bar{r}_0}{\bar{S}_u\bar{r}_u}\frac{h_0}{a} \tag{15.18}$$

where h_0 is the enddiastolic wall thickness, we see the influence of the heavy wall more clearly. Pressure p and fiber force f are related through the enddiastolic wall thickness and the instantaneous inner radius. Had we used the thin-wall approximation, we would have erroneously related p and fiber stress with the instantaneous values of both h and a.

In preparation for the next section we note that the volume V within the cylinder is not equal to the intraventricular blood volume. The papillary muscles and the trabeculae carneae (the "wrinkles" on the inner surface) take up a considerable volume. In man these amount to approximately

40 cc. Assuming an average enddiastolic blood volume of 160 cc, we can put

$$\frac{V}{V_0} = \frac{q + 40}{200} \tag{15.19}$$

where q is the intraventricular volume (cc).

Thus eqs. 15.19 and 15.13 or 15.17 define the static relation between the intraventricular pressure, volume, and wall forces.

C. PUMP PERFORMANCE OF THE ISOLATED VENTRICLE

The assumptions introduced here are very simple, essentially reflecting the state of the art of the early 1960s, yet the model allows us to judge to what extent we can explain interdependencies between the important observables on the basis of simple, fundamental knowledge of muscle properties. In particular, three basic sets of premises are presented in this study of the left ventricle. (See Beneken [15.8] for models including the right ventricle and the atria.) First, the pumping action must be defined by functions relating pressure, volume, and rate of volume change. Second, the entire myocardial mass can be represented by a three-element system comprising an actively contracting part, a parallel-elastic part, and a series-elastic part. Third, the active pressure source produces (*a*) increasing (peak isovolumetric) pressures when volume increases, and (*b*) decreasing pressures when rate of volume change increases; both of these relations can be approximated fairly well quantitatively. Further features of the model are elaborated below.

The characteristics of the three-element ventricle in the pressure–volume–flow (flow = rate of change of volume) space are given best by adopting the two-chamber model put forward originally by Robinson [15.14] and further tested by Grodins et al. [15.15]. In this (Fig. 15.3) we consider one large volume spanned by that part of the muscle which shortens actively; a second and much smaller volume is spanned by the effective series-elastic component of the muscle mass. In this model we neglect the particular details that determine how the individual fibers contribute, as a function of their position, orientation, and time, to the behavior of the whole ventricle. We therefore must also introduce an *effective* time course of active state. Separation into the two volumes is conceptually useful; it is needed, from the system analyst's viewpoint, if the model is to be studied as part of the cardiovascular system. The model would fail if such effects as dependence of contractile force on asynchrony of fiber excitation were to be investigated.

Both Beneken [15.8] and Robinson [15.14] attempted to bridge individual fiber action with the behavior of the ventricle. These "fibers," however, can only be "model fibers" or functional building blocks, if the procedure

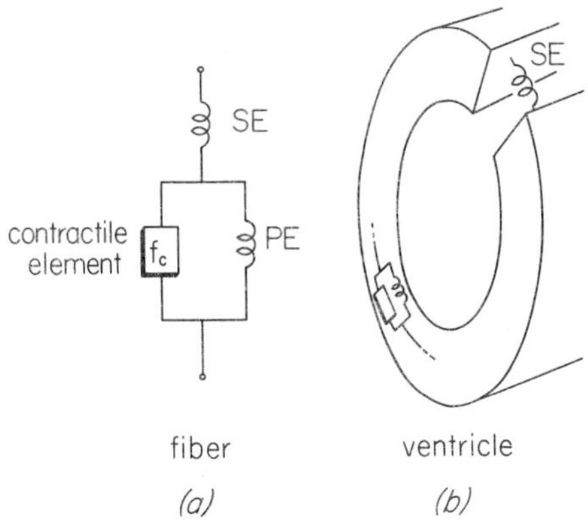

Fig. 15.3. Evolution of the three element model of the (cylindrical) myocardium. As we indicated in Chapter 10, most of the basic features of muscle performance are modeled by (*a*) the fiber system. Here f_c includes a force source whose output depends on length of *PE*, the parallel-elastic element, and its rate of change (shortening velocity). Total muscle length equals the length of *PE* and the (shorter) length of *SE*, the series-elastic element. The ventricle (*b*) shows the conceptual separation of ventricular volume into two compartments. All force sources and *PE*'s of all model fibers are lumped into one, as are all their series-elastic elements. This model describes only overall performance of the muscle mass regardless of the contributions of individual fibers. Generally, all springs shown exhibit nonlinear properties. The force sources (f_c) are active only during systole, and they are assumed to be freely distensible in passive muscle. Compare also with Fig. 15.7. After Robinson [15.14].

of these authors is adopted. This, again, is partly due to asynchrony of contraction of the real fibers and, more important, due to the highly nonuniform fiber arrangement [15.12].

1. Properties of the Ventricle

The characteristics of the *series-elastic element*, although studied by a few authors in the whole, intact ventricle, are here inferred from observations in excised papillary muscle. The kinds of errors so introduced are not known. Yeatman, Parmley, and Sonnenblick [15.16] found the stresses depending exponentially on length changes, that is,

$$\sigma_{SE} = \frac{A}{B}\left(e^{B\,L_{SE}/L_u} - 1\right) \tag{15.20}$$

where σ_{SE} = stress in series-elastic element of papillary muscle
L_{SE} = length of series-elastic element
L_u = resting length of muscle
A/B = 40 to 60 g*/cm^2
B = 32

Following Robinson [15.14] we approximate the exponential stress curve by a straight line. At 100 g*/cm^2, L_{SE} is 3.5% of L_0. But 100 g*/cm^2 is approximately 24% of peak isometric stress. When converting these figures into pressure and volume, we can assume (as is explained later) that peak active isometric pressure is 320 mmHg. Volume change in the series-elastic component at 24% of this (i.e., 77 mmHg) therefore is 7.1% of the total, normal average enddiastolic volume. If the latter is assumed to be 20 cc in a 10-kg dog, the compliance of the series-elastic compartment becomes

$$C_{SE} = \frac{20 \times 0.071}{77} = 0.0185 \frac{\text{cc}}{\text{mmHg}} \tag{15.21}$$

Robinson [15.14] computed $C_{SE} = 0.0256$ cc/mmHg from other values published by Sonnenblick and his co-workers. He also showed that the influence of C_{SE} on cardiac output and other pump characteristics (see below) is small. Therefore, linearization of the exponential characteristics does not cause serious errors, although it must have considerable influence on some details in the waveforms in the system, notably on dp/dt in the ventricle. However, the basic findings are noteworthy. The series-elastic component contributes 5 to 20% of the total ventricular volume, independently of inotropic state (but not independent of temperature). All these data were inferred from observations at small fiber distensions, to avoid interference by the highly nonlinear parallel spring.

The *parallel-elastic component* can be determined relatively easily for papillary muscle from its passive behavior, but it probably would be unwise to extrapolate from this to the parallel-elastic compartment of the ventricle (i.e., the properties of the large compartment in Fig. 15.3 during diastole). Robinson [15.14] used the direct determinations of p–V curves of the relaxed ventricle by Brecher et al. [15.17] and Ullrich et al. [15.18], and scaled them to a 10-kg dog (Fig. 15.4). (Blinks and Koch-Weser [15.19] quote additional studies in this area.) It has become clear that the so-called parallel spring in the myocardium contributes a considerable fraction (10–20%) of the total tension at volumes where the peak isometric tension added during systole is maximal (i.e., at "resting length" of the fibers). This is not so in skeletal muscle. Curve P_d in Fig. 15.4, giving data used by Robinson in his model study, may not indicate this clearly enough. It is for lack of more precise data on this important pressure–volume relation of whole ventricles that Robinson adopted this curve. Influences exerted by the pericardium or by blood pressure in the coronary artery are not taken into account.

It is important to note that the viscous forces (not shown in Fig. 15.4) in the relaxed, or relaxing, ventricle are not negligible. Robinson estimated these by dividing the maximum pressure driving flow into the ventricle during diastole by an estimate of peak inflow. He found a diastolic myocardial

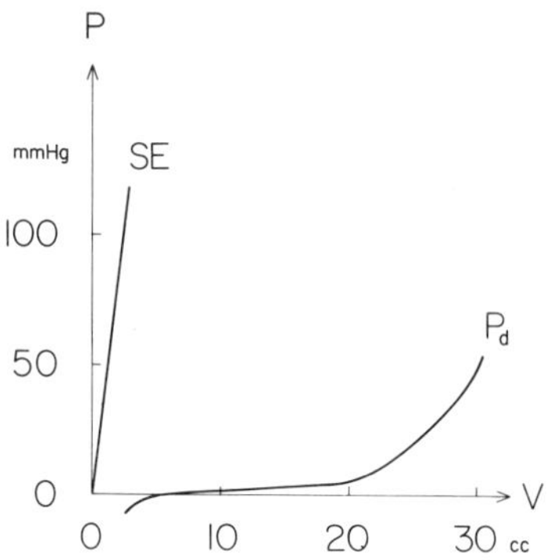

Fig. 15.4. Elastic properties of passive heart muscle. To deal with the entire ventricle, the "spring" properties (cf. Fig. 15.3) are expressed as pressure–volume relations. The volume contributed by all series-elastic elements is given by *SE*, the line is an approximation to the exponential characteristics; see text. The passive pressure–volume relation of all "parallel-elastic" elements of the ventricle of a 10-kg dog is represented by P_d. Negative pressures are needed to decrease the volume below 5 cc. At large volumes the elastic elements become stiffer; nonlinearities are encountered. After Robinson [15.14].

resistance $R_d = 0.2$ mmHg/(cc/sec), which he had to adjust to 0.1 mmHg/(cc/sec) to obtain normal filling in the whole model. These values are small with respect to the viscous terms during systole, which certainly influence total myocardial viscosity during early diastole. It is also interesting to note that rheological behavior of relaxing papillary muscle [15.20] does seem to indicate the presence not only of fast time constants (as computed by $R_d(\Delta V/\Delta p)_{20\text{cc}}$, where $(\Delta V/\Delta p)_{20\text{cc}}$ is the slope of the diastolic pressure–volume curve at 20 cc) but *also* of much longer ones. Thus the data used in this study to characterize passive ventricular behavior rest on very simple considerations.

The two-compartment ventricle is defined completely if we add, to the quantities already given, the pressure–volume curve of the *active* myocardium, its viscous property, and the time course of active state. Robinson [15.14] approximated by the curve $P_{CE}(V)$ in Fig. 15.5 the data reported by Ullrich et al. [15.18]. It shows the typical increase of peak isometric pressure–volume over the operating region below 20 cc of volume. The total available isometric pressure (P_S) is the sum of $P_{CE}(V)$ and $P_d(V)$, that is, during (or shortly after) *maximal* activation, approximated by

$$P_S(V) = P_d(V) + 320\left[1 - \left(1 - \frac{V}{20}\right)^2\right] \tag{15.22}$$

Internal viscosity causes ventricular pressure to fall considerably below P_S as soon as the active portion contracts at velocities occurring during ejection. By dividing $(P_S - p_{\text{ventricle}})$ by the outflow velocity at different times during systole, Robinson [15.14] estimated the myocardial "viscosity" as

$$R_S \simeq 3\,\frac{\text{mmHg}}{\text{cc/sec}} \tag{15.23}$$

which, interestingly, does not seem to vary much between 10 and 20 cc of

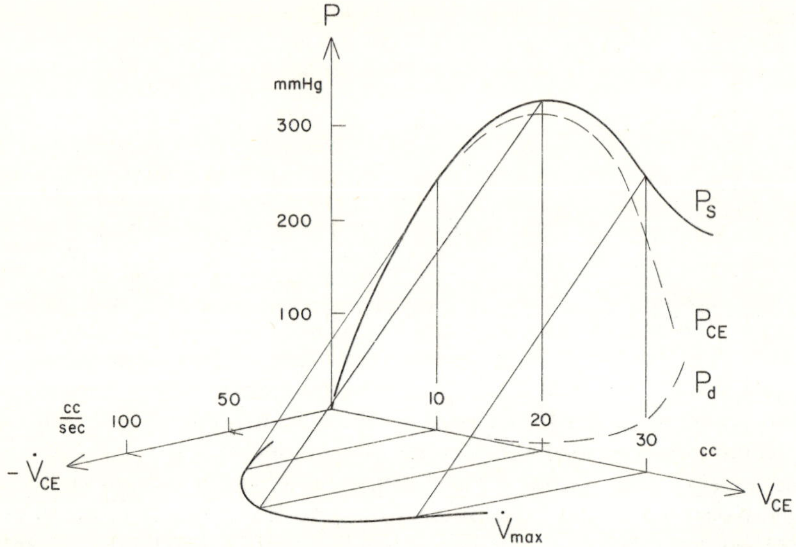

Fig. 15.5. Pressure–volume–volume (out)flow characteristics of the contractile element (*CE*) of the ventricle; P_{CE} gives the isometric (peak)pressure–volume curve of the actively contracting muscle elements. To this, P_d of the passive muscle is added to give P_s, the total mid-systolic peak pressure (isometric) as a function of volume. The entire system is scaled to a 10-kg dog. Auxotonic pressures become much lower than isovolumetric ones because of internal pressure drops due to viscous and rate limiting effects. The slope of the P–$\dot{V}$ lines is 2.5 mmHg/[(cc/sec)]. During peak activation, the state of the contractile element is determined by the oblique surface. Contractility therefore is best defined in terms of the particular shape of the surface. Inotropic interventions (e.g., change in sympathetic tonus) change the slope and/or the location of the contractility surface. To obtain complete ventricle performance, the (small) volume under *SE* (see Fig. 15.4) must be added to all volumes at all pressure and volume flows given here. After Robinson [15.14].

ventricular volume. The value must be adjusted to 2.5 mmHg/(cc/sec) to achieve normal operation of the entire model. This resistance determines, in Fig. 15.5, the slope of the oblique lines connecting P_S and the v_{max} curve. These lines are analogous to the Hill hyperbolas of skeletal muscle. It is interesting to note that v_{max} of papillary muscle is of the order of 2.5 rest lengths/sec [15.16] at 37°C. In the cylindrical model, with circumferential fibers only, this would result in a maximum volume change of 6.25 rest volumes/sec. From this we might compute $R_S = 320/(6.25 \times 20)$ $= 2.56$ mmHg/(cc/sec), which is close to the values just given. But remember that the peak velocity of papillary muscle (2.5 rest lengths/sec) was obtained at slow rates of stimulation (12/min) and that it probably assumes higher values when stimulus rate is increased.

Figure 15.5 gives the surface to which the output variables of the contractile part (*CE*) of the ventricular volume are bound during peak activation. In principle this is the same type of operating surface given for the *CE* of skeletal muscle (Fig. 10.13), defined by the isometric force–length curve and Hill's hyperbola. The surface of Fig. 15.5, however, is not fixed in the p–V–$\dot{V}$ space. It changes with time, as Fig. 15.6 shows, according to the status (and spatial distribution) of active state. It also changes under neurogenic control, as mentioned. Positive inotropic interventions, for example, move $P_S(V)$ and $\dot{V}_{max}(V)$ to larger values.

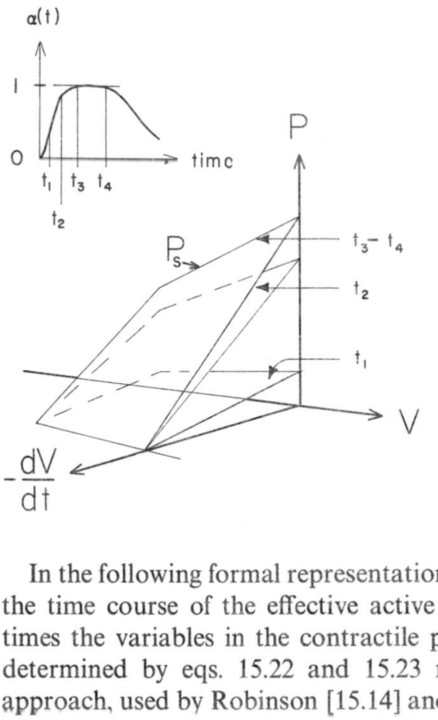

Fig. 15.6. Activation function $\alpha(t)$ and its action on a contractility surface. Part of the p–V–$\dot{V}$ surface of Fig. 15.5, together with a typical $\alpha(t)$ (insert) during systole and early diastole, is presented schematically. With the onset of activation, the peak isometric pressure is made to approach P_s (the mid-systolic value), while the slope of the surface increases correspondingly; t_1, t_2, t_3, t_4 are arbitrarily chosen instants during systole. Owing to the complexity of fiber orientation within the myocardium and asynchrony of fiber activation, $\alpha(t)$ is thought to represent a *net* active state for the entire myocardium. It may therefore differ from the time course of active state observed in papillary muscle. In Robinson's model [15.14] $\alpha(t)$ was assumed to reach unity in a very short time ($t_3 = 0$); $t_4 = 0.2$ sec; and $\alpha(t)$ for $t > t_4$ was assumed to decay exponentially with a time constant of 0.05 sec.

In the following formal representation of the model we use $\alpha(t)$ to represent the time course of the effective active state. As is seen in Fig. 15.6, at all times the variables in the contractile part are bound to a surface which is determined by eqs. 15.22 and 15.23 multiplied by $\alpha(t)$ (see below). This approach, used by Robinson [15.14] and particularly also by Beneken [15.8], has the advantage that inotropic influences can be made to act on $\alpha(t)$ (i.e., its amplitude and its time course), thus changing contractility.

Figure 15.7 enables us to derive the differential equations governing pump performance in a straightforward manner. Following Robinson [15.14], with $\alpha(t) = 1$ during systole we have,

$$\alpha(t) = \begin{Bmatrix} 0 & \text{for} \quad P_S(V) \\ e^{-t/\tau} & \text{for} \quad R_S \end{Bmatrix} \quad \text{during diastole} \qquad (15.24)$$

($t = 0$ at beginning of each diastole)

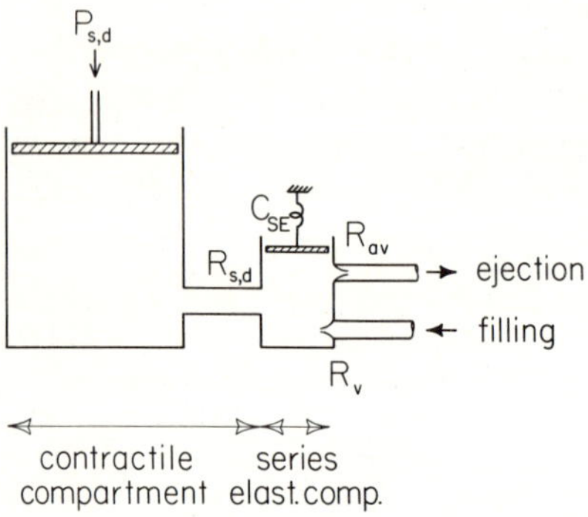

Fig. 15.7. Principal elements of a ventricle model, including internal "resistance," which limits outflow velocity. The main compartments at the left (*contractile compartment*) assume $P_d(V)$ in diastole, as given in Fig. 15.4, and $P_s(V)$ during systole, as in Fig. 15.5. The rate of change of V in systole is determined by internal rate limiting elements (Hill's hyperbola approximated) represented by R_s, a resistance term; in diastole it assumes a much lower value R_d (see text). The *series-elastic compartment* comprises a small volume; C_{SE} is approximated by curve *SE* in Fig. 15.4; R_{av} (aortic valve) and R_v (mitral valve, resistance to inflow) are small resistances needed in the derivation of the overall performance (eqs. 15.27–15.29). After Robinson [15.14].

The differential equation for ventricular pressure (p) is

$$C_{SE}\frac{dp}{dt} = \frac{dV}{dt} - \frac{1}{R_S}[p - P_S(V)] \quad \text{systole} \tag{15.25}$$

$$C_{SE}\frac{dp}{dt} = \frac{dV}{dt} - \frac{1}{(R_S - R_d)e^{-t'/\tau} + R_d}[p - P_d(V)] \quad \text{diastole} \tag{15.26}$$

where $\frac{dV}{dt}$ = outflow during systole, inflow during diastole

C_{SE} = series-elastic compartment volume distensibility, eq. 15.21

R_S = slope of internal pressure drop with $\dot{V}$, eq. 15.23

R_d = term representing viscous forces in completely relaxed (contractile) compartment

$P_S(V)$ = maximum available (isovolumetric) pressure, see Fig. 15.5

$P_d(V)$ = static pressure–volume curve of actively contractile portion of ventricle, when relaxed, see Figs. 15.4 and 15.5

t' = time determining decay of ventricular "viscosity" (i.e., $t' = 0$ at beginning of each diastole)

τ = time constant for decay of ventricular "viscosity" after systole; $\tau = 0.05$ sec, approximately the accepted values of isometric relaxation

The heart rate is kept fixed at 2 Hz; the ratio of systolic to diastolic duration remains at 2:3.

2. Arterial Load and Venous Inflow Conditions

To test the model from the system analyst's viewpoint, we need to connect the ventricle to pulmonary veins (atrial action is neglected in this study) and to the systemic arterial circulation. The latter is assumed to comprise, in a first-order approximation, a lumped compliance ("Windkessel") in parallel with the peripheral resistance. In addition, the resistance to flow of the aortic valve is taken into account. Hence for the arterial system in *systole* we have

$$-\frac{dV}{dt} = C_a \frac{dp_a}{dt} + \frac{p_a}{R_a}$$

$$-\frac{dV}{dt} = \frac{1}{R_{av}}(p - p_a)$$

or

$$R_a C_a \frac{dp}{dt} + p + R_a R_{av} C_a \frac{d^2V}{dt^2} + (R_a + R_{av})\frac{dV}{dt} = 0 \qquad \text{in systole} \qquad (15.27)$$

and

$$R_a C_a \frac{dp_a}{dt} + p_a = 0 \qquad \text{in diastole} \qquad (15.28)$$

where $-\dfrac{dV}{dt}$ = outflow of ventricle = inflow into arteries

p_a = arterial pressure
C_a = effective arterial distensibility
R_a = peripheral resistance
R_{av} = aortic valve resistance

The Windkessel model is applicable to this study because we emphasize ventricular performance and its dependence on input and on arterial load. The input impedance into the Windkessel does in fact approximate the aortic impedance fairly realistically at low frequencies. It does not *explain*, however, the behavior of the impedance as a function of frequency, as the discussion in Chapter 14 shows. Thus it is feasible to use the Windkessel model, as defined by eq. 15.27, as the *load* of the ventricle even though the waveforms of p_a obtained in the model are quite unnatural. Robinson [15.14] chose the following values for the system's parameters:

$$R_a = \frac{\text{mean arterial pressure}}{\text{cardiac output (10-kg dog)}} = \frac{100 \text{ mmHg}}{24 \text{ cc/sec}} = 4.2 \frac{\text{mmHg}}{\text{cc/sec}}$$

$$C_a = \frac{\text{time constant causing decay from 120 to 80 mmHg in 0.3 sec}}{R_a} = \frac{0.74}{4.2}$$

$$= 0.18 \frac{\text{cc}}{\text{mmHg}}$$

$$R_{av} = 0.033 \frac{\text{mmHg}}{\text{cc/sec}} \qquad \text{causing a pressure drop of 2 mmHg at 60 cc/sec}$$

The venous source for ventricular filling can be approximated by a pressure source (6 mmHg in closed-chest animal) and a resistance representing losses in veins, atrium, and mitral valve. The equation governing inflow is

$$\frac{dV}{dt} = \frac{1}{R_v}(p_v - p) \qquad \text{in diastole} \tag{15.29}$$

where p_v = filling pressure in the pulmonary venous reservoir
R_v = resistance to inflow

We can estimate R_v by assuming a pressure drop of 1 mmHg during peak inflow of 63 cc/sec; hence R_v is 0.016 mmHg/(cc/sec).

3. Integrative Performance of the System

When solving eqs. 15.25 to 15.29 simultaneously on an analog computer, Robinson [15.14] carefully investigated the *parameter sensitivity* of the model. He found that enddiastolic volume, stroke volume, and stroke work are remarkably insensitive to changes in individual parameters such as R_d, R_s, C_{SE}, or C_a. For example, a change of +50% in R_s causes stroke volume to change by less than −20%. Furthermore, a change in heart rate by 25% does not cause any change in cardiac output at all. These observations not only justify the choice of a very simple model, but also indicate that the ventricle has a certain inherent stability of performance. If changes must occur, several parameters must vary concurrently—for example, $P_s(V)$ together with $\alpha(t)$ [15.8, 15.14]. Stability with respect to changes in single parameters is particularly striking when load sensitivity is studied. Dynamically, arterial compliance determines the load of the ventricle. Yet it can vary greatly (e.g., −75%) without influencing the pump variables much at all (stroke volume, −6%; enddiastolic volume, +1%). This is largely explained when we compare R_s, the coefficient of ventricular viscosity, with the magnitude of the load impedance at 2 Hz (the basic heart rate used in this study). At 2 Hz we have (with $j = \sqrt{-1}$, $\omega = 2\pi \times$ heart rate)

$$|Z_{\text{load}}| \simeq \frac{R_a}{|1 + j\omega R_a C_a|} = 0.45 \frac{\text{mmHg}}{\text{cc/sec}} \tag{15.30}$$

which is 18% of R_s.

Thus from the dynamic viewpoint, the ventricle essentially acts as a flow source. The normal myocardium contracts as fast as its internal frictional elements allow, nearly independently of $|Z_{load}|$. In passing we note that the value of $|Z_{load}|$ given in eq. 15.30 is approximately 10% of R_a, the peripheral resistance. As is seen in Chapter 14, this is quite realistic, at least at 2 Hz, even though it is obtained, in this study, from the Windkessel model.

The performance of the ventricle in response to hemodynamic stresses should be tested in two ways, as Robinson [15.14] pointed out. A normal work trajectory of the isolated, paced ventricle appears in Fig. 15.8. We now should ask how the work loops, or the dependent variables such as end-diastolic volume and stroke volume, change when input conditions and load

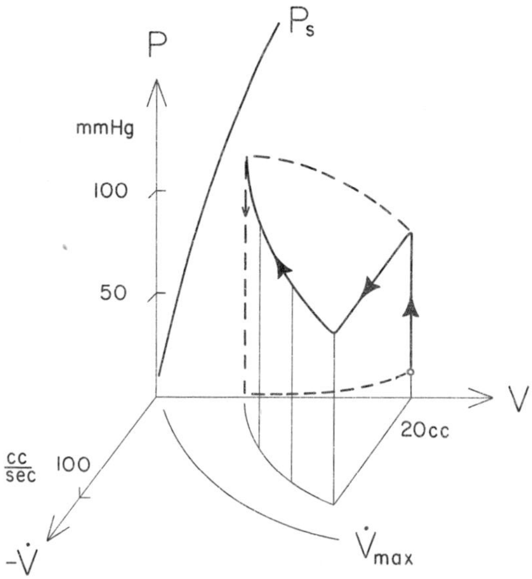

Fig. 15.8. Trajectory of the state variables of the *whole* ventricle in the P–V–$\dot{V}$ space, as produced by Robinson's model [15.14]. The heavy line, beginning (in time) at point O, traces one systole. Its first vertical part corresponds to isovolumetric contraction. As soon as the valve opens, the myocardium contracts approximately isotonically. Hence the velocity of ejection is essentially given by the contractility surface (cf. Fig. 15.5). The trajectory moves onto this surface; velocity decreases rapidly while ventricular volume decreases also. The upper dashed curve is the projection of the trajectory during ejection onto the P–V plane. In reality, onset of relaxation prevents the trajectory reaching the upper left corner of the figure. The trajectory, however, drops vertically during early diastole until it reaches the lower dashed curve (corresponding to P_d of Fig. 15.5). Diastolic filling follows; its projection on the P–V plane is the lower dashed curve returning to point O. Note that this figure does not reveal all details of the action of the contractile element (CE). Certainly during isovolumetric contraction, CE works against the series-elastic element, at finite velocity. Hence *its* trajectory is outside the P–V plane.

are varied. Venous filling pressure can be varied conveniently while peripheral resistance is kept constant, and vice versa. Robinson termed the first a "flow run" and the second a "pressure run." With these, of course, filling pressure and peripheral resistance are considered to be independent variables, and enddiastolic volume, stroke volume, mean aortic pressure, and stroke work become the dependent variables.

In *flow runs* it is found that steady-state stroke volume increases linearly from 12 to 14 cc when p_v is varied from 14 to 25 mmHg ($p_v = 14$ mmHg, normally, in dog with open chest). For the same change in p_v, mean arterial pressure increases from 100 to 115 mmHg, so that stroke work increases from 1600 to 2100 g*cm. This is indeed what is expressed by the Frank-Starling law. The finding is explained by the positive slope of $P_s(V)$ for $V < 22$ cc. In other words, the volume-dependent isometric pressure curve of active heart muscle ensures heterometric autoregulation.

On the other hand, the isolated ventricle performs poorly in *pressure runs.* When peripheral resistance is increased such that arterial pressure rises from 100 to 200 mmHg, incomplete emptying causes enddiastolic volume to go from 20 to 21 cc. Stroke volume, however, decreases from 12 to 6 cc such that stroke work remains approximately constant. Stroke work would decrease if R_a were increased further. These observations pertain to the isolated model ventricle. Experimental findings in animals agree in some reports and differ in others. Robinson [15.14] added a control loop to the model in order to explain the findings reported by Sarnoff and his co-workers [15.21] concerning homeometric autoregulation. If $P_s(V)$ is enhanced proportionately to a tension-time index, the ventricle indeed can be made to increase stroke work when peripheral load (or arterial pressure) is increasing. This represents a positive inotropic intervention. It is still a matter of debate to what extent this is found in the living isolated heart.

Thus the Robinson model clarifies several points. The Frank-Starling mechanism is successfully explained, when flow runs are considered, by the characteristics of active heart muscle which are used in connection with a simple three-element model suggested by investigators in muscle physiology. The large internal pressure drops, to a certain extent analogous to the force "drop" in skeletal muscle (as illustrated by Hill's hyperbola), render cardiac output relatively independent of aortic impedance. Contractility of the ventricle cannot be described by one particular quantity. Rather it is given by a surface (Fig. 15.5) in the flow–volume–pressure space and also by the time course of active state. Moreover, and perhaps most important, it becomes evident that further experimental work is needed if the many unresolved questions are to be attacked. The viscous properties of passive muscle, the decay of active state (both with respect to P_s and R_s) in early diastole, the behavior of enddiastolic volume in pressure and flow runs, and the question of whether the conceptually useful active state function is constant in a flow run, are but a few of the topics that should be studied before the model can be improved.

D. REFERENCES

15.1. Campeau, L., L. Pospisil, P. Grondin, I. Dyrda, and G. Lepage. Cardiac catheterization findings at rest and after exercise in patients following cardiac transplantation. *Amer J Cardiol* **25**: 523–528, 1970.

15.2. Guyton, A. C. *Textbook of Medical Physiology*, 4th ed. W. B. Saunders Co., Philadelphia, 1971.

15.3. Sonnenblick, E. H. Determinants of active state in heart muscle: force, velocity, instantaneous muscle length, time. *Fed Proc* **24**: 1396–1409, 1965.

15.4. Koch-Weser, J. and J. R. Blinks. The influence of the interval between beats on myocardial contractility. *Pharmacol Rev* **15**: 601–652, 1963.

15.5. Rushmer, R. F. *Cardiovascular Dynamics*, 2nd ed. W. B. Saunders Co., Philadelphia, 1961.

15.6. Fry, D. L., D. M. Griggs, and J. C. Greenfield. Myocardial mechanics: tension–velocity–length relationships of heart muscle. *Circ Res* **14**: 73–85, 1964.

15.7. Levine, H. J. and N. A. Britman. Force–velocity relations in the intact dog heart. *J Clin Invest* **43**: 1383–1396, 1964.

15.8. Beneken, J. E. W. and B. DeWit. A physical approach to hemodynamic aspects of the human cardiovascular system, pp. 1–45 in *Physical Bases of Circulatory Transport*. E. B. Reeve and A. C. Guyton (Eds.). W. B. Saunders Co., Philadelphia, 1967.

15.9. Davila, J. C. and M. E. Sanmarco. An analysis of the fit of mathematical models applicable to the measurement of left ventricular volume. *Amer J Cardiol* **18**: 31–42, 1966.

15.10. Sandler, H. and H. T. Dodge. Left ventricular tension and stress in man. *Circ Res* **13**: 91–104, 1963.

15.11. Mirsky, I. Left ventricular stresses in the intact human heart. *Biophys J* **9**: 189–208, 1969.

15.12. Streeter, D. D., Jr., H. M. Spotnitz, D. P. Patel, J. Ross, and E. H. Sonnenblick. Fiber orientation in the canine left ventricle during diastole and systole. *Circ Res* **24**: 339–347, 1969.

15.13. Hefner, L. L., L. T. Sheffield, G. C. Cobbs, and W. Klip. Relation between mural force and pressure in the left ventricle of the dog. *Circ Res* **11**: 654–663, 1962.

15.14. Robinson, D. A. Quantitative analysis of the control of cardiac output in the isolated left ventricle. *Circ Res* **17**: 207–221, 1965.

15.15. Grodins, F. S. and J. F. Buoncristiani. General formulation of the cardiovascular control problem—mathematical models of the mechanical system, pp. 61–75 in *Physical Bases of Circulatory Transport*. E. B. Reeve and A. C. Guyton (Eds.). W. B. Saunders Co., Philadelphia, 1967.

15.16. Yeatman, L. A., W. W. Parmley, and E. H. Sonnenblick. Effects of temperature on series elasticity and contractile element motion in heart muscle. *Amer J Physiol* **217**: 1030–1034, 1969.

15.17. Brecher, G. A. and A. T. Kissen. Relation of negative intraventricular pressure to ventricular volume. *Circ Res* **5**: 157–162, 1957.

15.18. Ullrich, K. J., G. Riecker, and K. Kramer. Das Druckvolumendiagramm des Warmbluetlerherzens. *Pflügers Arch Ges Physiol* **259**: 481–498, 1954.

15.19. Blinks, J. R. and J. Koch-Weser. Physical factors in the analysis of the actions of drugs on myocardial contractility. *Pharmacol Rev* **15**: 531–599, 1963.

15.20. Feigl, E. O. Effects of stimulation frequency on myocardial extensibility. *Circ Res* **20**: 447–458, 1967.

15.21. Sarnoff, S. J., J. H. Mitchell, J. P. Gilmore, and J. P. Remensnyder. Homeometric autoregulation in the heart. *Circ Res* **8**: 1077–1091, 1960.

CHAPTER SIXTEEN

The Cardiovascular System and its Controls

The cardiovascular system is primarily designed to transport oxygen and metabolites, as well as carbon dioxide and wastes. It is indispensable as a heat exchanger and for distribution of materials essential in nervous and digestive activities, in the growth process, and in protection against disease. The distribution of blood flow to different parts of the body and to different organs must be patterned to local demands. These demands may vary with increased metabolic rate because of increased muscular activity, with changes in ambient temperature, increases in digestive activity, and many other factors. On the other hand, distribution of pressures throughout the system must guarantee that capillary pressure stays close to the level optimal for diffusion of oxygen, carbon dioxide, and metabolites. The hydraulic pressure required in most capillary beds is between 15 and 30 mmHg and should be maintained even when flow changes drastically. Capillary equilibrium is especially important in the brain and spinal cord. Furthermore, blood volumes must be shifted to avoid excessive pooling in one part and to supply other parts of the system when need arises.

These and other aspects of the performance and regulatory controls of the circulation have been studied more intensively and longer than many other physiological systems. Yet they are still imperfectly understood, even though more data on blood flow, equally significant as observations of pressures, have been collected since the advent of commercially available flowmeters. Without flowmeters we would be in a situation comparable to an electrical engineer studying active networks but unable to measure currents or impedances. Systems analyses of the baroreceptor action on the heart and on the peripheral resistances have been carried out in anesthetized animals, as we discuss later in this chapter. On the other hand, many other regulatory mechanisms in the circulatory system have not been investigated from the systems engineer's point of view. Thus we cannot yet present a detailed analysis of all mechanisms involved in shifting blood volumes between skin, viscera, and muscle such as occurs in exercise. Details of autoregulation of flow in arterial beds, control of local flow according to blood oxygen concentration, and central mechanisms involved in exercise are also poorly understood. It is impossible, then, to present a complete model of the cardiovascular system, alone or coupled with thermocontrol and with regulation of electrolyte balance and blood gas pressure. These are areas of research where strong give-and-take relations between physiologists, medical scientists, and engineers are needed.

In this chapter we discuss the phenomena and regulatory systems whose main tasks are the control of blood pressure and flow. Some of the tasks of the circulation as a transport system are covered in subsequent chapters.

A. CONTROL OF SYSTEMIC ARTERIAL PRESSURE

Pressure in the capillaries depends on the central, aortic pressure and on venous pressure as well as on resistance to flow at the arteriolar level. When slow variations in pressure are considered, the large systemic arteries can be thought of as one elastic reservoir. Here the mean pressure normally is much higher (80–100 mmHg) than required at the capillary level. The largest pressure drop is observed at the precapillary level, across the arterioles (see [2.2], or any standard physiology text book). Because of the high resistance of these vessels, small fluctuations in central pressure do not appear at the capillary level. For the same reason the central pressure cannot change rapidly if one of the peripheries dilates. Hence the inherent delays in the regulatory mechanisms keeping central pressure constant are less influential. The particular distribution of pressure throughout the vascular system, therefore, is determined by many different factors. Some of these can be defined by studying the interplay of the elements of the uncontrolled, *passive* vascular system which we will discuss in Section 16.A.1. It will then be clearer to what extent and in which situations *control mechanisms* are needed if the circulation is to fulfill the variety of its tasks.

1. The Passive, Uncontrolled System

For elucidating basic relationships governing mean pressure distribution, a greatly simplified model of the entire circulation, originally proposed by Grodins [16.1], and later discussed in great detail by the same author [16.2], can be analyzed. Consider, as in Fig. 16.1, the two sides of the heart to be

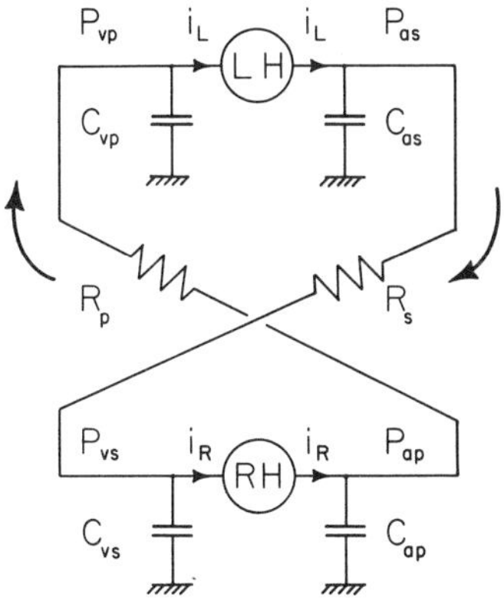

Fig. 16.1. Electric analog of the uncontrolled cardiovascular system. The two pump systems of the heart are separated (LH = left heart, RH = right heart); both are assumed to be flow sources. The symbols used are: *i*, mean blood flow; *p*, mean blood pressure; *R*, (hydraulic) resistance; *C*, net compliance of entire vascular bed. The subscripts are defined as follows: *L*, left; *R*, right; *as*, arterial systemic; *vs*, venous systemic; *ap*, arterial pulmonary; *vp*, venous pulmonary. This system can only be used to model slow changes and steady states.

flow sources, i_L (left ventricle) and i_R (right ventricle). (Throughout this chapter we employ the notation used in electric network theory, modified by using p for pressure or voltage; hence i stands for current or blood flow, and q for blood volume.) The systemic circulation is represented by the (arterial) storage capacity C_{as}, peripheral resistance R_s, and venous distensibility C_{vs}. Similarly, C_{ap}, R_p, and C_{vp} model the pulmonary circulation. The important feature of this model is that we can study phenomena seen in a closed circuit. Blood returning from the venous sides must be driven, by the ventricles, to the arterial sides. The obvious shortcoming is that we consider only steady flows and steady pressures. Furthermore, we certainly do not expect to study, with this model, phenomena such as shift of blood volumes from intestines to muscles. We do gain insight, however, into some aspects of the performance of the system, the distribution of volume between systemic and pulmonic beds, and so on, as given by the distribution of elasticities (i.e., ratios of capacities in Fig. 16.1) and by the large difference between systemic and pulmonary circuits. Analysis of the circuit in Fig. 16.1 yields the following linear, nonhomogeneous system of differential equations, written in matrix form:

$$\frac{d}{dt} \cdot \begin{vmatrix} \tau_{as} & p_{as} \\ \tau_{vs} & p_{vs} \\ \tau_{ap} & p_{ap} \\ \tau_{vp} & p_{vp} \end{vmatrix} - \begin{vmatrix} -1 & 1 & 0 & 0 \\ 1 & -1 & 0 & 0 \\ 0 & 0 & -1 & 1 \\ 0 & 0 & 1 & -1 \end{vmatrix} \cdot \begin{vmatrix} p_{as} \\ p_{vs} \\ p_{ap} \\ p_{vp} \end{vmatrix} = \begin{vmatrix} R_s & i_L \\ -R_s & i_R \\ R_p & i_R \\ -R_p & i_L \end{vmatrix} \tag{16.1}$$

where $\tau_{as} = R_s C_{as}$, $\tau_{vs} = R_s C_{vs}$, $\tau_{ap} = R_p C_{ap}$, $\tau_{vp} = R_p C_{vp}$.

Restricting attention to the homogeneous part of the equations, we see that the four pressures are coupled through the offdiagonal terms in the matrix on the left side. There are no terms, however, that make pressure changes in the systemic side of the circulation dependent on pressures in the pulmonary side, and vice versa. Hence, considering the homogeneous equations, we note that coupling exists between arterial and venous halves, but not between systemic and pulmonary sides of the circulation.

The vector on the right side of eq. 16.1, the forcing function, depends on the assumed characteristics of the flow sources. Note that the steady-state responses of the systemic and pulmonary pressures are not decoupled (e.g., dp_{vs}/dt depends on i_R, the flow in the pulmonary system). To explore this point, we rewrite eq. 16.1 (left side in abbreviated matrix notation) as follows:

$$\frac{d}{dt}\mathbf{TP} - \mathbf{MP} = \begin{vmatrix} R_s \\ -R_s \\ R_p \\ -R_p \end{vmatrix} i + \begin{vmatrix} -R_s \\ -R_s \\ +R_p \\ +R_p \end{vmatrix} \Delta i \tag{16.2}$$

where $i = (i_L + i_R)/2$ mean flow
$\Delta i = i - i_L = i_R - i$ deviation from mean flow

In this form it is clear that the mean flow, in i, determines average steady-state pressures. The time courses of the pressures reaching their mean values, due to i imposed at $t = 0$, can be found by letting Δi be zero.

$$\begin{aligned}
p_{as} &= P_0 + \frac{R_s C_{vs}}{C_{as} + C_{vs}}(1 - e^{-t/\tau_s})i \\
p_{vs} &= P_0 - \frac{R_s C_{as}}{C_{as} + C_{vs}}(1 - e^{-t/\tau_s})i \\
p_{ap} &= P_0 + \frac{R_p C_{vp}}{C_{ap} + C_{vp}}(1 - e^{-t/\tau_p})i \\
p_{vp} &= P_0 - \frac{R_p C_{ap}}{C_{ap} + C_{vp}}(1 - e^{-t/\tau_p})i
\end{aligned} \tag{16.3}$$

where $1/\tau_s = 1/\tau_{as} + 1/\tau_{vs}$
$1/\tau_p = 1/\tau_{ap} + 1/\tau_{vp}$
P_0 = pressure assumed to exist in all compartments at $t = 0$ and $i = 0$, the so-called mean circulatory pressure

Mean pressures are reached when the exponential terms in parentheses in eq. 16.3 are much smaller than unity. These time courses appear in Fig. 16.2. By judicious choice of network parameters *and* the level of i, these pressures are made to assume their physiological levels.

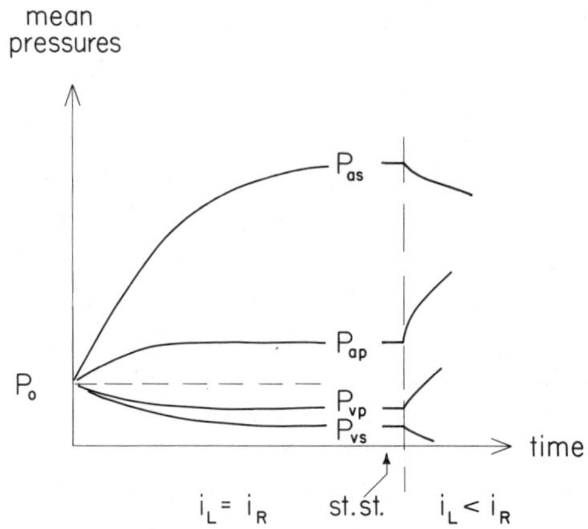

Fig. 16.2. Transients of the pressures p, in the uncontrolled circulation, in response to stepwise changes in flows (i_L and i_R). For definition of symbols see Fig. 16.1. So long as balance between flows ($i_L = i_R$) is maintained, the pressures seek their steady-state values (*st. st.*), which depend on the resistances R_s and R_p. On the right side of the dashed line, the catastrophical time courses of pressures appear as they result from an imbalance in flow ($i_L < i_R$). Adapted from Grodins [16.2].

If Δi, the deviation from the mean cardiac output, takes on a nonzero value, we obtain transients of a different nature. Let Δi assume a nonzero value at t_1 after all pressures have reached their normal values (when $i_L = i_R = i$, $0 \leqslant t \leqslant t_1$). The time courses of the pressures, due to a Δi, an imbalance in flow, are

$$\begin{aligned}
p_{as} &= P_{as} - K_{11}R_s(1 - e^{-t/\tau_s})\,\Delta i - (K_{12}\,R_s\,\Delta i)t \\
p_{vs} &= P_{vs} - K_{21}R_s(1 - e^{-t/\tau_s})\,\Delta i - (K_{22}\,R_s\,\Delta i)t \\
p_{ap} &= P_{ap} + K_{31}R_p(1 - e^{-t/\tau_p})\,\Delta i + (K_{32}\,R_p\,\Delta i)t \\
p_{vp} &= P_{vp} + K_{41}R_p(1 - e^{-t/\tau_p})\,\Delta i + (K_{42}\,R_p\,\Delta i)t
\end{aligned} \tag{16.4}$$

where P_{as}, P_{vs}, P_{ap}, P_{vp} are the pressures in the four compartments at time t_1 and

$$K_{11} = 2\frac{\tau_{vs}^2}{(\tau_{vs} + \tau_{as})^2} \qquad K_{12} = \frac{2}{\tau_{vs} + \tau_{as}}$$

$$K_{21} = 2\frac{\tau_{as}^2}{(\tau_{vs} + \tau_{as})^2} \qquad K_{22} = K_{12}$$

$$K_{31} = 2\frac{\tau_{vp}^2}{(\tau_{vp} + \tau_{ap})^2} \qquad K_{32} = \frac{2}{\tau_{vp} + \tau_{ap}}$$

$$K_{41} = 2\frac{\tau_{ap}^2}{(\tau_{vp} + \tau_{ap})^2} \qquad K_{42} = K_{32}$$

These equations, shown for an imbalance in flow in favor of the pulmonary circulation $[\Delta i = (i_R - i_L)/2 > 0]$, indicate the serious consequence of any such asymmetry in flows. The last terms of eqs. 16.4 represent what is intuitive: If flow in the pulmonary system i_R is larger than i_L, pressures in the pulmonary vessels keep increasing indefinitely and those in the systemic arteries and veins keep decreasing. This is schematized in Fig. 16.2.

The circulation in the living system has to be so regulated that flows on the two sides are, on the average, equal to each other. If Δi, half of the difference between pulmonary and systemic flows, deviates from zero and becomes positive for a certain limited period of time, pressures must be brought back by a Δi of opposite sign. Since the left and right heart pumps contract together at the same rate, the stroke volumes must be adjusted so that, on the average, $i_L = i_R$. Even without neural feedback, the ventricles are capable of fulfilling this task remarkably well. As we learned in Chapter 15, the properties of the myocardium ensure larger stroke volumes if they are filled more than normally (the inverse also is true) and if their hydraulic loads are normal. If the enddiastolic fiber lengths are larger than normal, the isometric ventricular pressure that can be developed is larger, too. Conversely, if the hydraulic load of one ventricle increases, incomplete emptying is the result. The next systole therefore is initiated from a larger enddiastolic volume causing more forceful ejection. This by itself may insure that the ventricular volume does not build up indefinitely. It does not guarantee, however, that aortic pressure is kept constant.

Neural feedback therefore is needed in response to many situations—venous pooling upon change in posture, a decrease of R_s due to increased metabolic rate in one area of the body or, certainly, severe stress. Thus any complete circulatory model should incorporate both the mechanical self-regulation of the heart and neural control mechanisms. Grodins [16.2] chose to model the ventricles by linear elastic chambers during diastole and by sources of flow fulfilling, a fortiori, the law of Starling and Frank in the following idealized form:

$$P_{as} q_s = S q_{ed} \tag{16.5}$$

where

q_s = stroke volume
q_{ed} = enddiastolic volume
S = constant of proportionality

We also refer the reader to Guyton's [16.12] extremely lucid analysis of coupling between cardiac output and venous return.

The more complete heart model proposed by Robinson [15.14] takes into account many structural and physiological details. This then can be used in connection with a more realistic representation of the vascular systems and a more complete model of neural reflex mechanisms, presented in the following section.

2. Baroreceptor Reflexes

From the preceding discussion it is clear that the physiologically required pressures can be maintained in a hydraulically closed system if the products of flow and peripheral resistance and ratios of vascular distensibilities have certain fixed values (see eq. 16.3). Hence if flow should increase, as in exercise, resistance must decrease proportionally. Flow can be increased by a faster heart rate, by larger stroke volumes, or both. Similarly, if the total blood volume is below normal, either flows or resistances must be augmented.

Long-term pressure control is achieved by adjusting blood volume to the level required. Fluid can be expelled or taken up at the capillary level; this effect can show within approximately 30 min or less (see Section 16.C). The kidney can change its water and electrolyte output, and it also plays an important role in pressure control over long periods.

Rapidly occurring pressure transients in the central, large arteries are sensed by tension-sensitive nerve endings within the wall of the aorta (at the aortic arch) and in the wall of the carotid sinus (areas just distal of the bifurcations of the common carotid arteries). Wall tension depends on the local blood pressure; these sensors therefore are called baroreceptors. Increased activity from these pressure sensors upon positive pressure changes excites the vagal center and inhibits the sympathetic area of the medullary vasomotor centers. Efferent fibers from these centers, both of the sympathetic and parasympathetic types, innervate the heart and smooth muscles in the walls of arteries and veins.

The heart rate is tightly controlled by information from the vasomotor center carried to the heart by way of the vagus nerves and by the sympathetic nervous system. Increased vagal activity decreases heart rate, and vice versa. Normally, basic vagal activity keeps the frequency below the level it would attain without parasympathetic innervation. The sympathetic nervous system influences the entire cardiovascular system. In particular, increased sympathetic activity increases heart rate and also enhances contractility of heart muscle. (Furthermore, coronary blood flow is caused to increase.) Thus high pressures and large pulse amplitudes in the aorta and in the carotid arteries reflexly bring about reduction in heart rate and in contractility, and vice versa, so that pressure is brought back almost to its normal, required value.

a. Input–Output Relation of Baroreceptors. The firing rate in the afferent fibers from the pressure sensors depends on mean arterial pressure and on pressure changes (dynamic response). The *static response* (to mean pressure) of one transducer cell, as seen on one afferent fiber, exhibits a (variable) threshold, a rather steep increase in firing rate over a range of approximately 20 mmHg; beyond this point it reaches a maximum firing rate (of the order of 20–50 impulses/sec) (Landgren [16.3]). The thresholds of various receptors differ considerably, and progressively more fibers are recruited with increasing pressure. This increases the dynamic range of the system. The overall static afferent firing rate, when plotted versus mean pressure, is S-shaped, leveling off at approximately 200 mmHg. The overall threshold appears to be near 40 mmHg, and the S curve has its maximum slope at about 100 mmHg. Whether the observed saturation at high pressures is of purely neural origin or due to limited stretch within the aortic wall (or both) is not decided.

The *dynamic response* of the baroreceptors shows directional rate sensitivity. Pressure increments elicit larger changes in firing rate than negative pressure changes. Warner [16.4] gave the following relation for the discharge rate $n(t)$ observed in an *individual* afferent fiber:

$$n(t) = C_1\left[(C_2 - \bar{p})\frac{dp^+}{dt} + (\bar{p} - C_3)\frac{dp^-}{dt} + \underbrace{C_4\frac{p(t) - (C_5\bar{p} + C_6)}{C_7 + p(t) - (C_5\bar{p} + C_6)}}_{n_3}\right] \tag{16.6}$$

where dp^+/dt, dp^-/dt = positive and negative pressure changes, respectively

$\bar{p}$ = pressure average over the past few beats

$p(t)$ = instantaneous pressure

The dependence of firing rate $n(t)$ on positive and negative pressure changes and on pressure $p(t)$ is expressed by eq. 16.6. It has been observed that $(C_2 - \bar{p}) > (C_3 - \bar{p})$; that is, $n(t)$ increases more with a (positive) dp^+/dt than it decreases with (negative) dp^-/dt. Changes in firing rate due to such pressure variations, however, become smaller at higher mean pressures. The third term in eq. 16.6 similarly shows saturation. If $p(t) = \bar{p}$, the steady-state contribution n_3 to $n(t)$ from the third term is

$$n_3 = C_4\frac{\bar{p}(1 - C_5) - C_6}{C_7 + \bar{p}(1 - C_5) - C_6} \qquad \text{steady state} \tag{16.7}$$

Hence we can conclude that $C_6/(1 - C_5)$ must be the steady-state threshold of the baroreceptor. The derivative of n_3 with respect to $\bar{p}$ is

$$\frac{dn_3}{d\bar{p}} = C_4 \frac{(1 - C_5)C_7}{[C_7 + \bar{p}(1 - C_5) - C_6]^2} \qquad \text{steady state} \tag{16.8}$$

From this equation it is seen that the increase of $n(t)$ with $\bar{p}$ decreases with larger $\bar{p}$'s.

When using eq. 16.6 for a description of the baroreceptor transfer function, it should be noted that the third term $n_3(p, \bar{p})$ also contributes to the phasic response. If pressure $p(t)$ increases from $\bar{p}$ to $\bar{p}_1$ in a step, n_3 is initially

$$n_3 = C_4 \frac{\bar{p}_1 - (C_5\bar{p} + C_6)}{C_7 + \bar{p}_1 - (C_5\bar{p} + C_6)} \tag{16.9}$$

After a few beats $\bar{p}$ becomes equal to $\bar{p}_1$; then

$$n_3' = C_4 \frac{\bar{p}_1(1 - C_5) - C_6}{C_7 + \bar{p}_1(1 - C_5) - C_6} \tag{16.10}$$

The difference $(n_3 - n_3')$ is a measure of the overshoot; it is

$$\Delta n_3 = C_4 \frac{C_5\, C_7(\bar{p}_1 - \bar{p})}{[C_7 - C_6 + \bar{p}_1(1 - C_5)]\,[C_7 - C_6 + \bar{p}_1(1 - C_5) + C_5(\bar{p}_1 - \bar{p})]} \tag{16.11}$$

The product $C_5\, C_7$ can be considered a measure of *adaptability*. The amount of overshoot, Δn_3, also decreases when pressure increases.

Since threshold, sensitivity (dn/dp), and level of saturation vary from fiber to fiber, the net afferent firing rate, due to all (recruited) baroreceptors at any given time, would have to be computed as a sum of many expressions of the form of eq. 16.6. Whether the temporal pattern of this sum of all discharges during a heart beat is important with respect to control of heart rate and vasometer tone is not yet resolved. Douglas et al. [16.5] found aortic pressure to be independent of *patterns* of afferent stimulation when the afferent fibers were severed and artifically stimulated. The net number of discharges during one beat seemed to be the significant factor. On the other hand, there is some evidence (Koepchen [16.6]) that the timing within the heart cycle of an afferent discharge influences its effect.

Poitras et al. [16.7] simulated nerve firing patterns, observed in *multifiber* preparations, in response to natural pressure pulses and to artificially intro duced pressure steps. Their model was much simpler than indicated by eq. 16.6. These authors related net firing rate $\bar{n}(t)$ to pressure $p(t)$ as follows:

$$\left[1 + (\tau_1 + \tau_2)\frac{d}{dt} + \tau_1\tau_2\frac{d^2}{dt^2}\right]\bar{n}(t)$$

$$= \left\{(a_1 + a_2) + [\tau_1(a_2 + b_1) + \tau_2(a_1 + b_2)]\frac{d}{dt} + \tau_1\tau_2(b_1 + b_2)\frac{d^2}{dt^2}\right\} p(t) \tag{16.12}$$

In the Laplace domain this becomes

$$\bar{N}(s) = \left[\frac{a_1 + b_1\tau_1 s}{1 + \tau_1 s} + \frac{a_2 + b_2\tau_2 s}{1 + \tau_2 s}\right]P(s) \tag{16.13}$$

This linear model was followed by another *nonlinear* one exhibiting threshold and saturation. The entire system predicted behavior of the baroreceptors satisfactorily, especially in response to relatively high-frequency pressure changes: τ_1 was approximately 5 msec and τ_2 was approximately 5 sec; all other constants were also optimized using a computer. Figure 16.3 presents a typical baroreceptor response to the natural pulse and the prediction by the model.

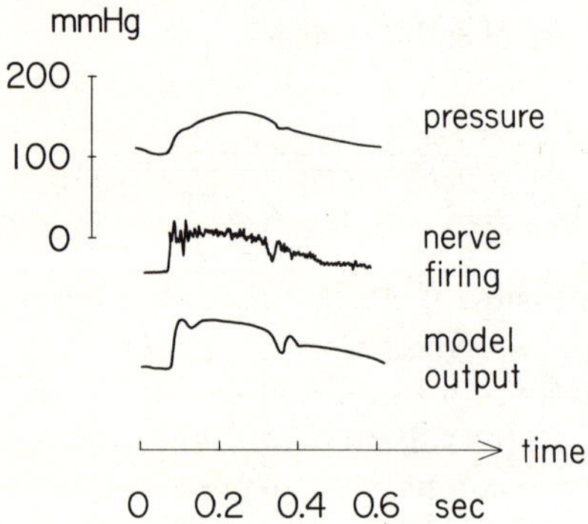

Fig. 16.3. Baroreceptor response (middle trace, multifiber preparation) to a typical blood pressure pulse (top curve) obtained in a dog, together with the model response, as reported by Poitras et al. [16.7]. The Poitras model relates afferent nerve activity to aortic pressure through the differential equation (16.12).

Katona [16.8] found it sufficient, for a model of the entire baroreceptor–heart rate loop, to compute the *approximate* number of total nerve discharges for each beat. Over a limited range, the number of firings for the kth beat could be expressed in terms of peak systolic pressure (p_{sk}) and enddiastolic pressure (p_{dk}) as follows:

$$n_k = C_1\left[C_2(p_{sk} - p_{dk}) + \frac{p_{sk} + p_{dk}}{2} - C_3\right] \tag{16.14}$$

where $p_{sk} - p_{dk}$ = pulse amplitude of kth beat

$(p_{sk} + p_{dk})/2$ = estimate of mean pressure during kth beat

Equation 16.14 neglects the contribution of negative pressure changes (dp^-/dt) within the kth beat. Calculated by a computer, n_k changes stepwise with every beat at peak systolic pressure (i.e., when p_{sk} is available). Negative pressure slopes, although not explicit in eq. 16.14, are also sensed by this model. If the enddiastolic pressure of the $(k + 1)$th pulse (i.e., $p_{d(k+1)}$) falls below p_{dk} (the enddiastolic pressure of kth beat), $n_{(k+1)s}$ will differ from n_{ks}. In other words, changes in negative pressure slopes during one beat will be signaled to the vasomotor area with a slight delay. Beneken [16.9], who adopted Katona's model of eq. 16.14 in his analog computer simulation of the entire circulation, let $C_2 = 1.5$, and $C_3 = 40$ mmHg. The size of C_2—larger than unity—expresses the finding that baroreceptors show large phasic responses.

In summary we note that the phasic response of the baroreceptor system carries information about the current "state of affairs" in the large arteries (i.e., about the pressure fluctuation during the current heart cycle). The static response, on the other hand, depends on previous history. Mean pressure is given by peripheral resistance, flow, and elastances during the past several heart beats (eq. 16.3).

b. ***Effects on the Heart of Efferent Autonomic Stimulation.*** Parasympathetic fibers, part of the vagus nerves, originate in the medullary vasomotor area and innervate the atria of the heart. Increased discharge rate results in increase of acetylcholine release at the nerve endings. Its direct effect is to slow down the rate of discharge of the sinoatrial (SA) node and to decrease the excitability of the atrial–ventricular junction. (Artificial, strong stimulation of the vagus nerve can arrest the heart for a limited period of time, i.e., 4–10 sec.) Warner's formulation [16.10] of this depressing action of the parasympathetic system is given below.

Let m be the fraction of all available vesicles at the nerve endings able to release acetylcholine. Then $(1 - m)$ is the fraction that previously has expelled acetylcholine and is being replenished. If the rate at which vesicles are discharged is proportional to both the number of available vesicles and the vagal firing frequency f_p, we can write

$$\frac{dm}{dt} = k_1(1 - m) - k_2 m f_p \tag{16.15}$$

where k_1 = rate constant of replenishing, ca. 2.75/sec
k_2 = constant of proportionality

The rate of change of concentration [A] of acetylcholine, A, at the sinoatrial node, is proportional to mf_p (rate of acetylcholine release) minus the rate at which acetylcholine is hydrolyzed, which is proportional to the concentration [A] at its site of action. Hence

$$\frac{d[\mathrm{A}]}{dt} = k_3 m f_p - k_4[\mathrm{A}] \tag{16.16}$$

where k_4 = rate constant of acetylcholine destruction; ca. 0.36/sec

Finally, Warner [16.10] assumed the heart period T to be linearly dependent on acetylcholine concentration up to a certain level $[\mathrm{A}_0]$, above which the heart ceases to contract; that is,

$$\begin{aligned} T &= T_0 + k_5[\mathrm{A}] \qquad [\mathrm{A}] < [\mathrm{A}_0] \\ T &= \infty \qquad\qquad\quad [\mathrm{A}] > [\mathrm{A}_0] \end{aligned} \tag{16.17}$$

The steady-state response of this system can be computed from eqs. 16.15 to 16.17 by letting $dm/dt = 0$ and $d[\mathrm{A}]/dt = 0$.
Then

$$T = T_0 + \frac{k_1 k_3 k_5}{k_2 k_4} \frac{f_p}{k_1/k_2 + f_p} \qquad \text{steady state} \qquad [\mathrm{A}] < [\mathrm{A}_0] \tag{16.18}$$

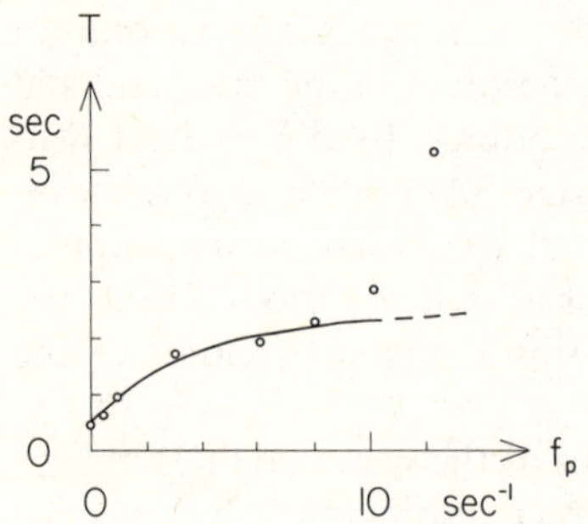

Fig. 16.4. Steady-state relation between the period (T) of the heart cycle and the net efferent firing rate of vagus nerve. The circles are observations obtained in a dog; the solid line (eq. 16.18) represents prediction by Warner's model. Note that f_p strictly represents the rate of externally applied stimulus. From Warner [16.10].

The nonlinear function $T(f_p)$ is plotted in Fig. 16.4, where the prediction of the model is compared with experimental results. In this case k_1/k_2 is of the order of 4.0/sec; and $k_3 k_5/k_4 \simeq 0.7\ \mathrm{sec}^2$ per unit of concentration. The main shortcoming of the model seems to be its abrupt transition from finite to infinite T. However, eq. 16.18 demonstrates why the heart rate depends more strongly on f_p at initially high heart rates than it does at initially low ones.

The dynamic response of heart rate to changes in f_p, the efferent vagal firing rate, is also quite well predicted by this model. As Fig. 16.5 indicates, we observe directional rate sensitivity. Heart rate is decreased faster upon increased vagal activity than it increases when vagal influx diminishes.

One reason for this is that the time constant of eq. 16.15 is $(k_1 + k_2 f_p)^{-1}$. Thus an increase in f_p means a short time constant, and a decrease in f_p lets m decrease slowly. Note that, if $k_2\ \Delta f_p < k_1$, this system is linear and essentially first-order.

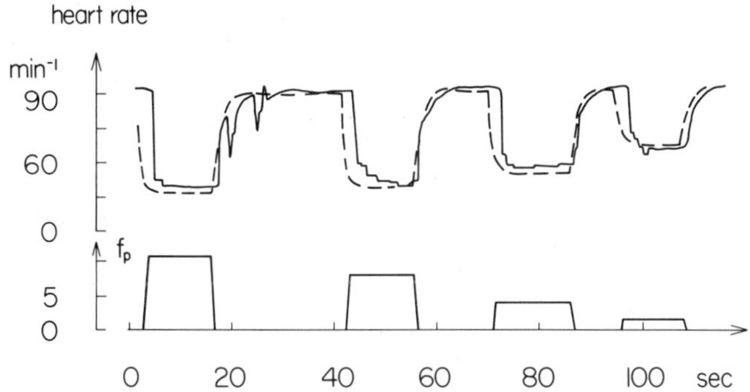

Fig. 16.5. Responses of the heart rate (upper curves) to stepwise changes in vagal stimulation (lower curve), as functions of time. The solid lines are experimental data; the dashed line represents the predictions by Warner's model. Note the directional rate sensitivity in the responses; heart rate is decreased faster than it is increased. From Warner [16.10].

The action of the *sympathetic* system, innervating atria and ventricles of the heart, is twofold. Both heart rate and contractility increase with larger sympathetic influx. Warner [16.10] sought to describe the heart rate control in detailed form, taking into account diffusion of noradrenalin released at the sympathetic end plates, into the nodal cells, and in and out of the circulation (eq. 16.19). Furthermore, noradrenalin is thought to react with another substance, by a second-order reaction (eq. 16.20 and 16.21); then heart rate is influenced by the product of the reaction (eq. 16.20). Formally, the rate of change of concentration $[B_0]$ of noradrenalin B at the site of release is

$$\frac{d[B_0]}{dt} = s_1 f_s + s_2\{[B_1] - [B_0]\} + s_3\{[B_2]-[B_0]\} \qquad (16.19)$$

where s_1, s_2, s_3 = rate constants
f_s = firing rate on sympathetic fibers
$[B_0]$ = concentration of noradrenalin near nerve endings
$[B_1]$ = concentration of noradrenalin in blood
$[B_2]$ = concentration of noradrenalin at SA node

If reaction between noradrenalin B and another substance D is governed by rate constants s_4 and s_5 as follows:

$$B + D \underset{s_5}{\overset{s_4}{\rightleftarrows}} BD$$

we can write
$$\frac{d[BD]}{dt} = s_4[B_2][D] - s_5[BD] \tag{16.20}$$

But then, concentration of noradrenalin at the SA node changes according to

$$\frac{d[B_2]}{dt} = s_3\{[B_0] - [B_2]\} - \frac{d[BD]}{dt} \tag{16.21}$$

Finally, the heart rate v ($= T^{-1}$) is made to depend linearly on [BD]:

$$v = v_0 + s_6[BD]$$

Warner optimized all constants so that this model, containing nonlinearity due to the second-order chemical reaction, matched his observations.

Thus in both his models Warner included mechanisms and chemical reactions that attempt to explain the relation between efferent stimuli and heart rate. Other methods of attack are available, as discussed in the following sections. These are useful when the observed control action is incorporated in a model of the whole circulatory system.

c. ***Transformation of Afferent Baroreceptor Signals into Heart Rate in Anesthetized Dogs.*** Katona, Barnett, and Jackson [16.8, 16.11] optimized a phenomenological model simulating the observed transfer function between afferent nerve activity, as estimated by eq. 16.14, and heart rate. Their model therefore includes the action of the medullary vasomotor control center in a particular state of the animal (chloralose-anesthetized dogs). Interestingly enough, it was necessary to incorporate in this model two paths of signal processing. One could be associated with the parasympathetic, the other with the sympathetic system. When artificial pressure disturbances were created within the aorta (by inflating a small balloon or by withdrawing blood) two, and sometimes three, modes of heart rate control were observed. (Innervation paths were kept intact, i.e., data were collected from the "closed-loop" system.)

Heart rate changes induced by aortic pressure perturbations were small and slow if, initially, heart rate was high (3 Hz) and pressure low. In this response the sympathetic system seems to dominate. Second, at slower heart rates (2 Hz) and higher pressures heart rate was influenced quite quickly and to a larger degree by pressure fluctuations. Here the parasympathetic system is most influential. A representative record showing a mixture of these two

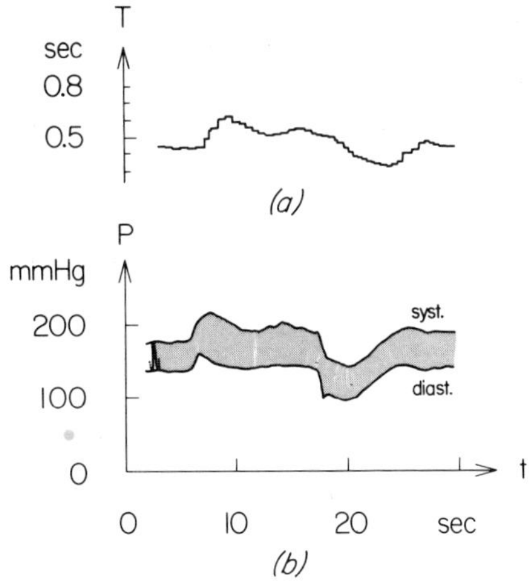

Fig. 16.6. Action of baroreceptor feedback system (closed loop) on heart period (T) observed in a dog. (a) Changes in the duration of the heart cycle in response to the pressure fluctuations, plotted in (b), caused by inflation and deflation of an intraarterial balloon. For simplicity, systolic and diastolic blood pressure values are connected by continuous lines. (b) The pressure pulses lie within the shaded area, as indicated on the left. Note the fairly quick response of the system to increases in pressure. After Katona [16.11].

modes is reproduced in Fig. 16.6. Third, action of the reflex loop was very strong and complex if the initial heart rate was below 1.7 Hz and if the initial (mean) aortic pressure was of the order of 180 mmHg. In the third domain, heart rate was not uniformly affected as in the other cases. Upon a positive disturbance, from approximately 175 to 200 mmHg, the period of a subsequent beat was almost twice the value it had had previously. This caused the enddiastolic pressure to fall back to 175 mmHg or lower, in turn creating command signals that contradicted the original one. The next beat was of shorter duration, upon which another long heart cycle may have followed, and so on. In this domain of discontinuous behavior, both average pressure and pulse amplitudes and variations during *one* cycle reflexly control heart rate. This probably occurs by way of the parasympathetic system and must be due, in part, to the dynamics of the ventricle.

The first two domains could be simulated elegantly by the model presented in Fig. 16.7. The input to the model is the estimated number of afferent receptor nerve discharges per beat as computed by eq. 16.14. This is thought to represent signals from sensors in both carotid sinus and aortic arch. The lower path, active for small signals ($n_k < \mu$, a kind of threshold) contains a linear overdamped second-order system. The output, the heart *period*, of this section yields a "smeared" but symmetric (with respect to time axis) transformation of the input. Such responses, seen in the first domain at low pressures and initially short heart periods, are thought to occur when sympathetic control of the myocardium dominates.

The second mode of regulation, occurring when vagal activity dominates, is modeled by the upper path of the model in Fig. 16.7. Response is governed by the nonlinear first-order system. Capacitor C charges rapidly through the

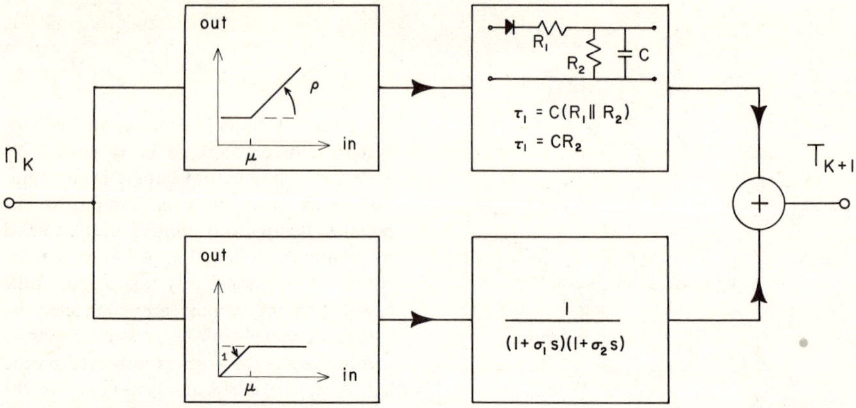

Fig. 16.7. Katona's model for the overall relation between the measure n_k, indicative of afferent firing rate of the baroreceptor system (eq. 16.14), and the heart period T. Heart period of the $(k+1)$th beat is determined by n computed from the kth beat. The upper pathway is active for large mean n's; it exhibits directional rate sensitivity and simulates parasympathetic control. The lower pathway, incorporating both parasympathetic and sympathetic controls, comes into action when mean pressure is relatively low. From Katona [16.11].

diode but discharges slowly through resistor R_2, causing asymmetry in the response. Output signals generally are large. This path, however, is exclusively engaged only if $n_k > \mu$.

Values of the parameters, optimized in each experiment to simulate the experimental animal responses, varied somewhat from case to case. The time constant τ_1 ranged from 0.5 to 2.5 sec, τ_2 was always larger and varied from

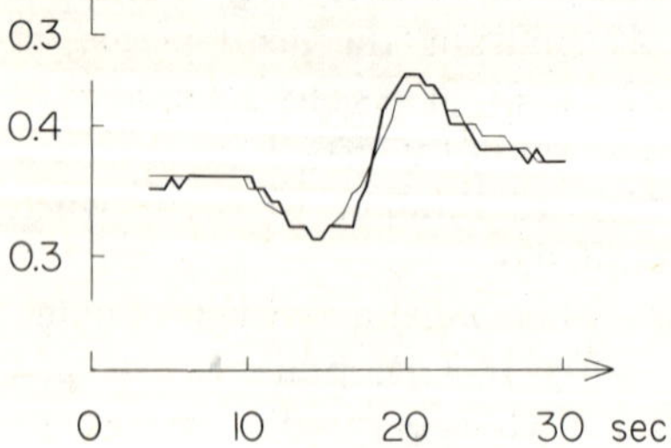

Fig. 16.8. Performance of Katona's model in two typical cases. The actual time course of the heart period T (heavy line), observed in a dog whose arterial pressure was disturbed in the manner of Fig. 16.6, is compared with the prediction of the model (fine line). In this animal, $\tau_1 = 1.5$ sec, $\tau_2 = 4.5$ sec. After Katona [16.8].

2.5 to 10.0 sec; σ_1 is of the order of 1.0 sec, σ_2 of 2.0 sec. Performance of the model (Fig. 16.8) can be called satisfactory in the first two low-pressure regions.

The result of this reflex loop can be seen in the example in Fig. 16.6. After the pressure is artificially raised by approximately 20 mmHg, the mean pressure is brought back 3 sec later to within perhaps 5 mmHg of the original level due to slowing of the heart. The same pressostat action is seen when there exists a negative pressure disturbance. Generally, such pressure transients may occur when a peripheral vascular bed constricts or dilates because of local flow adjustment. The action of the parasympathetic nervous system is a powerful depressor in the former case. The sympathetic nervous system, important in the latter case, increases heart rate as described previously, at least if initial pressure is relatively low. In Chapter 15 it was noted that increased activity on the sympathetic efferent tracts also has a positive inotropic effect on the myocardium (increase in maximum isometric tension or peak-shortening velocity, or both). Additional systems also controlled by the sympathetic outflow from the vasomotor areas are discussed in the following section.

d. Sympathetic Control of Fluid Resistance of Systemic Arterioles. Resistance to steady flow of blood is a function of the geometry of the vessel tree. The pressure drop in the arterioles, the precapillary vessels, normally amounts to approximately 75% of the total pressure drop between aorta and capillaries (i.e., the total arteriolar resistance is larger than at any of the other levels). Since lengths of the arterioles are fixed, their resistance to volume flow is, in a first-order approximation, inversely proportional to the fourth power of the radius. Therefore, not much contraction of smooth muscle in the arteriolar wall is required to change fluid resistance. It appears that smooth muscle is most sensitive to concentrations of oxygen, carbon dioxide, and nutrients in the blood. If there is a surplus of carbon dioxide the arterioles dilate, allowing more blood to flow [16.12]. This locally induced "reflex" is important for local flow regulation and thus for distribution of flows in the entire system. It may override the central, sympathetic control of peripheral resistance. Action of the sympathetic nervous system is mediated by norepinephrine released at some nerve endings within the vascular walls, causing smooth muscle to contract. (Other sympathetic fibers are cholinergic and may induce vasodilation.) This direct action of the sympathetic system may be supported indirectly by norepinephrine and epinephrine circulating in the blood when released by the adrenal medulla.

Beneken [16.9] suggested a straightforward approach to model the central control of resistance by way of adrenergic sympathetic fibers responding to afferent information from the baroreceptors. The firing rate n_k from the pressure sensors, calculated from eq. 16.14, is assumed to be transformed by the vasomotor areas to influence the vascular resistance R as follows (in response to a step change in n_k):

$$\frac{1}{R} = \frac{1}{R_0}\left[1 + k\left(\frac{n_k}{n_{k0}} - 1\right)(1 - e^{-t/\tau_s})\right] \tag{16.22}$$

where R_0 = normal, resting vascular resistance
n_{k0} = normal, resting afferent firing rate
τ_s = time constant of response, of the order of 12 sec

According to eq. 16.22, which is the solution of a first-order differential equation for $1/R$, increased nervous influx, caused by high aortic pressure (or positive pressure changes), decreases the partial peripheral resistance R. Approximately 25 sec (i.e., $2\tau_s$ or more) elapses before the new level of R is approached. This is in accord with observations in animals [16.12].

The static relation between R and mean aortic or carotid pressure is given in Fig. 16.9. The shape of this curve is like that used by Grodins [16.2] but it is not of the sigmoid shape discussed by Grodins and also by Scher [16.13]. The curve in Fig. 16.9 is not extended to zero pressure because of the assumed threshold of the baroreceptors. In reality the curve in Fig. 16.9 is modified greatly by additional inputs to the vasomotor centers. The chemoreceptors in the carotid and aortic bodies, as well as sensitivity to pCO_2, pO_2, and pH on the part of the epithelial cells of the vasomotor area itself, reflect on the peripheral vessels. Emotion, mental activity, and temperature changes are other stimuli to the control centers. Information from the motor cortex also influences the medullary vasomotor center.

Equation 16.22 is thought to apply to the resistance concentrated at the arteriolar level. Sympathetic efflux also affects the larger arteries. It is

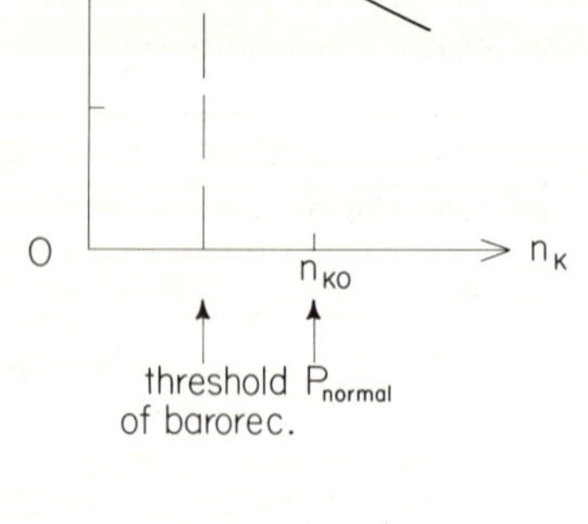

Fig. 16.9. Steady-state values of peripheral resistance R, normalized by its resting level R_0, as a function of n_k, the measure of afferent baroreceptor activity (eq. 16.14); n_{k0} exists at the normal resting level of arterial pressure. For the dynamic relation see eq. 16.22. This figure gives the influence on R of baroreceptors only—many other factors can influence R concomitantly. After Beneken [16.9].

interesting that these changes in arterial radius, while not contributing significantly to variation in total resistance, probably bring about changes in overall elasticity [16.14]. Equation 16.3 shows that this latter effect bears on the value of steady-state pressure as well as on the time constant of transients. Quantitative studies of these interrelations have not been made.

B. VENOUS RETURN

1. Control of Systemic Venous Volume

Braunwald et al. [16.15] observed, in an "open-loop" preparation, significant influences of carotid pressures on venous volume. These investigators used extracorporeal pumps and an oxygenator to perfuse the systemic circulation of dogs at a constant rate. Carotid arteries were cannulated separately, which allowed for independent variations of pressures at the carotid sinus. Aortic pressure could be used as a measure of systemic resistance, since flow (output of pump) to trunk and hindlimbs was kept constant. If venous return varied, the volume within an extra reservoir of the extracorporeal pumping system changed also. The latter can be assumed to be, essentially, a measure of the change in venous volume. Thus a positive pressure step in the carotid, which decreased vascular resistance (in caudad circulation) by an average of 40 %, increased total venous volume by approximately 250 cc. This pooling effect, beginning after a delay of about 40 sec, appeared to follow an exponential law with a time constant of approximately 3 min. Thus venous return decreased transiently.

After reaching a steady state, the venous distensibility ($\Delta q/\Delta p$) was approximately 70 % of its original value. In the intact animal, these effects would support the heart's task of decreasing cardiac output so that pressure could be maintained. The dynamics of this particular baroreceptor reflex, however, have not been worked out in detail. From the engineering point of view, the "parametric forcing" of both capacitance and resistance to flow, due to diameter changes, is of interest.

Humoral control of venous capacity is also important. Braunwald et al. [16.15] demonstrated the constricting effect of norepinephrine and the dilation induced by trimethaphan.

2. Blood Flow through Veins

It is rather difficult to discuss rigorously the laws governing flow of blood through the venous systems, although some useful experimental studies (Noordergraaf and Kresch [16.16]) are available. In veins, pressure gradients are very small, and since transmural pressures, at least in the larger veins, are small or even negative, the vessels may collapse. Moreover, valves distributed throughout the venous circulation (except in neck, head, and lungs) prevent centrifugal flow, and muscular activity helps in maintaining venous return through external "massage" of the veins. Furthermore, upon inspiration the lowering of pressure in the thoracic cavity below atmospheric level periodically supports all forces returning venous blood.

Beneken's approach [16.9] in modeling the venous system takes into account several of these complicating factors. The resistive and elastic properties of segments of the venous vascular tree may be concentrated, considering an electric analog, into lumped network elements. (Inertial forces are neglected. Although venous flow may be intermittently oscillatory,

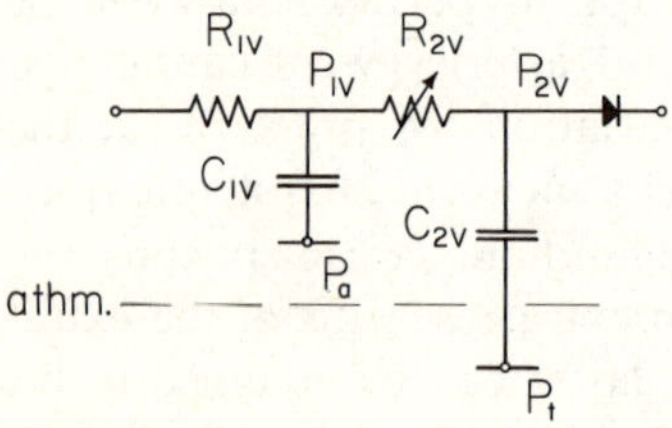

Fig. 16.10. Representation, in the form of an electric analog, of the abdominal veins (subscript 1) and inferior vena cava (subscript 2) by two "lumped" segments; P_a is intraabdominal pressure, P_t is intrathoracic pressure. (Cf. eq. 16.23). The R_2 depends on *mean* transmural pressure; collapsing may occur. After Beneken [16.9].

accelerations are relatively small.) Figure 16.10 shows the analog of two such segments for which the following relations between flow and pressure gradients can be given. Consider the sections of the inferior vena cava in the abdomen (index 1) and the one in the thorax (index 2); flow between the two segments is then determined by

$$
\begin{aligned}
p_{1v} - (p_{2v} + p_t - p_a) &= R_{2v} i_{2v} \qquad \text{if} \quad (p_{2v} + p_t - p_a) > 0 \\
p_{1v} &= 10 R_{2v} i_{2v} \qquad \text{if} \quad (p_{2v} + p_t - p_a) < 0
\end{aligned}
\tag{16.23}
$$

where

p_a = pressure in abdomen

p_t = pressure in thoracic cavity

$p_{1v} + p_a$ = absolute pressure in abdominal segment

$p_{2v} + p_t$ = absolute pressure in thoracic vena cava

$p_{2v} + p_t - p_a$ = transmural pressure in the system at a point between the two segments; if negative, vein partly collapses

R_{2v} = resistance of thoracic segment, when open

These equations introduce the transmural pressures at each level and the collapsible nature of the vessel. If the transmural pressure at a point between the two segments is negative, the inferior vena cava is made to collapse partly so that flow essentially is independent of the "downstream" pressure, while resistance to flow increases tenfold. Obviously we have not attempted to analyze the dynamics of a collapsible vessel as such. Equation 16.23 enables us, however, to introduce some of the important features of the venous beds into a model of the entire circulation.

Beneken [16.9] included stress relaxation in the vascular wall in the relation between volumes q_v and pressures (the additional equations needed). For each venous segment we can write

$$p_v = \frac{1}{C_v}(q_v - q_{v0}) + R_v \frac{dq_v}{dt} \tag{16.24}$$

and

$$q_{v0} = q_{v0}(0) + \Delta q_{v0}(1 - e^{-t/\tau_{\text{rel}}}) \tag{16.25}$$

where q_{v0} = volume within vein at zero pressure
q_v = blood volume in excess of unstressed volume
Δq_{v0} = additional volume allowed to collect due to stress relaxation
τ_{rel} = time constant of stress relaxation, ca. 12 min

These two equations express not only that pressure in a venous segment depends on instantaneous volume (as calculated by integrating the difference between inflow and outflow) and on temporal volume changes (due to wall viscosity, second term in eq. 16.24), but also that volume within the segment is a function of its previous history. If volume is increased stepwise, pressure at first rises too. Subsequently a slow decrease is observed (time constant τ_{rel}), even if segmental volume is kept constant, because stress relaxation occurs. This is equivalent to letting the unstressed volume q_{v0} be larger; that is, $[q_{v0}(0) + \Delta q_{v0}]$. In eq. 16.24 C_v, the capacity of the vein, is assumed to be a constant, which probably is an approximation. Time t in eq. 16.25 has its origin whenever the total volume within the vein is changed. It is noteworthy that inverse stress relaxation also occurs after an instantaneous volume decrease.

An advantageous feature of this model is that effects of hydrostatic pressure may be introduced. For a given position of the body, q_{v0}, and perhaps Δq_{v0}, of every venous segment can be made dependent on the height of the fluid columns exerting the static head. Furthermore, according to our previous discussion of neural control of venous volume, the term q_{v0} also should be made a function of the firing rate of the sympathetic fibers innervating the veins. However, there is not enough experimental data available to decide how much C_v, the distensibility, varies with static pressure changes and with neural stimulation.

Beneken [16.9] designed this model as part of an analog circuit of the entire circulation with which he studied mainly regulatory responses to Valsalva maneuvers and hemorrhage (see below). Especially for the latter it was important that stress relaxation (for which at least some quantitative data are available) be incorporated. At the same time, such a simple model can offer some guidance to the experimenter investigating vascular mechanics of the veins. Not only are simultaneous observations needed on pressure p, volume q (i.e., cross-sectional area), and distensibility $\Delta q/\Delta p$, but also on nonlinear phenomena such as stress relaxation and creep (i.e., volume change at constant pressure).

C. TRANSPORT OF FLUID THROUGH THE CAPILLARY WALL

As mentioned in the introduction of Section 16.A.2, fluid can be taken up or given off at the capillary level. By this fluid shift the body controls the local capillary pressure quite independently of mean, central pressure and vascular resistance. Guyton [16.12] summarizes the necessary pressure balance in the capillaries as follows:

(capillary fluid pressure) – (plasma colloid osmotic pressure)
= (tissue pressure) – (tissue colloid osmotic pressure)

If, for example, the fluid pressure in a capillary bed is too high, more liquid is expelled than reabsorbed so that tissue fluid pressure is increased, capillary pressure is reduced, and balance is maintained. Effective resistance to flow through the capillary wall is high; volume changes occur slowly. Hence this mechanism plays a role in the long-term regulation of volume and pressure (as do sodium retention and fluid transport by the kidney).

Fluid loss can be calculated as follows. Let p_a, p_c, and p_v be mean aortic, capillary, and venous pressures, respectively. With R_s as arteriolar resistance, R_c as resistance to flow through capillary wall, and R_v as resistance to flow through venules, we can write generally

$$p_a - p_c = R_s i \tag{16.26a}$$

$$p_c - p_v = R_v(i - i_c) \tag{16.26b}$$

$$p_c - 25 = R_c i_c \tag{16.26c}$$

where i = flow through arteries
i_c = flow through capillary wall
25 = interstitial fluid pressure for which $i_c = 0$, in the same units as p_a, p_c, and p_v

Hence, upon elimination of i and i_c from eqs. 16.26, we obtain an equation for p_c in terms of p_a and p_v:

$$p_c = \frac{R_v p_a + R_s p_v + 25 R_s R_v / R_c}{R_v + R_s + R_s R_v / R_c} \simeq \frac{R_v p_a + R_s p_v}{R_v + R_s} \quad \text{since} \quad R_c > R_a, R_v \qquad (16.27)$$

From the normal condition, such as $p_a = 90$ mmHg, $p_v = 6$ mmHg, and $p_c = 25$ mmHg, we can obtain the ratio $R_s/R_v = 65/19$. When this value is introduced into eq. 16.27, we find

$$p_c \simeq \left(\frac{19}{84} p_a + \frac{65}{84} p_v \right) \qquad (16.28)$$

If p_c, as given by eq. 16.28, deviates from 25 mmHg, i_c begins to flow, so that

$$i_c = \frac{p_c - 25}{R_c} = \frac{1}{R_c} \left(\frac{19}{84} p_a + \frac{65}{84} p_v - 25 \right) \qquad (16.29)$$

Here R_c is of the order of 200 mmHg/(cc/sec), since a capillary pressure change is reduced, by fluid loss, to about half its value within 90 min [16.9, see also 16.17]. Equation 16.29 enables us to compute outflux i_c through the capillary wall. Its integral is, of course, the blood volume change. The latter influences, in the complete model, such integration constants as P_0 in eq. 16.3. These considerations are correct, assuming that osmotic pressures are normal.

D. PERFORMANCE OF BENEKEN'S MODEL OF THE CONTROLLED CIRCULATION

To summarize the previous discussion of the different control loops, we reproduce the circuit diagram of Beneken's model [16.9] in Fig. 16.11. The equations governing the heart's performance are given in Chapter 15. The *systemic arterial tree* is represented by seven R–L–C–R sections, labeled A_0 to A_6; R simulates the viscous effects due to blood viscosity. In series with C, the analog of the distensibility of the arterial tree segment, we introduce a resistor that incorporates viscosity of the vascular wall. Clearly these circuit elements lump the properties of whole subsections of the vessel tree, which are named in Fig. 16.11. Although Beneken assumed equal proportional action of the vasomotor control center on peripheral resistance, the particular partition he chose might be used in a study of shifts of blood volumes from viscera to muscle in the extremities. The model, however, is not capable of controlling skin blood flow separately. For completeness we repeat the equations that must be solved in each of these arterial subsystems called A_0 to A_6. For the kth arterial segment we have

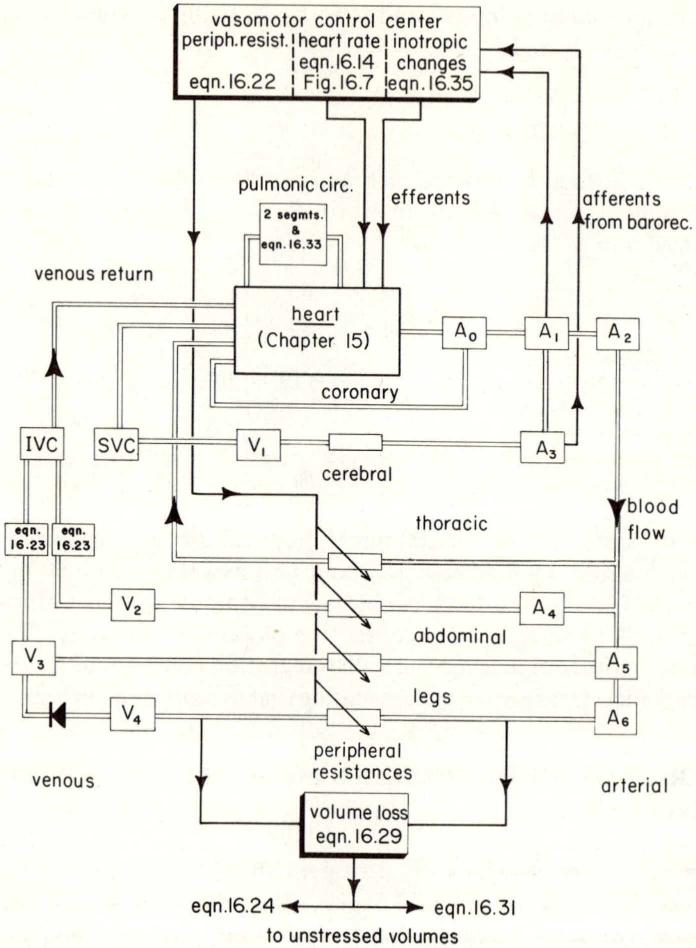

Fig. 16.11. Beneken's model of the entire cardiovascular system, showing compartments used to simulate controlled circulation: heart; seven arterial (A_0–A_6) and six venous (V_1–V_4, *IVC* and *SVC*) segments on the systemic side; two segments and a fairly complex pulmonary resistance (see eq. 16.33) on the pulmonary side. Double lines indicate paths of blood flow. The diode indicates valves in the venous circulation. Peripheral resistances are shown as rectangles; *IVC*, inferior vena cava; *SVC*, superior vena cava. Neural control circuits (heavy single lines) involve afferent paths from baroreceptors to vasomotor centers and efferent pathways to control peripheral resistances (arrows) and ventricular performances, as indicated in the box on top. Blood volume change depends on pre- and postcapillary pressures (bottom) and influences the state (i.e., "unstressed" volumes) of all vascular segments. The heart is described, in this model, by a system comparable to the one discussed in Chapter 15, to which Beneken added many refinements. After Beneken [16.9].

$$\underbrace{p_k - p_{k+1}}_{\text{Pressure drop across segment}} = \underbrace{R_k i_k}_{\text{Frictional pressure drop}} + \underbrace{L_k \frac{di_k}{dt}}_{\text{Inertial, accelerative force}} \tag{16.30}$$

$$q_k = \underbrace{q_{k0}}_{\text{Unstressed volume}} + \int (i_k - i_{k+1})\, dt \tag{16.31}$$

$$p_{k+1} = \underbrace{\frac{q_k - q_{k0}}{C_k}}_{\text{Elastic force}} + \underbrace{r_k \frac{dq_k}{dt}}_{\text{Viscous force in wall}} \tag{16.32}$$

The *systemic venous system* is represented by six segments of the kind shown in Fig. 16.10 and described by eqs. 16.23 to 16.25. The corresponding vessels are labeled in Fig. 16.11. In addition, Beneken put valves (i.e., diodes) at the places indicated in Fig. 16.11.

The *pulmonary circulation* was considerably simplified by Beneken [16.9]. He took into account the critical closing pressure ($p_{cc} \simeq 7$ mmHg) of the capillaries. If the pulmonary venous pressure (p_{pv}) is below p_{cc}, flow (i_{pulm}) is determined by

$$i_{\text{pulm}} = \frac{1}{R_{\text{pulm}}} (p_{pa} - p_{cc}) \qquad \text{if} \quad p_{pv} < p_{cc} \tag{16.33}$$

where R_{pulm} = a constant

p_{pa} = pulmonary artery pressure

If $p_{pv} > p_{cc}$, then we can use the common law

$$i_{\text{pulm}} = \frac{1}{R_{\text{pulm}}} (p_{pa} - p_{pv}) \tag{16.34}$$

The points to be emphasized here are the *control loops* incorporated in Beneken's model. It is seen in Fig. 16.11 that four mechanisms were simulated:

1. Baroreceptor reflex action on heart rate.
2. Baroreceptor reflex action on maximum isometric force of heart muscle (see Chapter 15).
3. Baroreceptor action on peripheral vascular resistance.
4. Blood volume control by fluid transport through capillary wall.

These mechanisms, discussed in detail in subsections of this chapter, are those for which quantitative data were available when Beneken was constructing the model. An additional feature of the model is that the pressure receptor reflexes do not influence the circulation in the brain, since this is regulated primarily by carbon dioxide concentrations in the blood. Furthermore, the duration of systole was made a function of heart rate.

Beneken also included indirect coupling between heart rate changes as induced by parasympathetic stimulation and contractility. If heart rate is stepped up, vigor of ejection decreases transiently, after which it climbs to a new, higher level (Koch-Weser and Blinks [16.18]). This can be introduced into the model by multiplying α, the activation factor, by a function H depending on heart rate. The factor α, introduced in Chapter 15, assigns a time-varying fraction of the maximum isometric force to the heart muscle (i.e., its contractile element). During maximum activation $\alpha = 1.0$. Beneken used the following equation for H as a function of the heart rate.

$$H(t) = a + bv - cve^{-t/\tau'} \tag{16.35}$$

where v = heart rate
t = time, referred to instant at which v changes
a, b, c, τ' = constants

On the other hand, many other factors are omitted from the model of Fig. 16.11. In particular, processing of neural information in the medullary vasomotor center is simulated only as far as Katona [16.8] observed it in anesthetized animals. Even Katona's approach did not take into account that the afferent neural information is used to act on the heart and on periphery in reference to a set point. This reference would make mean aortic pressure attain its resting level (e.g., 90 mmHg). In exercise this set point is shifted so that mean central pressure is higher than normal. More subtle weighting of sympathetic and parasympathetic action has been described by Robinson et al. [16.19]. Furthermore, there exist connections between the thermoregulatory center in the hypothalamus and the vasomotor area, in addition to the coupling with the motor cortex. Equally powerful interactions occur between the respiratory control center and the medullary vasomotor areas. Nevertheless, even if Beneken's model [16.9] can be checked only against limited experimental data (those for which factors assumed to be constants were kept reasonably constant or noninfluential), it is an important development, opening up new questions and calling for more investigations.

Beneken simulated the response of the cardiovascular system to the Valsalva maneuver. He also directly observed the time course of heart rate in healthy subjects asked to blow, with open glottis, into a reservoir so that air pressure was maintained at 40 mmHg for 12 sec. During this interval, intrathoracic and intraabdominal pressures are close to the same high level. Hence it is easy to perform a "Valsalva maneuver" on the computer. The abdominal and thoracic pressures (respectively p_a and p_t of eq. 16.23) are raised to a value analogous to 40 mmHg, for 12 sec.

Generally these high extravascular pressures impede venous return, upon which blood pools in the systemic veins. This causes a decrease in stroke volume, whereupon rate and vigor of contraction of the heart as well as peripheral resistances are reflexly affected to restore the mean central pressure. Variations in all these variables can be easily observed in the model and compared with physiological responses. Some information on blood pressure is available from Rushmer [16.20] and Sharpey-Schafer [16.21]. The neural signals from the aortic pressure sensors must be affected also by the high intrathoracic pressure during the Valsalva maneuver.

On the other hand, we might expect the carotid sinus receptors to sense the true deviations from normal blood pressure. The relative weights with which the afferent signals from these two receptor fields enter the reflex loops is not exactly known. Observations by Guyton [16.12] suggest equal weights. Beneken [16.9] also found with his model that at least the heart rate responded realistically when the neural information of the two receptors was made to control the effectors with equal weights. Even if this finding is considered to be a theoretical prediction, applicable perhaps only to a special case and subject to further, experimental verification, this is a good example of the possible application of a careful model study. If the existing elements of a system and their responses are known, but if their interconnections are less well established, the latter can be changed conveniently in a model until the entire analog conforms to observations. The task then remains to look for evidence in the living system to support or disprove the predicted interconnections.

Figure 16.12 shows an experimental response of heart rate during the Valsalva maneuver, together with heart rate as computed by the model. As

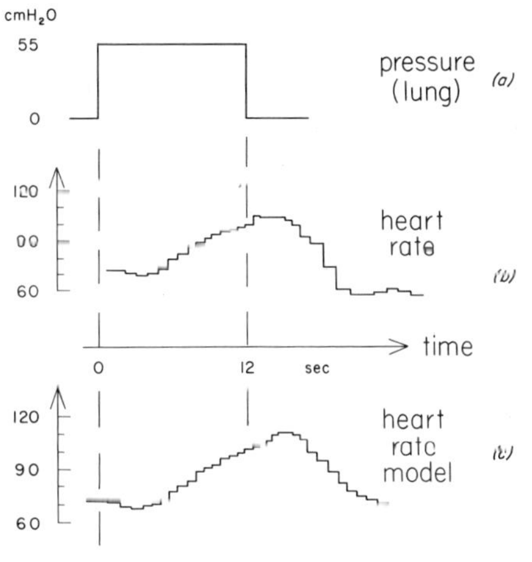

Fig. 16.12. Heart rate (beats/min) in response to a Valsalva maneuver: (*a*) and (*b*) as observed in man; (*c*) as predicted by the model discussed in this chapter. The Valsalva maneuver is controlled in that the test subject maintains approximately 55 cm of water in his airways and lungs during 12 sec (*a*). The model, as worked out by Beneken, predicts response best (*c*) if the signals from aortic and carotid baroreceptors are taken into account, by the vasomotor centers, with equal weights. After Beneken [16.9].

stated earlier, signals from baroreceptors in both segments A_1 and A_4 were made to affect heart rate, contractility, and peripheral resistances equally strongly. The mean aortic pressure thus is brought back, after 10 sec, very close to its original value, whereas stroke volume is considerably smaller, for venous return has diminished. At the very beginning of the experiment, the ventricle, having been filled normally, empties incompletely because of high aortic pressure. This causes the next few strokes to be larger than normal, which (together with the transient increase in mean pressure) causes the heart rate to decrease slightly during the first few seconds. Both this initial phase and the pressure regulation setting in after 5 to 10 sec are modeled quite satisfactorily. This, by itself, clearly is not a complete test of the applicability of the model. However, Beneken [16.9] also demonstrated that the responses of his model to hemorrhage (and infusion) were quite similar to the ones observed in nature.

It can be stated quite generally that, in addition to tests pertaining to physiological responses, a model itself should be carefully investigated. Since many of the subsystems were simplified, and several parameters approximated, it is important to investigate to what extent the model's responses depend on the various assumed parameters (parameter sensitivity). Let us illustrate this with an example. The systemic vascular tree was lumped into seven segments. Hence the input impedance, faced by the model's ventricle, cannot correspond exactly to the one in nature [16.9]. Thus the aortic pressure waveforms differ from reality and the afferent neural information from the baroreceptors may not equal what is needed in the baroreceptor reflex loop. However, this fault may not be severe if it can be shown that, for the particular problem, regulatory performance changes insignificantly when the phase of the aortic impedance better approximates that in animals.

We think it worthwhile to repeat that Beneken's model, designed to investigate some basic components of baroreceptor reflexes, does not include all existing mechanisms available for circulatory regulation. The peripheral resistances should be made to react to local concentration of oxygen, carbon dioxide, and other metabolic products which would require additional independent variables (i.e., metabolic rates) in each compartment. Furthermore, it has not been determined whether Beneken's choice of compartments (of the systemic circulation) would suffice to study complex reactions, such as those to exercise.

In moderate exercise, flow to active muscles and coronary flow increase, whereas splanchnic and renal perfusions may diminish. Blood flow to passive muscles may decrease, so does flow to the skin at the beginning of exercise. Cerebral flow stays constant in exercise unless vigorous ventilation causes a fall in arterial carbon dioxide concentration. Central blood pressure is higher than normal in exercise, and thus flows to muscle and skin can increase more than permitted by local vasodilation. As pointed out earlier, much simpler tasks of the cardiovascular systems, such as maintenance of pressure after assumption of upright posture, may also require more complete modeling of change in venomotor tone [16.19].

Clearly, certain aspects of pressure control by the baroreceptor reflexes can be studied elegantly with Beneken's analog [16.9]. Understanding of the coupling with the thermocontrol system, respiration, and the motor system, as well as local regulation of flow and shifts of blood volume will gain by further collaborative efforts of physiologists and systems engineers.

E. REFERENCES

16.1. Grodins, F. S. Integrative cardiovascular physiology: a mathematical synthesis of cardiac and blood vessel hemodynamics. *Quart Rev Biol* **34**: 93–116, 1959.

16.2. Grodins, F. S. *Control Theory and Biological Systems*. Columbia University Press, New York, 1963.

16.3. Landgren, S. On the excitation mechanism of the carotid baroreceptors. *Acta Physiol Scand* **26**: 1–34, 1952.

16.4. Warner, H. R. Simulation as a tool for biological research. *Simulation* **1**: 57–63, 1964.

16.5. Douglas, W. W., J. M. Ritchie, and W. Schaumann. A study of the effect of the pattern of electrical stimulation of the aortic nerve on the reflex depressor responses. *J Physiol (London)* **133**: 232–242, 1956.

16.6. Koepchen, H. P., H. D. Lux, and P.-H. Wagner. Untersuchungen über Zeitbedarf und zentrale Verarbeitung des pressoreceptorischen Herzreflexes. *Pflügers Arch Ges Physiol* **273**: 413–430, 1961.

16.7. Poitras, J. W., N. Pantelakis, C. W. Marble, K. R. Dwyer, G. O. Barnett, and P. G. Katona. Analysis of the blood pressure–baroreceptor nerve firing relationship. *Proc Annual Conf on Engineering in Medicine and Biology* **8**: 105 (no. 15.6), 1966.

16.8. Katona, P. G. Computer simulation of the blood pressure control of heart period. Sc.D. thesis, Massachusetts Institute of Technology, Cambridge, Mass., 1965.

16.9. Beneken, J. E. W. and B. DeWit. A physical approach to hemodynamic aspects of the human cardiovascular system, pp. 1–45 in *Physical Bases of Circulatory Transport*. E. B. Reeve and A. C. Guyton (Eds.). W. B. Saunders Co., Philadelphia, 1967.

16.10. Warner, H. R. Control of the circulation as studied with analog computer technics, pp. 1825–1841 in *Handbook of Physiology, Sec. 2 Circulation*, Vol. 3. W. F. Hamilton and P. Dow (Eds.). American Physiological Society, Washington, D.C., 1965.

16.11. Katona, P. G., G. O. Barnett, and W. D. Jackson. Computer simulation of the blood pressure control of the heart period, pp. 191–199 in *Baroreceptors and Hypertension*. P. Kezdi (Ed.). Pergamon Press, New York, 1967.

16.12. Guyton, A. C. *Textbook of Medical Physiology*, 4th ed. W. B. Saunders Co., Philadelphia, 1971.

16.13. Scher, A. M. and A. C. Young. Servoanalysis of carotid sinus reflex effects on peripheral resistance. *Circ Res* **12**: 152–162, 1963.

16.14. Bergel, D. H. The dynamic elastic properties of the arterial wall. *J Physiol (London)* **156**: 458–469, 1961.

16.15. Braunwald, E., J. Ross, Jr., R. L. Kahler, T. E. Gaffney, A. Goldblatt, and D. T. Mason. Reflex control of the systemic venous bed: effects on venous tone of vasoactive drugs, and of baroreceptor and chemoreceptor stimulation. *Circ Res* **12**: 539–552, 1963.

16.16. Noordergraaf, A. and E. Kresch (Eds.). The venous system: characteristics and function, a biomedical engineering approach. *IEEE Trans Bio-Med Engin* **BME-16**: 235–338, 1969.

16.17. Landis, E. M. and J. R. Pappenheimer. Exchange of substances through the capillary walls, pp. 961–1034 in *Handbook of Physiology, Sec. 2 Circulation*, Vol. 2. W. F. Hamilton and P. Dow (Eds.). American Physiological Society, Washington, D.C., 1963.

16.18. Koch-Weser, J. and J. R. Blinks. The influence of the interval between beats on myocardial contractility. *Pharmacol Rev* **15**: 601–652, 1963.

16.19. Robinson, B. F., S. E. Epstein, G. D. Beiser, and E. Braunwald. Control of heart rate by the autonomic nervous system: studies in man on the interrelation between baroreceptor mechanisms and exercise. *Circ Res* **19**: 400–411, 1966.

16.20. Rushmer, R. F. Circulatory effects of three modifications of the Valsalva experiment. *Amer Heart J* **34**: 399–418, 1947.

16.21. Sharpey-Schafer, E. P. Effects of respiratory acts on the circulation, pp. 1875–1886 in *Handbook of Physiology, Sec. 2 Circulation*, Vol. 3. W. F. Hamilton and P. Dow (Eds.). American Physiological Society, Washington, D.C., 1965.

CHAPTER SEVENTEEN

Respiratory Mechanics and Gas Exchange

The history of respiratory biophysics is old, and understanding of the relationships between functionally important parameters and variables was already advanced by the turn of the century. Experimental substantiation of theories followed later, based on ingenious techniques of measuring important quantities such as pleural and alveolar pressures, and work in this area has progressed steadily, together with related mathematical studies. It therefore would be unjust to the many contributors to the field if we were to claim to offer a complete survey in this chapter. Excellent reviews [e.g., 17.1, 17.2] show how important physics, mathematics, and control theory are to the respiratory physiologist. Here we restrict attention to relations that are important for analysis of the main function of the uncontrolled system. Beginning with the "fundament" (i.e., gas exchange in the effective compartment), behavior of the building blocks of the system is described and interconnected to form the whole, including the respiratory muscles. Thus analysis is oriented toward the following task: pumping air, by way of the airway system, to and from the alveoli to ensure continuous exchange of oxygen and carbon dioxide. No additional roles of the lung—for example, in cardiovascular dynamics, for water balance, for control of respiratory infections [17.3]—are discussed.

The geometry of the airway system of the tracheobronchial tree is necessarily complex because the large total surface (60–100 m^2) of the alveolar membranes must be ventilated by small volumes of air. The complex architecture of the bronchial system, in itself interesting to anatomists (Weibel [17.4]) and to aerodynamicists [17.5], has two major consequences with respect to our specialized viewpoint.

1. The oxygen and carbon dioxide contents within alveolar air do not correspond to the oxygen and carbon dioxide contents of inspired and expired air, respectively. The residual volumes, large compared with the tidal volume, are responsible for " dilution " of the gas concentrations of interest. Hence it is necessary to incorporate anatomic and functional dead spaces into our calculations. This is not easy, strictly, since residual volumes depend in part on ventilation rate and on previous events.

2. Pressure gradient–flow relations in the airways, and mechanical properties of lung tissue and of pleural space, are not easily formulated. Airflow through muscular, viscoelastic bronchi is turbulent during a large fraction of the respiratory cycle. Surface tension phenomena in the fluid film lining the alveoli are nonlinear and introduce instabilities, countered by nonlinear tissue forces and variable cohesion forces in the surfactant. In other words, the lung's volume–pressure–flow relationships are inherently nonlinear and nonstationary.

We show in the following how these two sets of problems can be analyzed from the systems analyst's viewpoint.

A. RESPIRATORY EXCHANGE RATIO AND GAS EXCHANGE IN THE THREE–COMPARTMENT MODEL

The goal of this section is to develop relations between the following functionally significant variables: flow rates and partial pressures of gases in airways and blood. First this is done for the *effective* lung compartment. We then show how the *anatomic dead space* (tracheobronchial tree and upper respiratory tracts) and the *functional dead space* of the lung (alveoli ventilated but not perfused by blood) are taken into account by more comprehensive analysis. In this we assume familiarity with the basic physical properties of gases and fluids and with the interaction of gases with blood, as summarized in most physiology textbooks [2.2]. The following analysis, which is detailed and quite rigorous, pertains only to a steady state of the lung, however.

1. Gas Exchange in Effective Compartment [17.6]

Considering all alveoli that are both ventilated and perfused (i.e., the effective alveolar compartment of the lung), the basic relations for mass flow are

$$\dot{V}_{CO_2} = \dot{V}_E F_{E,\,CO_2} \tag{17.1}$$

$$\dot{V}_{O_2} = \dot{V}_I F_{I,\,O_2} - \dot{V}_E F_{E,O_2} \tag{17.2}$$

where $\dot{V}_{CO_2}$ = flow (cc/sec) of CO_2 component out of effective compartment
$\dot{V}_{O_2}$ = net inflow of O_2 to effective compartment
$\dot{V}_E$ = expiratory flow
$\dot{V}_I$ = inspiratory flow
$F_{x,y}$ = fractional concentration in *dry* gas of component y (O_2 or CO_2) during phase of respiration x (E or I)

In these equations we take into account the small difference between inspired and expired volumes. Obviously, the variables $\dot{V}_{O_2}$ and $\dot{V}_{CO_2}$ are the most important variables functionally. It is the main task of the lung to furnish oxygen and to remove carbon dioxide. Therefore the ratio R, called the *respiratory exchange ratio*, is introduced:

$$R \equiv \frac{\dot{V}_{CO_2}}{\dot{V}_{O_2}} = \frac{F_{E,\,CO_2}}{F_{I,\,O_2}(\dot{V}_I/\dot{V}_E) - F_{E,\,O_2}} \tag{17.3}$$

This can be made more useful by introducing a quantity easily measured when the subject simultaneously breathes a gas that is not exchanged across the lung. If the gas is nitrogen, then

$$\dot{V}_E F_{E,\,N_2} = \dot{V}_I F_{I,\,N_2} \tag{17.4}$$

Under this condition, we have

$$F_{E,\,O_2} + F_{E,\,CO_2} + F_{E,\,N_2} = 1 \tag{17.5a}$$

$$F_{I,\,O_2} + F_{I,\,N_2} = 1 \tag{17.5b}$$

Hence

$$R' = \frac{F_{E,\,CO_2}(1 - F_{I,\,O_2})}{F_{I,\,O_2}(1 - F_{E,\,CO_2}) - F_{E,\,O_2}} \tag{17.6}$$

The symbol R' indicates that the exchange ratio is calculated from observed quantities and plays the role of a parameter in the following. The gas fractions F are measured, in this case, at the input to the effective compartment. The tracheobronchial airways, however, are not actively engaged in gas exchange. We therefore should distinguish between total input quantities and ventilation and gas fractions of the alveolar region of the effective compartment ($\dot{V}^{ac}$ and F^{ac}, respectively). The law of mass conservation yields

$$\dot{V}_E F_{E,\,O_2} = \dot{V}_E^{ac} F_{E,\,O_2}^{ac} + (\dot{V}_E - \dot{V}_E^{ac}) F_{I,\,O_2} \tag{17.7}$$

$$\dot{V}_E F_{E,\,CO_2} = \dot{V}_E^{ac} F_{E,\,CO_2}^{ac} \tag{17.8}$$

(since carbon dioxide uptake is zero).

When F_{E, O_2} and F_{E, CO_2} from these equations are introduced in eq. 17.6, we find R' in terms of the alveolar gas fractions:

$$R' = \frac{F^{ac}_{E, CO_2}(1 - F_{I, O_2})}{F_{I, O_2}(1 - F^{ac}_{E, CO_2}) - F^{ac}_{E, O_2}} \tag{17.9}$$

Gas fractions, however, are easily converted into partial pressures, since

$$P_y = F_y(P_{baro} - P_{H_2O}) \tag{17.10}$$

where P_y, P_{H_2O} = partial pressures of gas y and water vapor, respectively
P_{baro} = barometric pressure

We now can find the relation between alveolar partial pressures of oxygen and carbon dioxide at a given respiration exchange ratio R'. Introducing eq. 17.10 into eq. 17.9 and solving for P^{ac}_{E, CO_2} gives

$$P^{ac}_{E, CO_2} = \left[F_{I, O_2} + \frac{1 - F_{I, O_2}}{R'}\right]^{-1}(P_{I, O_2} - P^{ac}_{E, O_2}) \tag{17.11}$$

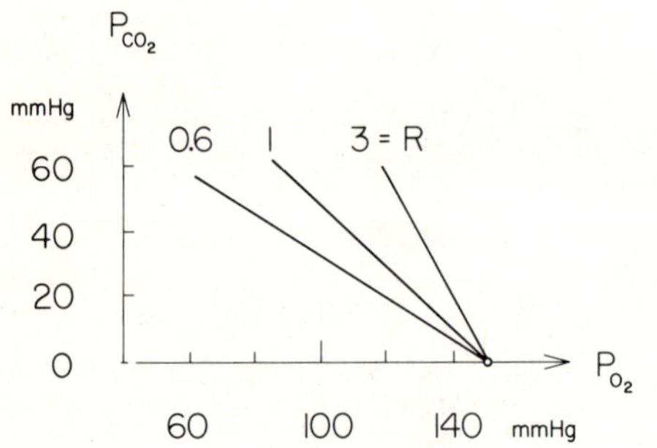

Fig. 17.1. Relation between partial pressures of carbon dioxide and oxygen in alveolar *air* of the effective compartment of the lung (eq. 17.11), for different respiratory exchange ratios R (valid when breathing normal air). This diagram pertains to the air spaces only and does not take into account gaseous exchange with blood. The lines, called isopleths, must be superimposed on the lower graph of Fig. 17.2*a* to yield the locus of intersections seen in Fig. 17.3. After Riley [17.6].

This equation is graphed in Fig. 17.1 showing different isopleths (i.e., for different values of R'). We also see from the foregoing equations that R is approximately equal to the ratio of P^{ac}_{E, CO_2} and $(P_{I, O_2} - P^{ac}_{E, O_2})$—that is, the ratio of pressure differences of inspired gas and the alveolar component of the expired gas.

Equation 17.11 clearly does not suffice for analysis of lung function because gas exchange with the blood has not been introduced. The gaseous and fluid phases are closely coupled (alveolar membrane thickness ca. 1 μm; transit time of blood particles through capillaries ca. 0.7 sec, in normal state). In fact, we can assume that no difference in partial pressures exists across the capillary membrane, at least on the pulmonary venous side. Thus the exchange ratio R computed earlier for the gas phase must also apply to pulmonary circulation, and R therefore establishes a relation between P_{O_2} and P_{CO_2} in blood, which must hold simultaneously with eq. 17.11.

The equations of mass transport through the pulmonary capillaries yield, for *blood* perfusing the effective alveoli,

$$\dot{V}_{CO_2} = \dot{Q}_a^{ac} C_{v,\,CO_2} - \dot{Q}_a^{ac} C_{a,\,CO_2}^{ac} \tag{17.12}$$

$$\dot{V}_{O_2} = \dot{Q}_a^{ac} C_{a,\,O_2}^{ac} - \dot{Q}_a^{ac} C_{v,\,O_2} \tag{17.13}$$

where $\dot{Q}_a$ = (arterial) blood flow rate (cc/sec)

C_a, C_v = arterial and venous gas concentrations, respectively

Hence R from the viewpoint of gas transport by blood must be

$$R = \frac{C_{v,\,CO_2} - C_{a,\,CO_2}^{ac}}{C_{a,\,O_2}^{ac} - C_{v,\,O_2}}$$

or

$$C_{a,\,CO_2}^{ac} = C_{v,\,CO_2} + C_{v,\,O_2} R - R C_{a,\,O_2}^{ac} \tag{17.14}$$

To determine P_{O_2} and P_{CO_2} that must exist at the venous end of capillaries, we use the dissociation curves to convert (graphically) concentrations into partial pressures. Figure 17.2 illustrates how results of eq. 17.14 are used to arrive at isopleths for blood.

Because equilibrium can be assumed for any one alveolus, the P_{O_2} and P_{CO_2} existing for any given value R must lie on the intersection of the corresponding gas and blood isopleths. The locus of these intersections, the so-called distribution curve, appears in Fig. 17.3. Different alveoli may lie at different points on the curve. An unperfused alveolus, belonging to the functional dead space, is at the left end of the curve. If an alveolar shunt exists in a part of the lung, its position is at the extreme right of the distribution curve. For the effective compartment of the lung, however, there is an average P_{O_2}–P_{CO_2} combination on the distribution curve. Partial pressures in expired gas, on the other hand, differ from these because of mixing with gases from anatomical and functional dead spaces.

2. Effect of Dead Spaces; the Three-Compartment Lung

The distribution curve of Fig. 17.3 relates the gas partial pressures within the effective alveolar compartment (volume V^{ac}) with the respiratory exchange ratio as a parameter. Because there always exist unperfused alveolar volumes (functional or alveolar dead space $V^{d\,alv}$) and anatomical dead

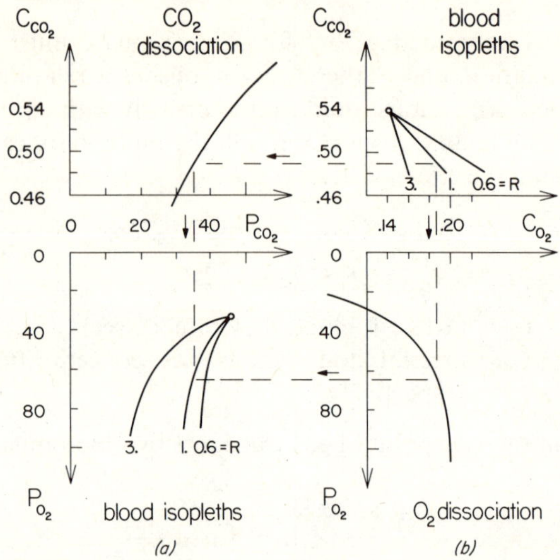

Fig. 17.2. Relation between concentrations of gases in *blood* (units, cc/cc of blood) (*b*), top, as given by the law of conservation of mass (eq. 17.14), shown for different *R*'s (respiratory exchange ratio). In order to match these data to Fig. 17.1, the relation between concentrations must be converted to that between partial pressures. For this the dissociation curves of blood for oxygen (*b*, bottom) and for carbon dioxide (*a*, top) are needed, as shown for normal hemoglobin content. Thus, by projection of any point on the concentration isopleth, the new point on the partial pressure isopleth is obtained. To yield the locus of the intersections of corresponding isopleths, as given in Fig. 17.3, we must superimpose (*a*, bottom) on Fig. 17.1.

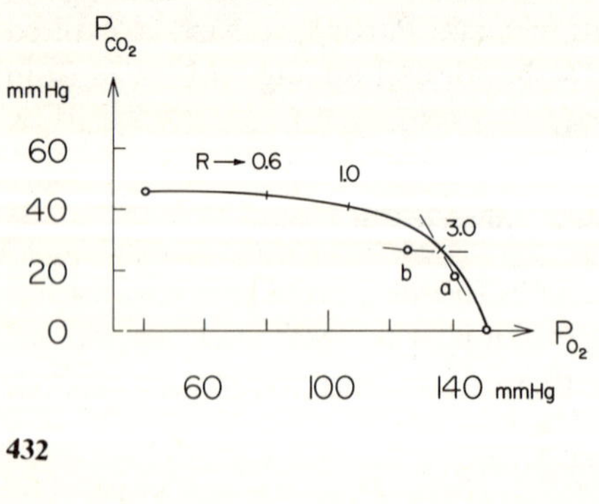

Fig. 17.3. The *distribution curve* of possible P_{O_2} and P_{CO_2} values in the *effective* compartment of the lung. The curve is the locus of the intersections of corresponding *R*-lines of Figs. 17.1 and 17.2; *R*, the exchange ratio, varies along the curve from zero at the left to infinity at the lower right end of the curve. Interrupted lines show the isopleths for air and blood ($R = 3.0$). The distribution curve gives partial pressures for equilibria between gaseous and fluid phases, and point *a* marks possible partial pressures in *air* resulting from mixing with air from dead spaces (see eq. 17.17). Similarly, point *b* marks the shift in partial pressures resulting from arteriovenous shunts in the alveolar capillaries. After Riley [17.6].

spaces (volume $V^{\text{d ana}}$), partial pressures in expired air differ significantly from those at the alveolar level. The relations are straightforward. Using the following symbolism:

$$V_E = V_E^{\text{e}} + V_E^{\text{d alv}} + V_E^{\text{d ana}} \tag{17.15}$$

where V_E = expired volume
V_E^{e} = expired volume from *effective* region of alveolar compartment
$V_E^{\text{e}} + V_E^{\text{d alv}} = V_E^{\text{ac*}}$, the total alveolar component volume

the partial pressure of carbon dioxide in expired air is related to that in the effective compartment:

$$\frac{P_{E,\,CO_2}}{P_{E,\,CO_2}^{\text{e}}} = \frac{V_{E,\,CO_2}^{\text{e}}}{V_{E,\,CO_2}} \equiv \frac{V_E^{\text{e}}}{V_E} \tag{17.16}$$

Experimentally $P_{E,\,CO_2}^{\text{e}}$ can be obtained by analyzing the last air leaving the mouth after a sudden forced expiration preceded by a series of normal respiratory cycles, so that steady state in $P_{CO_2}^{\text{e}}$ is reached. Equation 17.16 therefore can be employed when computing the volume of the dead spaces.

$$\frac{V_E^{\text{d alv}} + V_E^{\text{d ana}}}{V_E} = 1 - \frac{V_E^{\text{e}}}{V_E} \simeq 1 - \frac{P_{E,\,CO_2}}{P_{E,\,CO_2}^{\text{e}}} \tag{17.17}$$

To gain insight into the effects of dead spaces on (spatial) distribution of oxygen concentrations, we note that, in a first approximation,

$$V_E F_{E,\,O_2} = (V_E^{\text{d alv}} + V_E^{\text{d ana}})F_{I,\,O_2} + V_E^{\text{e}} F_{E,\,O_2}^{\text{e}} \tag{17.18}$$

or, when dividing by V_E and using eq. 17.17,

$$P_{E,\,O_2} = \left(1 - \frac{P_{E,\,CO_2}}{P_{E,\,CO_2}^{\text{e}}}\right) P_{I,\,O_2} + \frac{P_{E,\,CO_2}}{P_{E,\,CO_2}^{\text{e}}} P_{E,\,O_2}^{\text{e}} \tag{17.19}$$

Thus $P_{E,\,O_2}$ differs from $P_{E,\,O_2}^{\text{e}}$ by significantly more than the "dilution" factor V_E^{e}/V_E. The first term on the right side of eq. 17.19 is large due to oxygen storage in the dead spaces. Figure 17.3 relates the location of input signals $P_{E,\,O_2}$ and $P_{E,\,CO_2}$ to the distribution curve, although it pertains only to the effective alveolar part.

The three-compartment model allows for a satisfactorily complete description of lung function, although several approximations were made. Because we have neglected the fact that gases exhaled from the effective alveolar compartment reach the alveolar dead space during the following inspiration, eqs. 17.16 and 17.18 do not strictly apply. Neither can we be sure that $P_{E,\,CO_2}^{\text{e}}$, an important observable in the foregoing equations, can be obtained with satisfactory accuracy. In fact, large alveolar dead space and large arterio-venous shunts increase the difficulty in obtaining $P^{e}_{E,\,CO_2}$. Special care should be exercised in defining the steady state of the lung, one of the basic assumptions for the model. Steady state pertains to signals *as well as* to those system's properties which can be changed, for example, by one deep breath. Nevertheless, as elucidated by Riley [17.6], the model presented here is of great conceptual and tutorial value, allowing for satisfactory evaluation of the functionally important events.

It is impossible to predict, from information presented so far, the dynamics of the system governing transition from one state into another. For this, spatial pressure distribution, compliances, and resistances must be known. Moreover, important roles are played by the couplings between air spaces and parenchyma of the lung and surrounding spaces and tissue, as explained in the following sections.

B. PULMONARY MECHANICS; LARGE SIGNAL ANALYSIS

Let us continue to build up the respiratory system. First the dynamics of lung ventilation and the energetics of breathing must be explored. It then becomes possible to connect the functional aspects of gas exchange with those of air transport.

There are important consequences for the study of the aerodynamics in the anatomical finding that the tracheobronchial airway system is mechanically imbedded in the parenchyma and is in parts directly acted upon by intrathoracic pressure. These couplings make it too restrictive to model the system by an in-series arrangement of one airway resistance and one viscoelastic compartment representing alveoli and lung tissue. Results of clinical tests (maximum effort vital capacity; maximum expiratory flow), as well as observations in diseased lungs, could not be understood on the basis of such a simple system. This holds even if the model is made nonlinear by introducing turbulent flow and a compliance decreasing with increasing volume. We therefore follow Fry's [17.7] approach to characterizing the relation between transpulmonary pressure and flow. We find an essential nonlinearity in the pressure–flow relation and a compliance that is highly nonuniform (i.e., a function of volume and flow).

Consider the simple pneumatic system of Fig. 17.4*a* which incorporates the salient mechanical features of the lung. The dimensions of the air passages depend not only on their elastic properties and the spatial pressure distribution within as determined by flow, but also on the tension exerted on the outside by the surrounding viscoelastic mesh structure *and* by intrathoracic pressure P_t. (Note that P stands, henceforth, for air pressure.) During *inspiration*, the action of diaphragm and rib cage (i.e., movement to the left of the piston in the schema of Fig. 17.4*a*) causes inflow by lowering P_t and tends to widen the airways, especially the extrapulmonary ones. The pressure–flow relation ($P_t - \dot{V}$) is therefore concave toward the $\dot{V}$ axis (see Fig. 17.4*b*). We do not take into account, in particular, the resistance of the nasal airways, across which about 50% of the total pressure drop normally appears. Neither are Bernoulli pressure drops considered. These amount to but a few centimeters of water column in normal cases.

During *forced expiration*, large P_t's act on the outside of the air passages, directly or indirectly, to increase resistance to flow. In fact, resistance may increase so much that flow is limited, or even decreased, when P_t is increased

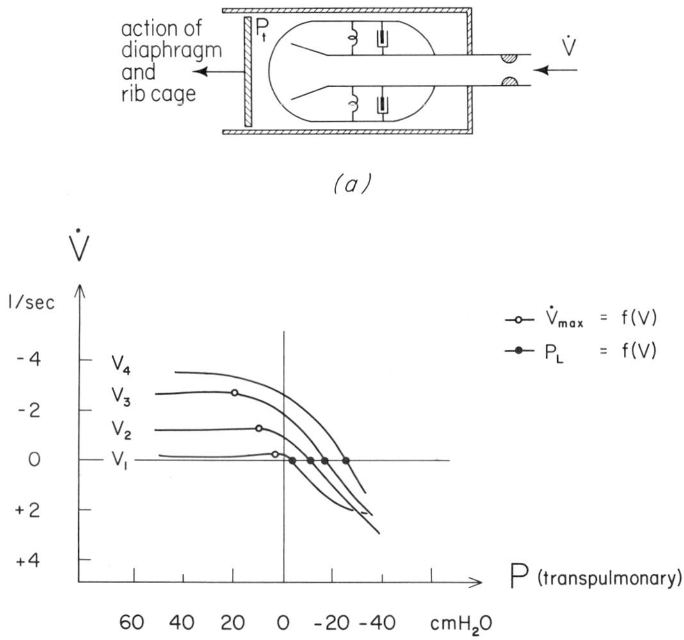

(b)

Fig. 17.4. (*a*) Simplified mechanical model of the lung, which explains (*b*) the pressure–flow curves. The most important couplings between pressure, flow, and systems properties are introduced into this system. In particular P_t, the intrathoracic pressure, influences directly the dimensions of the larger airways. Expiratory flow therefore is limited at $\dot{V}_{max}$ (*b*, top), especially at small lung volumes. The spring–dashpot combinations in (*a*) approximate only roughly the nonlinear rheological properties of lung tissue. Transpulmonary pressure at zero flow, P_L (*b*), which does not depend on lung volume in a simple proportional manner, must be obtained to estimate effective static lung compliance. See Fig. 17.5 for the complete description needed for study of the dynamic phenomena. After Fry and Hyatt [17.7].

beyond a critical level (see open circles in Fig. 17.4*b*). Flow maxima are particularly distinct in emphysematous lungs [17.7]. However, it is obvious that the level of critical pressure must depend on the state of the spring–dashpot system, that is, the distension of the lung. At large volumes, P_t must also be large if flow is to be limited. At small distensions, resistance may be increased by a smaller P_t to reach its flow-limiting value. This is seen in the curves of Fig. 17.4*b*, constructed from measurements of differences between esophageal ($\simeq$ intrathoracic) and tracheal pressures, flow, and volume above residual volume. A particular P_t–$\dot{V}$ curve can only be obtained by connecting the pairs of pressure and flow values that exist at a fixed volume. During one breath, several different P_t–$\dot{V}$ curves are touched, as we explain in more detail below.

This very simplified model permits prediction of natural phenomena occurring in a complex structure [17.4] and caused by complicated mechanisms involving many variables. It is also interesting to note that the classical expression [17.2] $\Delta P \propto (\alpha\dot{V} + \beta\dot{V}^2)$, used for turbulent flow in rigid pipes, seems to be applicable in a limited region for small volumes only (see curve V_1 in Fig. 17.4*b*, concave toward the *P*-axis in inspiration). At volumes larger than 0.5 liter, the factors explained previously dominate in determining "resistance." It should be remembered that viscous elements of lung tissue, significant in disease, also play a role in determining resistance because the driving pressure is P_t and not alveolar pressure.

Fry et al. [17.7] showed that dependence of peak flow (open circles in Fig. 17.4*b*) on volume yields clinically useful information on the state of the lung, particularly because the function is not influenced by the pressure drop in the upper respiratory tract.

Closed circles on the P–$\dot{V}$ curves in Fig. 17.4*b* mark pressures (P_L) necessary to distend the lung to the corresponding volumes at zero flow. As a function of volume, P_L is a nonlinear curve from which an incremental compliance ($\Delta V/\Delta P_L$) can be derived. Factors determining compliance are: the elastic components of lung tissue dependent on elastin and collagen structures, distensibility of the pulmonary vasculature, and the variable surface tension of the surfactant lining the alveoli.

In this study of overall lung peoperties, the degree of mechanical coupling between the elastic mesh of the lung and the inner surface of the visceral pleura is another important factor shaping the P_L–V curve. It also should be remembered that both uneven ventilation (i.e., pressures not equalizing throughout lung) and recruitment of alveolar spaces influence the P_L–V relation obtained in the manner described.

We refer the reader to the literature [17.1, 17.8] for discussions of interaction between alveolar distensibility and stress-dependent surface tension of the fluid linings, resulting in stabilization of alveolar volume. These phenomena are important at the level of the individual alveolus. In the systems analysis of this chapter they must not be introduced separately, although they can play significant roles in abnormal lungs (hyaline membranes; deficiency of lung surfactant, e.g., after prolonged cardiopulmonary bypass). Optimal insight into the overall behavior of the normal lung is gained by developing, after Fry [17.7], the three-dimensional surface of Fig. 17.5.

Figure 17.5 is a view of the surface in the V–P–$\dot{V}$ space to which the trajectory of any breathing maneuver is confined. The surface is bounded by the curve of a maximum effort vital capacity breath (heavy dashed curve). During inspiration it coincides with the maximum flow vital capacity trajectory [17.7]. The extreme segments of a maximum breathing capacity test are also given and are marked *mb*. Normal tidal breathing follows the small trajectory, *t*. The system obviously may be linearized in the very limited region of trajectory *t*. However, most other modes of breathing comprise nonlinear performance. Mechanical work done on the lung, and in particular energy put into the system (i.e., absorbed by frictional resistances) can be evaluated in projections of trajectories on the V–P plane. Loop *t* for tidal breathing appears as a small ovoid curve centered on *C*. The partial area of the circle on the inspiratory side of *C* approximately equals that on the expiratory side. Hence the energies expended are about the same, too. The situation is different during a maximum breathing capacity test, for which most of the energy is dissipated during expiration (curve *mb*).

Projections of trajectories on the P–$\dot{V}$ plane can yield insight into the ratio of power stored and power dissipated. For this we need to assume sinusoidal signals, which obviously can only exist over small regions of the V–P–$\dot{V}$ surface. Nevertheless, the state of the lung often is tested by applying artificial, sinusoidal respiration from which pneumatic input impedance can be calculated. In artificial respiration, the factors shaping the P–$\dot{V}$ trajectories differ from those listed earlier for spontaneous breathing. Driving pressure does not act directly on the outside of the tracheobronchial tree. In the inspiratory phase, forces needed to expand the thoracic cage must be balanced by stresses in the lung mesh structure. In any case, both resistive and reactive parts of an impedance determined externally depend on mean lung distension. In addition, only in normal lungs is resistance approximately independent of applied frequency.

Thus Fry's approach is powerful as a guide to analysis of functionally important mechanical events. The shape of the V–P–$\dot{V}$ surface defines the state of the lung. Not all regions of the surface, however, are known with equal precision. For example, the inspiratory segments of the P–$\dot{V}$ curves at low volumes have not been investigated thoroughly. Other problems not yet fully explored concern coupling between lung compliance, intrathoracic pressure, and breathing musculature, which is related to tests of compliance at large

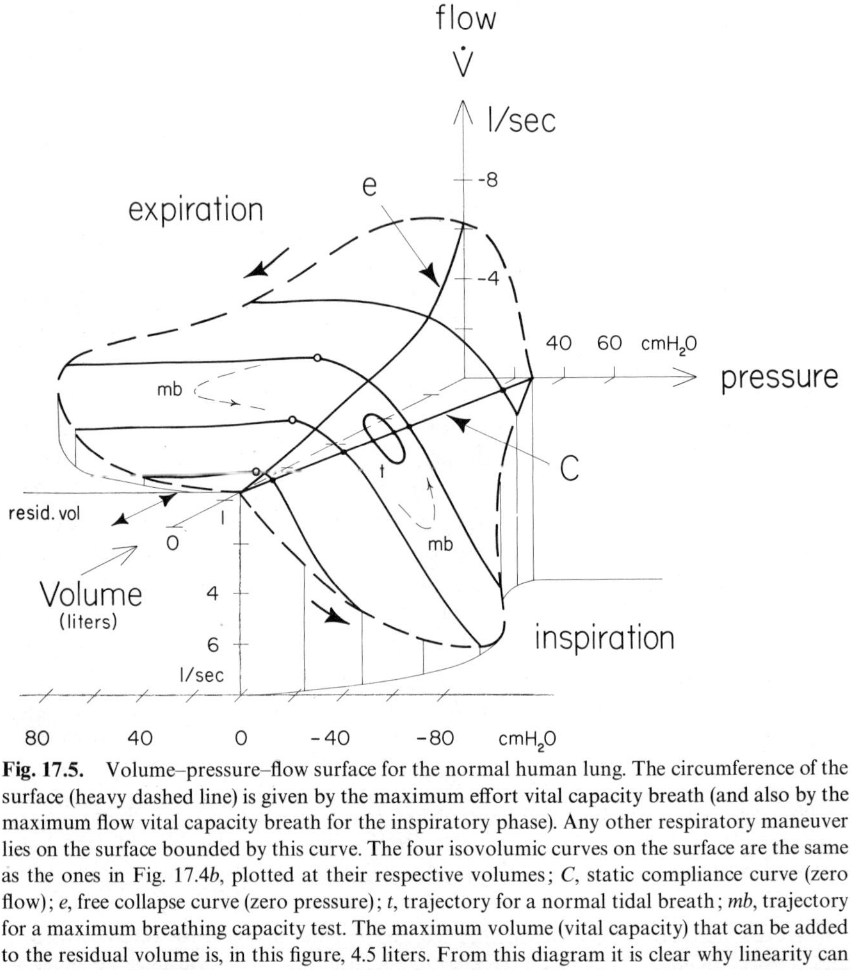

Fig. 17.5. Volume–pressure–flow surface for the normal human lung. The circumference of the surface (heavy dashed line) is given by the maximum effort vital capacity breath (and also by the maximum flow vital capacity breath for the inspiratory phase). Any other respiratory maneuver lies on the surface bounded by this curve. The four isovolumic curves on the surface are the same as the ones in Fig. 17.4*b*, plotted at their respective volumes; *C*, static compliance curve (zero flow); *e*, free collapse curve (zero pressure); *t*, trajectory for a normal tidal breath; *mb*, trajectory for a maximum breathing capacity test. The maximum volume (vital capacity) that can be added to the residual volume is, in this figure, 4.5 liters. From this diagram it is clear why linearity can be assumed only for tidal breathing (in the normal case). However, most breathing tests used clinically follow trajectories extending into the nonlinear regions. See [17.7] for a discussion of projections of various trajectories on P–V, P–$\dot{V}$, or V–$\dot{V}$ planes. After Fry and Hyatt [17.7].

volumes and maximum effort vital capacity. Moreover, position of the patient and previous breathing history influence the surface in the state space (Fig. 17.5).

To complete the uncontrolled respiratory system, the simple lung model must be connected to the mechanical system of rib cage, abdomen, and diaphragm with their muscles. Jodat, Horgan, and Lange [17.9] have made a beginning in analyzing the *entire system*, and in particular the small signal

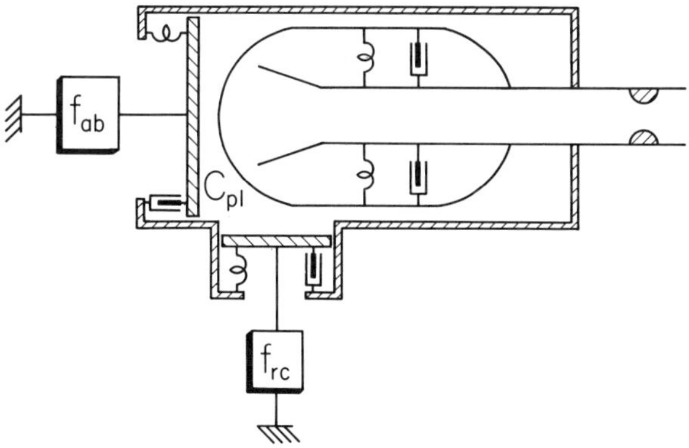

Fig. 17.6. Schematic model of the whole mechanical respiratory system. The action of the diaphragm and abdominal muscles is introduced separately from that of the rib cage. Here f_{ab} and f_{rc} represent the force sources of abdominal and rib cage muscles, respectively. The spring–dashpot systems connecting to the pistons represent elastic and velocity-limiting elements of these muscles; C_{pl} is the compliance of intrapleural cavity. The lung is represented by the system explained in legend of Fig. 17.4. After Jodat et al. [17.9]

performance of the model shown in Fig. 17.6. In the model, pressure acting on the outside of the lung depends on forces on two "pistons": forces determined by the intercostal muscles and those resulting from contraction of diaphragm and changes in abdominal volume. These force sources, in turn, are lumped into one ideal force generator in parallel with a dashpot and a spring. This is a satisfactory model of any muscle system, as was discussed in Chapter 10. Thus the equations governing behavior of the entire system can be found by inspection of Fig. 17.6:

For thoracic space compartment due to change in rib cage volume:

$$\rho_{rc}\dot{V}_{rc} + \frac{1}{\gamma_{rc}}V_{rc} = P_{rc} + P_{pl} \tag{17.20}$$

where P_{pl} = pleural pressure

For thoracic space compartment determined by position of diaphragm:

$$\rho_{ab}\dot{V}_{ab} + \frac{1}{\gamma_{ab}}V_{ab} = P_{ab} + P_{pl} \tag{17.21}$$

For lung volume:

$$\rho_{l}\dot{V}_{l} + \frac{1}{\gamma_{l}}V_{l} = -P_{pl} \tag{17.22}$$

where

$$V_{rc} + V_{ab} - V_l = V_{pl} \tag{17.23}$$

and the frictional component and the compliance are ρ and γ, respectively.

Note that linearity is assumed in eqs. 17.20 to 17.23. In particular, eq. 17.22 should incorporate the volume dependency of ρ_l and γ_l, as evident from Fig. 17.4*b*. The intrapleural space is assumed, in a first approximation, to behave like an elastic chamber:

$$P_{pl} = \frac{1}{C_{pl}} V_{pl} \tag{17.24}$$

We can express the interdependence of P_{rc} and P_{ab} elegantly, as follows [17.9]:

$$\begin{aligned} P_{rc} &= kP_m \\ P_{ab} &= (1 - k)P_m \end{aligned} \tag{17.25a}$$

where P_m = pressure due to total muscle effort
$k = 0.75$ (= 0.7 in exercise [17.10])

Jodat et al. [17.9] chose a time course for P_m which best fits available information:

$$\begin{aligned} P_m &= P_{\max}[1 - \cos(\alpha_1 t)] \qquad \text{in inspiration} \\ &= P_{\max}[1 - \cos(\alpha_2 t - \alpha_1 \Delta t_{\text{insp}})] \qquad \text{in expiration} \end{aligned} \tag{17.25b}$$

where $\alpha_1 < \alpha_2$.

Furthermore, mechanical coupling between abdominal and rib cage cavities should be considered. Changes in P_{rc}, called P_{fr}, due to variations in abdominal volume, occur because stretching (i.e., flattening) of the diaphragm lets the lower part of the rib cage expand. This happens especially during early inspiration. In the following we assume with Jodat et al. [17.9] a linear relation between P_{fr} and P_{ab}, but only up to a limiting value P_c. Formally, during inspiration, we have

$$\begin{aligned} P_{fr} &= mP_{ab} \qquad \text{if} \quad P_{ab} \leqslant mP_{ab,\max} \\ P_{fr} &= P_c \qquad \text{if} \quad P_{ab} > mP_{ab,\max} \end{aligned} \tag{17.26}$$

and during expiration

$$\begin{aligned} P_{fr} &= P_c - m(P_{ab,\max} - P_{ab}) \qquad \text{if} \quad P_{ab} \geqslant \frac{P_c}{m} - P_{ab,\max} \\ P_{fr} &= 0 \qquad \text{otherwise} \end{aligned} \tag{17.27}$$

Figure 17.7 presents the functional relationships between the variables governing performance of the whole system. The results of eqs. 17.26 and 17.27 appear graphically.

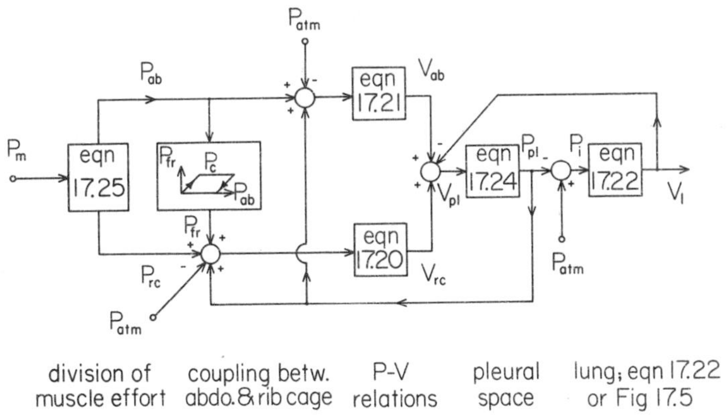

Fig. 17.7. Functional relationships of variables significant in respiratory mechanics. Some transfer functions are hypothetical, these may be determined by matching the overall performance of the model with experimental values, after which they should be verified separately and experimentally. See text. After Jodat et al. [17.9].

As Horgan and his co-workers indicated, this model has considerable value even though some transfer functions are hypothetical (e.g., eq. 17.24) and some constants are unknown (k and m). First, the development of the model has forced us to bring couplings between the subsystems into quantitative form. Second, by adjusting unknown parameters such that the overall performance simulates experimental findings, we can derive values for some unknowns. (This step requires, however, a careful parameter sensitivity study.) For example, Jodat et al. [17.9] were able to match curves in this manner fairly well, for tidal breathing, by choosing $P_{max} = 3$ cm H_2O (eq. 17.25b), $k = 0.2$, $P_c = 3$ cm H_2O. Third, a model study of this sort leads back to the laboratory, where the physiologist in collaboration with the engineer will want to verify the hypotheses and validate transfer functions that were adjusted by curve fitting. In this, the beauty of a model study is displayed best. Often, the functional importance of some variables and parameters is recognized only after theoretical *and* experimental investigations. It then becomes possible to attack the remaining problems and to introduce known nonlinearities and control mechanisms.

C. REFERENCES

17.1. *Handbook of Physiology, Sec. 3 Respiration*, Vol. 1. W. O. Fenn and H. Rahn (Eds.). American Physiological Society, Washington, D.C., 1964.

17.2. Mead, J. Mechanical properties of lungs. *Physiol Rev* **41**: 281–330, 1961.

17.3. Heinemann, H. O. and A. P. Fishman. Nonrespiratory functions of mammalian lung. *Physiol Rev* **49**: 1–47, 1969.

17.4. Weibel, E. R. *Morphometry of the Human Lung*. Springer, Berlin, 1963.

17.5. Olson, D. E., G. A. Dart, and G. F. Filley. Pressure drop and fluid flow regime of air inspired into the human lung. *J Appl Physiol* **28**: 482–494, 1970.

17.6. Riley, R. L. Gas exchange and transportation, pp. 761–787 in *Physiology and Biophysics*, 19th ed. T. C. Ruch and H. D. Patton (Eds.). W. B. Saunders Co., Philadelphia, 1965.

17.7. Fry, D. L. and R. E. Hyatt. Pulmonary mechanics: a unified analysis of the relationship between pressure, volume and gasflow in the lungs of normal and diseased human subjects. *Amer J Med* **29**: 672–689, 1960.

17.8. Mead, J., J. L. Whittenberger, and E. P. Radford, Jr. Surface tension as a factor in pulmonary volume–pressure hysteresis. *J Appl Physiol* **10**: 191–196, 1957.

17.9. Jodat, R. W., J. D. Horgan, and R. L. Lange. Simulation of respiratory mechanics. *Biophys J* **6**: 773–785, 1966.

17.10. Grimby, G., J. Bunn, and J. Mead. Relative contribution of rib cage and abdomen to ventilation during exercise. *J Appl Physiol* **24**: 159–166, 1968.

CHAPTER EIGHTEEN

The Respiratory Chemostat

A. INTRODUCTION

The respiratory control system is explained in most modern physiology books [2.3] using concepts of control theory. Simple observations tell us that the organization of the respiratory control system must indeed resemble that of a closed feedback system. Chemosensitive organs detect deviations from "normal" levels of gas pressures in different body structures. Information on gas pressures is conveyed, by neural pathways, to the respiratory control centers in the brain, which in turn adjust ventilation of the lungs to bring gas pressures back to normal levels. The system therefore is called the respiratory chemostat.

In one mode of action, the chemostat reacts to small transient increases in partial pressure of carbon dioxide (P_{CO_2}) in the blood (which may be caused by admixture of carbon dioxide with inspired air) by large increases in ventilation, such that blood P_{CO_2} is restored to its initial value.

In general, it appears to fulfill its function of transporting oxygen-rich air into and carbon dioxide-rich air from the lungs, according to the body's demand, by servo-controlling blood gases. This observation leads immediately to many interesting questions. Which of the (partial) gas pressures are sensed and regulated? Where are gas pressures being sensed? How is the normal P_{CO_2} level determined, or which element of the respiratory system corresponds to the "reference signal" (set point) of the classical feedback system [0.8]? What are the static and dynamic input–output relations of important parts such as the controlled portion (lungs, circulatory system, tissue), the sensors, and the control center (also called the controller)? And on the effector's side, what are the mechanisms by which homeostasis is achieved?

Information is available today which allows us to answer most of these questions, at least in part, well enough that many aspects of the system's performance can be explained, predicted, or both [18.1]. The respiratory control centers in the brain stem (i.e., the lateral reticular substance of the medulla and the lower part of the pons) are known to affect respiration. Regulation depends on (1) local gas pressures and pH in blood and in brain tissue (cerebrospinal fluid), allowing for direct "chemical" control; (2) neural information "fed back" from chemosensitive nervous structures in the carotid body and aortic arch, which allows for "neurochemical" control, and (3) afferent neural information from stretch receptors in the tissues of the lung, airways, and chest wall, allowing for "neuromechanical" control. Higher cortical centers also can influence the controller. Storage capacities of the main tissues and organs have been evaluated in detail for the different gases.

In view of the great bulk of knowledge on the controlled system and controller, it is not surprising that there exist many excellent quantitative analyses of static and dynamic aspects of the chemostat. Of the pioneering work we would mention the static, linearized controller equation by Gray [18.2], expressing ventilatory rate as a function (i.e., linear combination) of arterial P_{CO_2}, P_{O_2}, and pH. Later works by Grodins [18.1], and Longobardo and Cherniack [18.3, 18.4], analyze many aspects of the dynamic performance, the role of gas transport and storage, and even of blood flow distribution, concentrating on the integrative chemoreceptor action on ventilation. In these studies quantitative explanation is given for the chemostat's actions, and particularly for the finding that the level of ventilation is so adjusted that mean arterial P_{CO_2} is kept low and that the blood is always nearly fully oxygenated. Priban [18.5] suggested approaches to respiratory control at a higher system's level, introducing some optimization criteria of respiratory work on the one hand and gas exchange on the other. At a lower level of organization, the genesis of ventilatory patterns (respiration frequency and depth) and their control at the neural level are also of great interest to both systems theorist and physiologist (e.g., Burns [18.6]).

This chapter surveys some studies at the level of the chemostat, with respect to pathophysiological states (as opposed, e.g., to "artificial" carbon dioxide breathing reactions). Thus we try to prepare the ground for classes of problems that are less well understood: coupling of respiratory with cardiovascular control and—also very challenging—performance of the chemostat in exercise. As Schmidt [18.7] noted, the "real test of any concept of respiratory control is its ability to explain the known events of muscular exercise." Formulation of the necessary concepts indeed is not simple, even if the system is studied at rest (Fig. 18.1).

From the physiological viewpoint we should keep in mind that ventilation and cardiac output need to be so regulated as to satisfy the demands of the tissues depending on work done, metabolic rate, and so on. On the other hand, we arrive at a different feedback scheme if we consider blood gases to be the

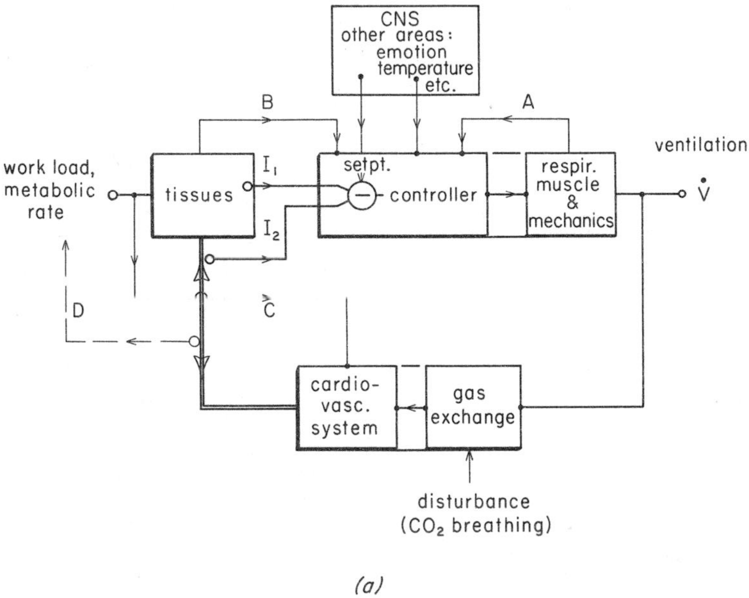

Fig. 18.1*a*. Functional block diagram of respiratory control system. The controlled variable, ventilation ($\dot{V}$), is adjusted such that the needs of the tissue (input) are satisfied. Chemoreceptive structures provide information on gas tension and pH in blood (I_2) and brain tissue (I_1) for the controller. Input–output relations of the controller have been obtained for certain states (rest, sleep, etc.), so that $\dot{V}$ can be expressed as a function of blood gases (see Section 18.C). A local control loop *A*, not discussed in this chapter, is involved in control of respiratory rhythm and depth, and probably in minimizing mechanical work [18.5]. Loop *B* represents afferent neural pathways from muscles that probably are involved in causing exercise hyperpnea [18.16]. Loop *C* indicates how the respiratory system is coupled with circulation. Pathway *D* indicates that metabolic rate also may depend on gas tensions. The model presented in this chapter does not take into account specifically loops *A*, *B*, *C*, or *D*. Acid–base regulation by the kidney, a slowly acting system also coupled to these loops, is omitted from the figure.

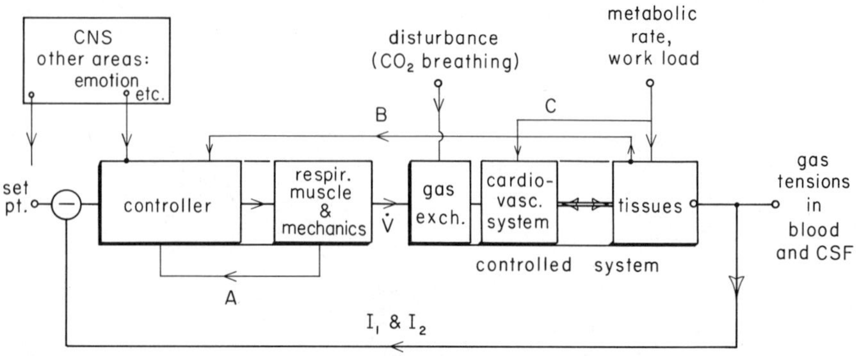

(b)

Fig. 18.1*b*. The system organization assumed by most authors [18.4, 18.8, 18.9] for discussion of the respiratory chemostat: gas tensions (output) are (nearly) kept constant by ventilation, as required by the setpoint, despite "disturbances" (varying exercise, gas compositions in inspired air, etc.). For definitions of loops $A \to C$ and I_1 and I_2 see Fig. 18.1*a*.

"output," which must be kept constant even when "disturbances" owing to stress occur. We prefer to adhere to the former approach, although the mathematical treatment of the whole system, at the level of organization assumed in this chapter, does not depend on the particular position chosen. However, the design of experiments probing deeper into the system does generally depend on one's viewpoint.

B. THE CONTROLLED SYSTEM

As Fig. 18.1*b* indicates, the so-called controlled system comprises lungs and all other tissues involved (passively) in gas exchange, uptake, release, and transport. Relatively simple laws relate signals that are important in blood gas control, as functions of ventilation and blood flow. They all rest on the law of conservation of mass and take into account the capacity of different tissues to store gases, as is expounded below.

Grodins' [18.8] first model study demonstrated that careful compartmentalization of the body is needed because different parts play different functional roles. One of the first considerations is that oxygen is stored mainly in lung and blood, whereas tissues and blood are the important stores for carbon dioxide. This suggests that, in a first approximation, lungs on the one hand and muscle and all other tissues on the other hand should be considered as separate compartments. The two compartments, "lungs" and "tissues," formed the regulated system in all Grodins' early work.

Defares [18.9] discussed in a very instructive paper the evolution of a more complex model useful in predicting responses of ventilation to carbon dioxide breathing. Brain tissue was treated as a third compartment, separate from lungs and all other tissues because of chemical control. We should ask, then, how finely different organs (muscle, viscera, etc.) should be differentiated in any model study. The answer depends on the information sought (response to carbon dioxide breathing, to exercise, to low oxygen pressure, or contribution to short-term acid-base balance). Many aspects of this problem were discussed quantitatively by Farhi and Rahn [18.10] and Cherniack et al. [18.4]. The latter authors have been concerned mainly with the dynamics of the chemostat and Cheyne-Stokes breathing [18.3], and we often assume their viewpoint. Even if the composition of the controlled system is specified in detail, many simplifying assumptions enter the model.

1. Assumptions

In this chapter we assume that storage of carbon dioxide in the following parts is important and should be treated separately: arterial blood together with the functional residual capacity of the lungs, muscles together with their venous blood volumes, and all tissues except muscle together with their venous blood stores (Fig. 18.2). Oxygen is mostly stored in two compartments: lungs and arterial blood, and venous blood.

Events *during* respiratory cycles are ignored, the lung is assumed to be ventilated by a continuous flow of air. Fluctuations lasting less than approximately 5 sec are ignored and neglected. (See [18.1] for a less restrictive model.) Similarly, cardiac rhythms are ignored; thus gas transport by the blood is assumed to take place at the mean flow velocity.

Within any of the compartments treated, gas tensions are thought to be uniform throughout. Thus it is possible to describe such a reservoir by one "lumped" store. This approach is feasible even if uneven distributions within a compartment exist, so long as the "unevenness" stays constant.

Furthermore, we assume that the gas tensions of expired air and of arterial blood leaving the lung are identical. Pulmonary dead spaces and shunts are not introduced. Carbon dioxide partial pressures in tissues are assumed to be equilibrated with those of venous blood leaving the compartments.

Oxygen uptake and production of carbon dioxide by tissues, assumed to proceed at the same rate, are made proportional to metabolic rate, which is kept constant (although metabolic rate may also depend, in turn, on blood gases). Figure 18.2 is a schematized diagram of the controlled system used in the study of Cheyne-Stokes breathing.

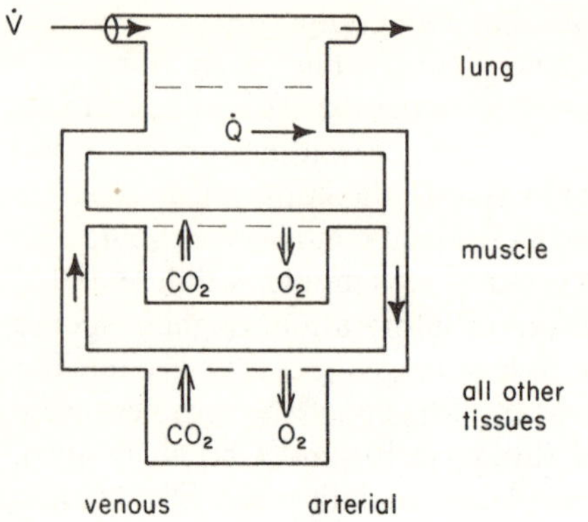

Fig. 18.2. Compartments of the controlled system. Oxygen is stored mainly in lungs and arterial blood, carbon dioxide is stored mainly in the tissues and venous blood. Rhythmic respiration is replaced by a steady $\dot{V}$, representing mean (alveolar) ventilation. More complex systems, including separate compartments for brain [18.9] and heart [18.10], may be needed depending on the processes studied. For study of Cheyne–Stokes breathing, this model seems to suffice.

In general investigations [18.1, 18.9] it is necessary to keep track of all gases circulating in the system: oxygen, carbon dioxide, and nitrogen. Furthermore, the concentration of H^+ ions must be considered [18.9], although it is not independent of P_{CO_2} (relation given by Henderson-Hasselbalch equation). The pH of cerebral blood and that of the cerebrospinal fluid influence the respiratory centers directly. Cherniack and Longobardo [18.3] found, however, that in Cheyne-Stokes breathing it suffices to consider P_{O_2} and P_{CO_2} only.

Here the gases are assumed to be bound by hemoglobin, so that their dissociation curves must be introduced when concentrations and partial pressures are related. The physically dissolved gases whose concentrations are proportional to partial pressure, taken into account by Grodins [18.1], are neglected in this chapter.

Blood flow (cardiac output) is treated as a constant during individual studies, although it is known that arterial P_{CO_2} influences cerebral flow. A transport delay, however, is taken into account: it is approximately 6 sec between lung and arrival at the controller. Grodins et al. [18.1] introduced relatively simple expressions to describe, when needed, known variations of cerebral blood flow and transport delays due to changes in blood gas tensions.

2. The Controlled System

Using the foregoing assumptions we can formulate differential equations for the partial pressures as follows. Applying the law of conservation of mass to the compartment including *lungs and arterial blood*, we have

For oxygen:

$$\frac{d}{dt}\left(V_a C_{a,O_2} + \frac{V_r}{P_B - 47} P_{a,O_2}\right) = \dot{Q}(C_{v,O_2} - C_{a,O_2}) + \frac{\dot{V}}{P_B - 47}(P_{I,O_2} - P_{a,O_2}) \tag{18.1}$$

where

V_a = arterial blood volume (storing oxygen); ca. 2.4 liters (see [18.3])

V_r = functional residual lung capacity; ca. 2.5 liters

$\dot{V}$ = alveolar ventilation, a variable dependent, in the *closed* system, on $P_{a,\,CO_2}$ and $P_{a,\,O_2}$

$C_{a,\,O_2}, C_{v,\,O_2}$ = arterial and venous concentration of oxygen, respectively (cc gas/cc blood)

$P_{a,\,O_2}, P_{I,\,O_2}$ = partial pressure of oxygen in arterial blood and inspiratory air, respectively (mmHg)

P_B = barometric pressure (mmHg)

47 mmHg = partial pressure of water vapor in (saturated) alveolar air

$\dot{Q}$ = blood flow (cardiac output)

With eq. 18.1 we state that changes in oxygen content in this store (left side, including oxygen in alveolar air and arterial blood) must be due, exclusively, to influx from the pulmonary artery and the outside air $[\dot{Q}C_{v,\,O_2} + \dot{V}P_{I,\,O_2}/(P_B - 47)]$, minus outflow toward tissues and to outside air $[\dot{Q}C_{a,\,O_2} + \dot{V}P_{a,\,O_2}/(P_B - 47)]$.

For carbon dioxide:

$$\frac{d}{dt}\left(V_a C_{a,\,CO_2} + \frac{V_r}{P_B - 47} P_{a,\,CO_2}\right) = \dot{Q}(C_{v,\,CO_2} - C_{a,\,CO_2}) - \frac{\dot{V}}{P_B - 47} P_{a,\,CO_2} \tag{18.2}$$

where $C_{a,\,CO_2}$ and $C_{v,\,CO_2}$ are the arterial and venous carbon dioxide concentrations, respectively.

In the range of P_{CO_2} encountered, the relation between carbon dioxide content and its partial pressure, given by the CO_2 dissociation curve, can be approximated by the linear equation (see Chapter 17)

$$C_{a,\,CO_2} = 0.25 + 0.0065 P_{a,\,CO_2} \tag{18.3}$$

Assume the following analogies:

P_{CO_2} = E, voltage

$V_r/(P_B - 47)$ = C_1, capacity of functional residual lung volume

$0.0065 V_a$ = C_2, capacity of arterial blood volume, given by slope of carbon dioxide dissociation curve

$(0.0065\dot{Q})^{-1}$ = R_1, "resistance" determining outflow (transport) by circulatory system

$[\dot{V}/(P_B - 47)]^{-1}$ = R_2, "resistance" determining outflow of carbon dioxide during expiration

The circuit in Fig. 18.3, then, is exactly analogous to the carbon dioxide store under discussion. It should be noted, however, that the resistive network elements do not correspond to hydraulic or pneumatic resistances: R_1 is inversely proportional to ventilatory rate, R_2 to cardiac output. In the entire (closed-loop) regulatory system $\dot{V}$ is a controlled *variable*. Thus, although the various "black boxes" of the system are arranged in the fashion of a typical feedback system, regulation actually is achieved by what is called *parametric* control; that is, one of the multiplying factors in the differential equation (18.2) is adjusted until the dependent variable $P_{a,\, CO_2}$ reaches the desired level.

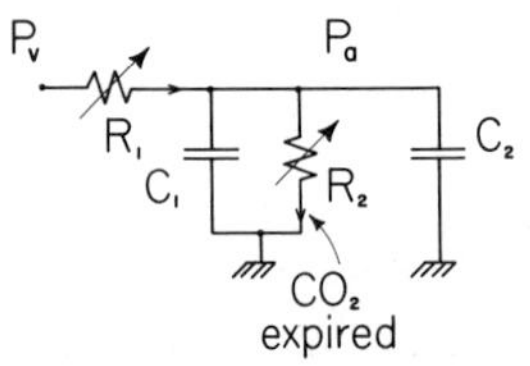

Fig. 18.3. Analog circuit for carbon dioxide system of lung and arterial blood; see eq. 18.2. The signals P_v and P_a are the venous and arterial partial pressures of carbon dioxide, respectively. R_1 is inversely proportional to cardiac output; R_2 is inversely proportional to alveolar ventilation; C_1 is the carbon dioxide storage capacity of the lungs, C_2 of the arterial blood. From the viewpoint of the electrical engineer, respiratory control of P_{aCO_2} is achieved by parametric control: R_2 is regulated.

The second major store for oxygen, as explained earlier, is the *venous blood*, for which we can write

$$\frac{d}{dt}(V_v C_{v,\, O_2}) = \dot{Q}(C_{a,\, O_2} - C_{v,\, O_2}) - \dot{V}_{m,\, O_2} \tag{18.4}$$

where V_v = volume of venous blood; ca. 3.6 liters, see [18.3]
$\dot{V}_{m,\, O_2}$ = metabolic comsumption (0.28 liters/min)

When solving eqs. 18.1 and 18.4 we require the oxygen dissociation curve. Various forms of analytical expressions have been suggested approximating the measured dependence of oxygen concentration on partial pressure [18.1, 18.9]. Here we follow Cherniack's suggestion [18.3] and approximate the sigmoid curve by three segments of straight lines, as in Fig. 18.4 where the mathematical expressions are given.

Carbon dioxide is stored in the lungs, and in addition in *muscle* and in all *nonmuscle tissue*. The equations governing the concentrations of this gas are

For muscle tissues:

$$\frac{d}{dt}(V_m C_{m,\, CO_2}) = \dot{Q}_m(C_{a,\, CO_2} - C_{v,\, CO_2}) + \dot{V}_{m,\, CO_2} \tag{18.5}$$

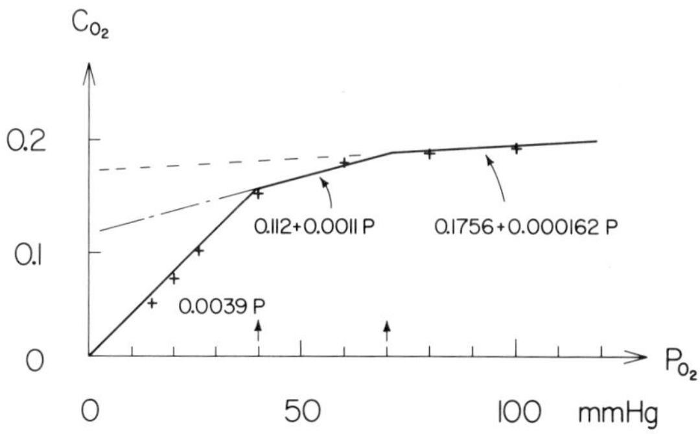

Fig. 18.4. The oxygen dissociation curve for normal hemoglobin content. Three straight-line segments, whose equations appear in the figure, are used to approximate the true curve. The crosses are experimental points on the sigmoid curve (see also Chapter 17). The break points of the piecewise linear approximation are marked by arrows on the abscissa.

For all remaining tissues:

$$\frac{d}{dt}(V_n C_{n,\,CO_2}) = \dot{Q}_n(C_{a,\,CO_2} - C_{v,\,CO_2}) + \dot{V}_{n,\,CO_2} \tag{18.6}$$

where

V_m, V_n = volumes of muscular and nonmuscular tissues with their venous blood (29 and 12 liters)

$C_{m,\,CO_2}, C_{n,\,CO_2}$ = concentrations of carbon dioxide in muscular and nonmuscular tissues, respectively

$\dot{V}_{m,\,CO_2}, \dot{V}_{n,\,CO_2}$ = rate of metabolic carbon dioxide production by muscular and nonmuscular tissues, respectively (0.038 and 0.19 liter/min)

Two remarks are in order concerning the dissociation curves of carbon dioxide (eq. 18.3) and oxygen (Fig. 18.4). Both these approximate the observed values well. However, in the first place the Haldane effect is neglected. In reality, the blood carbon dioxide concentration, at a fixed P_{CO_2}, decreases when P_{O_2} increases. Second, the so-called Bohr effect is also neglected, which means that C_{O_2}, at a fixed P_{O_2}, decreases when P_{CO_2} increases [2.2]. The Bohr effect can be large for $P_{O_2} < 75$ mmHg. It is not known to what extent these simplifications alter the overall performance of this model.

C. THE CONTROLLER

As explained briefly in the introduction of this chapter, the respiratory control centers must comprise several hierarchically linked controllers processing many different types of inputs. From the periphery, information is available from stretch receptors on lung inflation and deflation (Widdicombe [18.11]), and from chemosensors on blood gas tensions [18.2]; the concentration of H^+ ions in the cerebrospinal fluid (CSF) also represents input information. Furthermore, stimuli from nociceptors can elicit special breathing patterns. Different higher centers irradiate the neural structures of the respiratory controller, emotional states (Barchilon [18.12]) influence breathing movements, and speech and singing require special breathing patterns. Obviously, conscious control over respiration is also possible, at least within limits.

In spite of this complexity, we can analyze quantitatively control performance at certain levels of organization, while interactions with other centers or reorganization of the control loops are kept minimal. Studies of spontaneous breathing, such as presented in this chapter, form the basis of further investigations. From fixed states of the controller system, together with dynamic changes when the system goes from one mode into another (e.g., onset of exercise), more general rules of biological control can be sought.

In particular, analysis of chemoreflexes has been studied so thoroughly that the functionally important actions of the controller appearing in Fig. 18.1 can be predicted as functions of the three input variables. This holds especially for steady-state (static) chemostat action in response to these three inputs. Neural information on arterial P_{O_2}, P_{CO_2}, and H^+ ion concentration is processed in certain cells on the surface of the medulla (Mitchell et al. [18.13]). Defares [18.9] discussed carefully why H^+ concentration must be considered a controlled variable, although it is not directly affected by ventilation. Alveolar ventilation ($\dot{V}$) is taken to be the output of the controller. Thus the model, which is based on Gray's [18.2] and Nielsen and Smith's [18.14] work, does not include explicitly mechanisms by which changes in $\dot{V}$ are brought about. In particular, processes involved in the coupling between blood P_{CO_2} and H^+ concentration in the CSF are "absorbed" into one overall input–output relation. Lloyd [18.15] showed that Gray's data as well as his own (Fig. 18.5) are approximated best by

$$\dot{V} = a\left[1 + \frac{\alpha}{P_{a,\,O_2} - \gamma}\right]\left[\log\,(P_{a,\,CO_2} \cdot H^+) + \frac{c}{a}\right] \tag{18.7}$$

where
$\alpha = 10–30$ mmHg
$\gamma = 28–40$ mmHg
$a = 3–4.2$ liters/sec

$c/a = 5.85–5.97$
$P_{a,\,O_2}$ = alveolar P_{O_2}, close to arterial P_{O_2} (mmHg)
$P_{a,\,CO_2}$ = endtidal alveolar P_{CO_2}, close to arterial P_{CO_2} (mmHg)
H^+ = arterial H^+ concentration

The second bracket of eq. 18.7 obviously includes a linear combination of the logarithms of $P_{a,\,CO_2}$ and H^+, whereas the first bracket is a multiplier that becomes larger the smaller $P_{a,\,O_2}$ is. Over a limited range it is possible to simplify eq. 18.7 by linearization, as we show below. In any case we deal with an empiric curve fit to the observed $\dot{V}$-dependence on blood gases and pH. Equation 18.7 takes on a different form if we introduce the Henderson-Hasselbalch relation

$$\dot{V} = 0.5a\left[1 + \frac{\alpha}{P_{a,\,O_2} - \gamma}\right][2 + \log HCO^-_{3_\beta}] \log \left(\frac{P_{a,\,CO_2}}{\beta}\right) \qquad (18.8)$$

where β is the level of $P_{a,\,CO_2}$ at which apnoe occurs for a given $HCO^-_{3_\beta}$ concentration.

Lloyd [18.15] found an empiric relation between β and HCO_3^- such that eq. 18.8 simplifies, for any given acid-base condition, to

$$\dot{V} = 0.5a\left[1 + \frac{\alpha}{P_{a,\,O_2} - \gamma}\right] \log \left(\frac{P_{a,\,CO_2}}{\beta}\right)\left[\frac{c}{a} + 2 \log \beta - 5.62\right] \qquad (18.9)$$

However, an important question has not been raised. In many cases it may not be sufficient to have arterial pH as one of the independent variables in eq. 18.7. Grodins et al. [18.1] showed that a *combination* of arterial pH and of CSF–pH gives very satisfactory results for controller performance. Furthermore, we still do not know whether venous blood gases contribute to the input of the controller; we are not certain of the role of possible alveolar–arterial gradients; and not all factors are known that might influence the values of the constants in eq. 18.9 (Fig. 18.5).

From the systems analyst's viewpoint it is desirable to have a concise relation between the relevant inputs and outputs, such as eq. 18.7, which can be used together with the set of equations describing the controlled part of the system. It is, however, very limiting if only static characteristics are known, since fast transients do occur in nature. Slow changes apparently are sufficiently well described by these controller equations, as the studies of Cheyne-Stokes breathing by Longobardo et al. [18.3] demonstrated. For their model, these workers approximated the slightly curved ventilation response curves by straight lines:

$$\dot{V} = \left[1.81 + \frac{23.5}{P_{a,\,O_2} - 32.4}\right][P_{a,\,CO_2} - 31] - 15 \qquad (18.10)$$

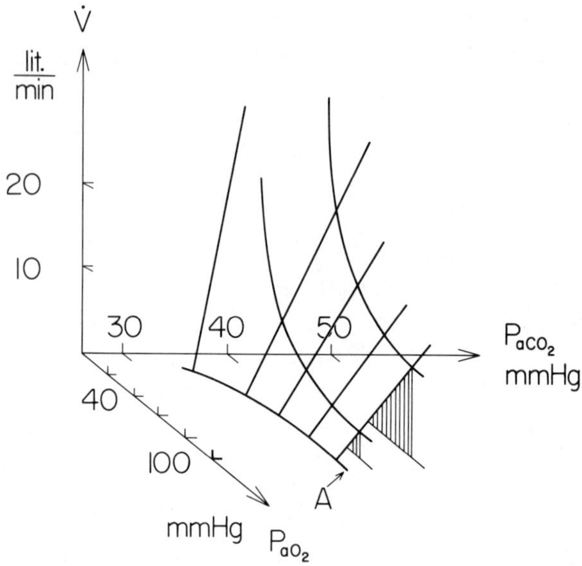

(a)

Fig. 18.5*a*. The controller surface used in the model of Longobardo and Cherniack [18.3] giving the steady-state ventilatory rate ($\dot{V}$) as a function of arterial oxygen tension (P_{aO_2}) and carbon dioxide tension (P_{aCO_2}); see eq. 18.10. The P_{aO_2} dependence at constant P_{aCO_2} is given by the hyperbolic curves at 40 and 45 mmHg of P_{CO_2}. The P_{aCO_2} dependence, at constant P_{aO_2} (20, 40, 60, 80, 100 mmHg), is given in this model by straight lines whose slopes increase with decreasing P_{aO_2} (i.e., there is an increased carbon dioxide sensitivity at hypoxia). At P_{aO_2}'s above 110 mmHg the slopes remain essentially constant. Curve *A* gives the apneic points which, however, are not always found in normal conscious subjects, as in Fig. 18.5*b*.

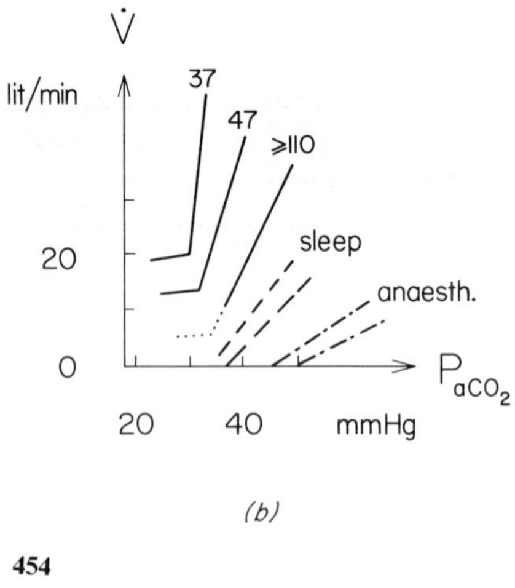

(b)

Fig. 18.5*b*. Experimentally obtained $\dot{V}$–P_{aCO_2} relations. The solid lines, summarizing the data by Nielsen and Smith [18.14], show the increased carbon dioxide sensitivity at low P_{O_2}'s seen also in Fig. 18.5*a*. At low P_{aCO_2} the curves exhibit knees such that $\dot{V}$ essentially depends on P_{aO_2} only. At very high P_{aCO_2} (not shown) the curves bend to the right (see eq. 18.9). During sleep and (halothane) anesthesia the controller action given by the dashed lines is less pronounced and, in addition, apnoe may occur [18.19]. Note that great variability is found in the data and that the relations here give averages only.

This, on the one hand, includes the increased carbon dioxide sensitivity with decreasing P_{O_2}, like eq. 18.7, but possible areas (at low P_{CO_2}) in which $\dot{V}$ is carbon dioxide independent (Fig. 18.5) are not taken into account. Neither does it incorporate the transport delays between pulmonary veins and chemosensitive areas, which can be incorporated quite elegantly as Grodins et al. [18.1] showed.

D. CLOSED-LOOP PERFORMANCE; REMARKS ON EXERCISE

Simultaneous solutions of the equations for the controlled system (18.1–18.6) and the controller (18.10), under different stresses, not only yield insight into the nature of transients, but also into all phenomena associated with Cheyne-Stokes respiration. Longobardo and Cherniack [18.3] investigated the stability of the chemostat and the kinds of abnormalities needed to cause the recurrent phases of apnea and hyperpnea. As long as cardiac output is normal, hypocapnia following hyperventilation (to $P_{a,\,CO_2}$ of 15 mmHg) causes only transient Cheyne-Stokes rhythms, which eventually die out. Stability of the normal system breathing low-altitude air is remarkable! This is possible because hyperpnea decreases gradually during the Cheyne-Stokes episode.

The model presented here does not explain the performance of the respiratory system in *exercise*. In particular, the level of hyperventilation under stress is predicted only partly by eq. 18.9. Cunningham [18.16] summarized in a very lucid paper the various factors contributing to exercise hyperpnea predicted by the controller characteristics at rest. The carbon dioxide acidity in exercise, in particular, shifts the controller surface to the left, thus eliciting a level of hyperventilation amounting to approximately 60% of that observed at steady state. The remaining 40% may be due to reflexes from the exercising muscle which, according to Kao's findings (see Defares [18.9]), may shift the controller surface even more. In addition, as Yamamoto [18.17] suggested, rapid fluctuations in P_{CO_2}, larger in exercise than at rest, may in fact also contribute to the unexplained 40%.

It seems likely that the dynamics—perhaps along the line of Clynes's unidirectional rate sensitivity (see Chapter 7)—of the controller come into play when variations are larger. This hypothesis, however, has never been tested. It is not known either whether such unidirectional rate sensitivity elements have short or long time constants. Long time constants might cause rapid steps of respiratory increases at the onset of heavy exercise.

The observation made in some cases that hyperpnea begins well before any humoral agents can become effective (see, e.g. [2.2]), has given rise to speculations about direct neural influence on the respiratory controller by the motor cortex. More recent careful analysis of respiratory transients

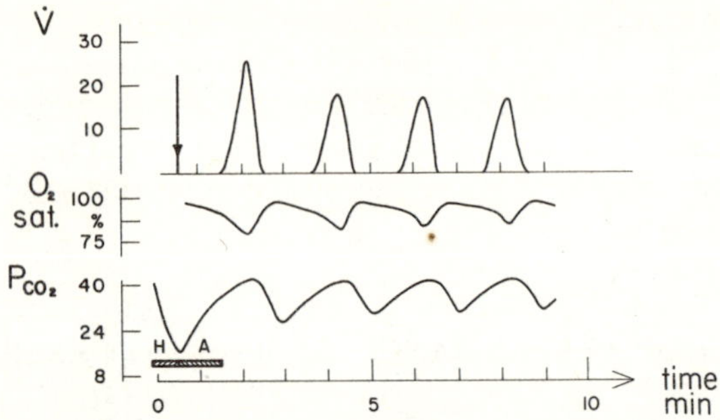

Fig. 18.6. Cheyne–Stokes breathing, by the model, in a hypothetical case: $\dot{Q} = 2.8$ liters/min, circulation delay between lung and brain = 30 sec. Hyperventilation (*H*), bringing P_{aCO_2} to 18 mmHg, followed by 1-min apnea (*A*), triggers the phenomenon. In a normal subject, Cheyne–Stokes breathing is not elicited by the maneuver. Cherniack and coworkers also studied periodic breathing in dogs [18.20] and found the model satisfactory. These curves, however, are based on data obtained from man. The Cheyne–Stokes cycle of 2 min also approximates that seen in man in certain conditions. Units are liters/min for $\dot{V}$ and mmHg for P_{CO_2}. After Longobardo and Cherniack [18.3].

[18.18], at the beginning and end of exercise, indicates that the abrupt changes reported earlier are not seen often. More frequent are gradual changes, resembling exponential curves or those of P_{CO_2} and heart rate.

Data are too sparse to conclude definitively on this matter. However, the long circulation times associated with low cardiac output during heart failure present a labile situation. Once $P_{a,\,CO_2}$ reaches 25 mmHg (following hyperventilation), the system of this example can go into apnea, after which the associated build-up of carbon dioxide causes hyperventilation. There follows a rapid drop of $P_{a,\,CO_2}$ which becomes large enough to stop respiration, because the effects of hyperventilation cannot be sensed quickly enough by the respiratory centers due to long circulation times. Afterward, the cycle starts again (Fig. 18.6). The cycle time thus becomes four times the abnormally long lung-brain circulation time. Metabolic alkalosis is another case in which Cheyne-Stokes breathing may occur because the controller surface is shifted. Stability is restored, however, after $P_{a,\,O_2}$ has reached normal levels.

E. REFERENCES

18.1. Grodins, F. S., J. Buell, and A. J. Bart. Mathematical analysis and digital simulation of the respiratory control system. *J Appl Physiol* **22**: 260–276, 1967.

18.2. Gray, J. S. *Pulmonary Ventilation and Its Physiological Regulation.* Charles C. Thomas, Springfield, Ill., 1950.

18.3. Longobardo, G. S., N. S. Cherniack, and A. P. Fishman. Cheyne-Stokes breathing produced by a model of the human respiratory system. *J Appl Physiol* **21**: 1839–1846, 1966.

18.4. Cherniack, N. S. and G. S. Longobardo. Oxygen and carbon dioxide gas stores of the body. *Physiol Rev* **50**: 196–243, 1970.

18.5. Priban, I. P. and W. F. Fincham. Self-adaptive control and the respiratory system. *Nature (London)* **208**: 339–343, 1965.

18.6. Burns, B.D. The central control of respiratory movements. *Brit Med Bull* **19**: 7–9, 1963.

18.7. Schmidt, C. F. Six-score years of chemoreflexes. *Ann NY Acad Sci* **109**: 631–633, 1963.

18.8. Grodins, F. S., J. S. Gray, K. R. Schroeder, A. L. Norius, and R. W. Jones. Respiratory responses to CO_2 inhalation. A theoretical study of a nonlinear biological regulator. *J Appl Physiol* **7**: 283–308, 1954.

18.9. Defares, J. G. Principles of feedback control and their application to the respiratory control system, pp. 649–680 in *Handbook of Physiology, Sec. 3 Respiration*, Vol. 1. W. O. Fenn and H. Rahn (Eds.). American Physiological Society, Washington, D.C., 1964.

18.10. Farhi, L. E. and H. Rahn. Dynamics of changes in carbon dioxide stores. *Anesthesiology* **21**: 604–614, 1960.

18.11. Widdicombe, J. G. Respiratory reflexes from the lungs. *Brit Med Bull* **19**: 15–20, 1963.

18.12. Barchilon, J. Emotions and respiration. *Ann NY Acad Sci* **109**: 619–630, 1963.

18.13. Mitchell, R. A., H. H. Loeschcke, J. W. Severinghaus, B. W. Richardson, and W. H. Massion. Regions of respiratory chemosensitivity on the surface of the medulla. *Ann NY Acad Sci* **109**: 661–681, 1963.

18.14. Nielsen, M. and H. Smith. Studies on the regulation of respiration in acute hypoxia. *Acta Physiol Scand* **24**: 293–313, 1952.

18.15. Lloyd, B. B. The chemical stimulus to breathing. *Brit Med Bull* **19**: 10–14, 1963.

18.16. Cunningham, D. J. C. Some quantitative aspects of the regulation of human respiration in exercise. *Brit Med Bull* **19**: 25–30, 1963.

18.17. Yamamoto, W. S. Mathematical analysis of the time course of alveolar CO_2. *J Appl Physiol* **15**: 215–219, 1960.

18.18. Beaver, W. L. and K. Wasserman. Transients in ventilation at start and end of exercise. *J Appl Physiol* **25**: 390–399, 1968.

18.19. Fink, B. R., E. C. Hanks, S. H. Ngai, and E. M. Papper. Central regulation of respiration during anesthesia and wakefulness. *Ann NY Acad Sci* **109**: 892–900, 1963.

18.20. Cherniack, N. S., G. S. Longobardo, O. R. Levine, R. Mellins, and A. P. Fishman. Periodic breathing in dogs. *J Appl Physiol* **21**: 1847–1854, 1966.

CHAPTER NINETEEN

Body Temperature and its Control

Specially evolved mechanisms maintain, in most mammals and birds, a steady state between heat production by the body and its loss of heat to the environment so that (deep) body temperature stays near the "normal" level within narrow limits. All organs produce metabolic heat due to exothermic reactions even if the body is resting. Heart, brain, and viscera (notably the liver) put out as much as 50% of the entire heat. In exercise, large amounts of heat are produced more peripherally by the muscles. The circulation of blood tends to equalize the temperatures within a central core of the body while it carries heat to the body's surface, where it is given off. The size of the body core, in which temperature is more or less uniform, depends on the particular state of the system and its environment. Figure 19.1 shows two possible temperature distributions within the body. Compared with the body

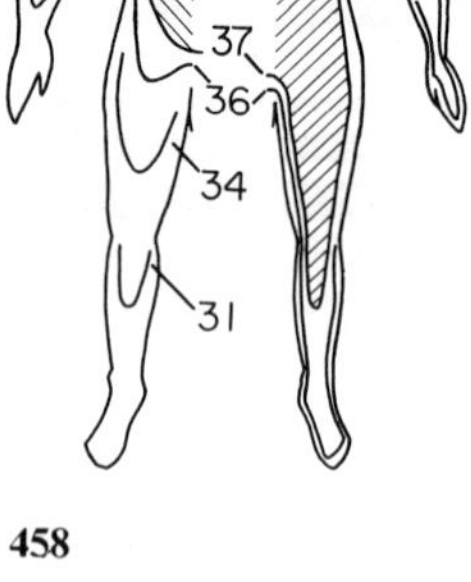

Fig. 19.1. Isothermal lines within the body in cold and in warm environments (left and right halves of figure, respectively). Curves are labeled in °C. Note that the region maintained at 37°C comprises the larger parts of head and trunk even in the cold environment. After Aschoff and Wever, quoted in Ruch and Patton [19.2].

in a warm environment, a relatively smaller core is near the normal temperature in cold exposure. Skin temperature is most often between that of the body core and the ambient temperature, except when evaporative heat losses are great. However, powerful mechanisms are available for maintenance of the body core temperature, that is, for protection of internal organs against large temperature changes.

Well-defined structures in the hypothalamus process information on local brain temperature, from temperature-sensitive nerve endings in the skin and perhaps from deep body temperature sensors outside the brain, and accordingly exert control over all mechanisms involved in heat loss, production and preservation: vasomotor activity, metabolic rate, sudomotor activity. This

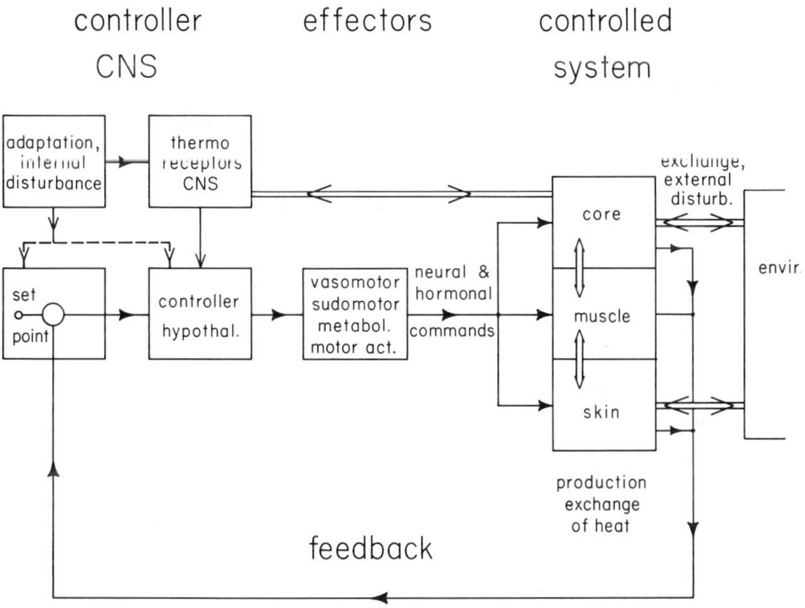

Fig. 19.2. Schematic diagram of feedback control of body temperature, illustrating the main components of the system and identifying concepts used in the discussion in this chapter. Information on body temperature is carried by neural pathways to the center (line "feedback" and connection between "thermoreceptors CNS" and "controller hypothalamus"), where it apparently is compared with a reference level ("setpoint"). The controller then triggers any of the available mechanisms (vasomotor, sudomotor, metabolic, motor activity), either to increase heat loss or to increase heat production and decrease heat loss, depending on the type of challenge to the system. Within the different compartments of the controlled system, heat is exchanged (double-line arrows). The environment is an important part of the system, to be considered simultaneously with homeothermic control mechanisms. Adaptation can take place in cold climates; internal disturbances may mean fever. The model discussed in this chapter involves many more details, as explained in the text

"control center" has a fine sensitivity to temperature deviations from "setpoints." We can draw a general block diagram (Fig. 19.2) indicating the principal components of the controller, the controlled part of the thermoregulatory system, and their main connections. The construction of feedback systems modeling this thermostat has proved to be useful, especially with regard to detailed action of the controller and modification of the setpoint, as in fever. In the following discussion, we disregard coupling between this control system and regulation of respiration and of cardiac output. Stolwijk, Hardy [19.1], and their co-workers, whose work forms the basis of this chapter, have shown that this is permissible for specific tasks of the thermoregulation in resting man. This method of attack is necessary as long as the significance of the choice of compartments is to be established. Similarly, whether the controller functions according to a weighted sum or proportionally to a weighted product of central and peripheral information, is a question that must be resolved by a model study *and* by experiments, using elements and couplings dominantly concerned with thermoregulation.

A. THE CONTROLLED SYSTEM

Heat is *produced* throughout the body by basal metabolic processes. Active muscles, as was discussed in part in Chapter 9, are additional heat sources. Thus shivering induced by way of the neuromuscular system produces heat useful for protection against cold, although the increased circulation to the shivering muscle brings more blood into the periphery, partly counteracting the increased heat production through larger heat losses. The rhythmically contracting heart is also a heat source; energy dissipated in the circulation due to viscosity of blood appears as a small amount of heat. In vigorous exercise heat production of the body may be 10 to 20 times the basic, resting level. Hence the body requires a very efficient thermostat.

Increased activity in the adrenergic autonomous nervous system can, under central control, boost the metabolic rate throughout the body. This then constitutes an additional source of heat possibly called upon in cold environment. Long-term effects may also include increased secretion of thyroxin by the thyroid gland, which in turn stimulates metabolic rate (after a delay of several days). This mechanism is important in adaptation to cold climate.

The specific heat capacity of tissue C_{th} relates the temperature change ΔT in an (isolated) mass m with the heat produced ΔQ as follows:

$$\Delta Q = C_{th} \cdot m \cdot \Delta T \qquad (19.1)$$

Due to the high water content, C_{th} is relatively large:

$$C_{th} = 0.83 \frac{\text{kcal}}{°\text{C kg}} \tag{19.2}$$

At rest the human body produces approximately 70 kcal/hr. If the body were completely isolated and devoid of heat loss mechanisms, its temperature would build up, according to eq. 19.1, at the rate of 1.2 °C/hr.

Heat is *exchanged within the body* (i.e., between parts of the body) mainly by convection by the blood and also by conduction, although tissue is a relatively poor heat conductor: conductivity = 18 kcal cm/(hr m^2 °C). The circulatory system has therefore an important task in transporting excess heat from the centers at higher temperature to the skin. If body heat must be preserved, cutaneous arteries constrict to increase the thickness of the outermost layer of "insulation."

Heat is also exchanged between the venous and arterial sides of the systemic circulation wherever arteries and veins are close to one another. This can be very beneficial in cold exposure, since arterial blood is precooled by the cooler venous return. Precooling in turn diminishes the temperature difference between capillary blood and surroundings, thus decreasing heat loss. Furthermore, this countercurrent exchange also warms the venous blood, thus minimizing central cooling. Clearly, heat transport by the blood depends on vasomotor activity, which is controlled to satisfy metabolic needs of the tissues (Chapter 16) as well as to maintain body core temperature. Mathematically, heat flow (kcal/sec) from blood into (or from) tissue is proportional to blood flow, (nonspecific) heat capacity of blood, and temperature difference between blood and tissue.

Heat is *exchanged with the environment* by several mechanisms. Natural and forced convection of air carries heat from the surfaces of clothing and exposed skin. This heat loss is also proportional to temperature difference and to exposed area. A wind velocity of 3 m/sec can increase the constant of proportionality, a function of the square root of wind velocity [19.2] up to 20 kcal/(m^2 °C hr). The body also exchanges heat with its surroundings by radiation. Since the emissivity of the body is close to unity in the infrared region, radiation can change the temperature of the skin of the resting body considerably. The power lost by radiation can be given as ($H = dQ/dt$)

$$\begin{aligned} H_{\text{rad}} &= K_r A_r e_s (T_s^4 - T_{ra}^4) \\ &\simeq K_r' A_r e_s (T_s - T_{ra}) \end{aligned} \tag{19.3}$$

where

$K_r = 7$ kcal/(m^2 °C hr)
A_r = effective area
e_s = emissivity
T_s = surface temperature of body
T_{ra} = ambient radiant temperature

and $(T_s^4 - T_{ra}^4)/(T_s - T_{ra}) = (T_s^3 + T_s^2 T_{ra} + T_s T_{ra}^2 + T_{ra}^3)$ = approximately constant. (See also [19.2] for a summary of physical laws governing heat exchange.) The body can also lose heat by conduction, when in contact with chairs, and so on. An additional powerful mechanism of heat release is evaporation from the surface.

Whereas the first three factors (convection, conduction, radiation) are properties of the *passive* system that depend on skin temperature (and ambient temperature), secretion of sweat and cooling by evaporation depend strongly on *neural* activity and play an important role in thermoregulation. One of the tasks of the epidermis is to prevent loss of water, yet some fluid is always evaporating, even at rest, though at an "insensible" rate. However, eccrine glands, stimulated by the cholinergic fibers of the autonomous system, may secrete sweat up to 1 liter/hr in exercise. All this is evaporated if the surrounding air is dry and forcefully moved across the body. The energy lost in this "ideal" case is given by the amount of water secreted, multiplied by the latent heat of vaporization of water (580 kcal/liter at constant temperature). In practice, the partial water pressures of ambient air and skin surface have to be taken into account. We then can give the energy lost as follows:

$$H_{ev} = K_e A_m (P_{sH_2O} - P_{aH_2O}) \tag{19.4}$$

where K_e = a function of air motion, wind velocity near body
A_m = moist area exposed
P_{H_2O} = partial water pressures, at surface (s) and in ambient air (a)

Evaporative heat exchange through the lungs and respiratory tracts is another important component. Usually, inspired air is at a much lower vapor pressure than the air within the lungs. Expired air is saturated with water vapor. Hence the heat lost by this mode is proportional to the rate of ventilation, which in turn depends on oxygen consumption (i.e., metabolic rate) by the body. This mechanism is important for animals with heavy fur; in man, respiratory heat loss may amount to 10% of the total.

The modes of heat exchange between skin and environment depend, in part, on the insulation provided by clothing. This is not taken into account in a model such as Stolwijk's and Hardy's [19.1], since the experimental data could be obtained from nude subjects. On the other hand, in studies concerned with protection against hostile environments (space flight), insulation by clothing, and perhaps its heat capacity, will be of prime interest. (See, e.g., Brown [19.3]).

At rest, the outermost layer of tissue of the body can also serve as insulation. In cold exposure, the cutaneous blood vessels constrict so that heat reaches the surface mostly by conduction. However, unperfused tissue, as pointed out earlier, is a relatively poor conductor. When exposed to heat, peripheral vasodilation allows blood to circulate closer to the surface.

We can now define the significant *compartments of the "passive," controlled system*. As suggested by Stolwijk and Hardy [19.1], it is satisfactory to simplify the geometry of the body as in Fig. 19.3. (For a more detailed study see [19.4]). Head and trunk are represented by one cylinder each, and all extremities are lumped together to form one cylinder. Skin, muscles, and viscera are placed in coaxial shells within these cylinders. The central blood volume, contained by the heart and the large central arteries and veins, is considered to be a separate compartment. Dimensions of the cylinders are computed so that the ratios of their masses and their surface areas (end faces neglected) correspond to those in a man of average build (71 kg; 1.83 m^3 total skin area). Heat exchange *by conduction* can be assumed to occur radially only within each cylinder. Axial temperature gradients in any of the cylinders are neglected. Areas of contact between the three main compartments are small. Therefore we neglect heat exchange by conduction between the cylinders. Resistance to radial heat flow between concentric shells is computed, in a simplified manner, by the following expression. Thermal resistance R_{ij} between shell i and shell j is

$$R_{ij} \simeq \rho_{th} \frac{\Delta r}{A_{ij}} \tag{19.5}$$

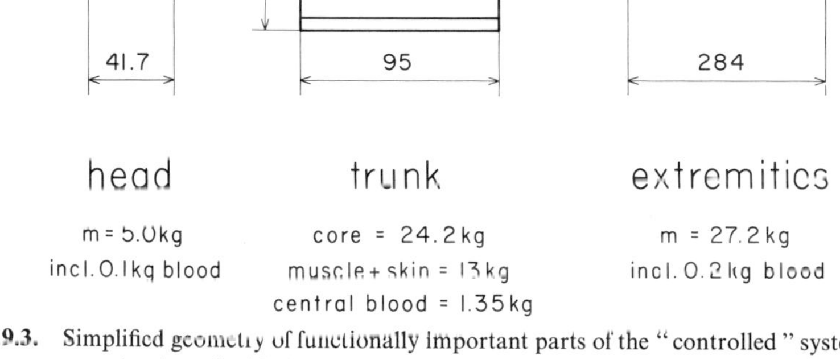

Fig. 19.3. Simplified geometry of functionally important parts of the "controlled" system. All parts are assumed to be cylindrical and are scaled to an average man weighing 71 kg. Measures are in centimeters. See text for assumption about conduction between the compartments. After Stolwijk and Hardy [19.1].

where ρ_{th} = thermal resistivity; ca. 0.05 m^2 hr °C/(kcal cm)
Δr = difference of radii of ith and jth shell (radii measured to the centers of the masses)
A_{ij} = area of contact between shells

Values calculated by Stolwijk [19.1] for the thermal resistances for this model appear in Fig. 19.4. In this figure, which gives an electric analog of the system, we also see how *convective* heat exchange by the circulation is introduced into the model. First, it can be assumed that there is little direct heat exchange between central blood volume and the core of the trunk. Most of the heat is exchanged at the capillary level in each compartment, varying with blood flow rate and the difference in temperature between blood and tissue. The temperature of the blood is assumed to be that of the central pool. The flow, being under control of the vasomotor centers, can and must vary significantly. Paths of convective heat flow therefore are represented in Fig. 19.4 as variable resistors. The only exceptions are the circulations to the brain and the viscera. Cerebral blood flow is known to be constant as long as the concentration of carbon dioxide in the blood is normal, and for the purpose of this study it is so considered. Furthermore, it can be assumed that blood flow to viscera and organs in the trunk is independent of thermoregulatory responses. (In exercise, however, blood flow may vary; the model therefore is inadequate for studying thermal responses to exercise.)

Values of the variable resistors, in the convective heat paths between central blood volume and the various compartments in Fig. 19.4, should reflect the following three factors determining heat convection by the circulation:

1. Maximum heat transfer rate = temperature difference between central blood and tissue compartment, multiplied by the thermal capacity of blood flowing (per unit time) to this compartment (assuming that blood reaches the tissue at central blood volume temperature).
2. Deviation from maximum heat transfer due to high flows, thereby preventing thermal equilibrium.
3. Deviation from maximum heat transfer due to countercurrent exchange causing arterial blood to be precooled.

Changes in factors 2 and 3 may partly cancel when flow is changing. Absolute values of these reductions in effectiveness of convective heat transport can be accounted for, in the model, by reducing blood flows to the different parts of the body. This, however, is only permissible because we restrict our attention to thermoregulation.

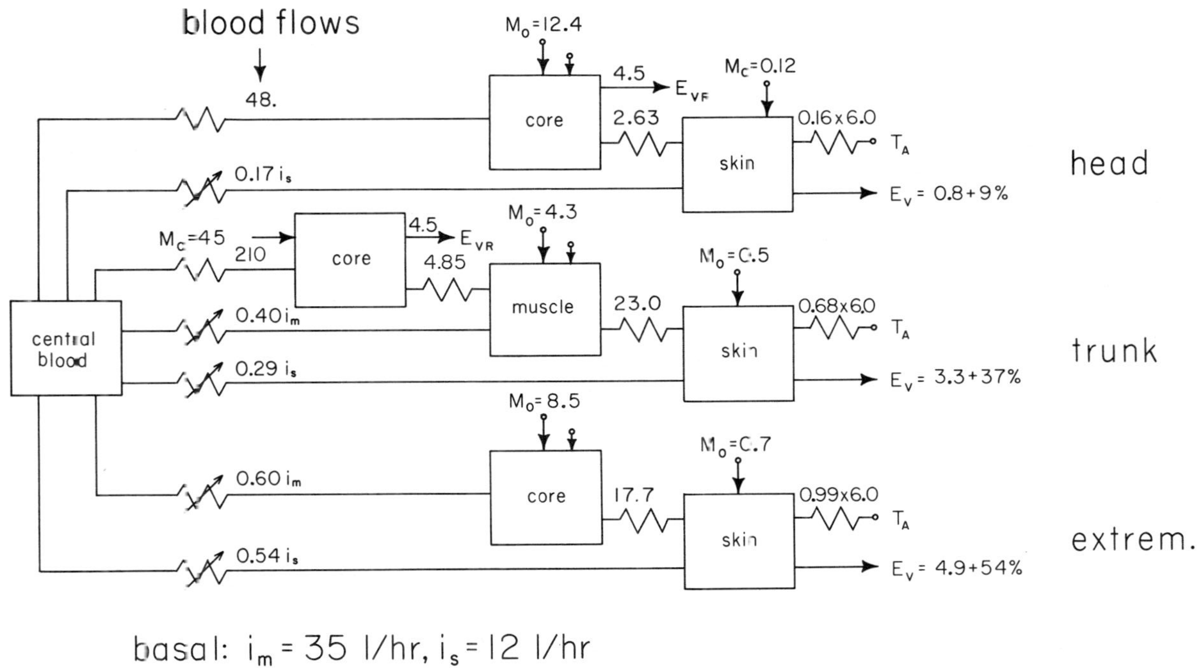

Fig. 19.4. Diagram giving, for all compartments, resting state values of metabolic heat production (M_0), heat-transfer coefficients between compartments (resistors between units, in the right half of the figure), heat-transfer coefficients between skin segments and environment (at temperature T_A), and insensible evaporative heat loss (E_V, marked by outward arrows). Metabolic heat production and evaporative heat loss are in kcal/hr, transfer coefficients in kcal/hr °C). Percentage listed together with basal E_{V0} is the fraction, assigned to the particular skin portion, of the total additional E_V brought into action when excess heat must be lost by evaporation of sweat. The left half of the figure shows the distribution of blood flows. The brain is perfused at 48, the core of the trunk at 210 liter/hr. All other flows may change according to demand. Their fractions of muscle (i_m) and skin blood flow (i_s) are assumed to be constants and are listed. The resting values of i_s and i_m, are given. The little arrows to the core of the head, the muscles of the trunk, and the core of the extremities indicate that extra heat inputs occur in shivering and in exercise. Furthermore, in exercise, it is assumed that 58% of the total heat originates in the core of the extremities and 38% in the muscles of the trunk. (Cf. eqs. 19.6–19.13.) The storage capacity of the compartments can easily be computed from their masses (see Fig. 19.3) and C_{th} (cf. eq. 19.1). After Stolwijk and Hardy [19.1].

Figure 19.4 also shows the basal *metabolic heat sources* and their assigned values for man at rest. Increased heat production by the muscle compartments due to shivering presents an additional input. This varies under central control. Stolwijk and Hardy [19.1] quote the physiological literature on the basis of which the values were assigned. Note that different compartments exhibit different ratios of resting blood flow to basal metabolic rate, the latter being made proportional to oxygen consumption. Indeed, organs like the kidney and the skin consume small quantities of oxygen compared with their blood flows, whereas in the brain, myocardium, and skeletal muscle the situation is reversed. The brain, perfused by about 12 to 15% of the total cardiac output, consumes 18 to 25% of the total oxygen uptake; on the other hand, the viscera of the trunk are assumed to obtain 70% of the total blood flow at rest while they consume 55% of the entire oxygen.

Paths of *heat exchange between skin and environment* also appear in Fig. 19.4. Heat losses by insensible evaporation, radiation, and convection, all made proportional to the difference in temperature ($T_{\text{skin}} - T_{\text{ambient}}$), are represented by currents flowing in one constant resistor. Heat loss by extra evaporation is a quantity under central control (i.e., it is considered to be an output of the controller).

In the following we list the differential equations derived by Stolwijk, determining thermal responses of the passive system of Fig. 19.4. Again, it is assumed that thermal capacities and resistances can be lumped into concentrated elements. The differential equations for temperature distributions then become ordinary first-order equations in which, for each compartment, the temporal temperature changes (times thermal capacity) are equated to the sum of heat inflows and outflows. The following symbols are used in these equations:

i = blood flow (liters/hr)
i_s = total skin blood flow (under central control)
i_m = total muscle blood flow (under central control)
h = environmental heat transfer coefficient for radiation and convection (suitable for experimental condition)
E_{V0} = insensible evaporative heat loss rate
E_{VR} = evaporative heat loss rate assigned to respiratory tracts
E_V = rate of total evaporative heat loss, induced by sweating
M_0 = basal metabolic heat production rate
ΔM = rate of extra metabolic heat production, due to shivering
α = factor ($\simeq 0.75$) accounting for countercurrent heat exchange
ρc = density × specific heat of blood (0.92 kcal/°C liter)
ε = fraction of total evaporative heat loss assigned to compartment (kcal/hr)

κ = thermal conductance, kcal/(hr °C); 1/(resistance)
σ = fraction of total shivering heat production rate assigned to compartment (kcal/hr)
β = fraction of total i_s assigned to compartment
γ = fraction of total i_m assigned to compartment

these quantities are subscripted to express the necessary scaling according to surface area, mass, metabolic rate, and so on, assigned to a particular compartment.

The subscripts are:

HC = head core
HS = head skin
TC = trunk core
TM = trunk muscle
TS = trunk skin
EC = extremities core
ES = extremities skin
CB = central blood volume
A = ambient

The system of differential equations governing the temperatures of the assumed, uncontrolled, compartments are:

For head core:

$$m_{HC}C_{th}\frac{dT_{HC}}{dt} - \overbrace{\alpha_{HC}\rho c i_{HC}(T_{CB} - T_{HC})}^{\text{Convective (blood)}} + \overbrace{\kappa_{HC-HS}(T_{HC} - T_{HS})}^{\text{Conductive (to skin)}}$$
$$= \underbrace{M_{0HC}}_{\substack{\text{Basal}\\ \text{metabolism}}} + \underbrace{\sigma_{HC}\Delta M}_{\substack{\text{Shivering}\\ \text{metabolism}}} - \underbrace{E_{VR}}_{\substack{\text{Evaporative}\\ \text{(airways)}}} \qquad (19.6)$$

For skin of head:

$$m_{HS}C_{th}\frac{dT_{HS}}{dt} - \overbrace{\kappa_{HC-HS}(T_{HC} - T_{HS})}^{\substack{\text{Conductive}\\ \text{(from core)}}}$$
$$= \overbrace{M_{0HS}}^{\substack{\text{Basal}\\ \text{metabolism}}} - \overbrace{E_{V0HS}}^{\substack{\text{Insensible}\\ \text{evaporation}}} - \overbrace{h_{HS}(T_{HS} - T_A)}^{\text{Radiation, etc.}}$$
$$+ \underbrace{\alpha_{HS}\rho c\,\beta_{HS}\,i_S(T_{CB} - T_{HS})}_{\text{Convective (blood)}} \quad \underbrace{c_{HS}E_V}_{\text{Evaporation}} \qquad (19.7)$$

For core of trunk:

$$m_{TC}\,C_{th}\frac{dT_{TC}}{dt} - \overbrace{\alpha_{TC}\,\rho c\, i_{TC}(T_{CB} - T_{TC})}^{\text{Convective (blood)}} + \overbrace{\kappa_{TC-TM}(T_{TC} - T_{TM})}^{\text{Conductive}}$$

$$= \underbrace{M_{0TC}}_{\substack{\text{Basal}\\\text{metabolism}}} + \underbrace{E_{VR,TC}}_{\substack{\text{Respiratory}\\\text{water vapor}}} \tag{19.8}$$

For muscles of trunk:

$$m_{TM}\,C_{th}\frac{dT_{TM}}{dt} - \overbrace{\kappa_{TC-TM}(T_{TC} - T_{TM})}^{\text{Conductive (from core)}} + \overbrace{\kappa_{TM-TS}(T_{TM} - T_{TS})}^{\text{Conductive (to skin)}}$$

$$= \underbrace{M_{0TM}}_{\substack{\text{Basal}\\\text{metabolism}}} + \underbrace{\sigma_{TM}\,\Delta M}_{\text{Shivering}} + \underbrace{\alpha_{TM}\,\rho c\,\gamma_{TM}\,i_m(T_{CB} - T_{TM})}_{\text{Convective (by blood)}} \tag{19.9}$$

For skin of trunk:

$$m_{TS}\,C_{th}\frac{dT_{TS}}{dt} - \overbrace{\kappa_{TM-TS}(T_{TM} - T_{TS})}^{\substack{\text{Conductive}\\\text{(from muscle)}}} = \overbrace{M_{0TS}}^{\substack{\text{Basal}\\\text{metabolism}}} - \overbrace{E_{V0TS}}^{\substack{\text{Insensible}\\\text{evaporation}}}$$

$$- \underbrace{h_{TS}(T_{TS} - T_A)}_{\text{Radiation, etc.}} + \underbrace{\alpha_{TS}\,\rho c\,\beta_{TS}\,i_s(T_{CB} - T_{TS})}_{\text{Convective (blood)}} - \underbrace{\varepsilon_{TS}\,E_V}_{\text{Evaporative}} \tag{19.10}$$

For core (muscles) of extremities:

$$m_{EC}\,C_{th}\frac{dT_{EC}}{dt} + \overbrace{\kappa_{EC-ES}(T_{EC} - T_{ES})}^{\substack{\text{Conductive}\\\text{(from skin)}}} = \overbrace{M_{0EC}}^{\substack{\text{Basal}\\\text{metabolism}}} + \overbrace{\sigma_{EC}\,\Delta M}^{\text{Shivering}}$$

$$+ \underbrace{\alpha_{EC}\,\rho c\,\gamma_{EC}\,i_m(T_{CB} - T_{EC})}_{\text{Convective (blood)}} \tag{19.11}$$

For skin of extremities:

$$m_{ES}\,C_{th}\,\frac{dT_{ES}}{dt} - \overbrace{\kappa_{EC-ES}(T_{EC} - T_{ES})}^{\text{Conductive (from core)}} = \overbrace{M_{0ES}}^{\text{Basal metabolism}} - \overbrace{E_{V0ES}}^{\text{Insensible evaporation}}$$
$$- \underbrace{h_{ES}(T_{ES} - T_A)}_{\text{Radiation, etc.}} + \underbrace{\alpha_{ES}\,\rho c\,\beta_{ES}\,i_s\,(T_{CB} - T_{ES})}_{\text{Convective (blood)}} - \underbrace{\varepsilon_{ES}\,E_V}_{\text{Evaporative}} \tag{19.12}$$

For (central) blood volume (all terms are "convective"):

$$m_{CB}\,c\,\frac{dT_{CB}}{dt} + \alpha_{HC}\,\rho c\,i_{HC}(T_{CB} - T_{HC}) + \alpha_{TC}\,\rho c\,i_{TC}(T_{CB} - T_{TC})$$
$$= -\alpha_{TM}\,\rho c\,\gamma_{TM}\,i_m(T_{CB} - T_{TM}) - \alpha_{EC}\,\rho c\,\gamma_{EC}\,i_m(T_{CB} - T_{EC})$$
$$- \alpha_{HS}\,\rho c\,\beta_{HS}\,i_s(T_{CB} - T_{HS}) - \alpha_{TS}\,\rho c\,\beta_{TS}\,i_s(T_{CB} - T_{TS})$$
$$- \alpha_{ES}\,\rho c\,\beta_{ES}\,i_s\,(T_{CB} - T_{ES}) \tag{19.13}$$

In these equations we arranged, if possible, the dependent terms on the left of the equality signs. On the right are the "forcing functions." These functions are considered either to be constants (basal metabolic heat production rate M_0, insensible evaporative rate of heat loss E_{V0}, etc.) or to vary according to "command signals" from the controller (rate of sweating, i.e., evaporation, E_v; skin blood flow i_s; muscle blood flow i_m). If the same model were used to study responses to exercise, this set of differential equations would have to be modified. All equations governing the temperature in the compartments including muscle (eqs. 19.6, 19.9, 19.11) should include a term representing extra heat production. Furthermore, E_{VR}, the respiratory heat loss, should also be made to depend on the level of exercise (through the rate of ventilation), as well as i_m.

We refer the reader to [19.1] for numerical values of all variables that are not given in Fig. 19.4. It may be possible to obtain more precise figures for distribution and changes in blood flow from current literature.

The response of the model, without thermocontrol as defined by eqs. 19.6 to 19.13, must depend on the particular compartmentalization assumed. Crosbie et al. [19.5] demonstrated that a "slab geometry" arrangement, much simpler than the one of Fig. 19.4, satisfactorily predicts equilibrium temperatures of the body (steady-state responses). The model presented here performs more satisfactorily in *dynamic* tests, particularly when compared with results reported by Piironen et al. (quoted in [19.1]). These authors cycled sinusoidally the skin temperature of subjects completely immersed in a bath and measured head core temperature at the tympanic membrane.

The temperature of the head's core (T_{HC}), as predicted by the model, initially was approximately 0.4°C above the one observed. However, during the entire temperature cycle (whose amplitude at the skin was ±3°C, and ±1°C at the tympanic membrane) T_{HC} was predicted within 0.1°C of the measured temperature; the phase difference between the theoretical and observed sinusoid was less than 10° and was probably close to experimental errors. Stolwijk [19.1] found that introduction of the central blood volume improved the model significantly over those of Wissler [19.4] and of Crosbie et al. [19.5].

B. TEMPERATURE SENSORS, THE HOMEOTHERMIC CONTROL CENTER, AND MODES OF TEMPERATURE CONTROL

In the previous section we enumerated the *effectors* through which the autonomic reflex loops of the thermostat system can act to maintain deep body temperature within a physiologically optimal range. Cutaneous arteries constrict or dilate, metabolism is controlled to increase or decrease, and muscles may be rhythmically excited to shiver (which is also partly under voluntary control). Furthermore, special glands may be stimulated to secrete sweat so that large quantities of evaporation heat are given off by the body, whose heat losses by convection and radiation are limited.

Control of these effectors depends on the integrity of more or less well-defined areas in the hypothalamus (which has additional important functions in regulation of electrolyte balance, blood volume, blood pressure, etc.). There is evidence—though partly contradictory and therefore contested—that the anterior hypothalamus is the center of protection against overheating, that the posterior hypothalamus elicits the defenses against cooling, and that the states of these two areas have inhibitory influences on each other. The particular mode of thermocontrol elicited by activity within the hypothalamus depends on local, hypothalamic temperature *and* on afferent neural information from peripheral temperature-sensitive nerve endings.

The *local* temperature of the thermocontrol center is determined by that of the arterial blood, by cerebral blood flow, by the local metabolic rate, and by conductive heat losses to or inputs from other parts of the brain. Additional factors may be involved, as indicated by eq. 19.6. All this should be remembered in direct experimental work on the temperature sensitivity of the hypothalamus. Thus even the "thermoclamp" technique (Hammel et al. [19.6]), involving an array of thermodes, only partly avoids thermal gradients in the control area. Another consequence of the multiplicity of factors determining hypothalamic temperature is that the control center must be able to differentiate between local temperature changes needing regulation and all others, as has been argued by Bligh in his excellent review [19.7].

Direct observations of hypothalamic activity as a function of local temperature do not give clear-cut evidence that this sensitivity to thermal changes plays the same role in normal homeothermy. The study by Stuart et al. [19.8] on single-unit activity in conscious animals seems to indicate strongly that at least some cells exist in the anterior hypothalamus that increase their discharge rate upon local cooling, and that some units in the posterior hypothalamus respond to warming of the anterior hypothalamus. Not all investigators agree, however, on the existence of central cold receptors. It certainly is not established whether there are distinct hypothalamic thermosensors whose activity is integrated with afferent information, or whether local temperature only affects the mode in which afferent information is processed and routed to the different effectors. These are problems to be investigated when the structures, with their thermal sensitivities and interrelations, are studied with respect to neural efflux from the hypothalamus. We wish here to concentrate on the functional, integrative performance of this area, processing information on local and peripheral temperature as it is relevant to the task of thermal control.

Skin temperature is sensed by separate cold receptors and heat receptors. We refer the reader to a survey of their physiology by Zotterman [19.9]. Thermal nerve fibers are distributed over the entire body surface, although probably not uniformly. Their neural activities, showing rate sensitivity and also adaptation, are transmitted to the hypothalamus. It has not been established whether some of these structures, all of them, or different ones, are responsible for the equally important conscious awareness of "feeling cold" and "being warm." They may be partly engaged in spinal reflexes, yet central control must overwhelm local, peripheral action in order to protect the cores of head and trunk. Other peripheral temperature sensors may be found in the upper respiratory tract. The relative importance of these sensors, and of some other aspects, such as control of panting, may vary in different species of animals.

There is some evidence [19.7] that part of the neural influx to the hypothalamus may carry information from sensors in deep organs other than the brain. Thus it has been found that the temperature of the blood in the large veins has an influence on the thermostat, at least in sheep and dogs. We are, however, lacking direct verification of thermosensitive structures in the vascular walls.

Observations in the whole system (Benzinger [19.10]) indicate that the body's defenses against overheating and those against cooling probably fall into two distinct domains, almost without overlap. The former are brought into action only above a certain local temperature, the latter only for negative deviations from this normal core temperature. At this critical level, neural activities in the hypothalamus are such that neither (extra) heat-preserving mechanisms nor heat-loss mechanisms, other than in the resting state, are called on.

This situation has been likened to a feedback control system whose error signal is zero only if the controlled variable is exactly equal to the reference. The analogy has been useful in explanation of several phenomena. In exercise, the reference or setpoint is thought to be raised by influxes from the motor cortex. Also in fever the operating point of the system is shifted due to a higher setpoint. The exact mechanisms involved are poorly understood, yet the concept of the setpoint, introduced by Liebermeister in 1860 (as quoted by Bligh [19.7]) has aided our understanding.

The setpoint certainly is of nonthermal origin, although it is represented in any mathematical or analog model as the "reference temperature," and it should be regarded as hypothetical. Some observations do not seem to be explained. In exercise, where the setpoint is thought to be above normal body temperature, onset of sweating occurs sooner than calculated delays indicate. On the other hand, steady-state operating points in different environments and (as we see in the following) transient responses to simple thermal stresses are quite satisfactorily explained by a negative feedback control system, with a reference that normally is a constant. However, the system is discontinuous and nonlinear, and its range of operation is limited.

Stolwijk and Hardy [19.1] tested different hypothetical controller functions, which they simulated on an analog computer together with the passive, controlled system. They assumed that information on temperature in skin, in muscle compartments, and of course in head core is available to the controller. Information from the muscles, not directly proven to exist, may reflect either the coupling between motor cortex and hypothalamus or the possible thermoreceptors in the veins. *Each* of the three sets of sensors providing the afferent information on temperature profiles throughout the body is assumed to have its individual setpoint. Phasic (i.e., transient) responses of transducers are assumed to be of too short duration and are neglected. The *products* of the deviations from the setpoints were used as determinants for the effector actions. In defense against overheating (i.e., for positive deviations), the rate of *evaporative heat loss* under central control was assumed to be

$$E_{V1} = H_1(T_{HC} - 36.6)(\overline{T}_S - 34.1) \quad \text{if} \quad E_{V1} > 0 \quad \text{and} \quad (T_{HC} - 36.6) > 0 \tag{19.14}$$

where $\overline{T}_S$ = area-weighted *average* of skin temperatures

36.6 = setpoint of thermocontrol center (°C)

34.1 = setpoint of skin sensors (°C)

H_1 = experimentally determined rate of heat loss, in a given experimental situation; Stolwijk and Hardy let $H_1 = 68$ kcal/(hr °C²)

This function is presented in Fig. 19.5*a*. The nonlinearities of this type of controller are easily seen. The skin temperature must be above 34.1°C before any sweating can be caused by the thermostat. If $(\overline{T}_S - 34.1)$ is small, high body core temperatures are needed for a certain level of evaporative heat losses, which is not true if $(\overline{T}_S - 34.1)$ is large. This is quite similar to the observations reported by Benzinger [19.10].

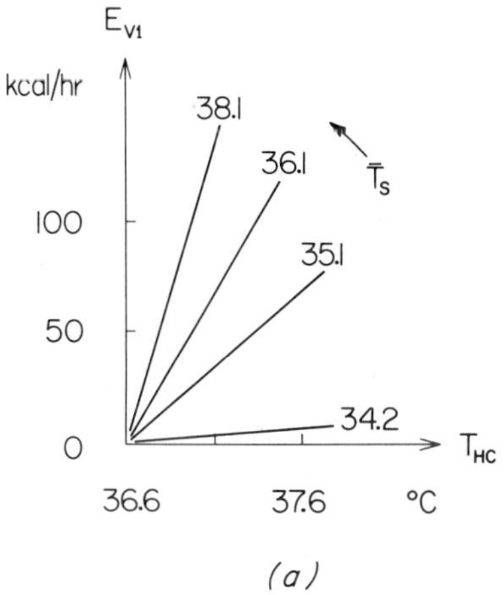

Fig. 19.5*a*. Controller action for high skin ($\overline{T}_s$) and brain (T_{HC}) temperatures, as assumed by Stolwijk and Hardy [19.1]. Above the setpoints, the sudomotor action and therefore the evaporative heat loss E_{V1} are proportional to the product of the deviations from the setpoints. Thus extra sweat secretion does not occur if $T_{HC} < 36.6$°C even if skin temperature is, for example, 38°C. (Cf. eq. 19.14.) Note that the complete model by Stolwijk also contains a term E_{V2} proportional to muscle temperature and to brain temperature (i.e., their deviations from their setpoints; eq. 19.15). See text for definition of $\overline{T}_s$.

Stolwijk [19.1] made evaporative heat losses also directly dependent on muscle activity. The following term can be calculated

$$E_{V2} = H_2(\overline{T}_m - 35.9)(T_{HC} - 36.6) \tag{19.15}$$

where

$\overline{T}_m$ = volume-weighted average of muscle temperatures

35.9 = setpoint of thermosensors in muscle (°C)

H_2 = rate of constant evaporative heat loss due to exercise; a value of 200 kcal/(hr°C²) approximates experimental values

Equations for both E_{V1} and E_{V2} show the same kind of nonlinearities. The total evaporative heat losses E_V that are needed in eqs. 19.6 to 19.13 to close one of the feedback loops are given as the sum

$$E_V = \underbrace{E_{V1}}_{\substack{\text{Evaporative loss}\\ \text{(high core temperature,}\\ \text{high skin temperature)}}} + \underbrace{E_{V2}}_{\substack{\text{Evaporative loss}\\ \text{(high core temperature,}\\ \text{high muscle temperature)}}} \tag{19.16}$$

Stolwijk included in his model the vasomotor action on skin blood flow. This flow increases in warm environment or when the core of the body has to be cooled, or both. In particular, this can be formulated as follows:

$$i_S = i_{S0} + \underbrace{\Delta i_{SC}(T_{HC} - 36.6)(\overline{T}_S - 34.1)}_{\substack{\text{Increase of skin blood flow}\\ \text{due to central action}\\ > 0 \text{ and } T_{HC} > 36.6^\circ\text{C}}} + \underbrace{\Delta i_{SL}(\overline{T} - 34.1)}_{\substack{\text{Increase or decrease,}\\ \text{local action}}} \tag{19.17}$$

where i_S = total skin blood flow, positive only

i_{S0} = resting level

$\Delta i_{SC}, \Delta i_{SL}$ = constants to be determined from experimental runs; $\Delta i_{SC} \simeq 36$, $\Delta i_{SL} \simeq 1$ liter/hr

We see that the thermostat can only raise the value of i_S. Changes due to local, peripheral action are needed in eq. 19.17, since they are real and since they in turn influence convective terms.

Power due to *shivering*, ΔM, was calculated as follows:

$$\Delta M = H_3(T_{HC} - 36.6)(\overline{T}_S - 34.1) \quad \text{if} \quad \begin{cases} \Delta M \geqslant 0 \\ (T_{HC} - 36.6) \leqslant 0 \end{cases} \tag{19.18}$$

where we know from experiments that H_3 is approximately 60 kcal/(hr °C^2).

This function is graphed in Fig. 19.5*b*. It can be introduced into the vasomotor action on *muscle blood flow*, as the next equation shows. Stolwijk [19.1] chose to express changes in metabolic demand of the muscle (increased oxygen consumption), which he equated to a fraction of the blood flow itself. To this must be added the shivering power and possibly the power consumed in exercise, if responses to exercise are of interest. When integrated, a measure of the extra blood flow necessary is obtained.

$$i_m = \underbrace{i_{m0}}_{\substack{\text{Basal}\\ \text{metabolism}}} + \alpha\int_{t_1}^{t_2} [\underbrace{\Delta M}_{\substack{\text{Flow increase}\\ \text{due to shiver}\\ \text{power}}} - \underbrace{\beta(i_m - i_{m0})]}_{\substack{\text{Change in flow}\\ \text{due to local activity}\\ \text{(oxygen demand)}}}\, dt \tag{19.19}$$

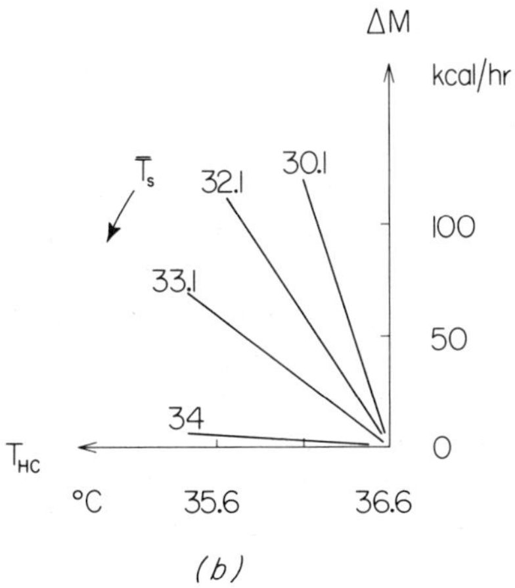

Fig. 19.5*b*. Controller action in defense against cooling. Extra heat production due to shivering ΔM is assumed to be a product of the deviations of brain temperature (T_{HC}) and of skin temperature ($\overline{T}_s$) from their setpoints. (Cf. eq. 19.18.) Note the very high gain with which the controller can cause increases in ΔM near the setpoints. These curves, however, give only static performance of the controller.

Changes due to oxygen demand (when shivering) should be included in i_m, since increased blood flow to muscles, be it in response to thermal stimuli or to others, influences the convective heat losses of some body compartments under consideration.

Summarizing the implications of the *controller equations* (19.16–19.19) we note the following. As long as the temperature in the core of the head is *below* its normally attained value of 36.6°C, evaporative heat losses are restricted to the insensible level (E_{V0}; $E_V = 0$ from eq. 19.16). Furthermore, if $T_{HC} <$ 36.6°C, skin blood flow deviations from the normal resting value are due only to local (i.e., peripheral) reflexes (eq. 19.17). Shivering and increased metabolism may begin as soon as $\overline{T}_S$ is below its normal setpoint of 34.1°C (eq. 19.18). If this occurs, blood flow to the muscles is increased, as expressed by eq. 19.19, which in nature as in the model may stop the shivering temporarily.

If, conversely, the temperature in the brain is *above* its setpoint, as may happen in a warm environment or in exercise, a new set of effectors is brought into play. Sweat glands are stimulated and E_V, the rate of evaporative heat loss, is increased as soon as the skin temperature is also higher than its setpoint. This occurs at high ambient temperature. Endomotor activity can be triggered, if the head core temperature is above 36.6°C, by active muscles that heat up above their operating levels. Blood flow to the skin is also under central control if $T_{HC} - 36.6 > 0$ and if the skin temperature is above its setpoint. (In the foregoing it is assumed that the skin temperature averaged over the body is the important measure).

If skin blood flow increases, the effective thickness of the insulation layer of the body is decreased and more heat is convected from the body core to the surface. As was stated earlier, the thermostat system is considered to be independent of cardiovascular regulation, so we assume that (skin) blood flow can increase when needed. This assumption does limit the applicability of the model, since only data obtained from resting or mildly exercising subjects can be studied.

Stolwijk [19.1] describes in detail the analog computer circuitry used to realize the discontinuous, two-domain controller. An important feature of this model is the multiplicative effect of the afferent "error signals" from the thermosensors. As Benzinger [19.10] has observed, larger changes in endomotor activity are brought forth (per °C deviation from central setpoint and especially near the brain's setpoint) when skin temperature is high than when it is low, and sweating stops entirely when the skin temperature is below a certain level (34.1°C in the case of this model). Equivalent phenomena are observed in defense against cooling. Below the setpoint of the skin the changes in (extra) metabolic rate (per dT_{HC} in the brain) increase with decreasing skin temperature, especially near the central setpoint.

In a first attempt to construct a simple model, we could approximate these characteristics by using "proportional" control of the following form:

$$E_V = c_1(T_{HC} - 36.6) + c_2(\overline{T}_S - 34.1) \tag{19.20}$$

where

$$E_V,\ (T_{HC} - 36.6),\ \text{and}\ (\overline{T}_S - 34.1) > 0$$

and also using

$$\Delta M = c_3(T_{HC} - 36.6) + c_4(\overline{T}_S - 34.1) \tag{19.21}$$

where both $(T_{HC} - 36.6)$ and $(\overline{T}_S - 34.1)$ must be negative.

Essentially, this was done by Crosbie et al. [19.5], who found that such a model accounts perfectly for the steady states the body assumes in different environments. But this proportional (i.e., linear) controller does not predict the transients satisfactorily.

The nonlinear thermostat with gain control is a more successful model. Here, the open-loop gains, between head core temperature and action exerted on effectors (e.g., dE_V/dT_{HC} and $d(\Delta M)/dT_{HC}$), depend directly on the deviation of the skin temperature from its setpoint. This feature is missing in eqs. 19.20 and 19.21, but it is given by eqs. 19.17 and 19.18. The latter equations let the defenses against overheating in warm environment, as elicited by peripheral sensors, be a function of central temperature. This is supported by the observation that the body does not begin to excrete sweat if it is placed, after prolonged undercooling, in a hot bath, so long as T_{HC} is below its setpoint, even if the skin temperature is 42°C.

The *discontinuous* and *nonlinear* behavior in terms of deviations from *setpoints* is indeed a significant feature which allows the system to react powerfully to temperature stresses and to preserve its deep temperature with remarkable stability. The setpoints, chosen to be 36.6, 35.9, and 34.1°C for head, muscle, and skin, respectively, should be considered to be average, representative values for resting man. It also should be remembered that the directional rate-sensitive nature of the peripheral thermosensitive structures has been neglected. These structures elicit phasic responses (overshoots) such as those reported by Benzinger [19.10], which last about 5 min. The model presented here is designed mainly to predict thermoregulatory responses over hours of time, and "fast" variations are ignored.

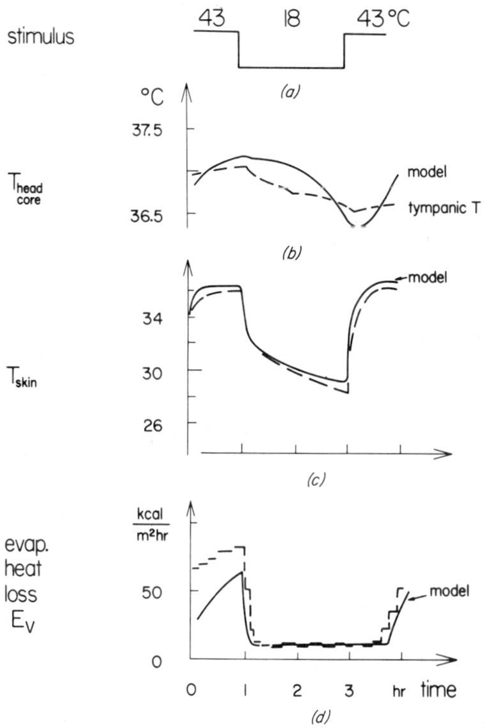

Fig. 19.6. Response of the closed-loop model, in comparison with experimental data obtained in a young nude man, to external thermal transients. (*a*) Ambient temperature showing stimulus to system; (*b*) computed head core temperature together with temperature measured at the tympanic membrane; (*c*) skin temperatures, computed and measured; (*d*) evaporative heat loss E_V computed in the model and calculated from loss of weight of the experimental subject. Note the excellent stabilization of T_{HC}, within 0.5°C, despite the 25°C of external change. From Stolwijk and Hardy [19.1].

Nevertheless, two kinds of transients are predicted quite well by Stolwijk's model. In a first run, a resting subject was exposed to stepwise changes in ambient temperature, and temperatures of skin and tympani were monitored. As Fig. 19.6 indicates, values obtained from the model compare well with observations, although initial conditions did not coincide completely. Second, Fig. 19.7 represents responses to an "ice cream experiment"—that is, to a forced transient in the trunk core temperature caused by ingestion of 420 g

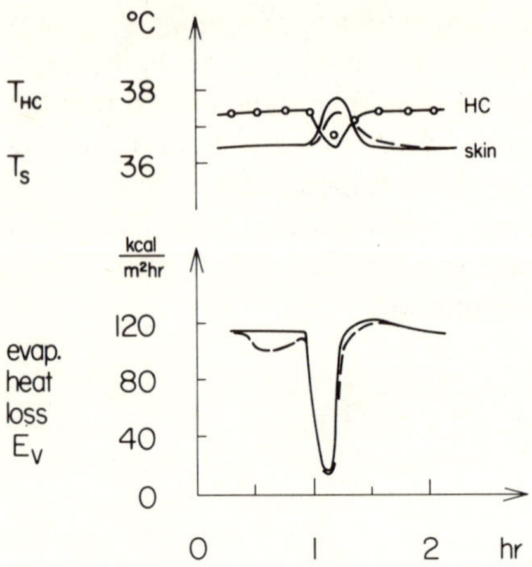

Fig. 19.7. The "ice cream experiment"—responses to rapid ingestion of 420 g of ice cream. The model predicts satisfactorily experimental data (circles and dashed curve) for head core temperature T_{HC}, skin temperature T_s, and the evaporative heat losses E_V. From Stolwijk and Hardy [19.1].

of ice cream. Here the decrease in evaporative heat loss is computed satisfactorily, whereas the extremes in skin and intracranial temperatures predicted by the model are too large.

Stolwijk also used his analog to explore shifts in the setpoints necessary to explain temperature patterns during artificial fever. Immune typhoid vaccine was injected into subjects exposed to 30°C and also to 38°C. It was found that, to approximate the model's responses to those observed, it was not only necessary to shift the "reference" of the cranial controller up by 2.5°C, according to a particular time course (as in Fig. 19.8), but the setpoint of the peripheral temperature sensors also had to be shifted by 5°C maximum. As Fig. 19.8 reveals, the model then simulates fever at least qualitatively quite well.

Model study indicates that it is likely that both the skin's and the brain's setpoints are affected by the pyrogens. This certainly calls for experimental verification.

Furthermore, since all variables in the model are conveniently followed, interplay between existing mechanisms and their relative importance in a complex situation can be delineated. In fever, it is of interest to see how the body adjusts its temperature to the new level dictated by the variable setpoints and how the adjustments depend on ambient temperature. For example, during onset of fever in a hot environment, skin blood flow and sweating decrease strongly. Metabolism does not increase as much as it does during the beginning of fever in a neutral environment. In the latter case this is the factor dominant over evaporation and skin blood flow. Later, when the action of the pyrogens wears off, body temperature is brought back to normal entirely by increases in sweating and skin blood flow.

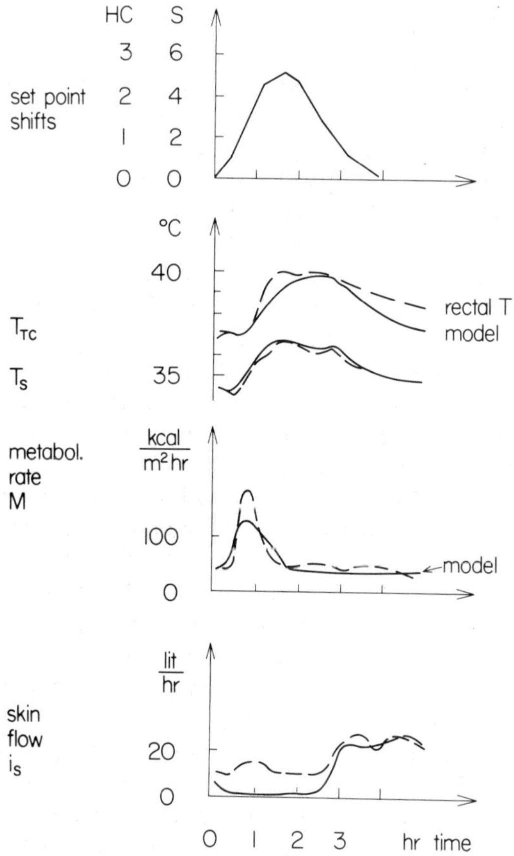

Fig. 19.8. Fever, induced by typhoid vaccine, in a young man exposed to 30°C ambient T_A. The temporal shifts in setpoints for T_{HC} and T_S, were determined empirically on the analog computer so that predicted values of temperature, M and E_V would match observations (dashed curves) best. The E_V (not shown) stayed constant at approximately 20 kcal/(m^{2}hr) for the first 2.7 hr; it then increased to 50 to 60 kcal/(m^{2}hr). Note that pyrogens probably do not act only on thermocontrol centers. From Stolwijk and Hardy [19.1].

C. CONCLUSION

The thermostat models discussed in this chapter give us good examples of the steps necessary in verifying and improving a set of mathematical equations used to gain insight into the relative importance of the many variables. Thus the convective heat transport within the body is found to be a significant factor in homeothermy. The models add also to our understanding of phenomena not easily observable directly (e.g., the variable gain controller is necessary for simulation of transients). Concurrently we have become aware of the limited nature of this model of the human thermostat.

Although the controller equations quite satisfactorily describe the limited amount of data, we must collect more detailed information. We can then investigate again the relative importance and ranges of dominance of central and peripheral thermoreceptors, which are by no means fully understood.

The effects of coupling between thermal and circulatory control systems have been largely disregarded. These doubtless come into play in exercise. Only some *local* reflexes on skin blood flow and on increased flow to shivering muscle have been included in eqs. 19.17 and 19.19. Partly this is because we need more simultaneous hemodynamic and "thermodynamic" data. As Bligh [19.7] points out, the simple notion of a shifting setpoint may not be sufficient to explain some of the information available.

Our discussion did not touch on extrahypothalamic control factors in homeothermy. Behavioral responses, small voluntary movements, and so on, tend to keep body temperatures within the range of the hypothalamic regulator, which lies approximately between 35.6 and 37.5°C in the core. How does the physiological system react to temperatures outside this range? There appears to be a centrally located, but not necessarily hypothalamic, secondary "broad-band" mechanism that calls on emergency measures to keep the body temperature within the bounds of 41 and 35°C. As quoted by Bligh [19.7], abrupt respiratory or sweating patterns are observed near the upper limit, unlike the ones discussed in this chapter, where we considered only the fine control mechanisms active *near* the normal levels. Other emergency mechanisms, such as massive shivering independent of skin temperature, may be responsible for keeping the deep body temperature above 35°C.

D. REFERENCES

19.1. Stolwijk, J. A. J. and J. D. Hardy. Temperature regulation in man—a theoretical study. *Pflügers Arch Ges Physiol* **291**: 129–162, 1966.

19.2. Ruch, T. C. and H. D. Patton (Eds.). *Physiology and Biophysics*, 19th ed. W. B. Saunders Co., Philadelphia, 1965.

19.3. Brown, A. C. Equations of heat distribution within the body. *Bull Math Biophys* **27**: 67–78, 1965.

19.4. Wissler, E. H. An analysis of factors affecting temperature levels in the nude human, pp. 603–612 in *Temperature*, Vol. III, part 3. J. D. Hardy (Ed.). Reinhold Publishing Corp., New York, 1963.

19.5. Crosbie, R. J., J. D. Hardy, and E. Fessenden. Electrical analog simulation of temperature regulation in man. *IRE Trans Bio-Med Electron* **BME-8**: 245–252, 1961.

19.6. Hammel, H. T., D. C. Jackson, J. A. J. Stolwijk, J. D. Hardy, and S. B. Strømme. Temperature regulation by hypothalamic proportional control with an adjustable set point. *J Appl Physiol* **18**: 1146–1154, 1963.

19.7. Bligh, J. The thermosensitivity of the hypothalamus and thermoregulation in mammals. *Biol Rev* **41**: 317–367, 1966.

19.8. Stuart, D. G., D. S. Maxwell, J. N. Hayward, M. D. Fairchild, W. R. Adey, and R. W. Porter. Unit activity in the hypothalamus. *Bol Estud Med Biol* **21**: 349–370, 1963.

19.9. Zotterman, Y. Thermal sensations, pp. 431–458 in *Handbook of Physiology, Sec. 1 Neurophysiology*, Vol. 1 J. Field and H. W. Magoun (Eds.). American Physiological Society, Washington, D.C., 1959.

19.10. Benzinger, T. H., C. Kitzinger, and A. W. Pratt. The human thermostat, pp. 637–665 in *Temperature*, Vol. III, part 3. J. D. Hardy (Ed.). Reinhold Publishing Corp., New York, 1963.

CHAPTER TWENTY

Hormonal Control: The Pituitary–Adrenocortical System

A. INTRODUCTION

The last chapter is devoted to a quantitative model of a hormonal system, developed from the viewpoint of systems control. Generally, hormones are recognized as prime controllers of tissue metabolism and tissue function. More specifically, hormones are vital information carriers within the multiply linked systems, regulating flow (exchange with environment or between tissue compartments), maintaining the *milieu intérieur*, and controlling tissue metabolism, growth, and development. Thus they maintain homeostasis and ensure procreation, as well as adaptation to, and protection against, disturbances and stimuli from the environment at the autonomic level.

It is possible to analyze hormonal activities at the level of the control systems in which they form important links, although their mode of action (at the molecular level) is not known in all cases. Some hormones act on genes, some play the role of enzymes or coenzymes, others may modify membrane properties. The aforementioned control systems comprise subunits of distinct forms: transport and diffusional systems, glandular secretion, and information transmission by the nervous system. Furthermore, many different couplings exist between the systems discussed here and those modeled in previous chapters. Water and electrolyte balance is one example in which functions of cardiovascular, respiratory, and hormonal systems should be treated concomitantly. Here lies one special challenge to the systems physiologist and biomedical engineer.

At the level of the subsystems, as we always find when a particular hormonal system is analyzed, static nonlinearities (e.g., due to saturation), dynamic nonlinearities (unidirectional rate sensitivity), adaptation (e.g., by stimulation of glandular growth), and parametric control (e.g., due to coupling with circulation or to hormone-driven chemical reactions) are ubiquitous. Thus many phenomena encountered in systems discussed in previous chapters must be treated in combined form, and as essential parts of hormonal control.

Many mathematical models already exist [20.1, 20.2]. Their applicability, however, is quite limited, since nonlinear effects too often have been linearized. Analyses of regulatory systems, which assume many different states and modes of action, frequently have been too restricted (e.g., including hyperglycemia but not hypoglycemia in blood glucose control). The limitations are partly due to lack of mathematical methods suitable for the analysis of biological control mechanisms. Much experimental data must still be obtained to allow formulation and validation of models at the level of metabolic control, electrolyte balance, and so on, all mediated by hormones.

B. GENERAL ASPECTS OF THE CORTISOL—CRF—ACTH SYSTEM

Glucocorticoid hormones—that is, cortisol and corticosteron—play important roles in the glucose metabolism and blood glucose control system (for which insulin, thyroxin, glucagon, and adrenalin also are essential) and in many other tissue functions. They have inhibitory effects on leucocytophagocytosis, inflammatory exudation, the development of fever, and the formation of antibodies. They inhibit the growth of fibroblasts important in wound healing, and they enhance blood clotting factors. Furthermore, they inhibit protein synthesis and influence calcium metabolism (e.g., by depressing renal calcium resorption). These are only some of the functions of the glucocorticoids, which are described more completely in physiology texts [2.2, 2.3] and, of course, in many specialized books [20.3].

For man, cortisol is the most important glucocorticoid. Because its concentration in plasma, and not its action, is regulated at constant level during rest, we might call this "indirect" control. In view of the hormone's many tasks, this seems to be an efficient "design" of nature. Cortisol level is subject to circadian rhythms, while (psychic) stress shifts the setpoint of the feedback system to higher values. The glucocorticoids are secreted by glands of the adrenal cortex, together with the mineral corticoids (important in electrolyte control) and the androgenic or estrogenic hormones. They are distributed by the circulatory system, in man mostly bound to globulins and erythrocytes (90%), or in their free active form (10%).

In certain structures of the hypothalamus, unbound cortisol influences production of a so-called corticotropin-releasing factor (CRF), which in turn stimulates production by the anterior pituitary gland of the adrenocorticotropic hormone (ACTH). As shown in Fig. 20.1, ACTH effects release of

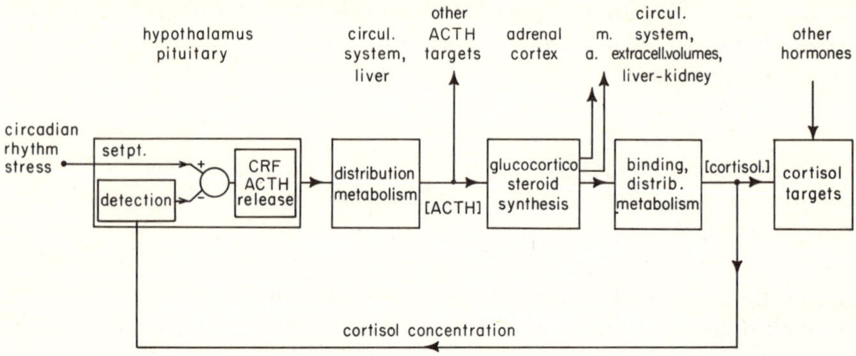

Fig. 20.1. Elements in control of adrenal cortical steroid secretion. The plasma concentration of cortisol, the most important glucocorticoid hormone in man, is sensed and controlled by negative feedback, as suggested by Yates and Urquhart [20.1, 20.2]. This loop forms part of many other hormonal control systems, as indicated here. ACTH, the adrenocorticotropic hormone, fulfills other functions besides being an important link in this loop. The glands of the adrenal cortex not only secrete glucocorticoid but also mineral corticoid (*m*) and androgenic (*a*) hormones. The functioning of cortisol also depends on other hormones (insulin, etc.). See Fig. 20.4 for details of the model of detectors, setpoint action, and ACTH release (box on upper left). The second stage, "distribution, metabolism," represents the stores and metabolic processes determining ACTH concentration in plasma. The fourth stage, "binding, distribution, metabolism," represents the two stores, the albumin to which part of the cortisol is bound, and metabolic processes: all these determine concentration of unbound cortisol (see text).

hormones by the adrenal glands, particularly the glucocorticoids. This closes the feedback loop which, for stabilization, exhibits negative feedback because blood–cortisol increases inhibit CRF and ACTH excretion, whereas decreases in cortisol allow for enhanced CRF and ACTH production.

Remember that the foregoing only summarizes the action of cortisol control, not the functions of the pituitary and adrenal glands. Much more must be considered. First, cortisol synthesis in the adrenal gland involves several enzyme-controlled reactions, where each enzyme can limit the rate of the entire chain. ACTH dominates one step of the synthesis and thus exerts very strong control over cortisol production. Furthermore ACTH, besides influencing release of mineral corticoids (aldosterone, etc.), stimulates growth of some of the excretory tissues of adrenal cortex (Fig. 20.1). This makes the system adaptive by morphokinesis. The secretive structures of the hypothalamus also produce so-called releasing factors other than CRF; at least one inhibitory and five hormone-releasing factors are known to originate there. The anterior pituitary, in turn, excretes at least six hormones, and one of these is ACTH. ACTH is not only involved in the regulation of the cortisol production; it also directly affects lipid metabolism and enhances the melanin formation involved in skin pigmentation ("other targets" in Fig. 20.1).

Release of ACTH may be stimulated by substances (e.g., angiotensin) other than cortisol.

Several pathological states can be explained using the feedback loop concept of the ACTH–cortisol system. In Cushing's disease, the setpoint seems to be shifted so that both ACTH and cortisol concentrations are higher than normal. In Addison's disease, with near-complete destruction of adrenal cortices, ACTH is high; the feedback loop is open.

In the following sections we describe some details of models for the cortisol–ACTH control system that we owe to Urquhart [20.1], Yates [20.2], and their co-workers. The models were designed, in most cases, to simulate the cortisol–ACTH system in dogs. The more physiological aspects are covered in [20.1], whereas the model of [20.2] takes recent findings more completely into account. For extensive presentations of experimental data, see [20.3].

C. THE ADRENAL GLANDS, CORTISOL EXCRETION AND DISTRIBUTION

We do not intend to give here anatomical or physiological discussions of the adrenocorticol glands [2.2, 20.3, 20.4]. This section shows first how Urquhart and Yates arrived at the system of differential equations governing the relation between plasma ACTH (the input) and cortisol at the output side. Then phenomena determining binding by globulins, distribution throughout the extracellular space, and metabolism (mostly in liver) are formulated.

The *synthesis* of cortisol from cholesterol comprises at least eight reactions. The following substances, synthesized at intermediate steps, are involved:

$$\underset{\text{Ch}}{\text{cholesterol}} \rightarrow \underset{\text{D}}{\Delta^5\text{-pregnenolone}} \rightarrow \underset{\text{E}}{\text{progesterone}}$$

$$\rightarrow \underset{\text{F}}{17\alpha\text{-OH-progesterone}} \rightarrow \underset{\text{C}}{\text{cortisol}}$$

Hereafter we assume that the overall kinetics of the entire biosynthetic chain are adequately represented by a series of four reactions, possibly the ones just given, of which only the first, Ch → D, is complex. The remaining three steps, D → E → F → C, are simple, irreversible first-order systems, each contributing one time constant. Figure 20.2 illustrates how the transform Ch → D proceeds (*a*) directly and ACTH-driven, and (*b*) by way of a branching pathway (via substance G, a cholesterol-derived precursor of D) which is rate limited *and* indirectly ACTH-driven. The reaction kinetics of all the

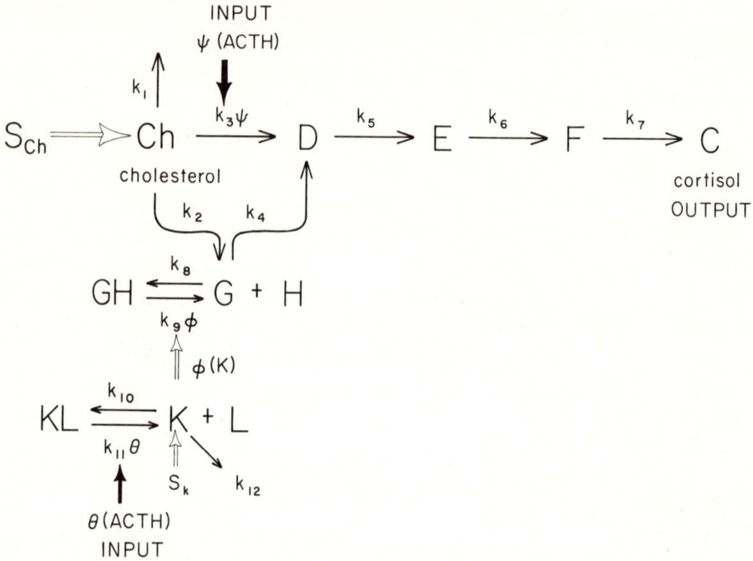

Fig. 20.2. Cortisol synthesis in adrenal cortex; system of (hypothetical) reactions. The first reaction of the direct path from cholesterol (Ch) to cortisol (C) is, on the one hand, driven by the ACTH input (ψ). On the other hand, a cholesterol-derived precursor G also contributes to D. G reacts with H, where decomposition of GH is driven by K. K is one of the reagents of a further reaction which is, in one direction, dependent on θ(ACTH), a function of plasma ACTH. S_{ch} is the source of cholesterol; S_K is a source for substance K, which itself disappears at the rate k_{12}. The last three steps, D → E → F → C, are simple unidirectional reactions. See text for differential equations governing the kinetics of the entire system. "Step responses" appear in Fig. 20.3.

stages between Ch and D are given by

$$\frac{d[\mathrm{Ch}]}{dt} = \dot{Q}_{\mathrm{Ch}} - k_1[\mathrm{Ch}] - k_2[\mathrm{Ch}] - k_3\,\psi[\mathrm{Ch}] \tag{20.1}$$

$$\frac{d[\mathrm{G}]}{dt} = k_2[\mathrm{Ch}] + k_9\,\phi[\mathrm{GH}] - k_8[\mathrm{G}][\mathrm{H}] - k_4[\mathrm{G}] \tag{20.2}$$

$$\frac{d[\mathrm{H}]}{dt} = k_9\,\phi[\mathrm{GH}] - k_8[\mathrm{G}][\mathrm{H}] \tag{20.3}$$

$$\frac{d[\mathrm{K}]}{dt} = \dot{Q}_{\mathrm{K}} + k_{11}\theta[\mathrm{KL}] - k_{10}[\mathrm{K}][\mathrm{L}] - k_{12}[\mathrm{K}] \tag{20.4}$$

$$\frac{d[\mathrm{L}]}{dt} = k_{11}\theta[\mathrm{KL}] - k_{10}[\mathrm{K}][\mathrm{L}] \tag{20.5}$$

$$\frac{d[\mathrm{D}]}{dt} = k_4[\mathrm{G}] + k_3\psi[\mathrm{Ch}] - k_5[\mathrm{D}] \tag{20.6}$$

where

$[\ ]$ = concentration of the substance within the brackets

k_i = reaction rate or disappearance rate constants ($i = 1, 2, 4, \ldots, 12$)

k_3 = rate constant proportional to blood flow to glands

$\dot{Q}_{\mathrm{Ch}}$ = internal source (rate) of cholesterol

$k_1[\mathrm{Ch}]$ = rate of disappearance of Ch by pathways other than shown in Fig. 20.2

G = cholesterol-derived precursor which contributes to D and is also involved in a reaction with H

ψ = ACTH-dependent rate parameter driving the direct Ch → D path; directly proportional to the delayed [ACTH] in plasma

$\phi(\mathrm{K}) = m[\mathrm{K}] - 1$ = rate parameter, influencing G + H → GH only; proportional to [K]

θ = rate parameter, driving K + L → KL, depending on [ACTH] in plasma: $\theta = p\psi + q$; p and q are constants

Additional conditions are that the total masses of H and L stay constant:

$$[\mathrm{GH}] + [\mathrm{H}] = \text{constant} \tag{20.7}$$

$$[\mathrm{KL}] + [\mathrm{L}] = \text{constant} \tag{20.8}$$

We refer the reader to [20.2] for numerical values of all constants. Note that the system is nonlinear, not only because of the parametric driving of three of the reactions, but also because [L] and ϕ are limited.

The remaining reactions E → F → C are determined by

$$\frac{d[\mathrm{E}]}{dt} = k_5[\mathrm{D}] - k_6[\mathrm{E}] \tag{20.9}$$

$$\frac{d[\mathrm{F}]}{dt} = k_6[\mathrm{E}] - k_7[\mathrm{F}] \tag{20.10}$$

$$\frac{d[\mathrm{C}]}{dt} = k_7[\mathrm{F}] - k_{cs}[\mathrm{C}] \tag{20.11}$$

where k_5, k_6, k_7 = rate constants
k_{cs} = rate constant determining secretion of cortisol (i.e., $\dot{Q}_c = k_{cs}[C]$)
$\dot{Q}_c$ = outflow rate of cortisol from adrenal glands

This system exhibits several interesting features, summarized in Fig. 20.3.

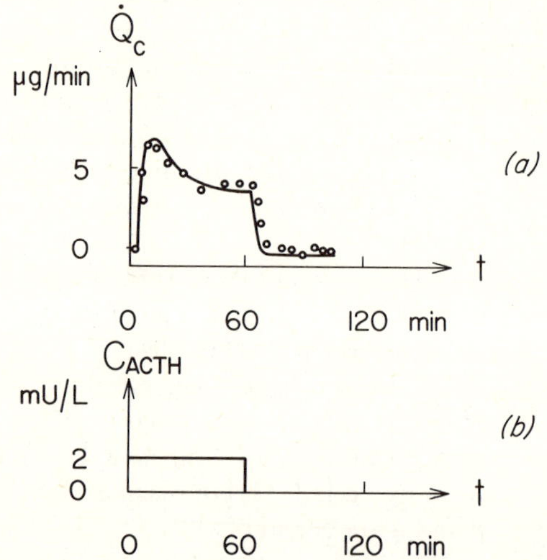

Fig. 20.3. Response of adrenal gland [(*a*) cortisol output flow $\dot{Q}_c$] to an impulselike change in plasma ACTH concentration [(*b*) C_{ACTH}]. The circles are Urquhart's observations; the solid line is computed with eqs. 20.1 to 20.11. Note the overshoot when C_{ACTH} increases and the simple decay at the end of the stimulus. The 2-min delay of the response was observed by Urquhart. More recent findings indicate that there may not be any time delay. From Yates and Brennan [20.2].

Those which bear mentioning here are:

1. Dynamic asymmetry, or unidirectional rate sensitivity (URS, see Chapter 7). Transients due to increasing ACTH steps show overshooting, approximately 65% above the steady-state level. Transients due to stepwise ACTH decrease following a decaying exponential.
2. Steady-state cortisol output is a nonlinear function of ACTH (approximately logarithmic); it obviously also depends on adrenal blood flow.
3. Minimum cortisol excretion (no ACTH) is very small; maximum output is limited (at 16 μg/min in dogs).

The *distribution* of cortisol can be formulated by taking into account the equilibria between the different globulin-bound fractions and the free cortisol, both in intravascular and extravascular compartments. Thus many different spaces are lumped into few—two, in this case. This procedure is successful, as examples in previous chapters indicate, so long as the compartments chosen comprise functionally important ones (e.g., in determining time course of transient concentration changes). Urquhart and Yates [20.1] showed that, in this case, compartmentalization into the two spaces yields satisfactory results. If we consider albumin-bound and transcortin-bound cortisol, besides free cortisol, we find the cortisol concentration in the plasma store (subscript 1) to vary as

$$V_1 \frac{d[\mathrm{C}_1]}{dt} = \dot{Q}_c - k_{20}[\mathrm{C}_1] - V_1\left\{\frac{d[\mathrm{CA}_1]}{dt} + \frac{d[\mathrm{CT}_1]}{dt}\right\} - V_2 \frac{d[\mathrm{C}_2]}{dt} \tag{20.12}$$

where
V_1 = plasma volume
$[C_1]$ = cortisol concentration in V_1
$[C_2]$ = cortisol concentration in V_2, the extravascular store
$\dot{Q}_c$ = cortisol production rate by adrenal glands; may include also cortisol injection
k_{20} = rate of excretion and metabolism in kidney and liver
CA_1 = albumin-bound cortisol in V_1
CT_1 = transcortin-bound cortisol in V_1

The bimolecular, reversible reaction of binding by transcortin is given by

$$V_1 \frac{d[\mathrm{CT}_1]}{dt} = k_{Ta}[\mathrm{C}_1][\mathrm{T}_1] - k_{Tr}[\mathrm{CT}_1] \tag{20.13}$$

and the monomolecular, reversible adsorption by albumin by

$$V_1 \frac{d[\mathrm{CA}_1]}{dt} = k_{Aa}[\mathrm{C}_1] - k_{Ar}[\mathrm{CA}_1] \tag{20.14}$$

where
k_{Ta}, k_{Tr} = rate constants of binding and of release by transcortin, respectively
k_{Aa}, k_{Ar} = rate constants of binding and of release by albumin, respectively
$[\mathrm{T}_1]$, $[\mathrm{A}_1]$ = concentration of transcortin and of albumin in V_1, respectively

The total amounts of cortisol, transcortin, and albumin are constants:

$$[\mathrm{C}_1] + [\mathrm{CT}_1] + [\mathrm{CA}_1] = \text{constant} \tag{20.15}$$

$$[\mathrm{T}_1] + [\mathrm{CT}_1] = \text{constant} = \sum \mathrm{T}_1 \tag{20.16}$$

$$[\mathrm{A}_1] + [\mathrm{CA}_1] = \text{constant} \tag{20.17}$$

We also need the restriction that the fraction of unoccupied cortisol sites $[T_1]$ is always nonnegative, that is, $\sum [T_1] - [CT_1] \geqslant 0$. (This inherently also limits $[A_1]$ to nonnegative values.)

In the "outer" compartment, of volume V_2, we have

$$V_2 \frac{d[C_2]}{dt} = k_{22}[C_1] - k_{21}[C_2] \tag{20.18}$$

where k_{21} and k_{22} are rate constants of cortisol transfer from V_2 to V_1, and from V_1 to V_2, respectively, and V_2 is to be considered a virtual volume, which can be larger than the extracellular spaces. (Some tissues seem to take up cortisol at higher concentrations than others.) See Yates's articles [20.1, 20.2] for methods of determining quantitatively k_{Ta}, k_{Tr}, k_{Aa}, k_{Ar}, k_{21}, k_{22}. The values of these constants are species specific.

It is worth summarizing the different features of the model. First, we deal with an approximative assumption when the sorption of cortisol by albumin is treated as monomolecular. Second, the metabolism and clearance of cortisol by liver and kidney are assumed to involve a first-order process. Nevertheless, the simplified system of reactions just presented, reproducing known clearance curves and their time constants as a function of the quantity injected, is satisfactory because of nonlinearities introduced by parametrically driven reactions. The terminal half-life of cortisol increases with the size of the stimulus and also with increasing availability of transcortin. Nonlinearity is also found in the steady-state relation between levels of infusion and excretion of cortisol [20.1].

D. COMPARATOR AND CONTROLLER ACTION

It is difficult to devise a model as explicit as that for cortisol excretion for the action of the centers in the hypothalamus and pituitary. Although the overall effects of stress, circadian rhythms, acoustic stimuli, morphine, and other substances are known [20.1], the sharing of functions and coordination among the centers (in rostral midbrain, hypothalamic median eminence, septal area, adenohypophysis) are not. Furthermore, it is experimentally difficult to observe directly ACTH excretion, a measurement needed for strict verification of a model. Yates [20.2], following the method of choice in all such situations, formulated the simplest plausible controller model, including as many data on pituitary function as possible and incorporating hypotheses, some debatable, to be proven or disproven.

The main assumptions for the model are:

1. The concentration of unbound cortisol and its positive time derivative (URS) are functionally important input variables.

2. A signal derived from these inputs is compared with setpoint signals. The difference between the two is the error signal, which determines how much ACTH should be produced to keep cortisol level normal. This is the simplest type of comparator yielding satisfactory results. Thus there is no reason, at this stage, for considering a controller based on a modulator (multiplier), which also might be plausible on neurophysiological grounds.

3. The relation between production rates of CRF and ACTH is unknown. It is therefore assumed that they change essentially simultaneously and are indistinguishable (1 : 1 relation).

4. Direct control by cortisol on the pituitary is omitted for lack of data.

5. There is experimental evidence [20.1] for multilevel comparison, probably anatomically distinct (median eminence of hypothalamus and septal region) as shown in Fig. 20.4 and as formulated in the following systems equations.

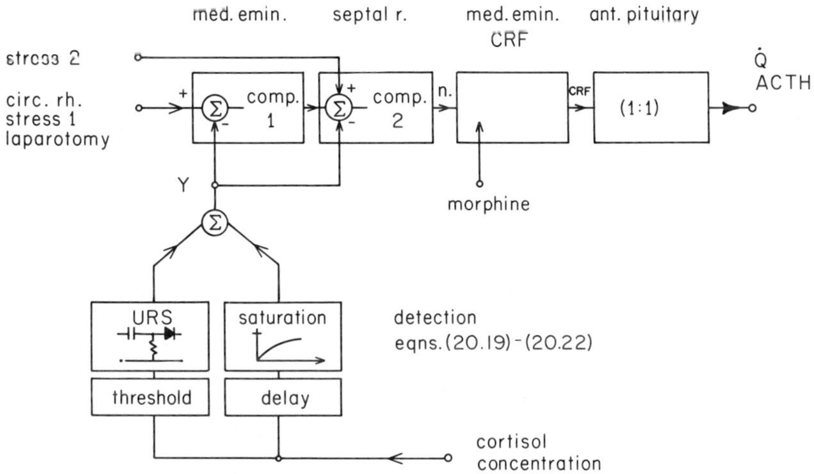

Fig. 20.4. Organization of central elements, cortisol "detectors," comparators, and ACTH secretion (cf. Fig. 20.1), as assumed in the model. The signal Y, containing information both on mean value (delayed) and on rate of rise of cortisol level, is probably compared at two different stages with "setpoints." Signals causing circadian rhythms originate within the median eminence of the hypothalamus. "Stress 1" may be due to histamine, scald, or laparotomy. The "error signal" from comparator 1 can only be positive. At comparator 2, the effects of acoustic stimuli, ether, or laparotomy are injected into the loop ("stress 2"). The output of comparator 2, restricted to be nonnegative, is a neural signal. The activity of the neurons (at median eminence) determining CRF secretion is influenced by stimuli such as morphine. The plasma concentration of ACTH is given by the secretion rate (output of this system) and by the volume over which it is distributed and the rate at which it is metabolized and excreted (eq. 20.27). After Yates and Brennan [20.2].

According to assumption 1, a "feedback" signal Y is computed comprising both a dynamic (Y_d) and a delayed static (Y_s) component; Y is compared, in the following form, with the setpoint parameters.

$$Y = Y_d + Y_s \tag{20.19}$$

where

$$Y_d = n_1[\mathrm{C}_1] - n_2 Z - \mathrm{Th} \tag{20.20}$$

$$\frac{dZ}{dt} = Y_d \qquad \text{positive only (unidirectional rate sensitivity)} \tag{20.21}$$

$$\mathrm{Th} = \text{threshold}$$

$$n_1, n_2 = \text{constants}$$

and

$$Y_s = Y_{s0}[1 - e^{-\eta[\mathrm{C}_1]}] \text{ delayed by 120 min} \tag{20.22}$$

where η is a constant.

Here Y_{s0} clearly is a limiting or saturation level. The exponential form of the static component may represent the basis of the often observed near-logarithmic (static) stimulus–response curve. Yates included in the exponent, in addition to $[\mathrm{C}_1]$, the concentration of injected dexamethose (synthetic glucocorticoid).

Assumption 5 leads to the following two comparators, stimulating (CRF and) ACTH release ($\dot{Q}_{\mathrm{ACTH}}$):

$$\dot{Q}_{\mathrm{ACTH}} = m_1(Y_{es} - Y_{em}) \tag{20.23}$$

$$Y_{em} = S_1 - Y \qquad \text{(median eminence of hypothalamus)} \tag{20.24}$$

$$Y_{es} = S_2 + Y_{em} - Y \tag{20.25}$$

where S_1 = "setpoint," subject to circadian rhythms and stress

S_2 = setpoint, depending on stress and influenced by some drugs

m_1 = constant determining "sensitivity" of controller

A restriction following some experimental evidence is imposed on the error signals [20.1]:

$$Y_{em}, Y_{es}, (Y_{es} - Y_{em}) \geqslant 0 \tag{20.26}$$

ACTH released is distributed in unbound form throughout a volume (V_{ACTH}) slightly larger than that of the blood plasma; it also is metabolized and cleared:

$$V_{ACTH}\frac{d[ACTH]}{dt} = \dot{Q}_{ACTH} - k_{ACTH}[ACTH] + \dot{C}_{ACTH} \qquad (20.27)$$

where k_{ACTH} = rate of ACTH metabolism
$\dot{C}_{ACTH}$ = concentration change of injected ACTH

The concentration of ACTH determines the parameters ψ and θ of eq. 20.2 and 20.4. Thus all components of the model system are defined.

E. CLOSED-LOOP PERFORMANCE

The cortisol–CRF–ACTH system as described by eqs. 20.1–20.27 inherently reproduces performances that were "put into the model," that is, those which served to identify the most important system parameters. Thus the static nonlinearity in the equilibrium relation between unbound and total cortisol, the cortisol-level-dependent time constant of cortisol disappearance, and the URS feature in cortisol distribution in V_1 and in the cortisol–ACTH relation are all responsible for static nonlinear and dynamic asymmetric responses of the closed-loop system.

Some of these features are apparent in Fig. 20.5. Curve *A* represents the continuous rise of plasma cortisol during a constant infusion of cortisol, not an exponentially rising curve. Curve *B* demonstrates the decay if infusion stops at $t = 120$ min. The disturbances *C* and *D* are caused by pulselike injections (over 20 min) of exogenous ACTH. If we try to elicit the same cortisol "pulses" by stress stimulation, during the same time intervals, it proves to be impossible. During the early phase, the rate of rise is large.

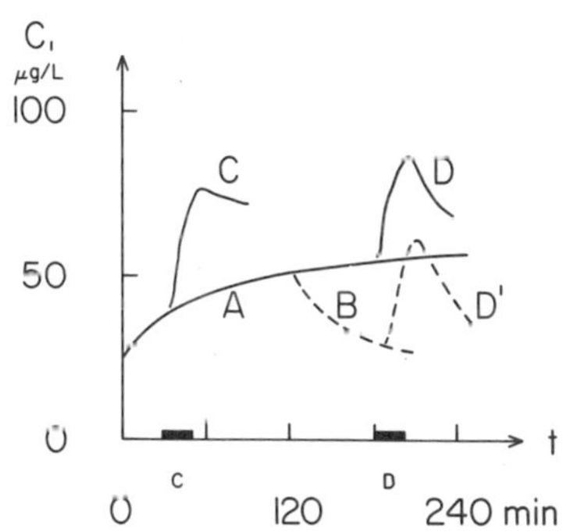

Fig. 20.5. Performance of the total ACTH–cortisol system, as predicted by the model: curve *A*, response to continuous infusion of cortisol (5 μg/min); curve *A–B*, response to pulselike infusion of cortisol during 2 hr; curves *C* and *D*, "pulses" elicited by injected ACTH pulses, lasting 20 min each, superimposed onto cortisol infusion at times *C* and *D* (heavy marks on abscissa); pulse *D'*, due to ACTH impulse after 2-hr infusion of cortisol. During times *C* and *D*, stresses (theoretically causing the same cortisol production as the ACTH pulses) cannot elicit cortisol production, since at time *C* the rate of rise is large and at time *D* the mean level of cortisol is large; both these factors dominate the system and therefore inhibit stress responses. This corresponds well to experimentally observed responses. After Yates and Brennan [20.2].

Negative feedback therefore is very strong, and stress signals cannot exert their effect (by eq. 20.26). During the late phase, information on high mean cortisol level reaches the controller through the proportional delay channel. Again, stress signals cannot influence ACTH release. (Only at intermediate times, at about 2 hr (Fig. 20.5), do the error signals depend on all inputs.) Yates showed that this is due to a time delay (rather than some integrative action) by demonstrating that the error signal cannot be affected even if cortisol is made to decrease as in curve *B*. Pulse *D'* cannot be elicited by a stress impulse, only by ACTH infusion.

Experimental work in animals corroborates these data quite satisfactorily. However, it is difficult to obtain comparable data from individual animals in consecutive runs or to estimate average responses from simultaneous runs in several animals. These questions could be attacked by new experimental methods. In any case, in the study of complex systems involving neurophysiological, cardiovascular, and biochemical subsystems, a working hypothesis is indispensable as a guide to thinking and working. The models by Yates and Urquhart are excellent examples.

F. REFERENCES

20.1. Yates, F. E., R. D. Brennan, J. Urquhart, M. F. Dallman, C. C. Li, and W. Halpern. A continuous system model of adrenocortical function, pp. 141–184 in *Systems Theory and Biology*. M. D. Mesarović (Ed.). Springer-Verlag, New York, 1968.

20.2. Yates, F. E. and R. D. Brennan. Study of the mammalian adrenal glucocorticoid system by computer simulation, in *Hormonal Control Systems*. E. B. Stear and A. H. Kadish (Eds.). *Mathematical Biosciences, Suppl* **1**: 20–87, 1967. American Elsevier Publishing Co., New York.

20.3. *The Hormones*, Vol. V. G. Pincus, K. V. Thimann, and E. B. Astwood (Eds.). Academic Press, New York, 1964.

20.4. Liddle, G. W. Physiological review of adrenocortical function, in *Hormonal Control Systems*. E. B. Stear and A. H. Kadish (Eds.). *Mathematical Biosciences, Suppl* **1**: 1–19, 1967. American Elsevier Publishing Co., New York.

Author Index

Numbers in brackets are reference numbers and show that an author's work is cited, although his name is not mentioned in the text. Numbers in *italics* indicate the pages on which full references appear.

Subject Index